A Text Book on

PRODUCTION ENGINEERING

(THIRD EDITION)

Useful for GATE / ESE / PSUs and other competitive examinations

Dr. Swadesh Kumar Singh
B.Tech. AMU Aligarh, Gold Medalist
M. Tech. and Ph.D. from IIT Delhi
Ex. IES officer, Govt. of India
Professor, GRIET, Hyderabad

MADE EASY Publications Pvt. Ltd.

Corporate Office: 44-A/4, Kalu Sarai (Near Hauz Khas Metro Station), New Delhi-110016
E-mail: infomep@madeeasy.in
Contact: 011-45124660, 8860378007
Visit us at: www.madeeasypublications.org

A text book on Production Engineering

© Copyright, by MADE EASY Publications
All rights reserved. No part of this publication may be reproduced, stored in or introduced into a retrieval system, or transmitted in any form or by any means (electronic, mechanical, photocopying, recording or otherwise), without the prior permission of the above mentioned publisher of this book.

First Edition: 2010
Second Edition: 2014
Third Edition: 2016
Reprint : 2017
Reprint : 2019

Reprint : 2020

MADE EASY PUBLICATIONS has taken due care in collecting the data and providing the solutions, before publishing this book. Inspite of this, if any inaccuracy or printing error occurs then **MADE EASY PUBLICATIONS** owes no responsibility. We will be grateful if you could point out any such error. Your suggestions will be appreciated.

To my teacher,

Who not only taught me Bhagavad-Gita as it is
but also trained me how to follow its teachings

FOREWORD

"Invention of tools to extend the capabilities of their bare hands has been one of the greatest advances made by the ancient humans that helped in the progress of civilization. Assembling individual tools having different functions into harmonious units called machines which could perform a range of complex functions was the next major step that ushered in an industrial society which brought unprecedented prosperity to a vast majority of people. Today's comfortable lifestyles which we so take for granted owe not a little to the manufacturing sector that produces not only the goods that are indispensable to our existence but also all the machinery that makes those goods. Production Engineering is the subject that deals with the techniques of converting a variety of raw materials into useful articles.

Metalworking is one of the oldest professions practised by human beings. From the ancient village blacksmithy to the present day robotic controlled automatic manufacturing, production technology has made mind boggling advances. Today's production technologies encompass a breath taking range and can satisfactorily address humongous challenges. Be it the fabrication of flawless individual components weighing a few hundred tons each such as the bulkheads of fusion reactors or nano-scale robots that could propel themselves through the bloodstream of a human being, be it the making of very-high temperature resistant nose cone of a ballistic missile or the layer-by-layer building-up of millions of transistors and other electronic components on a single crystal of silicon to make a VLSI chip of a few mm size, today's technologies have the answers.

Ancient man may have used only a simple stone to make his tools (the flint tools of prehistoric man), but the present day engineer has a bewildering variety of materials to work with. Although still most preoccupied with the forming of metals and alloys (especially steels) using traditional methods such as casting, forging, rolling, extrusion, machining and welding, he has to master the technologies required to shape a number of other materials such as ceramics (slip casting, sol-gel processing etc), polymeric materials (blow moulding, pultrusion etc), composites (co-extrusion, sintering, polymeric curing etc) and electronic materials (photogravuring, CVD, laser processing etc). Even a mere enumeration and brief description of the large number of production technologies is a daunting task. But a full understanding of these technologies can only come about by a study of the underlying science and material behaviour under different conditions of stress, temperature and environment. This latter part requires a thorough knowledge of the structure of materials from the atomic scale to the microscopic scale and its correlation to the externally observed properties and mechanical behaviour. Further, precise mathematical analysis of the response of materials to applied stress, strain rate and temperature is a necessary concomitant to a comprehensive description of the production processes. Study of the subject of Production Engineering is thus not only useful and practical but also highly exciting and rewarding in view of its multidimensional reach and approach."

Prof. V V Kutumba Rao

Formerly Professor Department of Metallurgical Engineering, IT BHU

Formerly Director, Jawaharlal Nehru Aluminium Research, Design & Development Centre, Nagpur

I heartily acknowledge my best wishes for the publication of the book on "Production Technology" by Dr. Swadesh Kumar Singh. He is an accomplished researcher and academician. I have read the book. Since I am in India, I know something about the competitive examinations like GATE and UPSC, there is a need for a book which can elucidate concepts and give the appropriate exercises for the students who appear for GATE/IES and other competitive exams. From this perspective, concepts are aptly presented and are absolutely lucid. Especially, the explanations on 'Material Science' and 'Metal Forming' are extremely impressive. I know him personally and his efforts for the welfare of the students are bountiful. My best wishes for his unique contribution to reach the many students who yearn for such wonderful compilations. Thank You!

Prof. Narsingh Bahadur Singh, FASM, FSPIE, FOSA, FRSC

University of Maryland, Baltimore County (UMBC), USA

PREFACE

I have great pleasure in writing the book of core subject of Mechanical Engineering viz. Production Technology/ Engineering. A thorough understanding of the concepts developed in this book will prepare the reader for more advanced course on the subject. The entire syllabus of Production Engineering is presented in a simple and lucid style to make it comprehensible to an average student.

In this book I tried to present the approach not only for competitive examinations like GATE, IES and IAS but also for university examinations of JNTU, UPTEC etc. especially on the subjects like Material Science, Metrology and Production Processes. I also understood by teaching various categories of students that it becomes easy for the students when things are explained by going through the fundamentals. So in the present book I tried to explain most of the topics through the basics. The questions which already appeared in competitive examinations like GATE, IES and university exams (both subjective and objective) are incorporated in each chapter.

I would like to thank All India Council for Technical Education (AICTE), Department of Science and Technology (DST) Govt. of India, Defence Metallurgical Research Laboratory (DMRL) and my present place of work GRIET for giving me opportunity to learn and keep in touch with recent advances in the production technology either by supporting the research or providing the forum of eminent personalities to discuss various technology related issues.

I would like to express my heartfelt regard and gratitude to my teacher from whom I have learned the subject matter and which gave me inspiration to write this book. I thank Mahesh, Apurv, Limbadri, Srinivasu, Gangadhar, Swathi and other students of mine for typing, drawing various figures and proofreading the text.

Dr. Swadesh Kumar Singh

CONTENTS

1. Metrology

1.1 Introduction 1
1.2 Accuracy and Precision 5
1.3 Interchangeability 6
1.4 Selective Assembly 7
1.5 Limits, Fits and Tolerances 8
1.6 Allowance 10
1.6.1 Width of Tolerance Zone 10
1.7 Tolerance Sink 17
1.8 Design of Limit Gauges 18
1.9 Violation of Taylor's Principle in Gauge Design 20
1.10 Gauge Material 20
1.11 Slip Gauges 22
1.12 Angle Blocks 23
1.13 Sine Bar 23
1.14 Precision Ball Measurement 24
1.15 Vernier Calipers and Micrometer 26
1.16 Types of Errors 26
1.17 Interferometry 27
1.17.1 Optical Flat as Comparator 29
1.17.2 NPL Flatness Interferometer 30
1.18 Straightness 30
1.19 Flatness 31
1.20 Comparator 31
1.20.1 Sigma Comparators 32
1.20.2 Mikrokator 32
1.20.3 Pneumatic Comparators 33
1.21 Surface Finish 37
1.21.1 Lay and Flaw 37
1.21.2 Methods of Establishing Datum 38
1.21.3 Numerical Values to Represent Surface Finish 39
1.21.4 Representation of Surface Finish 40
1.21.5 Talysurf 40
1.22 Screw Thread Metrology 40
1.23 Acceptance Tests 43
Practice Questions 47

2. Material Science

2.1 Metals 56
2.2 Non-metals 56
2.3 Organic Polymer 56
2.4 Ceramics 58
2.5 Crystal Structures of Ceramics 60
2.5.1 AX-Type Crystal Structures of Ceramics 60
2.5.2 Cesium Chloride Structure 60
2.5.3 Zinc Blende Structure 61
2.5.4 AmXP-Type Crystal Structures 61
2.5.5 AmBnXp-Type Crystal Structures 61
2.6 Composite Materials 62
2.7 Phase Diagrams 62
2.8 Experimental Methods of Constructing Phase Diagrams 68
2.8.1 Thermal Analysis Allotropy method 68
2.8.2 Metallographic Method 68
2.8.3 X-ray diffractionn Method 68
2.9 Fe-Fe3C Phase Diagram 68
2.10 Development of Microstructure in Fe-C System 71
2.11 Family of Cast Irons 72
2.12 Effect of Alloying Element on Iron 73
2.13 Relationship between the Phase Diagrams and Properties of Alloys 74
2.14 High Strength Low Alloy steel (HSLA) 75
2.15 Defects in the Materials 79
2.15.1 Point Defects 79
2.15.2 Surface Defects 80
2.15.3 Line Defects 80

2.16 Yield Point Phenomenon........................82
2.17 Strengthening Mechanism........................82
2.17.1 Alloying........................82
2.17.2 Grain Refinement........................82
2.17.3 Work (strain) Hardening........................83
2.18 Type of Bond........................85
2.19 Time Temperature Transformation (TTT) Diagram........................85
2.20 Heat Treatment........................89
2.20.1 Annealing........................89
2.20.2 Normalizing........................91
2.20.3 Hardening........................91
2.20.4 Tempering........................93
2.20.5 Case Hardening........................94
2.21 TTT Diagrams for Hypo and Hyper-eutectoid Steels........................99
2.22 Crystal Systems........................99
2.22.1 Simple Cubic Structure........................101
2.22.2 Body Centred Cubic (BCC) Structure........................101
2.22.3 Face Centred Cubic (FCC) Structure........................101
2.23 Hardenability Curves Jominy end Quench Test........................102
2.24 Mechanism of Crystallization........................102
2.25 Methods for Measurement of Grain Size........................103
2.26 Crystal Directions........................103
2.27 Crystal Planes........................104
2.28 Creep........................106
2.29 Tensile Test........................107
2.30 Hardness Tests........................107
2.31 Impact Test........................107
2.32 Aluminum Alloys........................108
2.32.1 Non Heat Treatable Al Alloys........................108
2.32.2 Heat Treatable Wrought Alloys........................109
2.32.3 Al – base Casted Alloys........................111
2.33 Brasses........................112
2.34 Bronzes........................113
2.34.1 Tin-bronzes........................114
2.34.2 Tin Free Bronzes........................115
2.34.3 Si-Bronzes........................115
2.34.4 Be-Bronzes........................115
2.34.5 Leaded Bronzes........................115
2.35 Introduction to Titanium Alloys........................116
2.36 Corrosion Definition........................117
2.37 Copper........................127
2.38 Nano-materials........................130
Practice Questions........................137

3. Casting

3.1 Design of Risers........................165
3.2 Riser Design for Steel Castings........................168
3.3 Gating System........................170
3.4 Green Sand Mould........................175
3.5 Types of Solidification........................176
3.5.1 Skin Forming........................176
3.5.2 Dandritic Growth........................176
3.5.3 Pouring Temperature........................177
3.6 Die Casting........................177
3.7 Centrifugal Casting........................179
3.7.1 True Centrifugal Casting........................179
3.7.2 Semicentrifugal Casting........................180
3.7.3 Centrifuging........................180
3.8 Slush Casting........................181
3.9 Sodium Silicate (CO_2) Moulding........................181
3.10 Blow Moulding........................181
3.11 Shell Moulding........................182
3.12 Investment Casting........................183
3.13 Types of Casting Defects........................184
3.14 Types of Furnaces........................184
3.15 Metallurgy of Cast Steel........................185
3.15.1 Composition........................185
3.15.2 Melting of Cast Irons (Cupola)........................185
3.15.3 Steps in Operation........................186
3.16 Methods of Purifying Material........................186
3.17 Continuous Casting........................187
3.18 Injection Moulding........................187
Practice Questions........................200

4. Metal Cutting

4.1 Terminology of Single Point Cutting Tool (As per ASA) 213
- 4.1.1 Back Rake Angle 213
- 4.1.2 Side Cutting Edge Angle 214
- 4.1.3 Side Rake Angle 215
- 4.1.4 Side Relief Angle 215
- 4.1.5 End Cutting Edge Angle 216
- 4.1.6 End Relief Angle (Clearance angle) 216
- 4.1.7 Nose Radius 216
- 4.1.8 ASA Tool Signature 216

4.2 British System 217

4.3 Normal or Orthogonal Rake System (ORS) 217

4.4 Types of Metal Cutting Process 217

4.5 Distribution of Heat in Metal Cutting 218

4.6 Types of Chips 219
- 4.6.1 Continuous Chips 219
- 4.6.2 Discontinuous Chips 219
- 4.6.3 Chips with Build up Edge 219
- 4.6.4 Chips with Unusual Strains 220

4.7 Merchant's Analysis for Chip Thickness Ratio 220

4.8 Determination of Cutting Forces 222

4.9 Determination of Area of Shear Plane and Shear Strength 223

4.10 Assumptions of Merchant's Circle 224

4.11 Determination of Shear Angle 225

4.12 Velocity Triangle 225

4.13 Ernest and Merchant Theory 226

4.14 Lee and Shaffer (slip line field theory) 227

4.15 Contact Length 238

4.16 Measurement of Temperature 238
- 4.16.1 Tool Workpiece Thermocouple 238
- 4.16.2 Inserted Thermocouple 239
- 4.16.3 Infrared Measurement 239
- 4.16.4 Hardness and Microstructure Studies 239

4.17 Cutting Fluids 239
- 4.17.1 Properties of Cutting Fluids 240
- 4.17.2 Selection of Cutting Fluids 240
- 4.17.3 Workpiece Material 240

4.18 Mechanism of Tool Wear 241
- 4.18.1 Types of Tool Wear 243
- 4.18.2 Taylor's Tool Life Equation 244

4.19 Effect of Parameters on Tool Life 249

4.20 Machinability 250

4.21 Economics of Machining 251

4.22 Tool Materials 257

4.23 Measurement of Cutting Forces 259

4.24 Grinding 259
- 4.24.1 Classification of Grinding Processes 260
- 4.24.2 Structure 262
- 4.24.3 Grade 263
- 4.24.4 ISO Designation of Grinding Wheel 263
- 4.24.5 Wheel Dressing 264

4.25 Unconventional/Non-traditional Machining Processes 264
- 4.25.1 Electro Chemical Machining 264
- 4.25.2 Electrode Discharge Machining 268
- 4.25.3 Ultrasonic Machining 270
- 4.25.4 Abrasive Jet Machining 271

4.26 Drilling 272
- 4.26.1 Drill Geometry 273
- 4.26.2 Important Angles of a Drill 275
- 4.26.3 Machining Time 276

4.27 Milling 277
- 4.27.1 Peripheral or Slab Milling Operation 277
- 4.27.2 Face Milling 277
- 4.27.3 End Milling 278
- 4.27.4 Gang Milling 278
- 4.27.5 Straddle Milling 278
- 4.27.6 Up Milling and Down Milling 279
- 4.27.7 Mechanics of Milling 279
- 4.27.8 Mean Chip Thickness 280

4.28 Powder Metallurgy 287

4.29 Jigs and Fixtures 289
- 4.29.1 Principle of Location 290
- 4.29.2 Steps in Designing the Jigs 291

4.30 Screw Thread Manufacturing 292

4.31 Gear Manufacturing 293
- 4.31.1 Materials for Gears 293
- 4.31.2 Forming 293
- 4.31.3 Machining 294

Practice Questions 297

5. Metal Forming

5.1 Wire Drawing 341

5.1.1 Defects of Wire Drawing 342

5.2 Tube Drawing 342

5.3 Extrusion 343

5.3.1 Forward Extrusion or Direct Extrusion 343

5.3.2 Backward Extrusion or Indirect Extrusion 344

5.3.3 Hydrostatic Extrusion 345

5.3.4 Impact Extrusion 346

5.4 Analysis of Wire Drawing and Extrusion 347

5.5 Extrusion Defects 356

5.6 Forging 356

5.6.1 Forging Defects 358

5.6.2 Analysis of Forging 359

5.7 Rolling 374

5.7.1 Mechanics of Rolling 375

5.7.2 Camber Rolls 378

5.7.3 Back up Rolls 378

5.7.4 Three High Mill 379

5.7.5 Planetary Mill 380

5.7.6 Tandem Mill 380

5.7.7 Rolling Defects 380

5.7.8 Test to Calculate the Defects 381

5.8 Punching and Blanking 383

5.9 Deep Drawing 385

5.9.1 Redrawing 387

5.9.2 Analysis of Deep Drawing 387

5.9.3 Measuring of Formability 391

5.9.4 Common Defects in Deep Drawing ... 391

5.10 Bending 392

Practice Questions 399

6. Welding

6.1 Metallurgy in Welding 414

6.2 Bead Geometry 416

6.3 Classification of Welding Process 417

6.4 Arc Welding 417

6.4.1 Shielded Metal Arc Welding (SMAW) 417

6.4.2 Transformer 418

6.4.3 Submerged Arc Welding (SAW) 429

6.4.4 Gas Tungsten Arc Welding (GTAW) or Tungsten Inert Gas Welding (TIG) ... 430

6.4.5 Arc Blow 431

6.4.6 Duty Cycle 432

6.4.7 Plasma Arc Welding (PAW) 432

6.4.8 Electro Slag/Electro Gas Welding ... 433

6.5 Resistance Welding 434

6.5.1 Spot Welding 435

6.5.2 Projection Welding 436

6.5.3 Seam Welding 436

6.5.4 High Frequency Resistance Welding (HFRW) 444

6.5.5 Flash Butt Welding 445

6.6 Solid State Welding processes 445

6.6.1 Friction Welding 445

6.6.2 Ultrasonic Welding 446

6.6.3 Explosive Welding 447

6.7 Radiant Energy Welding 448

6.7.1 Electron Beam Welding 448

6.7.2 Laser Beam Welding 449

6.8 Thermit Welding Process 450

6.9 Allied Processes 451

6.10 Gas Welding 452

6.11 Forces in Welding 455

6.12 Types of Metal Transfer 455

6.13 Types of Fluxes 456

6.14 ISO Designation 457

6.15 Welding Positions 458

6.16 Types of Joints 458

6.17 Types of Weld 459

6.18 Welding Symbols 460

6.19 Defects in Welding 461

Practice Questions 465

OOOO

CHAPTER 1

Metrology

1.1 INTRODUCTION

Metrology or industrial inspection involves inspection of components produced from production, a unit to check whether the dimensions fall within the acceptable limits assigned by the consumer. Inspection is divided into active and passive type. In active inspection parts are checked when it is being produced in any production system. It is also called online inspection i.e. if something goes wrong, correction can be taken. But in passive inspection parts are checked when it is already being produced and inspection, is meant for segregation of the components as good or bad. In industries generally active inspection are encouraged but it depends upon the type of application.

Parts which are produced from any manufacturing system follow a normal distribution. Machine environment can be considered as infinite universe and machined components are considered as random data collection. When data is collected randomly from an infinite universe, the data follows a normal distribution as shown in Fig. 1.1.

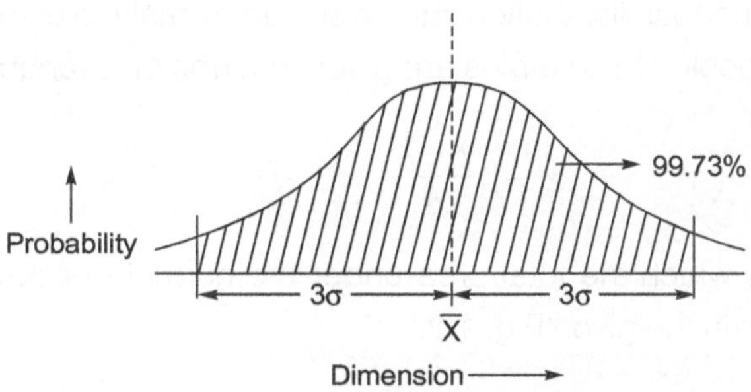

Fig. 1.1 (Normal Distribution)

Let us take one example to understand the concept. There is a lathe machine aimed at producing shafts of 25.0 mm diameter. As the shafts are coming out of the machine, no two dimensions will match and there will be some variation in the dimensions like 25.002 mm, 24.985 mm etc. It is because these dimensions are following a normal distribution. That is why it is said that what ever component you have made from your production systems, no two components will match and you cannot duplicate the part. In the above normal distribution $\overline{X}$ is mean or targeted value of the machine and σ is the process dispersion or standard deviation. Following are the factors that may change the process or targeted mean ($\overline{X}$) value in a typical machine environment :

1. Tool and die wear.
2. Excessive machine vibrations.
3. Wear or minor failure or loosening of machine parts.
4. Change of machine and process.
5. Measurement error or change in material property.

Following factors changes the standard deviation of the process.

1. Carelessness of operator.
2. New or inexperienced operator.
3. Deterioration in the condition of machine.
4. Frequent resetting or readjustment of the process.
5. Increase variability of the material.

99.73% area of the normal distribution takes between $\pm 3\sigma$ limits of this normal distribution. It means 99.73% components will be produced in 6σ range. This 6σ is called proves capability of machine. To move further with inspection let us understand normal distribution with following examples.

Example : 1.1

A lathe machine is targeted to produce shafts of 25.0 mm diameter. This conventional lathe is having process dispersion of 0.15 mm. Determine:

1. **How many shafts, machine is producing below 24.9 mm?**
2. **How many shafts, machine is producing above 25.2 mm?**
3. **Normally machine will produce shafts in what range?**
4. **If the acceptable limits of shafts by consumer is 25 ± 0.35 mm, how many defective shafts machine is producing?**

Solution :

Problems related to normal distribution can be solved by making use of normal distribution tables given at the end of book. These tables are given in terms of a variable Z which is

$$z = \frac{X - \overline{X}}{\sigma}$$

X is the value up to which the total area under the normal distribution from $-\infty$ to X is to be calculated as shown in the given Fig. 1.2.

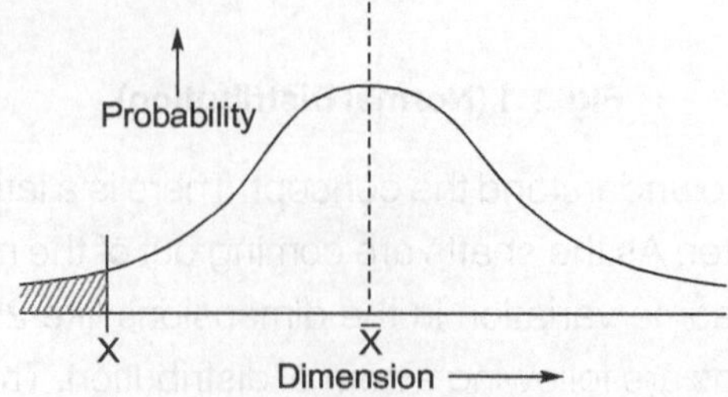

Fig. 1.2

1. In the given problem we have to calculate = $\int_{-\infty}^{24.9} P(X)dX$

where P(X) is the probability function

$$X = 24.9$$
$$\bar{X} = 25$$
$$\sigma = 0.15$$

So
$$z = \frac{X-\bar{X}}{\sigma} = \frac{24.9-25}{0.15} = \frac{-0.1}{0.15} = -0.666$$

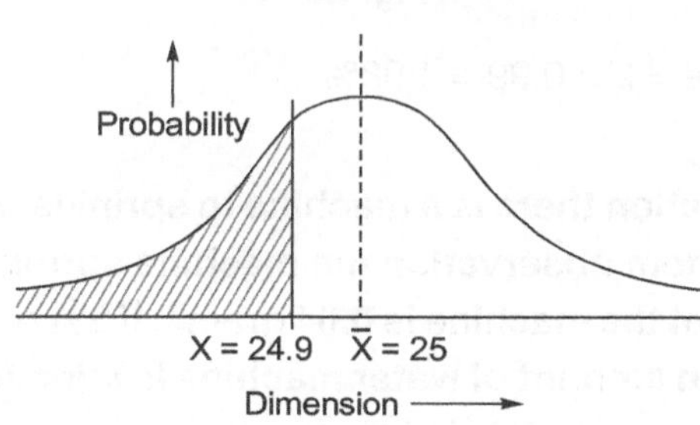

Fig. 1.3

Corresponding to this Z value, the area under normal distribution up to 24.9 mm from the table is 0.2546. So 25.46% shafts will be produced below 24.9 mm or 74.54% shafts will be produced more than 24.9 mm diameter.

2. In the above problem the shaded area has to be calculated as shown in Fig. 1.4.

$$z = \frac{X-\bar{X}}{\sigma} = \frac{25.2-25}{0.15} = \frac{0.2}{0.15} = 1.33$$

The area corresponding to this value from the table is 0.9082.
So the shaded area is (1 – 0.9082) = 0.0918. So 9.18% will be dimension more than 25.2 mm.

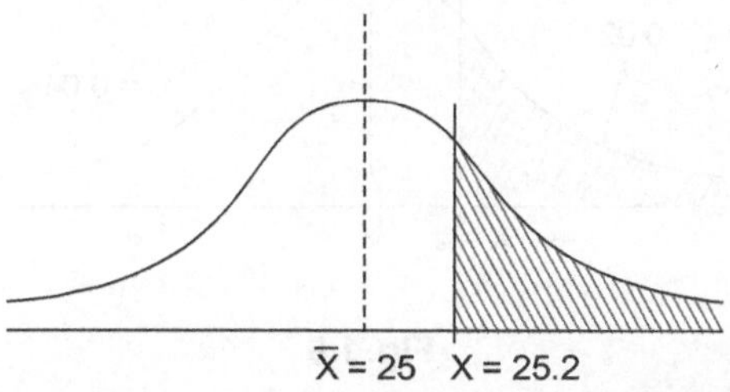

Fig. 1.4

3. Around 99.73% components will be produced within 25 ± 3(0.15) mm. So machine will produce components between 24.55 – 25.45 mm range.
4. Shaded area shows the defective components produced by the machine. Since the distribution is symmetrical, upon calculating area on one side defectives can be calculated by multiplying it by 2. Area under the distribution below 24.65 mm.

$$Z = \frac{24.65-25}{0.15} = -2.33$$

Area from the table corresponding to Z = – 2.33 is 0.0099. So shafts below 24.65 are 0.99%.

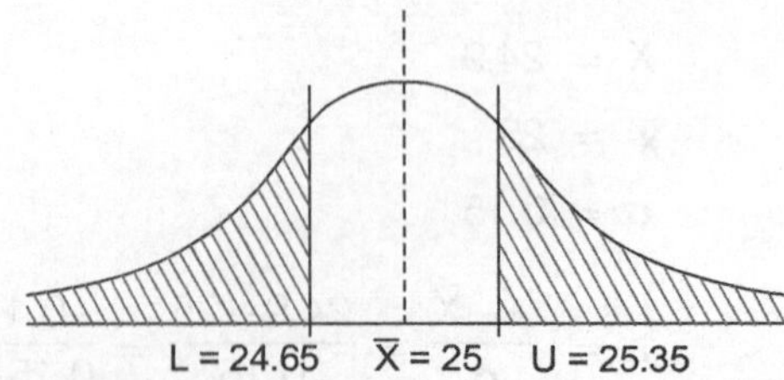

Fig. 1.5

Total defective percentage = 2 × 0.99 = 1.98%

Example : 1.2

To honor guests in a function there is a machine to sprinkle fragrant water on each guest's head. In a particular random observation the machine sprinkles 4 grams of fragrant water. The normal distribution of the machine is 0.04 grams. If 2% of time machine injects 4 grams of water what is the mean amount of water machine is injecting.

Solution :

$$Z = \frac{4-\overline{X}}{0.04}$$

The value of Z corresponding to 0.02 area = – 2.055 (from normal distributions tables)

$$-2.055 = \frac{4-\overline{X}}{0.04}$$

$$\Rightarrow \quad \overline{X} = 4.0822$$

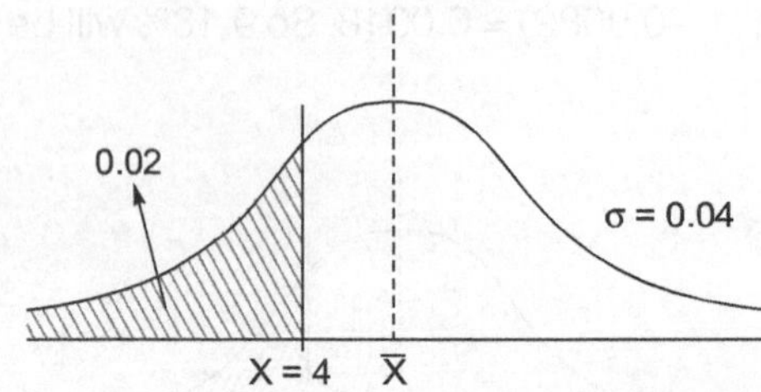

Fig. 1.6

Example : 1.3

A milling machine is programmed to produce round shape of 45 mm diameter. The machine has process dispersion of 0.2 mm. The customer has given specifications of product as 44.8 ± 0.3. If the product is over size it can be reworked and if it is undersize it will be scrapped. In the present settings how much percentage of scrap and rework machine is produced?

Solution :

Calculation of scrap

$$z = \frac{44.5-45}{0.2} = -2.5$$

Area corresponding to this value = 0.0062

So scrap = 0.62%

Calculation of rework

$$z = \frac{45.1-45}{0.2} = 0.5$$

$$\text{Area} = 0.6915$$

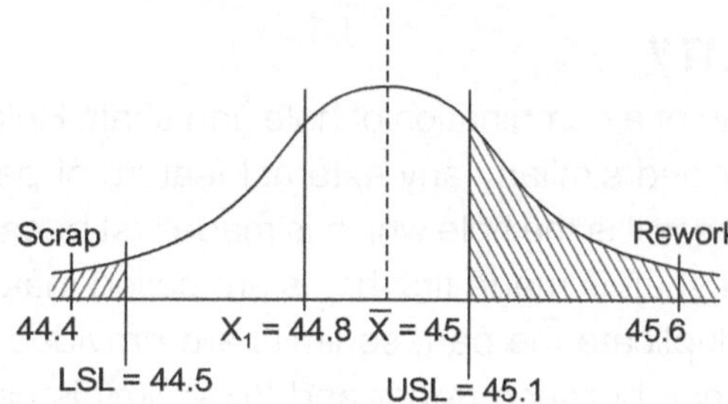

Fig. 1.7

So, percentage rework = 100 – 69.15 = 30.85%

1.2 ACCURACY AND PRECISION

Accuracy is defined as closeness to the exact value and precision is repeatability. Accuracy is something related to the mean value of the parts produced and precision is related to the process dispersion. In Fig.1.8 (i) mean value of the observations is very close to the targeted value but there are large fluctuations in the observations due to higher process capability, such systems are accurate but not precise.

In Fig.1.8 (ii) standard deviation of observations is low and also the mean is close to the targeted value, so such systems are precise and accurate.

In Fig.1.8 (iii) although the process dispersion of the system is low but due to the improper setting of the machine mean value is far from the targeted value. Such systems are called precise but are not accurate.

In Fig.1.8 (iv) the process dispersion is large and the mean is far from the targeted value. Such systems are said to be neither precise nor accurate.

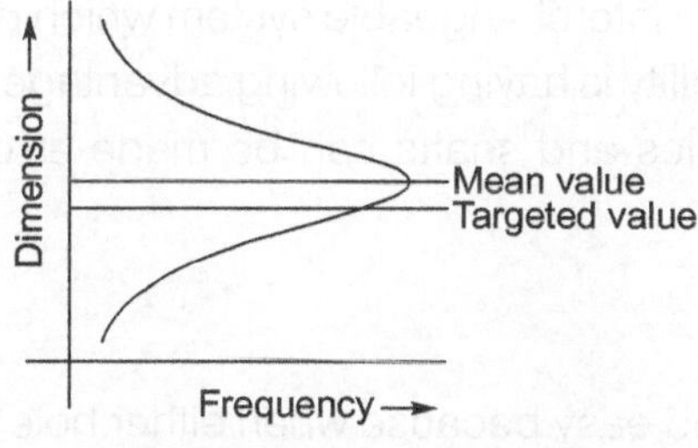

(i) Accurate but not precise

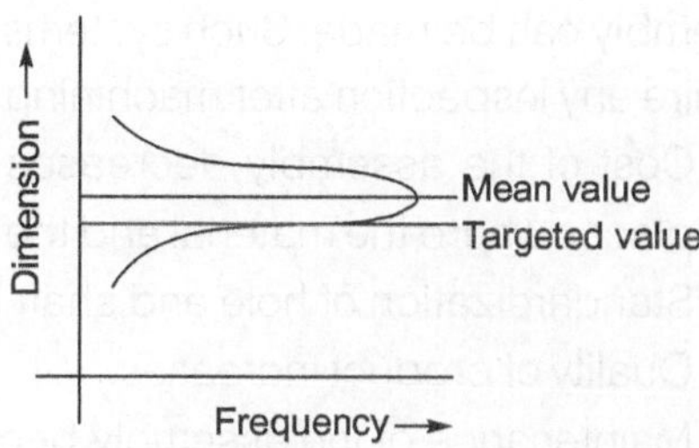

(ii) Accurate and precise

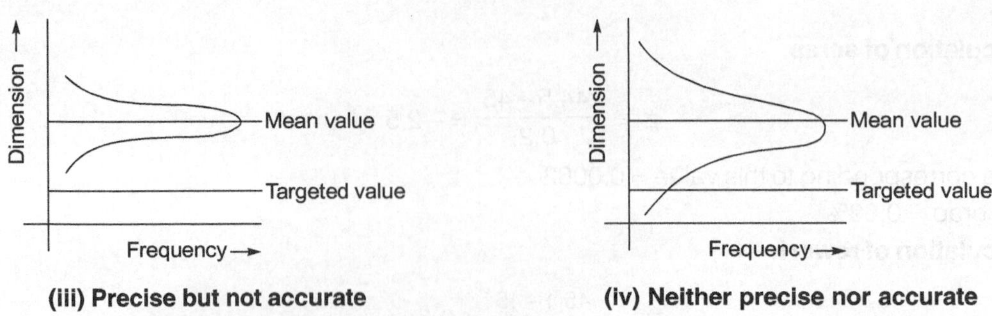

Fig. 1.8 : Accuracy Vs Precision

1.3 INTERCHANGEABILITY

Any assembly is made up of a combination of hole and shaft. Hole is any internal feature of part and may not be circular and similarly any external feature of part is called shaft, may not be circular. In most of the cases it is the hole which is made first because standard size of drills and reamers are available. If shaft is made first holes are called make to suit. As discussed earlier that it is impossible to duplicate the part, so limits are provided on the dimensions which are acceptable. This difference between upper and lower limit is called desired tolerance by the consumer. The targeted value of the hole and shaft is called basic size and nearest round number as per the standard is called nominal size.

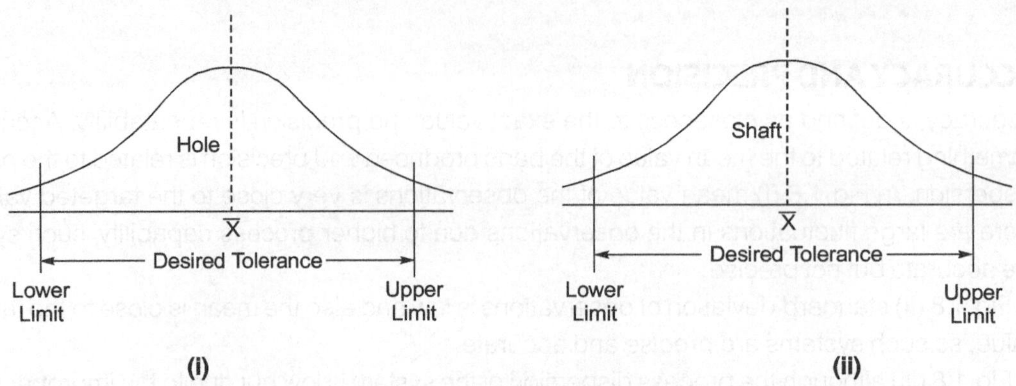

Fig. 1.9 : Full Interchangeability

Suppose there are two different machines producing holes and shafts. If it so happens that process capability is equal to the desired tolerance individually (as shown in Fig. 1.9 (i) and (ii)), if a hole is selected randomly from hole lot and a shaft is selected randomly from shaft lot, assembly can be made. Such systems are called a fully interchangeable system which does not require any inspection after machining. Interchangeability is having following advantages :

1. Cost of the assembly decreases because of holes and shafts can be made at different places where the material and the labour is cheap.
2. Standardization of hole and shaft is possible.
3. Quality of product increases.
4. Maintenance of the assembly becomes cheap and easy because when either hole or shaft fails one does not have to throw away the complete assembly and the failed parts only needs to be replaced.

1.4 SELECTIVE ASSEMBLY

To increase the quality of assembly, the desired tolerances for both hole and shaft should decrease but machines will have certain process capabilities. By converting conventional machines to NC, process capability decreases but desired tolerances are decreasing more than that. Selective assembly is a mathematical procedure to achieve full interchangeability although machines are not capable i.e. process capability is much larger than the desired tolerance. Let us understand it by discussing some cases.

Case : I

Process capability of hole and shaft are equal but more than the desired tolerance.

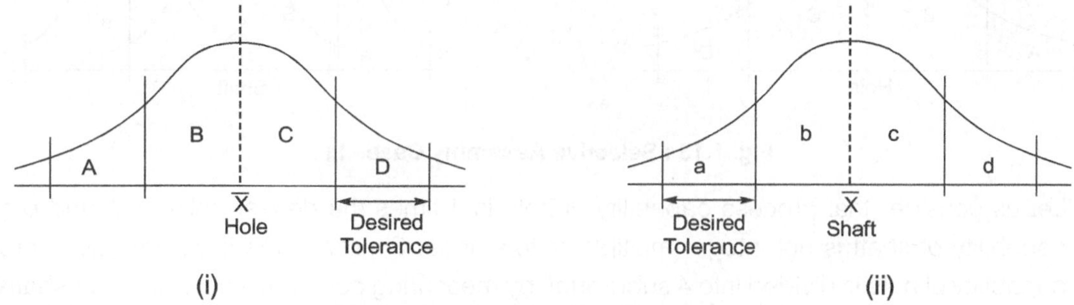

Fig. 1.10 : Selective Assembly Case - I

Let us assume that process capability is 4 times the desired tolerance of both hole and shaft. The process capability range of both hole and shaft is divided into 4 equal parts as shown in Fig. 1.10 (i) and (ii). This means in case of selective assembly each component has to be inspected after manufacturing and it is put into different subgroups according to its dimensions. The variation of dimensions with in one word the subgroup is equal to the desired tolerance. So when randomly a hole is selected from subgroup 'A' and a shaft is selected randomly from subgroup 'a', assembly can be made.

Case : II

Process capability of hole is equal to the desired tolerance and process capability of shaft is worse than the desired tolerance.

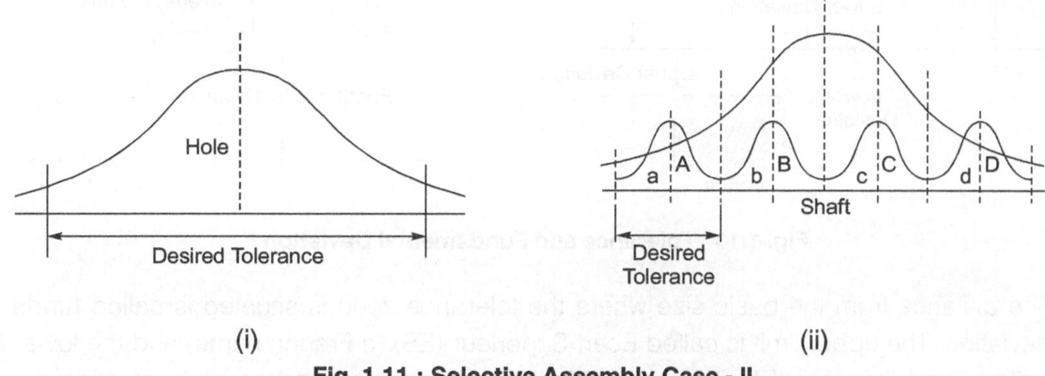

Fig. 1.11 : Selective Assembly Case - II

In this case process capability range of shaft is divided into 4 equal divisions so that variation with in the subgroup is equal to the desired tolerance. For producing holes of subgroup of shafts 'a', the machine is targeted at the mean of subgroup of shafts 'a' and the holes will be directly categorized as subgroup 'A' without any inspection because process capability of machine producing hole is equal to the desired tolerance. When variations are equal to the desired tolerances assembly can be made to give a quality of assembly.

Case : III

Process capability of both holes and shafts are different and worse than the desired tolerance.

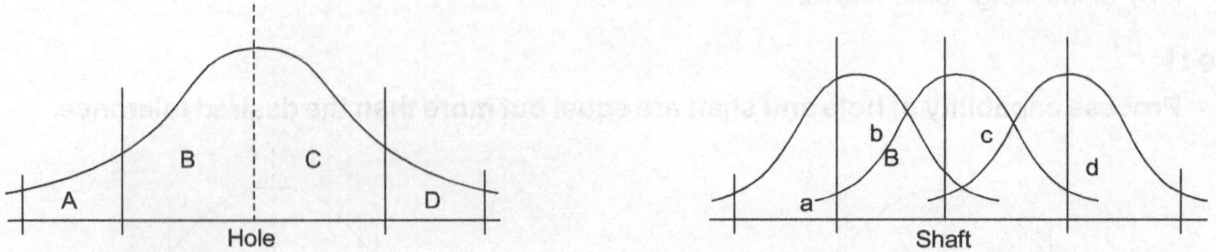

Fig. 1.13 : Selective Assembly Case - III

Let us consider that process capability of hole is 4 times the desired tolerance and process capability of shaft is not integer multiple of tolerance. Similar to the previous cases process capability of hole is divided into 4 subgroups by measuring each and every part. For shafts e.g. in subgroup 'b' shafts will be produced in two different settings. So in this case also each and every component must be inspected after manufacturing.

1.5 LIMITS, FITS AND TOLERANCES

As discussed earlier that due to process capabilities it is not possible to duplicate any part, so for assembly parts (either hole or shaft) should come within certain range of dimensions. This acceptable variation is called tolerance. Larger size is called upper limit and smaller size is called lower limit.

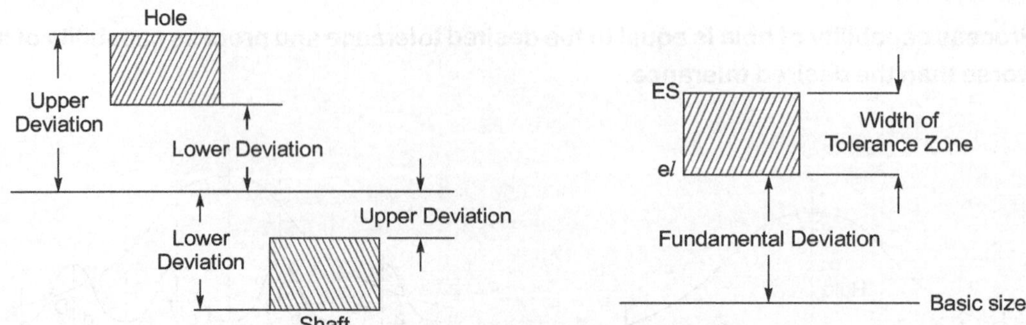

Fig. 1.13 : Tolerance and Fundamental Deviation

The distance from the basic size where the tolerance zone is situated is called fundamental deviation. The upper limit is called Ecart Superieur (ES) (a French name) and the lower limit is called Ecart Inferieur (EI). There are 25 different types of fundamental deviation as shown in Fig. 1.14.

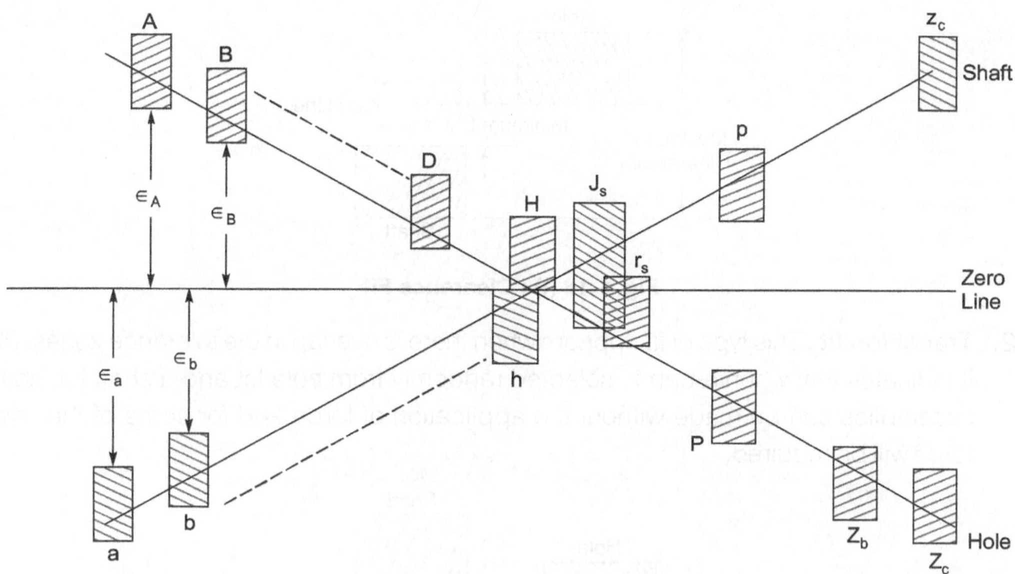

Fig. 1.14 : Variation of Different Fundamental Deviation types

Some of the alphabets are missing and some extra alphabets are there e.g. j_s, z_a, z_b and so on. Value of fundamental deviation for 'A' type hole will be same as 'a' type of shaft with a difference of sign and so on. All fundamental deviation has empirical formulas. An assembly is made up of a combination of hole and shaft. In this assembly if 'H' type of hole appears, the system is called **hole basis system** and if 'h' type of shaft is present it is called **shaft basis system**.

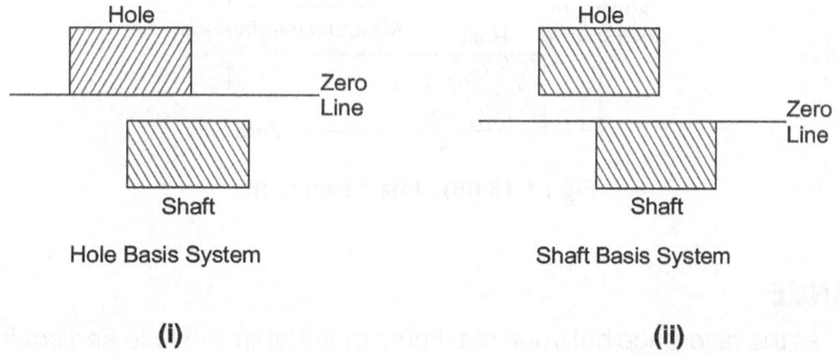

Fig. 1.15 : Hole and Shaft basis system

Fundamental deviation is also expressed as the distance between zero line and the limit closer to it i.e. either EI or e_s. Any information in capital letters refers to hole and any information given in smaller letters refers to shaft.

It is a common sense that large size shaft and small size hole will have more material. So these limits are called Maximum material limit and lower limits of shaft upper limit of hole is called Minimum material limits. A fit is defined as the relationship between hole and shaft before assembly.

Broadly there are three types of fits:

1. **Clearance fit** : If lower limit of hole is larger than the upper limit of shaft [as shown in Fig. 1.16 (i)]. It is considered as clearance fit.

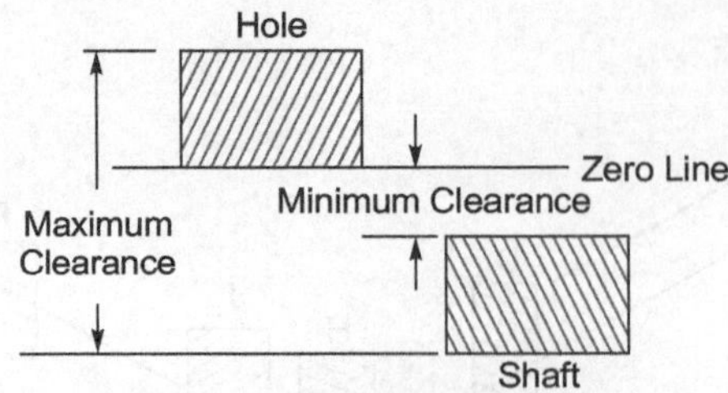

Fig. 1.16 (i) : Clearance Fit

2. **Transition fit :** This type of fit appears when there is overlap in the tolerance zones. Physically it indicates that when a part is selected randomly from hole lot and a shaft lot, some of the assemblies can be made without the application of force and for some of the assemblies force will be required.

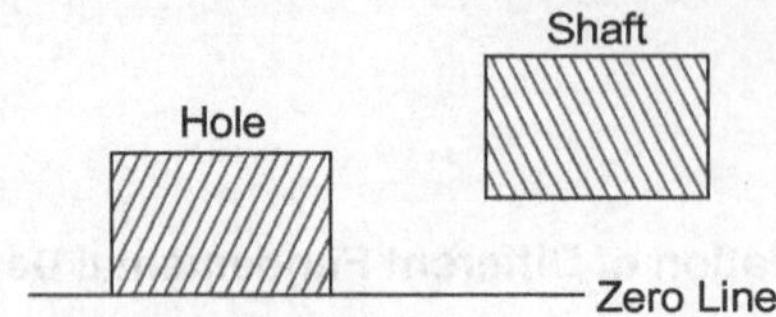

Fig. 1.16 (ii) : Transition Fit

3. **Interference fit :** If maximum size of hole is smaller than the minimum size of shaft then force has to be applied to make the assembly. Such fits are called interference fits.

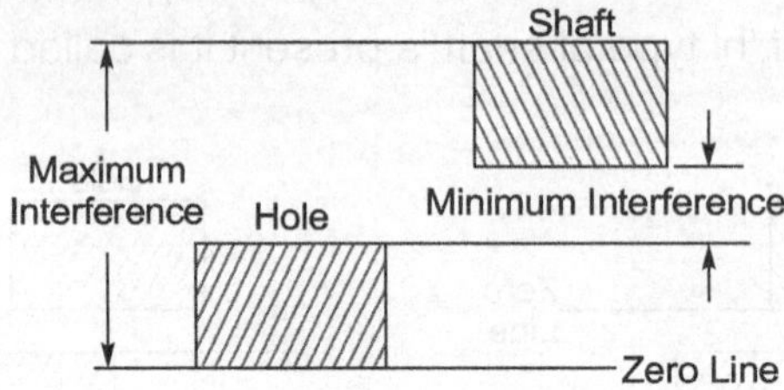

Fig : 1.16 (iii) : Interference fit

1.6 ALLOWANCE

It is defined as the difference between maximum material limit of hole and shaft. Depending upon the type of fit either it is equal minimum clearance or maximum interference.

1.6.1 Width of Tolerance Zone

There are three systems used in the world.

1. British Standard BS-4500-1969
2. International Standard ISO: 286-1988
3. Indian Standard IS-919

But the standards of fundamental deviation and width of tolerance zone are same. As discussed previously there are 25 different types of fundamental deviations that are standardized. For the tolerance width initial recommendation was to vary tolerance according to cubical expression.

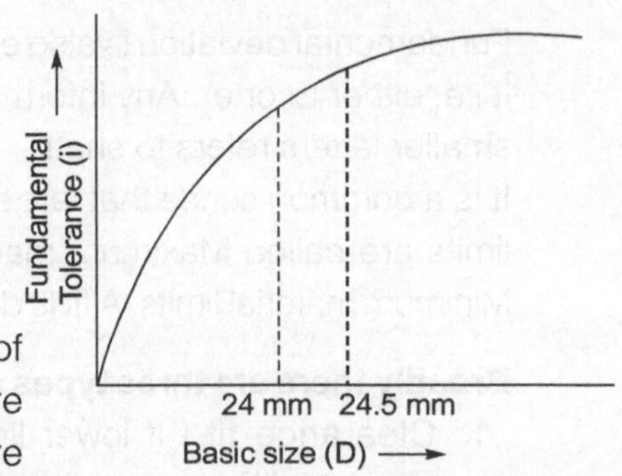

Fig : 1.17 : Fundamental Tolerance

$$i = 0.45\sqrt[3]{D} + 0.001D$$

It can be seen from the Fig. 1.17 that as the basic size increases the tolerances will also increase. For larger basic size it is not possible to control the tolerance preciously. But the major problem appeared when the basic size increases slightly e.g. from 24 to 24.5 mm, there is a change in tolerance values. So later on it was decided that there will be same tolerances between certain diameter range. These diameters are standardized as

Above (mm)	Upto and including (mm)	Above (mm)	Upto and including (mm)
—	3	80	120
3	6	120	180
6	10	180	250
10	18	250	315
18	30	315	400
30	50	400	500
50	80		

The curve of fundamental tolerance 'i' passes through the geometric mean of this diameter range as shown in Fig. 1.18. So in all the formulas of tolerance and fundamental deviation, the geometric mean diameter will be used. Initially there were 16 grades of tolerance zones. 2 more tolerance grades were added for precision industries.

These tolerances were arranged in the five step preferred series i.e. every fifth step will be multiple of 10 and the series is approximate geometric series. The calculation is shown in the Table.

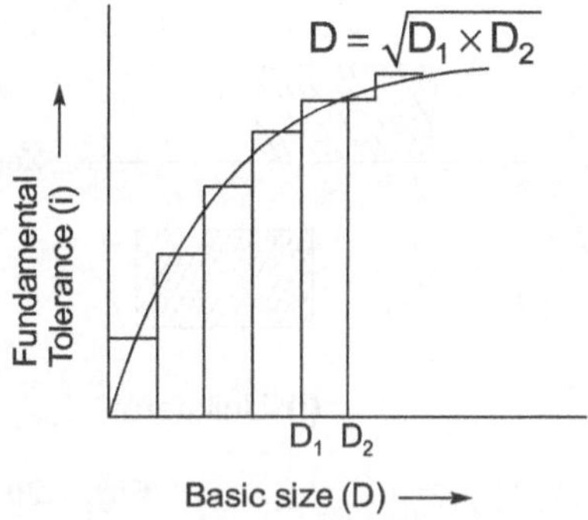

Fig : 1.18 : Modified Fundamental Tolerance

IT 01 $0.3 + 0.008\,D$	IT 0 $0.5 + 0.012\,D$	IT 1 $0.8 + 0.02\,D$ a	IT 2 ar	IT 3 ar^2
IT 4 ar^3	IT 5 $(10)^{4/5}\,i = 6.31\,i$ $ar^4 \Rightarrow 7i$	IT 6 $10\,i$	IT 7 $10(10)^{1/5}\,i =$ $15.8\,i \Rightarrow 16\,i$	IT 8 $10(10)^{2/5}\,i =$ $25.11\,i \Rightarrow 25\,i$
IT 9 $10(10)^{3/5}\,i = 40i$	IT 10 $10(10)^{4/5}\,i = 64i$	IT 11 $100\,i$	IT 12 $160\,i$	IT 13 $250\,i$
IT 14 $400\,i$	IT 15 $640\,i$	IT 16 $1000\,i$		

This series does not follow any rule that is why it is called modified 5 step preferred series. At present

IT 01 to IT 1 $\Rightarrow$ Empirical formula

IT 1 to IT 5 $\Rightarrow$ Exact geometric series

IT 5 to IT 16 ⇒ Preferred series

Suppose a hole and shaft assembly is designated by 25 H_7d_8. 25 indicates the basic size 'H' is the type of hole with width of tolerance zone IT 7. 'd' is the type of shaft with IT 8 is the width of tolerance zone. This can be represented by following fit diagram.

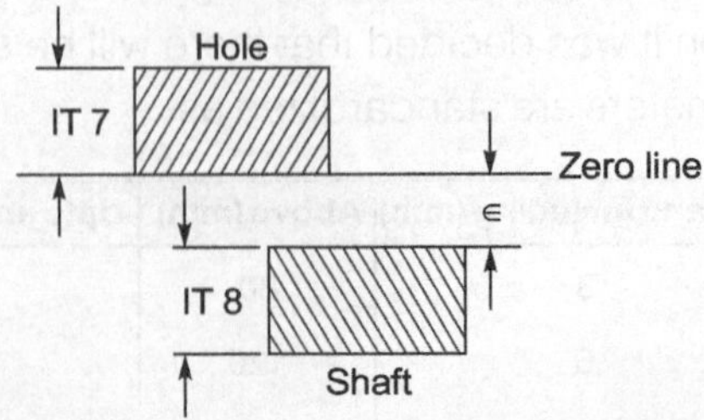

Fig : 1.19 : Fit Representation of $25H_7d_8$

When the tolerance zone is only on one side of zero line, it is called **unilateral**. When tolerance zone is on both side of zero line it is called **bilateral**.

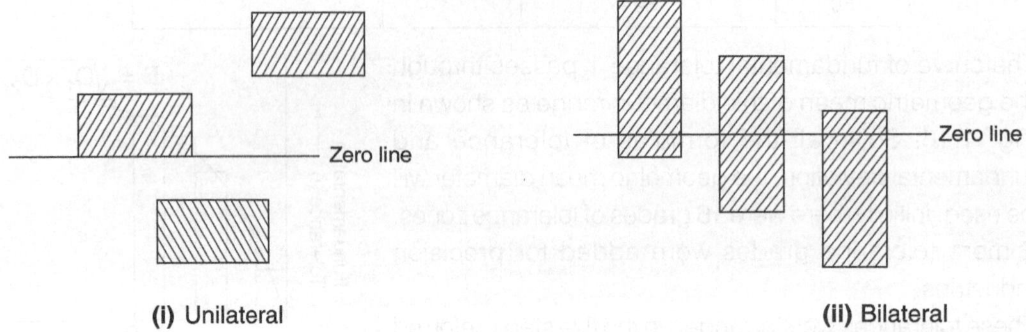

(i) Unilateral **(ii) Bilateral**

Fig. 1.20 : Unilateral and Bilateral Tolerances

Example : 1.4

A hole and shaft have a basic size of 30 mm and are to have a clearance fit with maximum clearance of 0.04 mm and a minimum clearance of 0.02 mm. The hole tolerance is to be 1.5 times the shaft tolerance. Determine limits for both hole and shaft using :

1. **A hole basis system**
2. **A shaft basis system.**

Solution :

1. **Hole basis system**

$$X + 0.02 + 1.5\,X = 0.04$$
$$X = 0.008 \text{ mm}$$
$$\Rightarrow \quad 1.5\,X = 0.012 \text{ mm}$$
$$\text{Hole} \rightarrow \text{Max} = 30.012 \text{ mm}$$
$$\text{Min} = 30.00 \text{ mm}$$
$$\text{Shaft} \rightarrow \text{Max} = 29.98 \text{ mm}$$
$$\text{Min} = 29.972 \text{ mm}$$

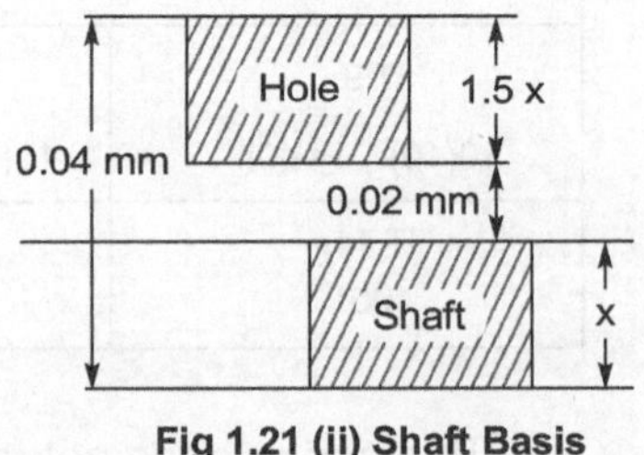

Fig 1.21 (ii) Shaft Basis

2. **Shaft basis system**

$$X + 0.02 + 1.5\,X = 0.04$$
$$X = 0.008 \text{ mm}$$
$$\Rightarrow \quad 1.5\,X = 0.012 \text{ mm}$$
$$\text{Hole} \rightarrow \text{Max} = 30.032 \text{ mm}$$
$$\text{Min} = 30.02 \text{ mm}$$
$$\text{Shaft} \rightarrow \text{Max} = 30.00 \text{ mm}$$
$$\text{Min} = 29.992 \text{ mm}$$

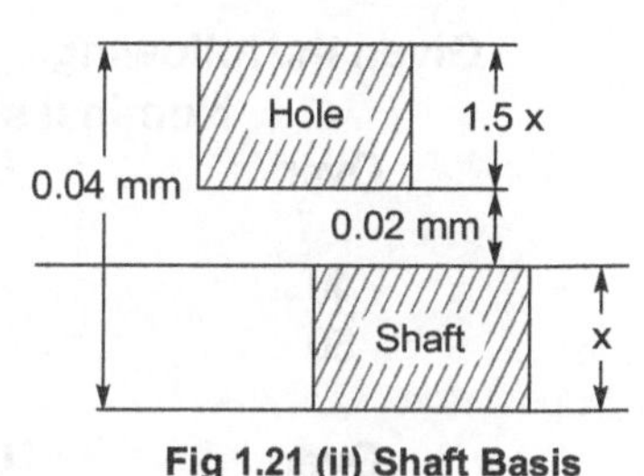

Fig 1.21 (ii) Shaft Basis

Example : 1.5

Calculate the fundamental deviation and tolerance and hence the limits of the size for the shaft and hole for the following fit 64 mm H_8f_7. The diameter steps are 50 mm and 80 mm. For shaft designation f, upper deviation is assumed as $-5.5\,D^{0.41}$:

Data :	**For**	**Tolerance**
	H8	**25 i**
	f7	**16 i**

Solution :

Since 64 falls between diameter range of 50 to 80 so,

$$D = \sqrt{50 \times 80}$$
$$= 63.24 \text{ mm}$$

Fundamental tolerance
$$i = 0.45\,D^{1/3} + 0.001\,D$$
$$= 1.856 \text{ microns}$$

We should note here that although input D is in mm but i is in microns.

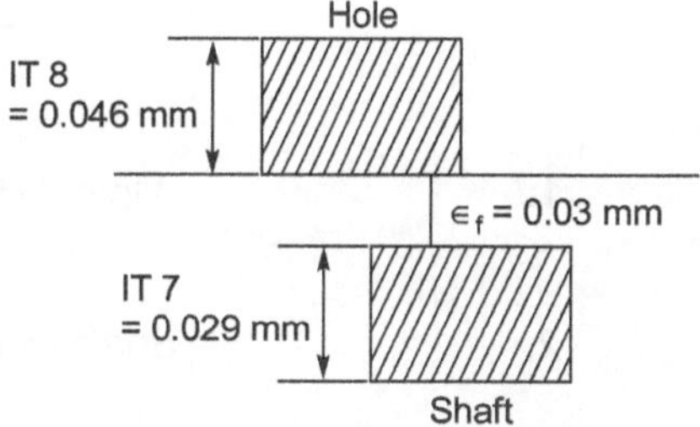

Fig 1.22: Fit diagram of 64 H_8f_7

$$IT\,7 = 16\,i = 29.69 \text{ microns} = 0.0296 \text{ mm}$$
$$IT\,8 = 25\,i = 46.4 \text{ microns} = 0.046 \text{ mm}$$

Fundamental deviation for 'f' type of shaft

$$\epsilon_f = -5.5\,D^{0.41} = -5.5(63.24)^{0.41} = -30.11 \text{ microns} = -0.03 \text{ mm}$$
$$\text{Hole} \rightarrow \text{Max} = 64.046 \text{ mm, Min} = 64.00 \text{ mm } 64^{+0.046}_{+0.0}$$
$$\text{Shaft} \rightarrow \text{Max} = 63.97 \text{ mm, Min} = 63.941 \text{ mm } 64^{-0.03}_{-0.059}$$
$$\text{Allowance} = 0.03 \text{ mm}$$

Example : 1.6

A medium force fit H_7p_6 is specified for an assembly of a hand wheel on a shaft nominal dia. Calculate

1. **The dimensions of hole**
2. **The dimensions of shaft**
3. **Maximum interference**
4. **Minimum interference**

Given the following:

Nominal size		Limits of tolerance for p6 shaft in microns	
Over	**Upto and including**	**es**	**ei**
		+	**+**
40	**50**	**42**	**26**
50	**65**	**51**	**32**

		Limits of tolerance for H7 hole in microns	
Over	**Upto and including**	**ES**	**EI**
30	**50**	**25**	**0**
50	**80**	**30**	**0**

Solution :

The fit diagram of above problem is

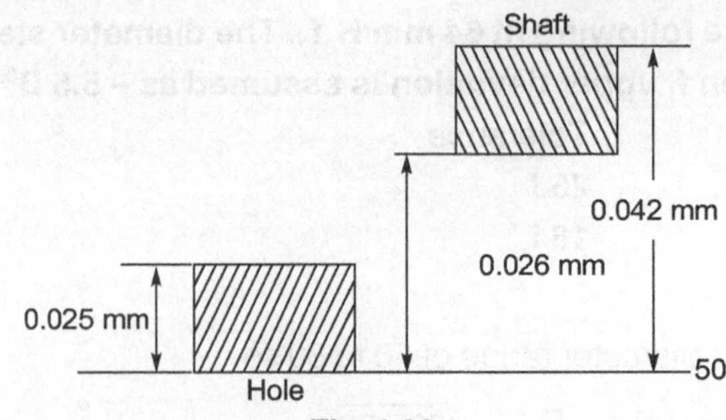

Fig. 1.23

1. Hole → Max. = 50.025 mm
   Min = 50.00 mm $50^{+0.025}_{+0}$
2. Shaft → Max. = 50.042 mm
   Min. = 50.026 mm $50^{+0.042}_{+.026}$
3. Maximum interference = 0.042 mm
4. Minimum interference = 0.026 − 0.025 = 0.001 mm

Example : 1.7

Determine and sketch the limits of tolerance and allowance for a 75 mm shaft and hole pair designated H_7s_8. The basic size lies in the range of 50-80 mm. The multipliers for grades 7 and 8 are 16 and 25 respectively. The fundamental deviation for 's' shaft is (IT 7 + 0.4D) microns.

Solution :

$D = \sqrt{50 \times 80} = 63.24$ mm

Fundamental tolerance

$i = 0.45D^{1/3} + 0.001\,D$

$= 0.45\,(63.24)^{1/3} + 0.001\,(63.24)$

$= 1.85$ microns

Hole → Max. = 75.029 mm

Min. = 75.0 mm

Shaft → Max. = 75.10 mm

Min. = 75.054 mm

Allowance = Maximum interference

= 0.054 + 0.046 = 0.10 mm

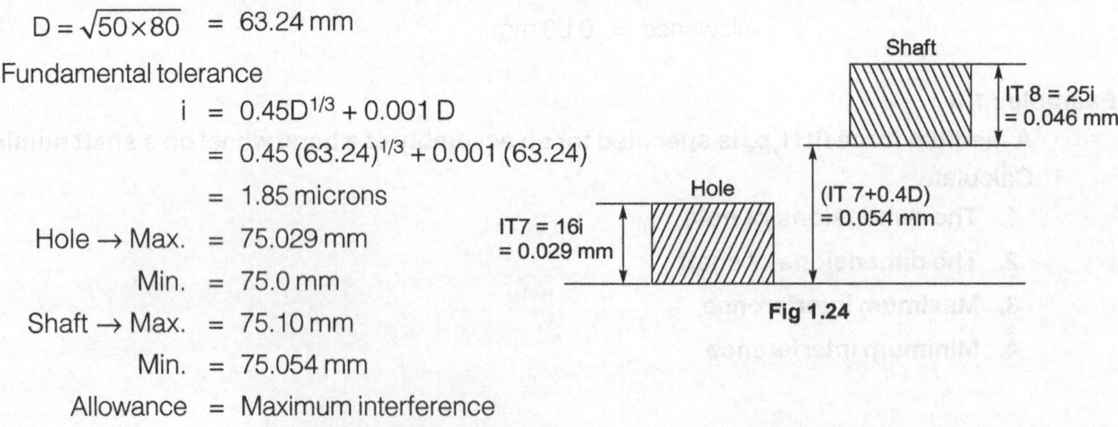

Fig 1.24

Example : 1.8

For each of the following shaft and hole pair, calculate shaft tolerance, hole tolerance and analyze whether the pair is,

1. **Clearance fit**
2. **Transition fit**
3. **Interference fit**

Pair 1: Hole - $50^{+0.50}_{+0.00}$ mm Shaft - $50^{-0.02}_{+0.005}$ mm

Pair 2: Hole - $50^{+0.25}_{+0.00}$ mm Shaft - $50^{+0.05}_{+0.005}$ mm

Pair 3: Hole - $50^{+0.04}_{+0.00}$ mm Shaft - $50^{+0.07}_{+0.04}$ mm

Sketch the three fits on the same zero line.

Solution :

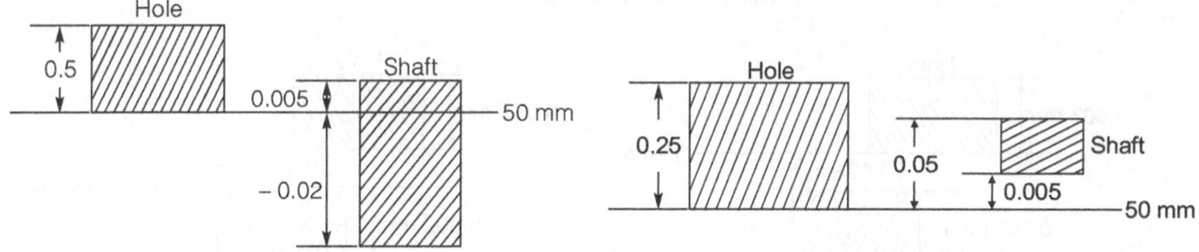

(i) Pair 1 : Transition fit **(ii) Pair 2 : Transition fit**

Hole tolerance = 0.5 mm

Shaft tolerance = 0.005 + 0.02 = 0.025 mm

Hole tolerance = 0.25 mm

Shaft tolerance = 0.05 – 0.005 = 0.045 mm

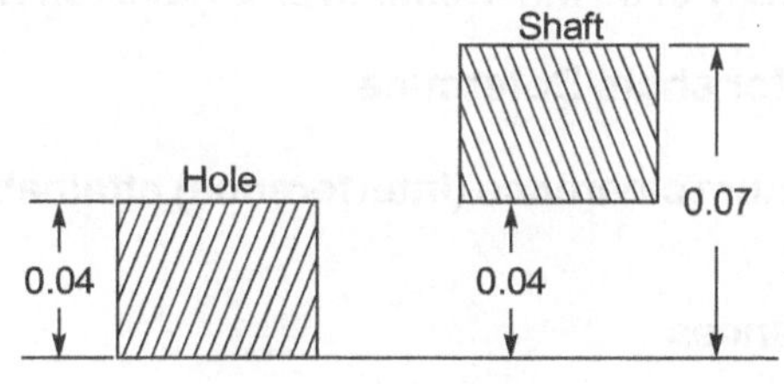

(iii) Pair 3 : Interference fit

Fig. 1.25

Hole tolerance = 0.04 mm

Shaft tolerance = 0.07 – 0.04 = 0.03 mm

Example : 1.9

A 35 mm diameter shaft and bearing are to be assembled with a clearance fit. The tolerance and allowance are as under

Allowance = 0.003 mm

Tolerance on hole = 0.007 mm

Tolerance on shaft = 0.002 mm

Find the limits of size for the hole and shaft if

1. **Hole basis system is used**
2. **Shaft basis system is used**

The tolerances are disposed of unilaterally.

Solution :

Allowance = Minimum clearance

1. **Hole basis system**

Hole → Max. = 35.007 mm
Min. = 35.0 mm
Shaft → Max. = 34.997 mm
Min. = 34.995 mm

2. **Shaft basis system**

Hole → Max. = 35.01 mm
Min. = 35.003 mm
Shaft → Max. = 35.0 mm
Min. = 34.998 mm

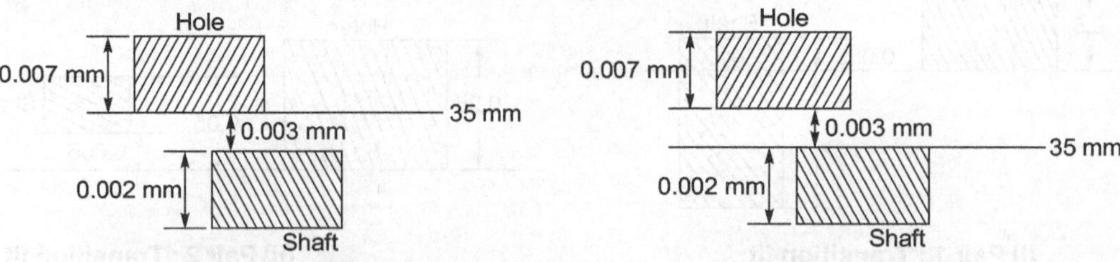

Fig 1.26 (i) : Hole basis **Fig 1.26 (ii) : Shaft basis**

Example : 1.10

The hole and shaft assembly of 30 mm nominal size have tolerances specified as $30^{+0.02}_{-0.00}$ mm for hole and $30^{-0.04}_{-0.07}$ mm for shaft. Determine

1. **Maximum and minimum clearance (interference) attainable**
2. **Allowance**
3. **Hole and shaft tolerances**
4. **Fundamental deviation**
5. **MML for shaft and hole**
6. **Type of fit.**

Sketch these values on a conventional diagram.

Solution :

1. Maximum clearance = 0.09 mm
   Minimum clearance = 0.04 mm
2. Allowance = Minimum clearance = 0.04 mm
3. Hole tolerances = 0.02 mm
   Shaft tolerances = 0.03 mm
4. Fundamental deviation of shaft = 0.04 mm
5. Maximum Material Limit for hole = 30.0 mm
   Maximum Material Limit for shaft = 29.96 mm
6. Clearance fit.

1.7 TOLERANCE SINK

Suppose a component of geometry shown in Fig. 1.28, has to be made from a round bar of 50 mm diameter and 200 mm length on a conventional lathe having process capability of ± 0.15 mm. After turning to ϕ40 mm point A is taken as reference and a cut is made at 50 mm length. The process capabilities will come as tolerances in the given section. After turning ϕ30 mm section by taking same point A as reference a cut is made at 190 mm. So some tolerances will appear in the overall length. 140 mm length section of the component will be the sink. A design engineer keeps one section of the part blank (without tolerance) so that production engineer can dump all the tolerances on that section which becomes most inaccurate dimension of the part. Position of sink can be changed by changing the reference point e.g. in the same problem, for turning 140 mm section if machine takes point B as reference, overall length will become sink.

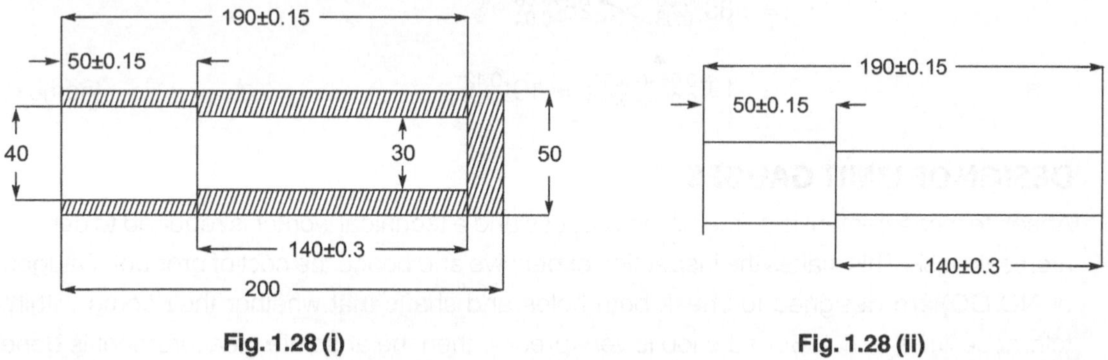

Fig. 1.28 (i) Fig. 1.28 (ii)

Tolerances for the sink is the cumulative sum of all the tolerances and only like minded tolerances can be added i.e. either equally bilateral or equally unilateral.

Example : 1.11

A part shown in Figure is machined to the sizes given below

$P = 35 \pm 0.08$

$Q = 12 \pm .02$

$R = 13^{+0.04}_{-0.02}$

With 100% confidence, the resultant dimension W will have the specification

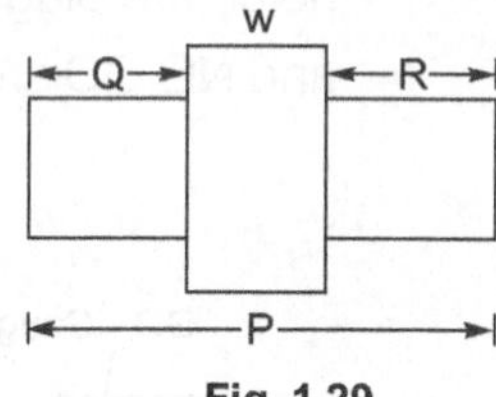

Fig. 1.29

Solution :

In the above problem dimensions P and Q are equally bilateral. So R also has to be converted into equally bilateral. This can be done by raising the basic size by 0.01 mm. Here section W is the tolerance sink. So

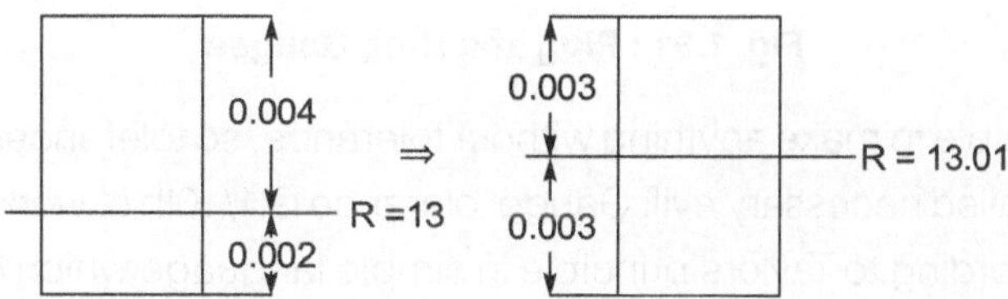

Fig. 1.30

$$\text{Basic size of W} = P - (Q + R)$$

$$= 35 - (12 + 13.01) = 9.99$$

$$\text{Tolerance} = \text{cumulative sum}$$
$$= 0.08 + 0.02 + 0.03$$
$$= 0.13$$

So W = 9.99 ± 0.13 same dimension can be written as by the same procedure by shifting basic size up by 0.01 mm.

$$W = 10^{+0.12}_{-0.14}$$

Method 2 : (Short cut)

$$W = P - (Q + R)$$

$$= 35^{+0.08}_{-0.08} - \left(12^{+0.02}_{-0.02} + 13^{+0.04}_{-0.02}\right) \text{ \{add upper and lower limits separately\}}$$

$$= 35^{+0.08}_{-0.08} \times 25^{+0.06}_{-0.04}$$

$$= 10^{+0.08-(-0.04)}_{-0.08-(+0.06)} = 10^{+0.12}_{-0.14} \quad \text{\{same result\}}$$

1.8 DESIGN OF LIMIT GAUGES

Absolute measurement is a time taking process and a technical worker is required to carryout the aforesaid task. This makes the inspection expensive and hence the cost of product. Gauges (GO or NO GO) are designed to check both holes and shafts that whether they come within their tolerance limits or not. But if the job is very precise then the absolute measurement is done and not gauged even if there are so many components.

The gauges used for shafts are called ring gauges [Fig. 1.31(ii)] and gauges used for checking holes are plug gauges [Fig. 1.31(i)]. When the basic size increases beyond a certain value GO and NO GO are provided on a separate handle [Fig. 1.31].

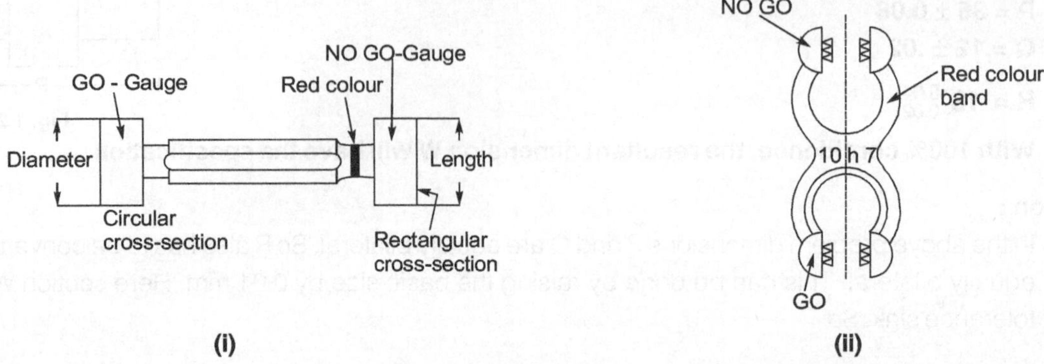

Fig. 1.31 : Plug and Ring Gauges

Since it is not possible to make anything without tolerance, so tolerances are also provided on the gauges and it is called necessary evil. Gauge tolerance is 1/10th of work tolerance. These gauges are designed according to Taylors principle in simple language which has two statements:

1. GO and NO GO gauges must be designed to check size as well as shape and shall be in the full form i.e. the full working length of hole or shaft.
2. NO GO gauges are designed to check one element at a time. e.g. suppose there is a hole which became oval during machining which one cannot see with naked eyes.

GO gauges with go completely inside the hole and cannot catch this ovality. NO GO gauge which is in the shape as shown in [Fig. 1.32(ii)] is kept in two positions as shown in Fig. 1.33. Along the major axis, NO GO gauge will go inside the hole it means there is ovality. Another element is suppose there is an unwanted cavity inside the hole. This cavity cannot be detected by the GO gauge. NO GO gauge as shown in Figure 1.33 is slightly tilted then it will go inside. At the position of cavity the axis of the NO GO gauge become horizontal.

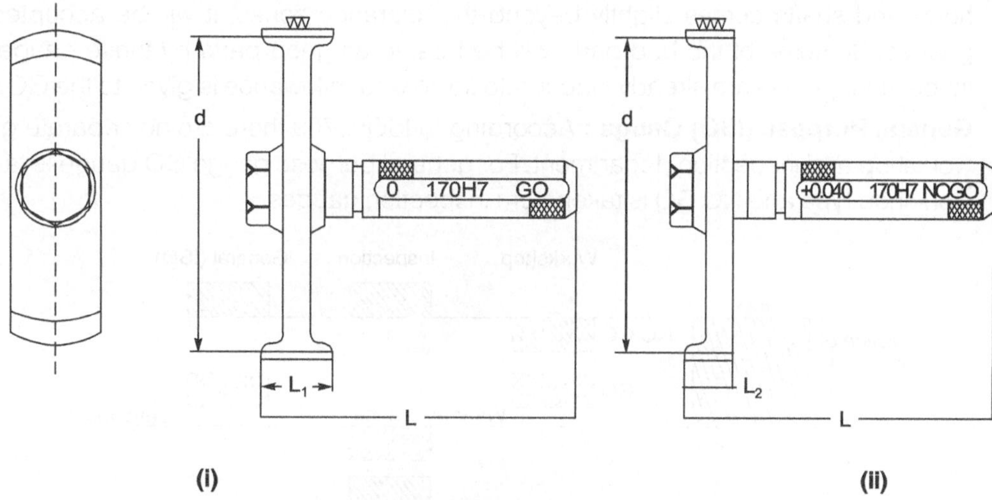

Fig. 1.32 : GO and NO GO on separate handle

Since Go gauge is moving in the complete working length to check size as well as shape of hole or shaft, there will be wear on the GO gauge. In some cases wear allowances which is 1/10th of gauge tolerance are provided on the gauge. Wear allowance is provided on the gauge only when work tolerance is greater than 0.1 mm. There are three types of gauges and the appearances of their tolerances are given in Fig. 1.33.

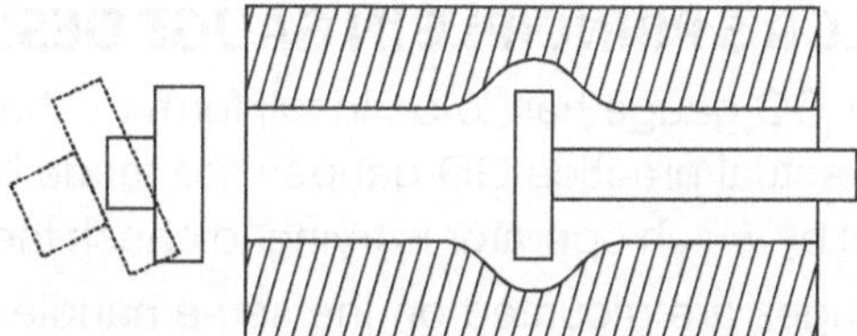

Fig 1.33: NO GO one element at a time

1. **Work Shop Gauges :** The philosophy of workshop gauge is that we will not produce anything defective and hence the tolerance on gauges will be towards the work tolerance. For hole there will be plug gauge and since it is repeatedly moving inside the hole, its diameter will keep on decreasing. So direction of wear will be in the downward direction. Ring gauge moves over the shaft so its diameter will keep on increasing so the direction of wear will be in the upward direction. A gauge can have dimension anywhere in tolerance zone. So when either hole or shaft which is within the tolerance limits near the boundaries inspected by these gauges it will be rejected. So by using these gauges some of the good parts will be treated as bad parts. If the GO gauge goes inside the hole it indicates that the dimension of hole is larger than the lower limit. If NO GO gauge does not go it means the size of hole is lower than the upper limit. If both GO and NO GO gauges goes inside the hole it means the hole is

oversize and when ever GO gauges does not go it means hole is undersize. By repeatedly using GO gauge its dimension changes. GO gauge is continuously being used until wear limit which is 4μ's is reached. Once GO gauge reaches wear limit it is thrown away and a new gauge is taken.

2. **Inspection Gauges :** The philosophy of inspection department is that already lot of effort has been taken to make the product so no good part should be rejected. So the tolerance of the gauges is away from the work tolerance as shown in Figure 1.34. It can be seen that when holes and shafts comes slightly beyond the tolerance zones, it will be accepted by the gauges. So some of the bad parts will be treated as good parts by these gauges. Since inspection gauges are already inaccurate so no wear allowance is given to the GO gauges.
3. **General Purpose (ISO) Gauge :** According to ISO: 2709 there are no separate gauge for workshop and inspection department. For general purpose gauge GO gauge is taken from workshop type and NO GO is taken from inspection gauges.

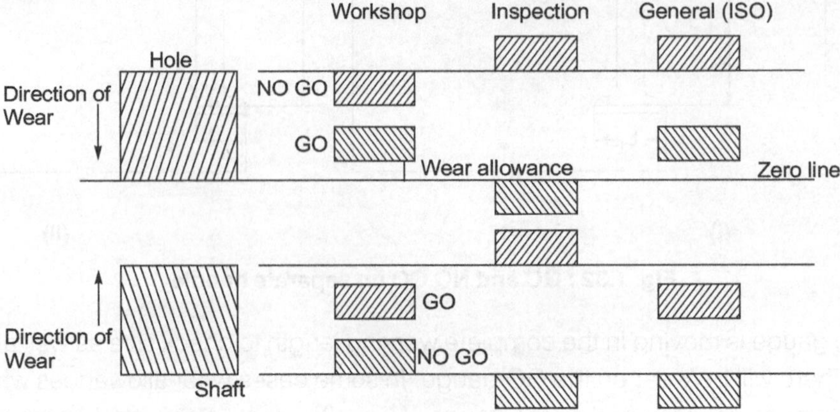

Fig. 1.34 : Tolerances of gauges

1.9 VIOLATION OF TAYLOR'S PRINCIPLE IN GAUGE DESIGN

1. As discussed earlier GO gauge has to be in full form to check shape as well as size. So it must be rigid but in actual practice GO gauges are made hollow so that it becomes light. Because it is the feel by which operator is trying to check the components.
2. GO and NO GO gauges are mounted on the same handle for simplicity which is another violation to Taylor's principle.
3. According to principle NO GO gauge must be in the form of sheet but to give life to the gauge some thickness is being provided to the gauge.

1.10 GAUGE MATERIAL

The properties of material required to make gauge out of it are :

1. **High hardness :** Since GO gauges moves repeatedly over the complete working length so the wear should be minimum.
2. **Lower coefficient of thermal expansion :** Since the difference in the GO and NO GO gauge is very small, even slight change in dimension will make the gauge incorrect.
3. **Machinability :** To finish it to exact size.
4. **Better corrosion resistance**

The commonly used materials are :

(a) En – 24
(b) High carbon steel
(c) Mild steel
(d) Inver (36% Ni) and Elinver (42% Ni)
(e) **Glass :** This material is having specific application in ordnance factories to check the size of gun. Sometimes by accident gauge falls from the hand of operator and produces micro burr which is not visible from naked eyes. When GO gauge with this burr moves inside the barrel it produces scratches which can be very fatel. So glass is the material which is used in such cases because when gauge falls down it breaks and the operator has to use another one.

Example : 1.12

Design general type GO and NO GO gauges for components having 25 H_8/f_9 fit. The basic size falls in the diameter range of 18-30 mm. The fundamental deviation for 'f' shaft = ($-5.5D^{0.4}$) microns. The multipliers for 8 and 9 grades are 25 and 40. Take wear allowance as 10% of gauge tolerance. Sketch the gauges with values.

Solution :

Fundamental tolerance

$$i = 0.45(D)^{1/3} + 0.001D$$

$$D = \sqrt{D_{max} \times D_{min}} = \sqrt{18 \times 30}$$

$$= 23.23 \text{ mm}$$

$$i = 0.45(23.23)^{1/3} + 0.001\,(23.23)$$

$$= 1.30\ \mu\text{m}$$

$$\text{Tolerance of hole} = 25i = 32.68\mu\text{'s}$$

$$= 0.032 \text{ mm}$$

$$\text{Tolerance on plug gauges} = 0.0032 \text{ mm}$$

$$\text{Tolerance of shaft} = 40i$$

$$= 0.052 \text{ mm}$$

$$\text{Tolerance on ring gauges} = 0.0052 \text{ mm}$$

Fundamental deviation of 'f' type of shaft is

$$\epsilon_f = -5.5(23.23)^{0.4} = 19.35\ \mu\text{m}$$

$$= 0.019 \text{ mm}$$

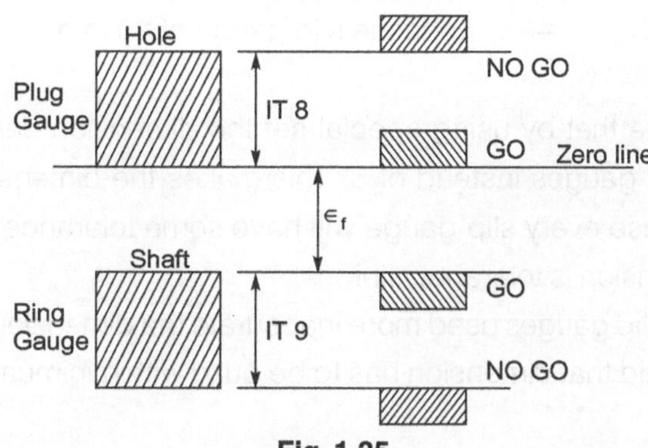

Fig. 1.35

Since work tolerance is less than 0.1 mm so no wear allowance will be given on GO gauge. So

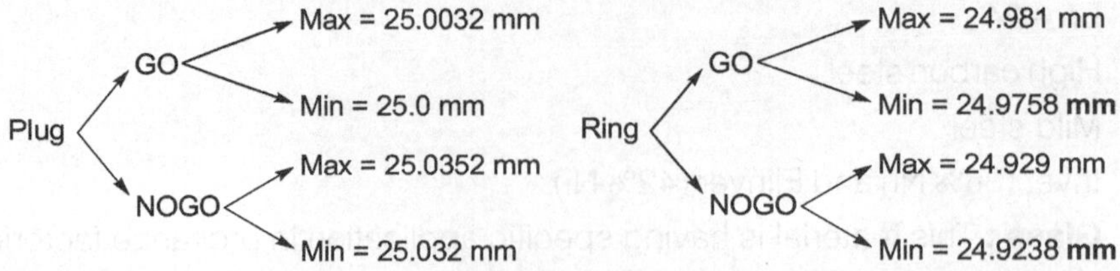

1.11 SLIP GAUGES

Slip gauges are rectangular pieces of block used to establish datum for length in industries. These gauges are available in 2 sets.

Normal set M 45		
Range (mm)	**Steps (mm)**	**No. of steps**
1.001 - 1.009	0.001	9
1.01 - 1.09	0.01	9
1.1 - 1.9	0.1	9
1 - 9	1	9
10 - 90	10	9
		Total = 45

Specific set M 87		
Range (mm)	**Steps (mm)**	**No. of steps**
1.001 - 1.009	0.001	9
1.01 - 1.49	0.01	49
0.5 - 9.5	0.5	19
10 - 90	10	9
1.005	–	1
		Total = 87

Let us build the dimension 25.832 mm by using normal and special sets. We begin from the last digit and move towards first digit.

Normal set :

25.832 → choose slip gauge of 1.002 mm ...1

25.832 – 1.002 = 24.83 → choose slip gauge of 1.03 mm ...2

24.83 – 1.03 = 23.8 → choose slip gauge of 1.8 mm ...3

23.8 – 1.8 = 22 mm → choose slip gauge of 2 mm ...4

22 – 2 = 20 mm → choose slip gauge of 1.8 mm 20 mm ...5

So by wringing these 5 gauges the required dimension can be build up. The faces of the slip gauges are ground and polished so that by wringing there will be fusion between the two surfaces.

Special set :

25.832 → choose slip gauge of 1.002 mm 1

25.832 – 1.002 = 24.83 → choose slip gauge of 1.33 mm 2

24.83 – 1.33 = 23.5 → choose slip gauge of 3.5 mm 3

23.5 – 3.5 = 20 mm → choose slip gauge of 20 mm 4

It can be noticed here that by using special set the dimension can be build by using only 4 gauges instead of 5. This makes the dimensions more accurate because every slip gauge will have some tolerance and overall build up dimension is tolerance sink.

So more is the no. of slip gauges used more inaccurate the dimension will be. That is why it is said that dimension has to be build with minimum no. of slip gauges.

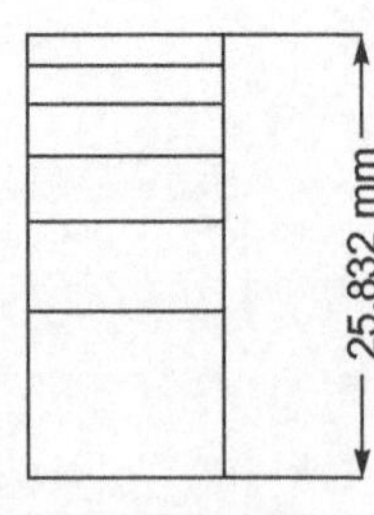

Fig: 1.36

1.12 ANGLE BLOCKS

Angle blocks are hardened steel blocks similar to slip gauges but it is used for angle measurement. Top and bottom surface of the block is ground and polished. Inclined surface defines angle from the flat surface as shown in Fig. 1.37. In a typical set of angle block following angles are present.

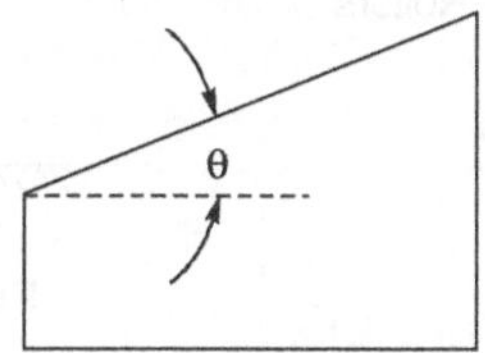

Fig : 1.37 : Angle Block

Table					
Degrees	1	3	9	27	41
Minutes	1	3	9	27	
Fractions of minutes	0.05	0.1	0.3	0.5	

By using these 13 blocks and an optical square angle between 0° to 360° can be made. e.g. If we want to make 4° angle surface 'a' has to be wringed on surface 'b''. This will make addition of angles as shown in Fig. 1.38 (i).

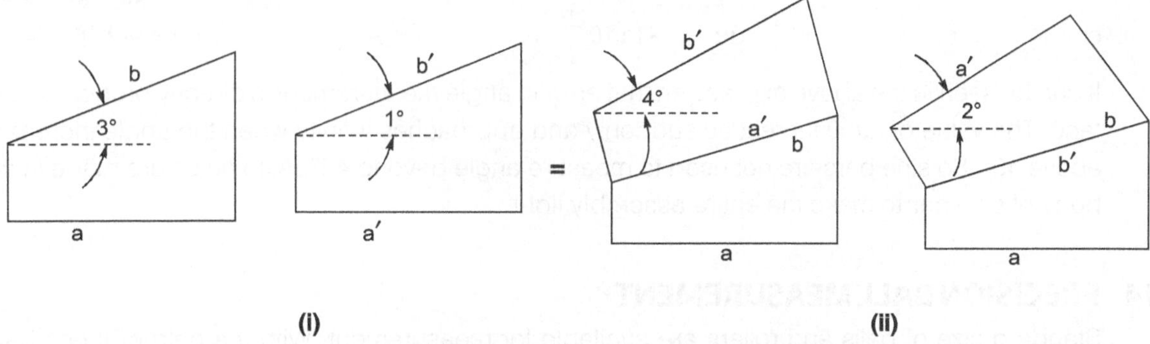

Fig. 1.38 : Addition and Subtraction of angles

If subtraction of angle has to be done (e.g. to make angle 2°) surface 'b' has to be wringed on surface 'b'' as shown in Fig. 1.38 (ii).

1.13 SINE BAR

It is used for indirect measurement of an angle of an machined surface. High carbon and high Cromium material are used for sine bar as it gives good surface finish (~0.2 μ's). Two rollers of exact size are attached to the body of sine bar. These exact size rollers are made by turning a shaft in a single operation and then dividing it into two halves.

For calculating angle θ initially the dimension is being build by using slip gauges (as shown in Fig. 1.39 (ii)]. These measurements are conducted by keeping the assembly on a surface plate.

In the given Figure θ can be calculated as $\sin\theta = \frac{h}{L}$

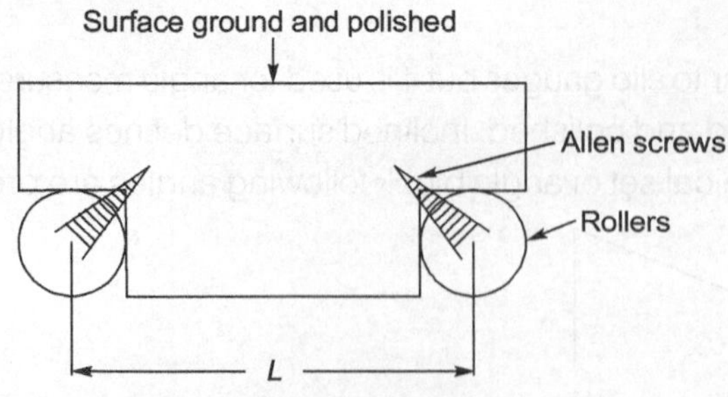

Fig. 1.38 (i) : Sine bar

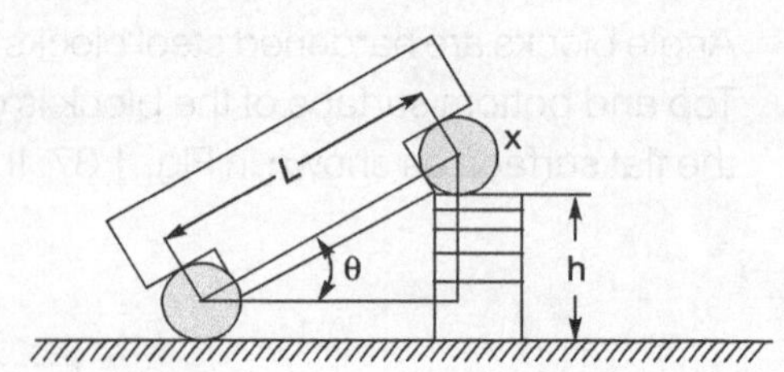

Fig. 1.38 (ii) : Assembly of Sine bar

Differentiating the above equation

$$\cos\theta \, d\theta = \frac{l\,dh - h\,dl}{l^2}$$

$$\cos\theta \, d\theta = \frac{dh}{l} - \frac{h}{l}\frac{dl}{l}$$

Where $d\theta$ is the error in angle measurement, dh is tolerance sink in slip gauge dimensional build up and dl is error in center and center distance of slip gauge. Since slip gauges are highly accurate dh can be neglected as compared to dl, so

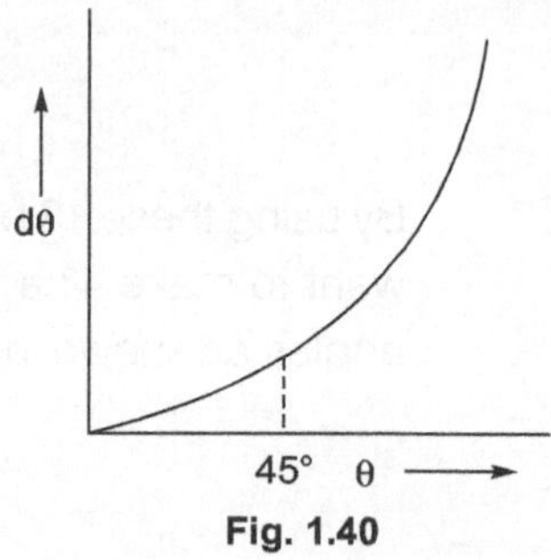

Fig. 1.40

$$\cos\theta \, d\theta = -\frac{h}{l}\frac{dl}{l} = -\sin\theta\frac{dl}{l}$$

or
$$d\theta = -\tan\theta\frac{dl}{l}$$

It can be seen in the above expression that error in angle measurement is directly proportional to tanθ. The value of tan θ increases suddenly and approaches infinity when the angle increases above 45°. So sine bars are not used to measure angle beyond 45°. Also holes are made in the body of sine bar to make the entire assembly light.

1.14 PRECISION BALL MEASUREMENT

Standard size of balls and rollers are available for measurement. With the help of these balls along with vernier, taper of any machined component can be measured. This procedure can be understood by the above examples.

Example : 1.13

Determine the taper of the V groove on a machined component as shown in Fig. 1.41.

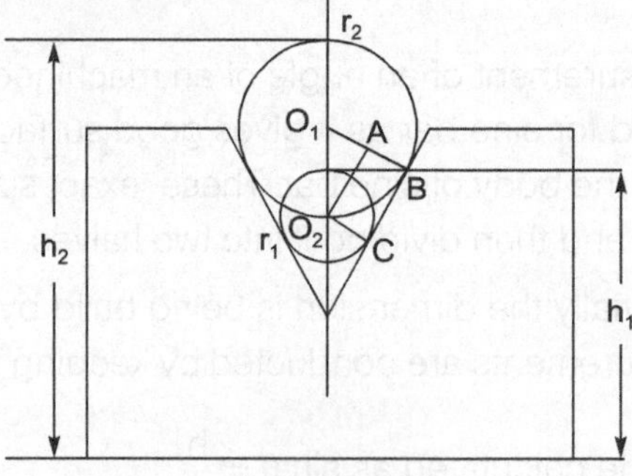

Fig. 1.41

Solution :

1. Initially keep a roller of radius r_1 in the groove and by using height gauge measure h_1 as the block is kept on the surface plate. So the reference for all the measurement will be same.
2. After removing the first roller, a large size rollers of radius r_2 is kept and height from the top surface h_2 (as shown in Fig. 1.41) is measured.

The angle θ is calculated by mathematical calculations as shown :

In $\Delta O_1 O_2 A$

$$O_1A = r_2 - r_1$$

$$O_1O_2 = (h_2 - r_2) - (h_1 - r_1) = (h_2 - h_1) - (r_2 - r_1)$$

$$\sin\theta = \frac{O_1A}{O_1O_2}$$

$$\operatorname{cosec}\theta = \frac{(h_2 - h_1) - (r_2 - r_1)}{r_2 - r_1} \Rightarrow \left(\frac{h_2 - h_1}{r_2 - r_1} - 1\right)$$

Example : 1.14

Identify the taper of the plug gauge as shown in Figure 1.42.

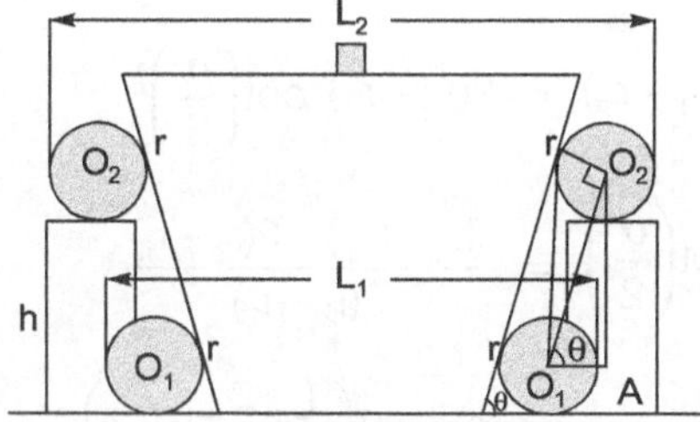

Fig. 1.42

Solution :

$$O_2A = h$$

$$O_1A = \frac{(L_2 - 2r) - (L_1 - 2r)}{2} = \frac{L_2 - L_1}{2}$$

$$\tan\theta = \frac{O_2A}{O_1A} = \frac{2h}{L_2 - L_1}$$

It is understood that there can be many ways to calculate these taper by using precision ball and rollers but all the methods should be practically feasible and mathematically correct.

Example : 1.15

Determine the taper of internal cavity of a machine part as shown in Fig. 1.43.

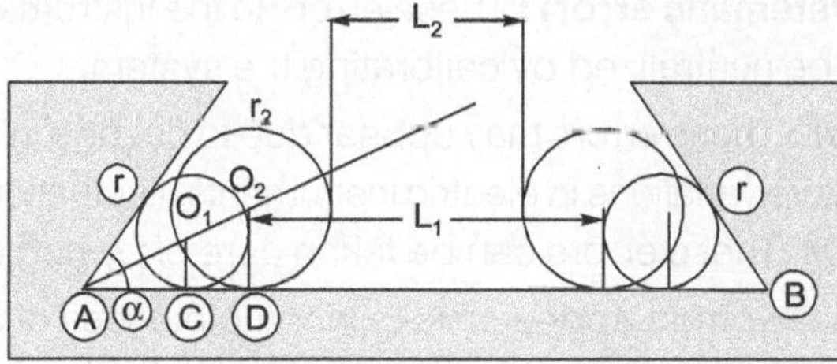

Fig. 1.43

Solution :

By putting 2 different size balls the length L_2 is calculated and in both the cases distance AB is equalized i.e. ΔAO_1C

$$\frac{r_1}{AC} = \tan\left(\frac{\alpha}{2}\right)$$

$$AB = L_1 + 2r_1 + 2AC$$

$$= L_1 + 2r_1 + 2r_1 \cot\left(\frac{\alpha}{2}\right) \quad ...(1)$$

ΔAO_2D

$$\frac{r_2}{AD} = \tan\left(\frac{\alpha}{2}\right)$$

$$AB = L_2 + 2r_2 + 2r_2 \cot\left(\frac{\alpha}{2}\right) \quad ...(2)$$

Equating equation (1) and (2)

$$L_1 + 2r_1 + 2r_1 \cot\left(\frac{\alpha}{2}\right) = L_2 + 2r_2 + 2r_2 \cot\left(\frac{\alpha}{2}\right)$$

$$(L_1 - L_2) + 2(r_1 - r_2) = 2(r_2 - r_1) \cot\left(\frac{\alpha}{2}\right)$$

$$\cot\left(\frac{\alpha}{2}\right) = \frac{(L_1 - L_2) - 2(r_2 - r_1)}{2(r_2 - r_1)}$$

$$\alpha = 2\cot^{-1}\left(\frac{L_2 - L_1}{2(r_2 - r_1)} - 1\right)$$

1.15 VERNIER CALIPERS AND MICROMETER

If a vernier calipers and micrometer is having same least count even then micrometer will be accurate one. It is because in vernier calipers, vernier scale is moving over a main scale and there is an area contact. So the reference plane of measurement keeps on shifting. If the contact points are more than 3, immediately it introduces redundancy because of change in reference plane. But in micrometer since there are balls which produces a point contact, the reference plane does not change. So micrometers are accurate.

1.16 TYPES OF ERRORS

There are three types of errors in measuring instruments.

1. **System error (systematic error) :** These errors in the instrument follow a particular pattern. These errors can be neutralized by calibrating the system.
2. **Short period error :** These errors may appear due to change in environmental conditions like humidity, temperature variations in electric instruments, fluctuations recorded while conducting tensile test on UTM. These errors can be taken care of by neglecting the irrelevant data.
3. **Erratic error :** These errors appear due to incompleteness of any link. Such error does not follow any pattern. Their magnitude is also uncertain. So these errors have to be eliminated by inspecting the instrument before starting the experiment.

There are 2 types of mathematical errors that appear during measurements :

1. **Sine Error :**

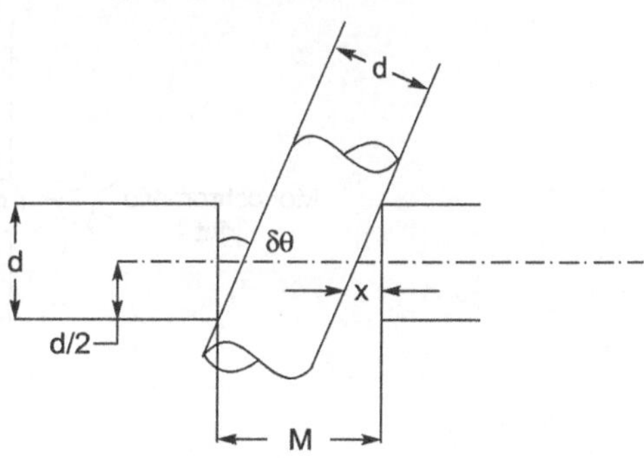

Fig. 1.44 (i) : Sine Error

$$\frac{X}{d/2} = \tan \delta\theta$$

$$\approx \delta\theta$$

$$X = \frac{d}{2} x \delta\theta$$

$$\text{Error} = M - D$$

$$= 2X$$

$$= 2 \times (d/2) \times \delta\theta$$

$$\text{Error} = d \times \delta\theta$$

Sine errors in the measurement are considerable.

2. **Cosine Error :**

$$M = L \cos \delta\theta$$

$$\text{Error in length} = L - M$$

$$= M \cos \delta\theta - M$$

Since $\delta\theta \to 0$ $\cos \delta\theta \to 1$

$$\text{Error} \approx 0$$

So cosine error has no significance in measurement.
It should also be noted down that to minimize errors in measurement; spherical ends should be used to measure flat surfaces and vice versa. It is because we are sure that measurement is done only between 2 points.

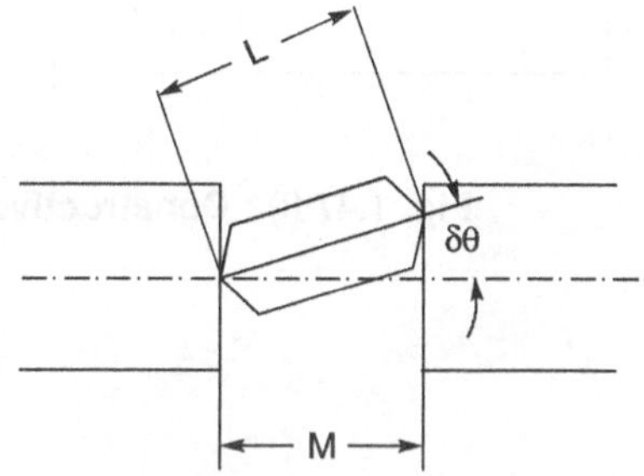

Fig. 1.44 (ii) : Cosine Error

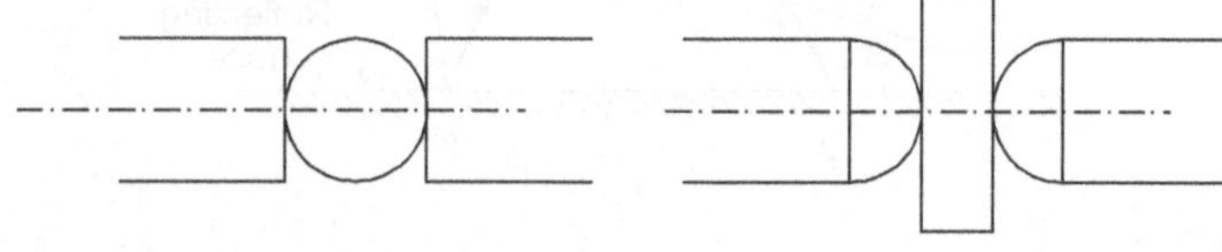

Fig. 1.45 : Measurement of Spherical and flat work pieces by micrometer

1.17 INTERFEROMETRY

A source of monochromatic light is kept at A. two rays are coming out through holes B and C and their interference is observed on the screen. At point O, the distance moved by both the rays is same so both the rays will be in phase [Fig. 1.47 (i)]. which results in constructive interference this produces white band at O. At a point M or N over the screen the distance covered by both the rays is path difference of λ/2. So both rays will be in opposite phase [Fig. 1.47 (ii)]. This results in destructive interference which produces dark band. So alternatively with every λ/2 path difference there will be dark and white fringes over the screen. This concept can be utilized in the measurement by using optical flat. Optical flat is a disc made of quartz with thickness of 8-10 mm and diameter between 25-300 mm. The top and bottom surfaces of the flat are parallel.

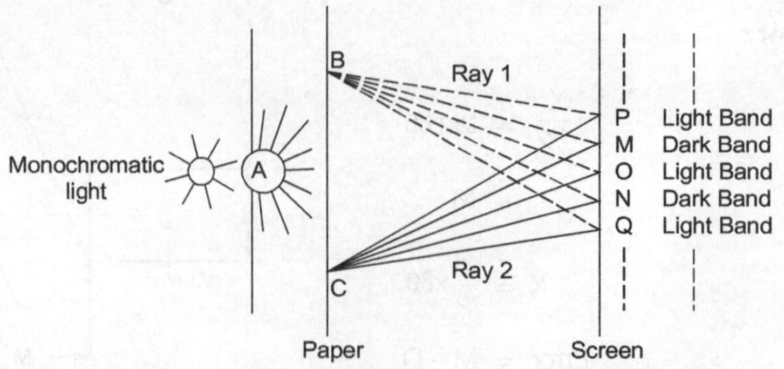

Fig. 1.45 : Interference

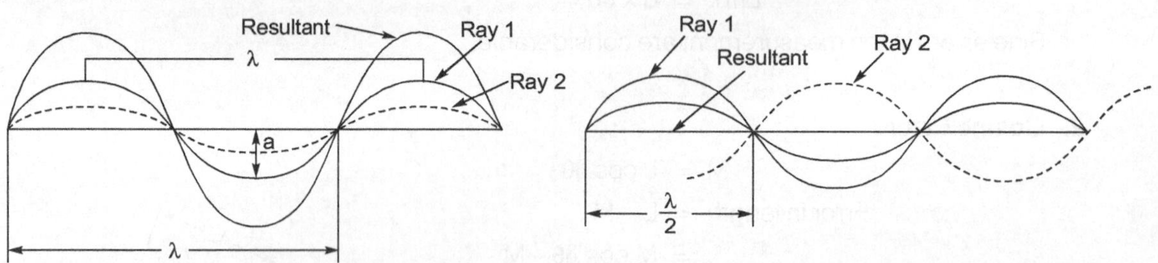

Fig. 1.47 (i) : Constructive Interference **Fig. 1.47 (ii) : Destructive Interference**

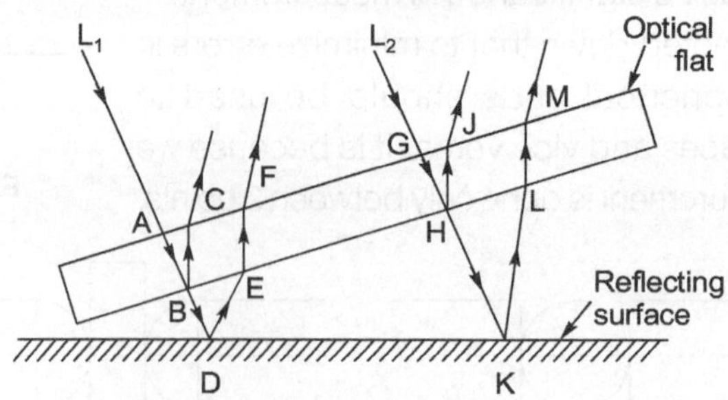

Fig. 1.48 : Optical flat over a surface

As shown in Fig. 1.48 an optical flat is kept over a tapered surface under examination which cannot be seen by naked eyes. L_1 is a ray of monochromatic light, a portion of this ray will be reflected from B and a portion will be transmitted. This transmitted ray will reflect back from the surface and come out of optical flat from F (as shown). Ray BC and EF, there will be path difference equal to (BD + DE). If this path difference is equal to $\lambda/2$ there will be a dark band and if this path difference is λ, a white fringe will appear. It should be noted here that when a rays reflects back from the surface, a path difference of $\lambda/2$ is automatically created. So on the other end of optical flat the total path difference is (HK + KL). Suppose there are n fringes seen in optical flat (we count no. of white fringes), then this path difference is equal to n λ.

$$HK \approx KL$$

$$\lambda = \text{wave length of monochromatic light}$$

So $$2HK = n\lambda$$

$$\Rightarrow \quad HK = \frac{(n\lambda)}{2}$$

Since the angle is very small so HK can be approximated to a vertical line. Suppose L is the length of surface $\frac{HK}{L} = \tan\theta \approx \theta$.

By this equation inclination angle of the surface can be calculated.

1.17.1 Optical flat as comparator

Using optical flat difference is the size of slip gauge can be calculated from a master reference. Suppose the difference Δh has to be calculated :

From the similar triangles

$$\frac{h}{L} = \frac{\Delta h}{G}$$

$$\Rightarrow \quad \Delta h = \left(\frac{h\lambda}{2}\right)\left(\frac{G}{L}\right)$$

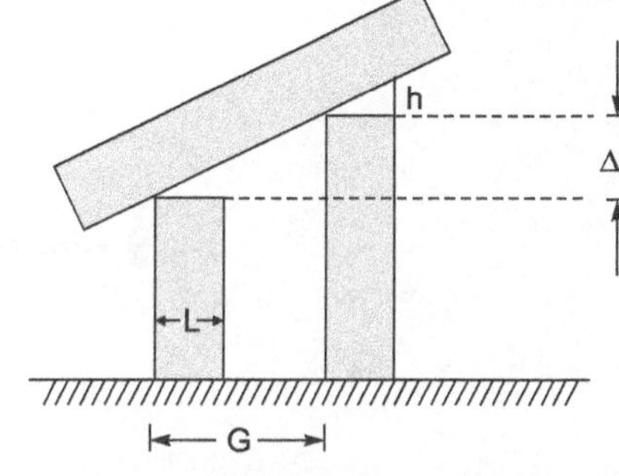

Fig. 1.49 : Flat as comparator of slip

{if n fringes appear in one slip gauge}

From the earlier discussion (Fig. 1.48) it can be understood that by increasing the angle between optical flat and surface, the wave lengths will quickly multiply by λ/2, so number of fringes over the surface will increase and their thickness will go down and opposite will be the case when the angle decreases.

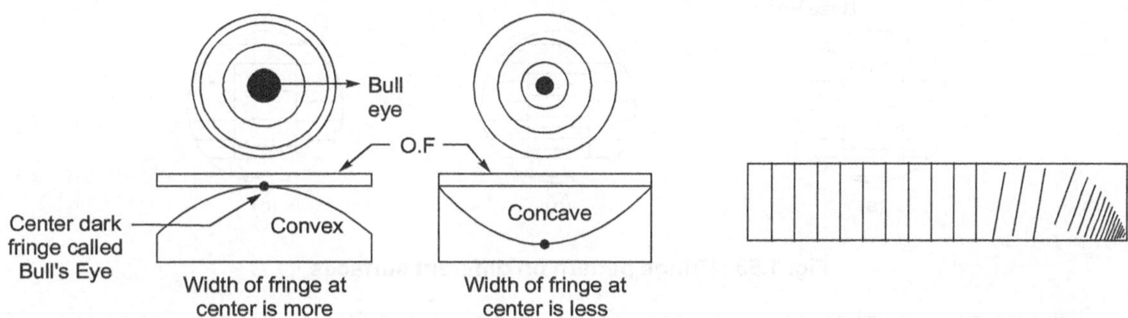

Fig. 1.50 : Optical flat over convex and concave surface

Fig. 1.51 : Wear on slip gauge

When the optical flat is kept over a convex surface, from the centre towards outward direction the angle between optical flat and surface will increase. So central fringes will be broad and in the outward direction the fringes will become thinner and thinner. Opposite will be the case when the optical flat is kept over a concave surface. Optical flats are having wide applications e.g. when it is kept over a slip gauge which has broken edge, the fringes will converge towards the edge as shown in Fig. 1.51.

1.17.2 NPL Flatness Interferometer

This instrument is used to check the flatness of any surface by comparing it with an optically flat surface i.e. base plate of interferometer. The instrument consists of a mercury vapour lamp whose radiations are passed through the green filter which gives less fatigue green monochromatic light. Light passes through a pinhole giving an intense point source. The beam is directed to the gauge to be tested via optical flat. Fringe pattern appears on the gauge to be tested as shown in Fig. 1.53.

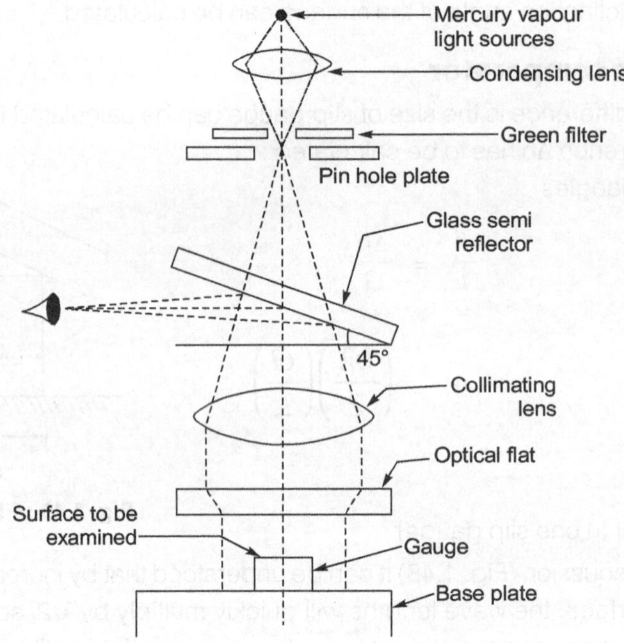

Fig. 1.52 : Interferometer

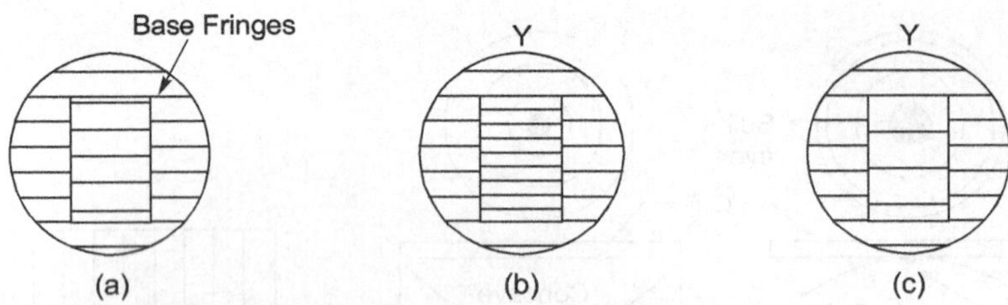

Fig. 1.53 : Fringe pattern on different surfaces

If the fringe appears parallel and symmetric as shown in Fig. 1.53 (a), it means that gauge surface is flat and parallel to base plate of interferometer. When fringes on the gauge are either more or less spaced, it means that gauge is at an angle from the base plate.

1.18 STRAIGHTNESS

It is defined as the deviation of surface from ideal straight line. This straightness can be measured in 3 ways.

1. **Straight edge :** Straight edge is a piece of block of which one surface is exactly straight. By keeping this surface on the machine part under investigation, the amount of light coming from the other side is being observed. If very small amount of light is coming from the interface, it means measured surface is straight.

2. **Spirit level :** Surface under examination is divided into number of segments equal to the size of spirit level. Spirit level is there kept from one segment to another and position of bubble in it is noted down. The deviations of bubble form the center position are recorded. Spirit is used in it due to its low viscosity.
3. **Autocollimator :** As shown in Fig. 1.54 the reflector of autocollimator is moved on the surface and the deviations $\delta\theta$ are recorded in terms of $\Delta S = 2f\delta\theta$. Hence the characteristic can be plotted on the paper. By joining first and last point, an straight line can be formed and the deviations of the surface from this ideal straight line can be noted down.

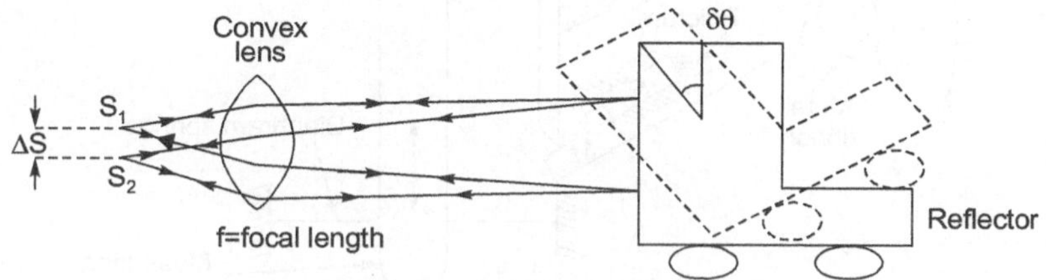

Fig. 1.54 : Autocollimator

1.19 FLATNESS

Flatness is defined as the departure of surface from ideal flat surface. Initially autocollimator or spirit level is moved along 3 different directions AB, BC and CA as shown in Fig. 1.55. These movements will identity 3 reference points A, B and C. Through these points a plane can be defined. Now autocollimator is now move practically in all the possible directions and characteristic of surface is noted down. The deviation of surface from the reference plane is called flatness.

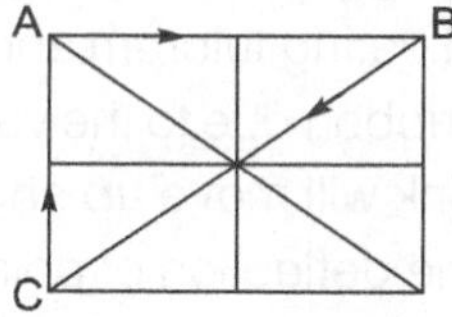

Fig : 1.55 : Movement of autocollimator over surface

1.20 COMPARATOR

Comparators are used for measuring the dimensions from a reference value. This datum is generally established using slip gauges. To maintain precision in the measurement comparators are not allowed to move large distances.

Mechanical comparators : The characteristic of mechanical comparators are :

(a) Usually cheaper than the other devices
(b) Does not require any external power supply to operate.
(c) Accuracy in inspection is independent of accuracy in manufacturing different linkages.
(d) Usually it has linear scale.
(e) More moving linkages which bring down the accuracy.
(f) Range of the instrument is limited.
(g) Due to the inertia of moving linkages makes the comparator sensitive to machine vibrations.

1.20.1 Sigma Comparators

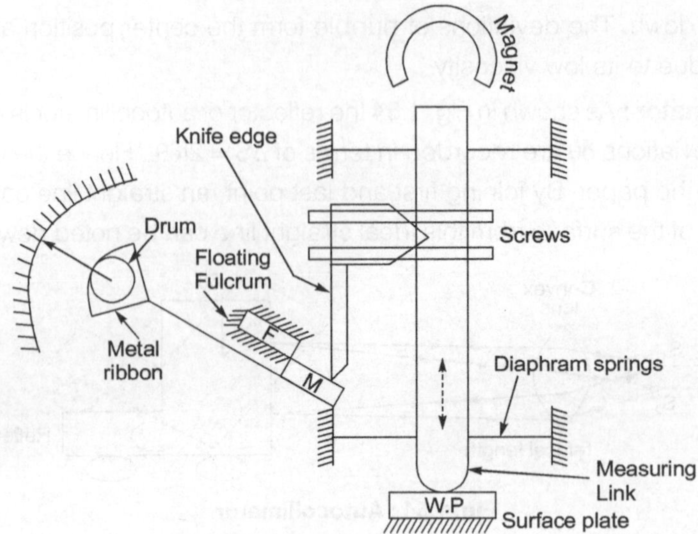

Fig. 1.56 : Sigma Comparator

Helical springs are not used in metrological instruments because residual stresses develops in the spring during deflection and it takes time for the system to come back to its original position. So diaphragm springs with C-slots are used as restoring device. These springs are in the form of a sheet and C-slots are being provided so that if any amount of residual stress develop, it is relieved. As shown in Fig.1.56 measuring link is connected with a knife edge to have a point contact. This edge resets over moving part of floating fulcrum. This ensures a positive contact between link is connected with the floating fulcrum and on the other side of which there is metal ribbon and a drum rotates over this ribbon due to the variations of this ribbon. Due to the variation of size of work pieces, measuring link will move up and down which will make the cylinder over ribbon to rotate and this results in the deflection of pointer. A magnet is placed above the link to have uniform pressure over it. By the movement of link an attractive force will develop between the magnet and measuring link and upon removing the work piece, sudden jerk will not appear over it.

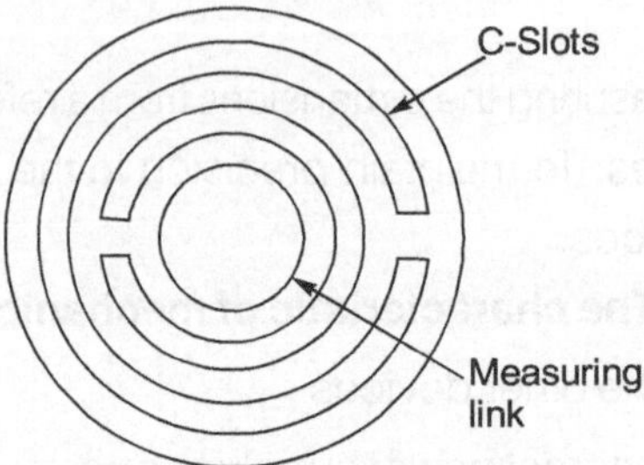

Fig. 1.57 : Diaphram Spring

1.20.2 Mikrokator

A thin metal strip is given left hand twist on one side and right hand twist on the other. The central portion will remain neutral. As shown in Fig. 1.58 due to pivot, movement of the measuring link converts into horizontal increment of the twisted spring. Due to that right hand side of the spring is either twisted more or twisted less. Due to this imbalance in the twist causes the central portion to rotate and the deflection can be noted down. The functions of the magnet here is similar to sigma comparator. The twisted spring is perforated all along its length so that if any amount of residual stresses appears it will be relieved.

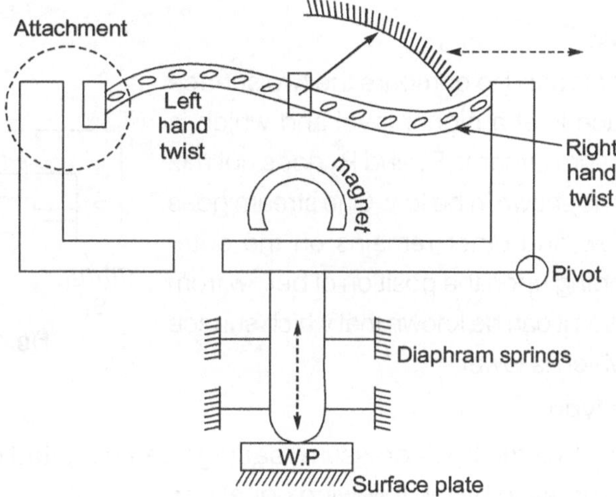

Fig. 1.58 : Mikrokator

1.20.3 Pneumatic comparators

Pneumatic comparators work on the principle that of an air jet. Upon decreasing the stand-off distance, the pressure on the back ground of jet will increase. This pressure is called back pressure and can be directly related to the measurement. These comparators can be further be divided into **three types :**

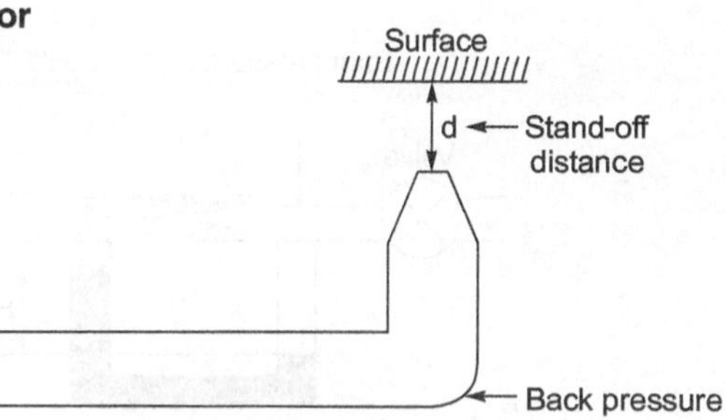

Fig. 1.59 : Principle of an airjet

1. **Flow type :**

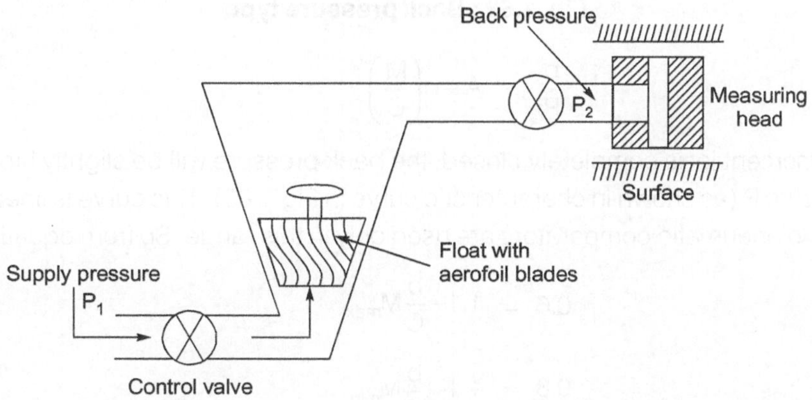

Fig. 1.60 : Rotameter

The design of head is decided by the geometric feature under measurement. As shown in

Fig.1.60 depending upon distance there will be back pressure 'P_2' and the difference between ($P_2 - P_1$) decides the position of the float. This float is having aerofoil shape blades so that due to the flow of air it rotates. If it is stationary, it will stick to the wall of rotameter and due to static friction between float and tube, the system will become insensitive to small change in back pressure 'P_2'. The tube is slightly tapered to maintain linear relationship in the measurement because air is compressible.

2. Differential type

This comparator is used to compare the two surface i.e. which surface is at a higher level and which is lower. Two pressure streams P_1 and P_2 does not mix with each other as shown in bellow. One stream goes inside the bellow and other remains on the outer surface. Depending upon the position of bellow from the neutral position it can be known that which surface is higher and which is lower.

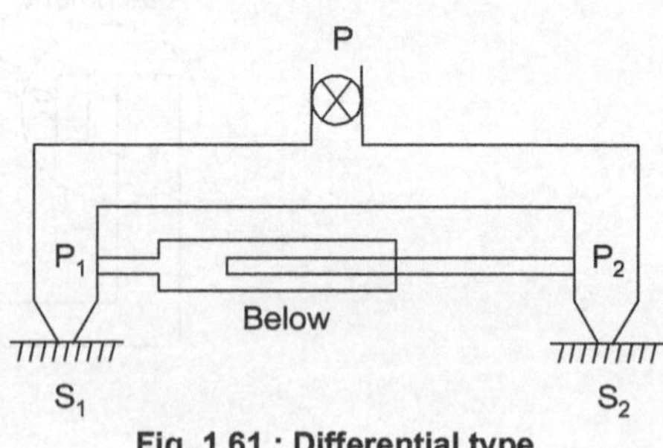

Fig. 1.61 : Differential type

3. Back pressure type

As shown in Fig. 1.62 the back pressure readings can be noted down from the manometer and it can be expressed by the following equation:

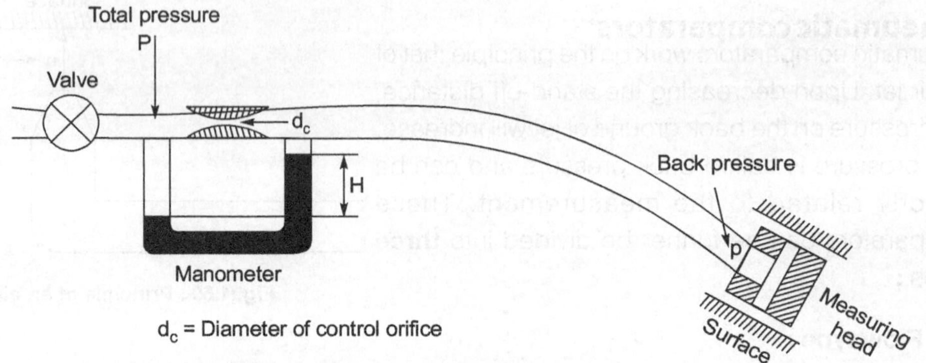

Fig. 1.62 : Back pressure type

$$\frac{p}{P} = A - b\left(\frac{M}{C}\right) \quad ...(1)$$

The moment jet is completely closed, the back pressure will be slightly higher then the total pressure P (as shown in characteristic curve in Fig 1.63). This curve is linear between 0.6 to 0.8. So pneumatic comparators are used only in this range. So from equation (1)

$$0.6 = 1.1 - \frac{b}{C} M_{max} \quad ...(2)$$

$$0.8 = 1.1 - \frac{b}{C} M_{min} \quad ...(3)$$

or $$\frac{b}{C} M_{max} = 0.5$$

$$\frac{b}{C}M_{min} = 0.3$$

Dividing these equations

$$\frac{M_{max}}{M_{min}} = \frac{5}{3} \quad ...(4)$$

Subtracting 1 from both the sides of this equation

$$\frac{M_{max}}{M_{min}} - 1 = \frac{5}{3} - 1$$

$$M_{max} - M_{min} = \text{Range} = \frac{2}{3}M_{min} \quad ...(5)$$

Adding 1 on both sides of equation (4)

$$\frac{M_{max}}{M_{min}} + 1 = \frac{5}{3} + 1$$

$$M_{max} + M_{min} = \frac{8}{3}M_{min}$$

Dividing by 2, we get $M_{avg} = \frac{4}{3}M_{min} = 2\text{ Range}$

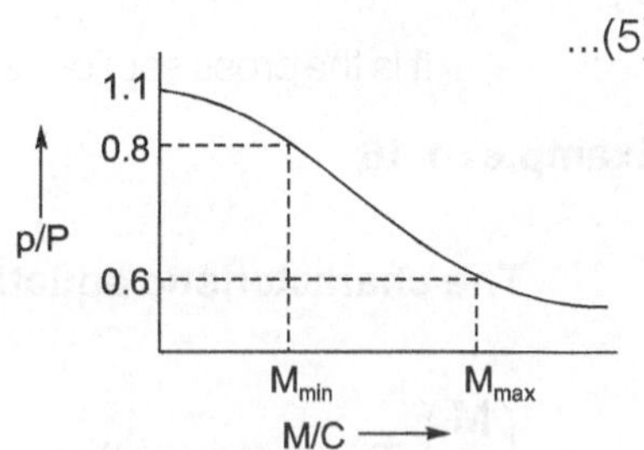

Fig. 1.63 : Characteristic equation of pneumatic comparator

Differentiating equation (1) we get

$$\frac{dp}{dM} = -\frac{b}{C}P \quad ...(6)$$

Taking average of (2) and (3) we get

$$0.7 = 1.1 - \frac{b}{C}M_{avg}$$

So $$\frac{b}{c} = \frac{0.4}{M_{avg}}$$ {putting this value in equation (6)}

$$\frac{dp}{dM} = -\frac{0.4P}{M_{avg}}$$

This is called pneumatic sensitivity. It is directly proportional to total pressure. So sensitivity of comparator will be better when total pressure is higher. But larger total pressure requires large size manometer which may not be suitable for ergonomics point of view. For small pressures, since it takes some time for the signal to reach manometer, it may be lost due to compressibility of air. So manometers are placed as close to measuring head as possible. The overall magnification is defined as output divide by input. So

$$\text{Magnification} = \frac{dR}{dl} = \frac{dR}{dp} \times \frac{dp}{dM} \times \frac{dM}{dl}$$

$$\frac{dR}{dp} = \text{Indicator sensitivity}$$

$$\frac{dp}{dM} = \text{Pneumatic sensitivity}$$

$$\frac{dM}{dl} = \text{Measuring head sensitivity}$$

Sensitivity is vector so '–' or '+' sign will come but magnification factor is scalar. If d_c is control orifice diameter and dm is measuring head then :

$$C = \text{Area of control orifice} = \frac{\pi}{4}d_c^2$$

$$M = \text{Measuring head area} = \pi\, d_m l$$

It is the cross section area through which compressed air is coming in contact with atmosphere.

Example : 1.16

The characteristic equation for a pneumatic comparator is represented by $\frac{p}{P} = 1.1 - 0.5\left(\frac{M}{C}\right)$. The curve is linear in the range of $\frac{p}{P} = 0.6$ to 0.8. The diameter of control orifice is 0.5 mm and diameter of measuring orifice is 1 mm. If indicator sensitivity is 30 mm per 1 kN/m², determine

1. **Range of linear measurement**
2. **Overall magnification of comparator if the total pressure behind control orifice is 3 bar.**

Solution :

1. The characteristic equation of pneumatic comparator is

$$\frac{p}{P} = 1.1 - 0.5\left(\frac{M}{C}\right)$$

Putting the linear range is the equation are get

$$0.6 = 1.1 - 0.5\left(\frac{\pi d_m l_{max}}{\frac{\pi}{4}d_c^2}\right) \quad ...(1)$$

$$0.8 = 1.1 - 0.5\left(\frac{\pi d_m l_{min}}{\frac{\pi}{4}d_c^2}\right) \quad ...(2)$$

where

$$d_m = 1 \text{ mm}$$

$$d_c = 0.5 \text{ mm}$$

Hence we get the linear range as

$$l_{max} - l_{min} = 0.025 \text{ mm}$$

2. Indicator sensitivity $= \frac{dR}{dP} = \frac{30\times10^{-3}}{1000}\frac{M}{N/m^2} = 30 \times 10^{-6}$ M/N/m²

Measuring head sensitivity

$$\frac{dM}{dl} = \pi\, d_m = \pi \times 1 \times 10^{-3}\, M$$

Pneumatic sensitivity

$$\frac{dp}{dM} = \frac{0.4P}{M_{av}}$$

$$= \frac{0.4 \times 300 \times 10^3}{\pi \times 10^{-3} \left\{\frac{l_{max} + l_{min}}{2}\right\} \times 10^{-3}} \qquad \because M_{av} = \pi d_m \left\{\frac{l_{max} + l_{min}}{2}\right\}$$

$$= \frac{2400}{\pi} \times 10^9\ N/m^4$$

$$\text{Overall magnification} = \frac{dR}{dp} \times \frac{dp}{dM} \times \frac{dM}{dl}$$

$$= (\pi \times 10^{-3}) \left(\frac{2400}{\pi} \times 10^9\right) \times (30 \times 10^{-6})$$

$$\text{Magnification} = 72{,}000$$

1.21 SURFACE FINISH

A machined surface is having 2 types of irregularities i.e. roughness and waviness. Small wavelength fluctuations are called roughness (Primary texture) and large wavelength deviations are called waviness (Secondary texture). Waviness is being produced by machine vibrations, chatter, errors in guide ways etc. Roughness appears due to improper selection of cutting fluid development of temperature on rake face etc.

1.21.1 Lay and Flaw

Predominant surface patterns produced by feed marks is called Lay and the surface irregularities present which are random is known as flaw.

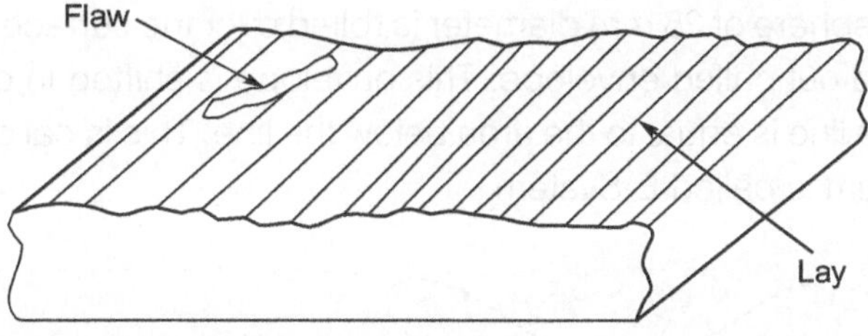

Fig. 1.64

Lay indicates the plane of projection of the view of symbol where finish is observed. The different types of lays are as under

Symbol	Diagram	Description
=		**Parallel lay :** Surface is produced by shaping planning etc.
⊥		**Perpendicular lay :** Surface is produced by shaping and planing
X		**Crossed lay :** Such surface can be produced by knurling operation
M		**Multidirectional lay :** Such surfaces are produced by grinding operation
C		**Circular lay :** The surfaces are produced by facing operation
R		**Redial lay**

1.21.2 Methods of Establishing Datum

1. **M – System :** After plotting the characteristic of any surface a horizontal line is drawn by joining 2 points. This line is shifts up and down in such a way that 50% area is below the line and 50% area in above the line. This line is called mean line which acts a datum in the measurement.

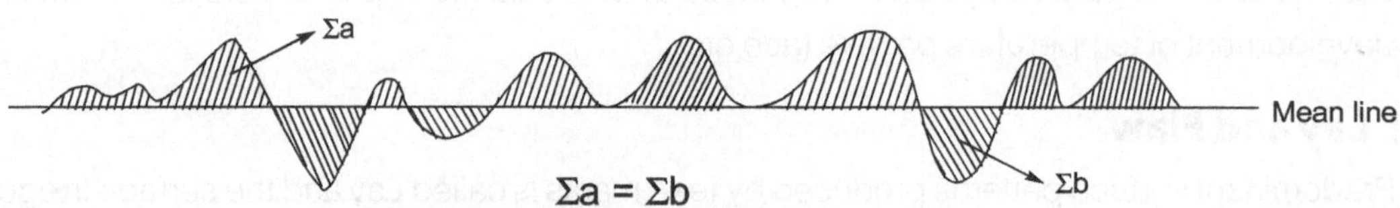

Fig. 1.65 : M-system

2. **E-System :** A sphere of 25 mm diameter is rolled over the surface and the locus of its center is being traced out called envelope. This envelope is shifted in downward direction till the area above the line is equal to the area below the line. This is called mean envelope and the system of datum is called E-System.

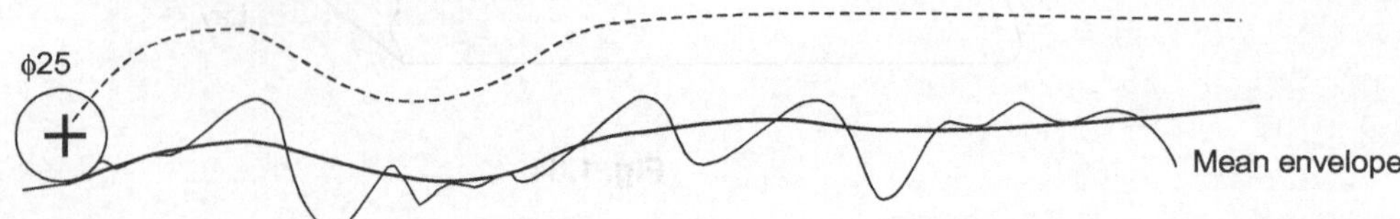

Fig. 1.66 : E-System

1.21.3 Numerical Values to Represent Surface Finish

1. **Peak to Valley height (R_t or R_{max}) :** It is the difference between highest peak and deepest valley. This measurement may not give the true characteristic of any surface because due to some sudden jerk peak to valley height is high in a very small region. Otherwise the entire surface is quite all right.
2. **Center line average value (CLA, R_a) :** n number of points are taken on hills and valleys and their distance from the mean line y_1, y_2, ... y_n is taken. Their average is called CLA value.

$$R_a = \frac{y_1 + y_2 + ... + y_n}{n}$$

Cutoff length is one in which measurement of roughness is being carried out (or length of travel of stylus). So R_a value can also be represented as

$$R_a = \frac{\Sigma a + \Sigma b}{L}$$

Where
Σa = area above the line
Σb = area below the line
L = cutoff length

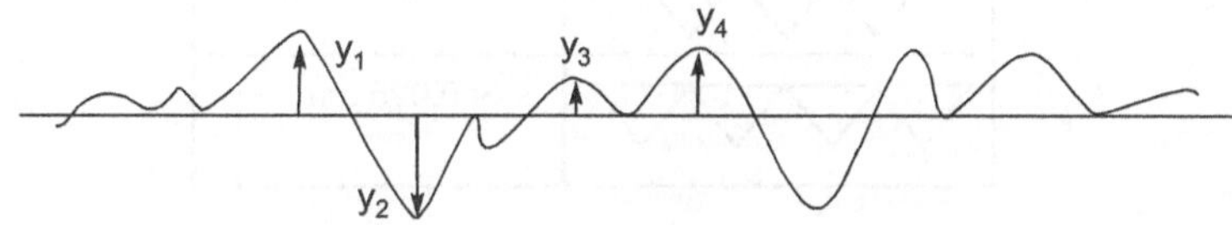

Fig. 1.67

If y = f(x) is the characteristic equation of the roughness, R_a value can also be expressed as

$$R_a = \frac{1}{L}\int_0^L y\,dx$$

The approximate value of R_a is $\frac{H_{max}}{4}$.

Maximum height of unevenness can also be expressed as

$$H_{max} = \frac{f^2}{8R}$$

If complete tool signature is given

$$H_{max} = \frac{f}{\tan\psi + \cot\psi_1}$$

where
f = feed
ψ = side cutting edge angle
ψ_1 = end cutting edge angle

3. Root mean square value (RMS, R_g)

$$R_g = \sqrt{\frac{1}{L}\int_0^L y^2 dx}$$

4. **10 point value (R_z).** It is the average of 5 highest peakes and 5 deepest valleys.

1.21.4 Representation of Surface Finish

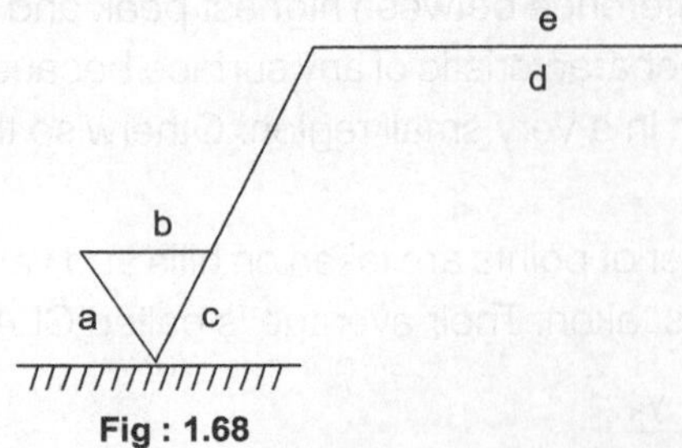

Fig : 1.68

a : Machining allowance
b : Roughness value
c : Direction of lay
d : Sampling or cutoff lengths
e : Production method

Symbols used in drawing	
Symbol	CLA value
∇	8 – 25 μm
∇∇	1.6 – 8 μm
∇∇∇	0.025 – 1.6 μm
∇∇∇∇	< 0.025 μm

1.21.5 Talysurf

Diamond point comes in contact with surface through a flexible skid for providing protection to the diamond point. As the diamond point moves up and down the position of core in the coil changes. This changes the effective resistance and it is measured by a wheat stone bridge.

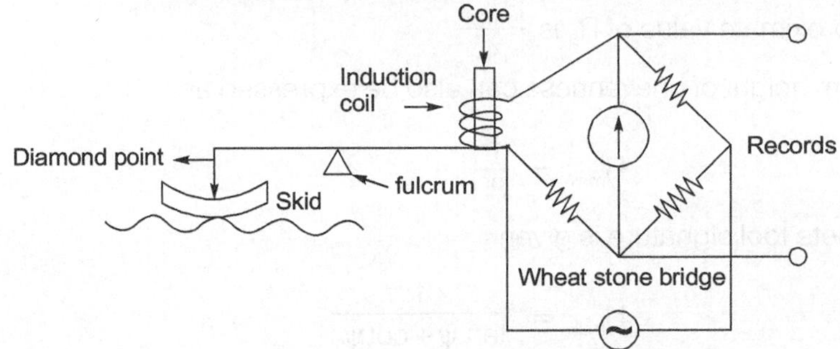

Fig. 1.69 : Principle of Talysurf

1.22 SCREW THREAD METROLOGY

The methods of measuring different parameters of screw thread are give as under :

1. **Major diameter :**
   (a) External Micrometer (b) Bench Micrometer
   (c) Optical Projector (d) Tool Makers Microscope
2. **Minor Diameter :**

(a) Optical Projector (b) Tool Makers Microscope
(c) Pointed anvil Micrometer

3. **Pitch :**
(a) Pitch gauges (b) Optical Projector

4. **Angle :** Optical Projector

5. Pitch circle diameter (effective diameter) can be measured by 2 or 3 wire (rolles) method. These wires are lapped within 0.0025 mm. Best size wire is one which coincide at pitch circle point but it will be a rare case when the wire used is best size wire but this has very little effect on the accuracy of measurement 3 wire method is used when pitch is very large 2 wires are held together as shown in Fig 1.70 d_w is the diameter of wire and it can be seen from Fig. 1.70 and 1.71 that :

$$T = M - 2\,d_w$$
$$PCD = T + 2CD$$
$$CD = OC - DO$$
$$DO = OB - DB$$

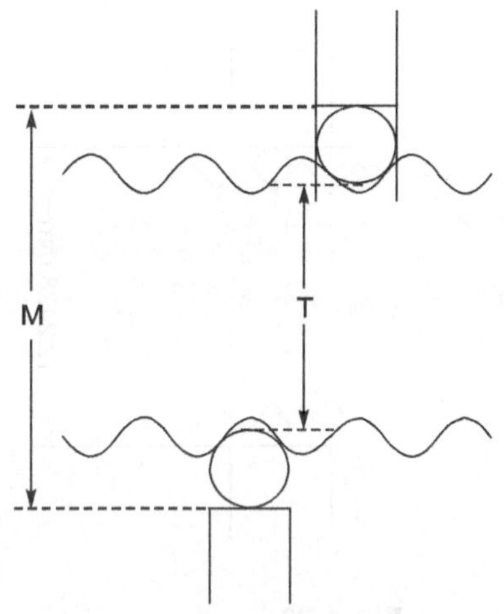

Fig. 1.70 : Two wire method

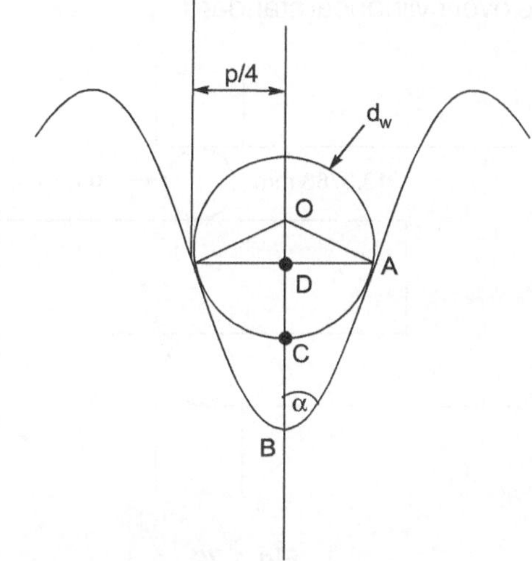

Fig. 1.71 : Calculation of effective diameter

ΔADB $$\frac{P/4}{DB} = \tan\alpha$$

$$DB = \frac{P}{4}\cot\alpha$$

ΔOAB $$\frac{OA}{OB} = \sin\alpha$$

$$\frac{d_w/2}{OB} = \sin\alpha$$

So $$OB = \frac{d_w}{2}\text{cosec}\,\alpha$$

So $$DO = \frac{d_w}{2}\text{cosec}\,\alpha - \frac{p}{4}\cot\alpha$$

$$CD = \frac{d_w}{2} - \left(\frac{d_w}{2}\text{cosec}\,\alpha - \frac{p}{4}\cot\alpha\right)$$

$$CD = \frac{d_w}{2}(1-\operatorname{cosec}\alpha)+\frac{p}{4}\cot\alpha$$

So effective diameter is

$$PCD = T + d_w(1 - \operatorname{cosec}\alpha) + \frac{p}{2}\cot\alpha$$

For best size wire from ΔOAD we get

$$d_w = \frac{p}{2}\sec\theta$$

Example : 1.16

While measuring the effective diameter of an external metric screw thread gauge of 3.5 mm pitch, a 30.5 mm diameter cylindrical standard and 2 mm diameter wires were used. The micrometer reading over the standard and wire was 13.3768 mm. The corresponding reading over the thread gauge and wire was 12.2428 mm. Calculate thread gauge effective diameter.

Solution :

Wire over cylindrical standard

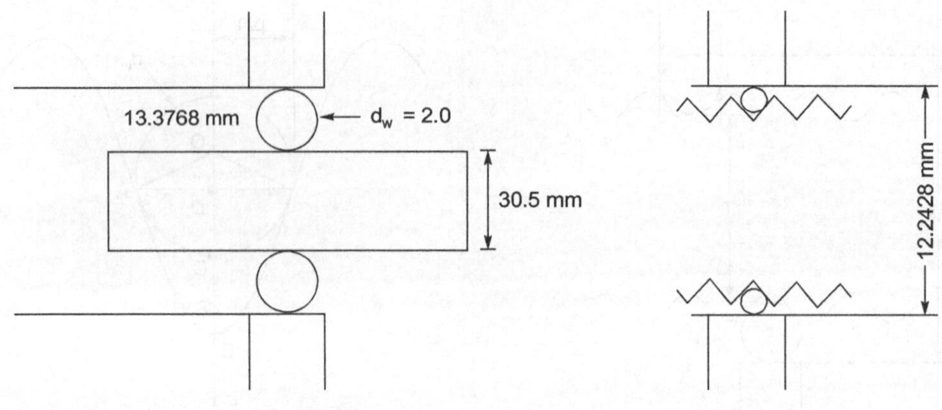

Fig. 1.72 Fig. 1.73

$$d_w = 2.0 \text{ mm}$$
$$\text{pitch} = 3.5 \text{ mm}$$

Since micrometer are used with some attachments correction factor must be considered to calculate the exact dimension. So from cylindrical standard are know that

$$\text{Correction} = (2.0 + 30.5 + 2.0) - 13.3768$$
$$= 21.1232 \text{ mm}$$

Actual value of m over thread gauge is

$$m = 33.366 \text{ mm}$$
$$T = m - 2d_w$$
$$= 33.366 - 2d_w$$
$$T = 29.366 \text{ mm}$$

$$PCD = T + d_w(1 - \operatorname{cosec}\alpha) + \frac{p}{2}\cot\alpha$$

Since it is metric thread and for metric thread

$$2\alpha = 60°, \quad \alpha = 30°$$

So $$PCD = 29.366 + 2(1 - \text{cosec } 30) + \frac{3.5}{2} \cot 30$$

$$PCD = 30.39 \text{ mm}$$

1.23 ACCEPTANCE TESTS

These tests are performed on the machine before inducting it for mass production. These tests are of two types :

1. **Dynamic Test :** Tests carried out over the machine in working condition. A work piece shown in figure is machined at 3 different places. Its diameter is equal to 1/8th of length. To adopt the machine is mass production machined portion should be cylindrical and error should not exceed 0.02 mm in 300 mm.

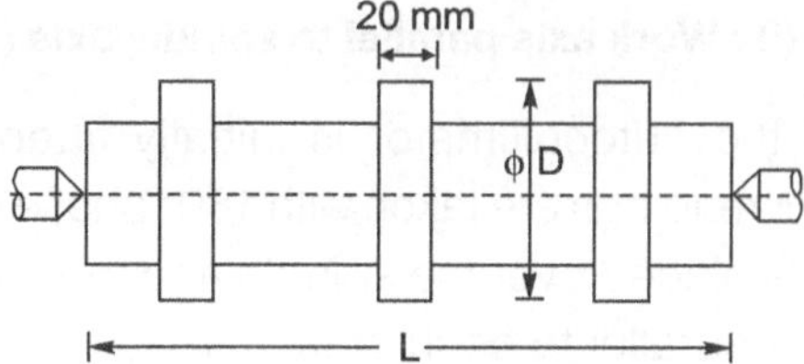

Fig. 1.74 : Test piece for practical tests on a lathe machine

2. **Static Test :** These tests will be discussed for 3 machines namely lathe, milling and drilling.

   (a) Acceptance test for lathe

   (i) Is bed true?

   The bed of the machine should be flat and to check it its surface is divided into segments and the segment size equal to the size of spirit level. Now the spirit level is kept from segment to segment to check flatness.

   (ii) Is spindle true?

   Work is held is the spindle so its axis should be parallel to bed. A mandrel is fitted into the spindle and properly centered. A dial gauge fixed over the carriage with its plunger touching the mandrel as shown in Fig. 1.75. By moving carriage towards spindle, it the axis is parallel there should not be any variation is dial gauge. It is not possible to have these errors zero. So there errors should not exceed beyond a certain value. These errors are permissible in upward direction and towards the tools.

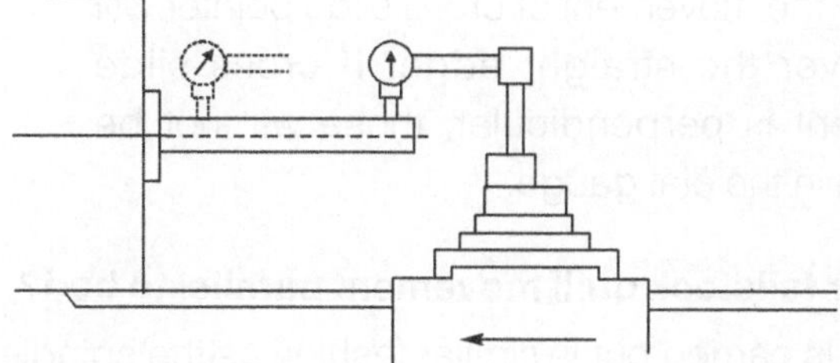

Fig. 1.75 : Spindle axis parallel to bed

(iii) Is line of centre is parallel to spindle axis?

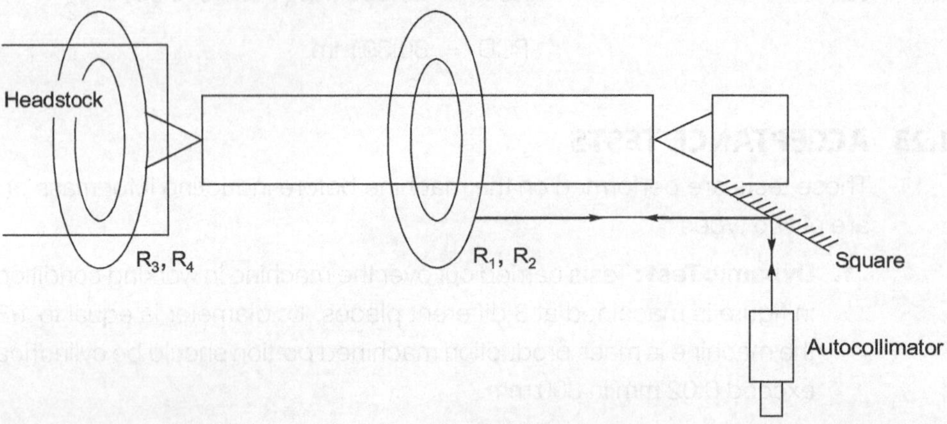

Fig. 1.76 (i) : Work axis parallel to spindle axis (Ist test)

Reflector of the autocollimator is initially fitted over the work piece and the autocollimator readings are taken with 180° phase difference i.e. R_1 and R_2. Later on the reflector is placed over the spindle and the same experiment was repeated. If both axes are parallel to each other, the autocollimator readings will match. The same test can be performed by placing mandrel between head stock and tailstock as shown in Fig.1.76 (ii).

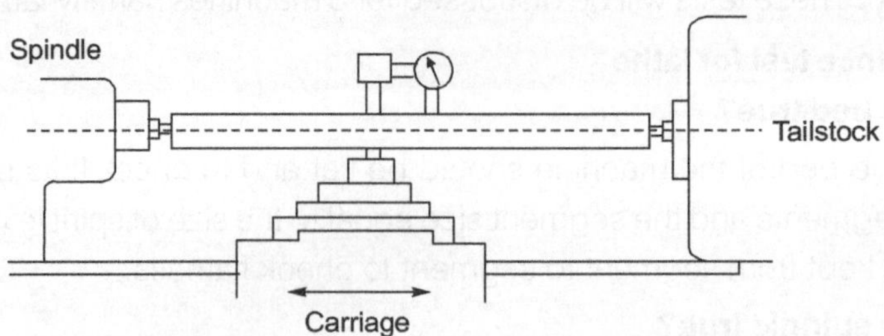

Fig. 1.76 (ii) : Work axis parallel to spindle axis (IInd test)

(iv) Whether cross slide movement is perpendicular?

Perpendicularity of cross slide is important because tool post is mounted over it. A straight edge is placed in the spindle. The plunger of dial gauge is touching on one side of straight edge and base is mounted on the cross slide. By the movement of cross slide pointer will move over the straight edge. If cross slide movement is perpendicular, these will not be variation in the dial gauge.

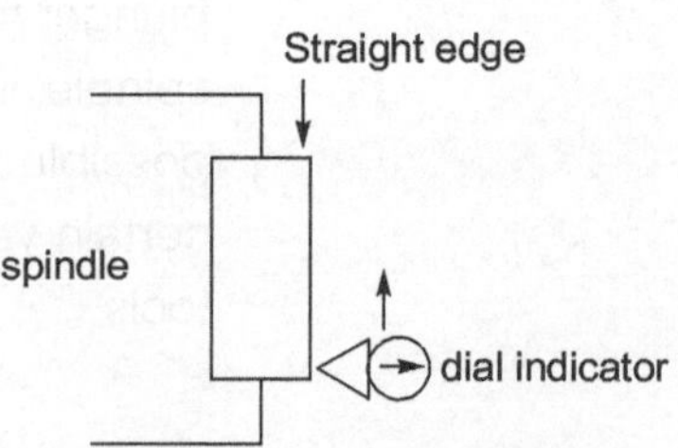

Fig. 1.77 Cross slide perpendicular to spindle axis

(v) Whether tailstock quill movement parallel to bed?

The test is carried out is similar fashion as the spindle movement parallel to bed.

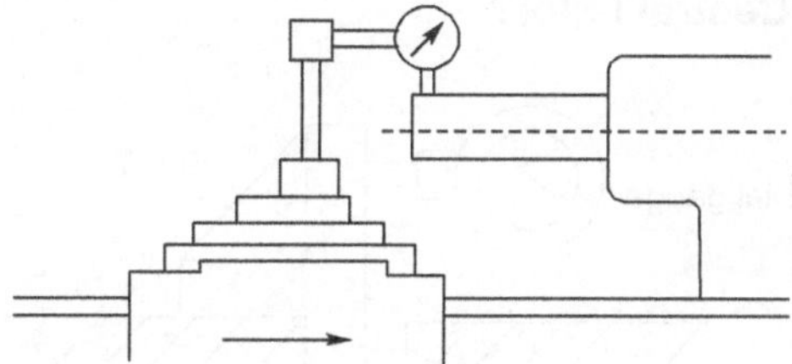

Fig. 1.78 : Tailstock quill

(b) Acceptance test of Milling

(i) Whether table surface square with the vertical movement?

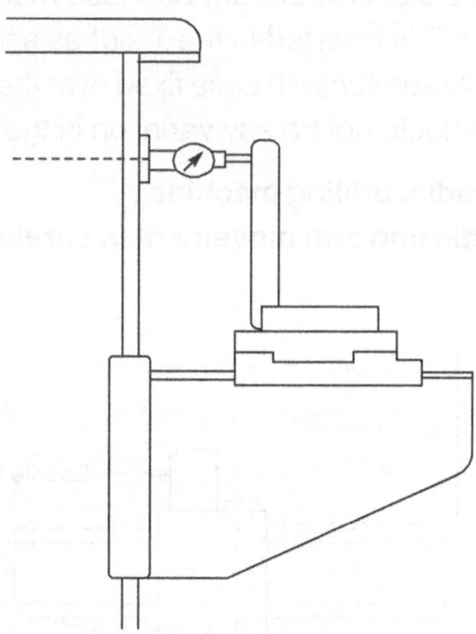

Fig. 1.79 : Testing squareness of table surface

A square is clamped on the table as shown in Fig. 1.79 the dial gauge based is fixed over the arbor with pointer touching one side of square. By giving vertical motion to the table, the variation in dial gauge should be within some permissible limits.

(ii) Whether table movement is parallel to cutter axis?

Base of the dial gauge is fixed over the arbor and plunger touching the table of milling machine. By giving table biaxial movement the variation on the dial gauge should be within permissible limits.

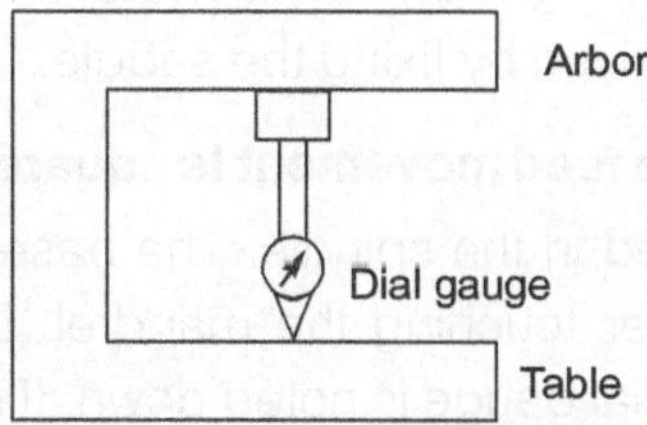

Fig. 1.80 : Table movement is parallel to arbor

(iii) Accuracy of Central T-slot?

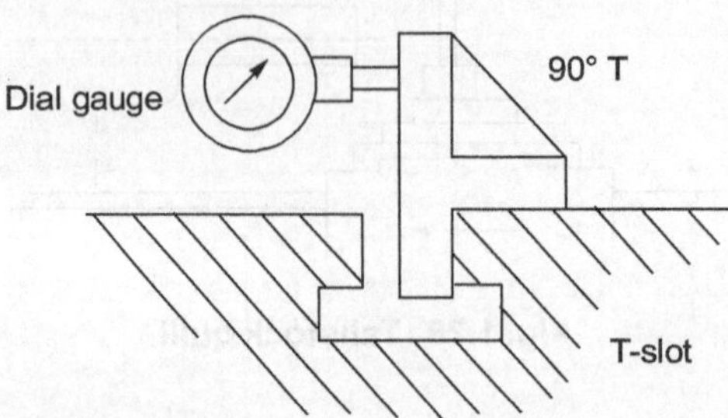

Fig. 1.81 : Central T-slot is perpendicular

Accuracy of the slot is important because machining fixtures are clamped in the central T-slot. A T is inserted in the T slot as shown in Fig. 1.81. The dial gauge is mounted over its surface with base fixed over the arbor. By giving vertical movement to table there should not be any variation in the dial gauge.

(c) Acceptance tests for radial drilling machine

(i) Whether saddle and arm movement is parallel to base plate?

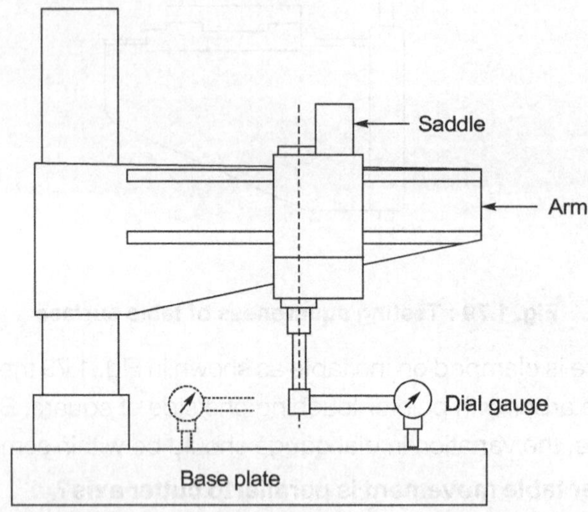

Fig. 1.82 : Saddle and arm movement parallel

The base of the dial gauge is kept over the saddle and initially by giving movement to the saddle, dial gauge variations are noted down. In the second set arm movements are given by fixing the saddle.

(ii) Whether spindle feed movement is square with the base?

A mandrel is fitted in the spindle. The base of dial gauge is fixed over the base plate with plunger touching the mandrel. By giving spindle vertical movement deviation is the dial gauge is noted down. If there is no variation in the dial gauge, it means spindle axis is perpendicular is base plate.

■■■■

PRACTICE QUESTIONS

1. A ring gauge is used to measure
(a) Outside diameter but not roundness
(b) Roundness but not outside diameter
(c) Both outside diameter and roundness
(d) Only external threads

2. Two slip gauges of 10 mm width measuring 1.000 mm and 1.002 mm a are Kept side by side in contact with each other lengthwise. An optical flat is kept resting on slip gauge as shown in the figure monochromatic light of wavelength 0.00058928 mm is used in the inspection. The total number of straight fringes that can be observed on both slip gauges is

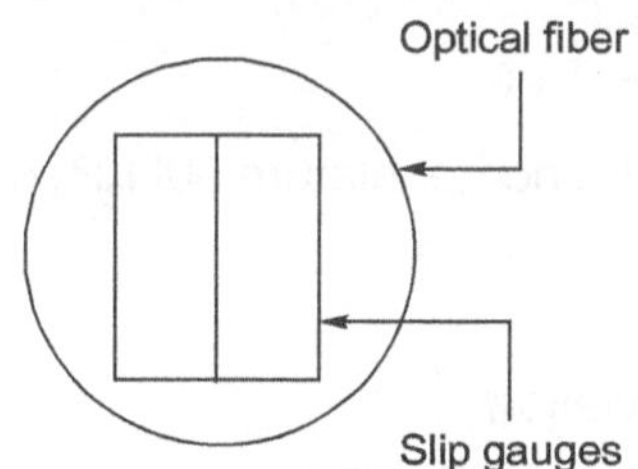

(a) 2 (b) 6
(c) 8 (d) 13

3. What is the dominant direction of the tool marks or scratches in a surface texture having a directional quality called?
(a) Primary texture (b) Secondary texture
(c) Lay (d) Flaw

4. A component requires a hole which must be within the two limits of 25.03 mm and 25.04 mm which of the following statement about the reamer size are correct
(i) Reamer size cannot be below 25.03 mm
(ii) Reamer size cannot be above 25.04 mm
(iii) Reamer size can be 25.04 mm
(iv) Reamer size can be 25.03 mm
Select the correct answer using the code given below
(a) 1 and 3 (b) 1 and 2
(c) 3 and 4 (d) 2 and 4

5. GO and NO-GO plug gauges are to be designed for a hole $20.000^{+0.050}_{+0.010}$ mm gauge tolerance can be taken as 10% the hole tolerance following ISO design of gauge design, sizes of GO and NO – GO will be respectively
(a) 20.010 mm and 20.050 mm (b) 20.014 mm and 20.046 mm
(c) 20.006 mm and 20.054 mm (d) 20.014 mm and 20.054 mm

6. In strain gauge dynamometers the use of how many active gauge makes the dynamometers more effective
(a) Four (b) Three
(c) Two (d) One

7. Match **List-I** (Measuring device) with **List-II** (Parameter measured) and select the correct answers using the codes given below the list:

List-I	**List-II**
A. Diffraction grating flat Surface	**1.** Small angular deviations
B. Optical flat	**2.** Online Measurement of Moving Parts
C. Auto Collimators	**3.** Measurement of Gear pitch
D. Laser scan of micrometer	**4.** Joining hydraraulic Piston rod for Agriculture Machinery
	5. Taper of any surface

Codes:

	A	B	C	D
(a)	5	4	2	1
(b)	3	5	1	2
(c)	3	5	4	1
(d)	5	4	1	2

8. Consider the following statement A nomenclature $50H_8P_8$ denotes that

1. hole basic size is 50mm
2. it is shaft basis system
3. 8 indicates fundamental deviation

Which of the statements given above are correct

(a) 1,2 and 3 (b) 1 and 2 only
(c) 1 and 3 only (d) 1 only

9. The value of surface roughness 'h' obtained during turning operation at a feed f with a round nose tool having a radius 'r' is given as

(a) f/8r (b) $f^2/8r$
(c) $f^3/8r$ (d) $f^3/8r^2$

10. A threaded nut of M 16,ISO metric type, having 2 mm pitch with a pitch diameter of 14.701 mm is to be checked for its pitch diameter using two or three number of balls or rollers of following size

(a) rollers of 2 mm φ (b) rollers of 1.155 mm φ
(c) balls of 2 mm φ (d) balls of 1.155 mm φ

11. which one of the following is a clearance fit

(a) $\phi\ H50^{+0.015}_{+0.005}\ h50^{-0.010}_{+0.000}$ (b) $\phi\ H50^{+0.015}_{+0.010}\ h50^{+0.025}_{+0.015}$

(c) $\phi\ H50^{-0.015}_{+0.000}\ h50^{+0.025}_{+0.005}$ (d) $H50^{-0.010}_{-0.000}\ h50^{+0.030}_{+0.005}$

12. In tolerance specification 25 D 6 the letter D represents

(a) garage of tolerance (b) upper deviation
(c) lower deviation (d) type of fit

13. what symbol is used to indicate surface roughness

(a) = (b) $\sqrt{}$

(c) 0.1 (d) ∇

14. $a^{+0.005}_{-0.001}$ $b^{+0.01}_{-0.01}$ $c^{-0.002}_{-0.003}$ $d^{-0.0}_{-0.001}$ $e^{+0.005}_{0.0}$ X will have the following tolerance

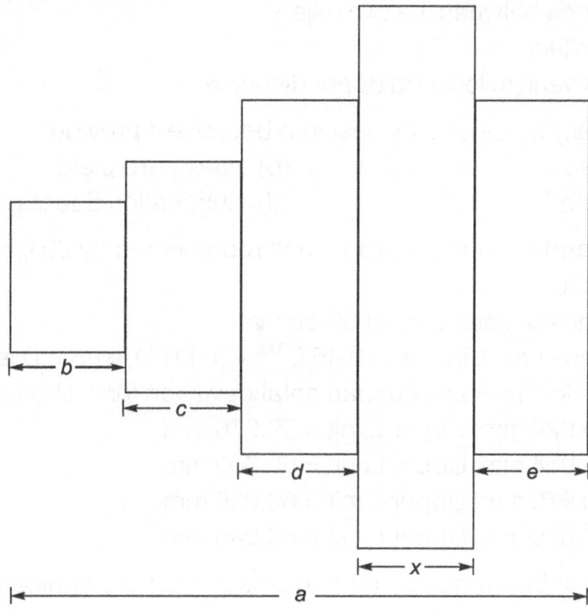

(a) $X^{+0.018}_{-0.007}$ (b) $X \pm 0.0125$
(c) $X^{+0.019}_{-0.014}$ (d) $X^{-0.012}_{+0.005}$

15. The M and E systems in metrology are related to measurements of
(a) Screw thread (b) flatness
(c) angularity (d) surface finish

16. A hole is specified as $40^{+0.050}_{+0.000}$ mm the mating shaft has a clearance fit with minimum clearance 0.01 mm the tolerance on shaft is 0.04 the maximum clearance in mm between hole and shaft is
(a) 0.04 (b) 0.05
(c) 0.10 (d) 0.11

17. Tolerances are specified
(a) to obtain desired fits
(b) because it is not possible to manufacture a size exactly
(c) to obtain higher accuracy
(d) to have proper allowances

18. Optical flats are made of
(a) quartz (b) glass
(c) plastic (d) steel

19. Repeatability of measuring equipment is
(a) the closeness with which a measurement can be read directly from a measuring instrument
(b) difference between measured value and actual value
(c) the smallest change in measurand that can be measured
(d) the capability to indicate the same reading again and again for a given measurand

20. External taper can be accurately measured with the help of
(a) sine bar and slip gauges (b) dividing head
(c) precision balls and height gauge (d) combination set

21. A sine bar is specified by
(a) its total length
(b) the centre distance between the two rollers
(c) the size of the rollers
(d) the distance between rollers and upper distance

22. Surface plate is usually made of grey cast iron because it provides
(a) non wearing plate (b) very hard plate
(c) easy to cast plate (d) lubrication due to graphite flakes

23. What are the upper and lower limits of the shaft represented by 60 f_8?
Use the following data :
Diameter 60 lies in the diameter step of 50-80 mm.
Fundamental tolerance unit, i, in mm = $0.45\ D^{1/3} + 0.001\ D$, where D is the representative size in mm; Tolerance value for IT8 = 25i. Fundamental deviation for 'f' shaft= $-5.5\ D^{0.41}$
(a) Lower limit = 59.924 mm, Upper Limit = 59.970 mm
(b) Lower limit = 59.954 mm, Upper Limit = 60.000 mm
(c) Lower limit = 59.970 mm, Upper Limit = 60.016 mm
(d) Lower limit = 60.000 mm, Upper Limit = 60.046 mm

24. A displacement sensor (a dial indicator) measures the lateral dispalcement of a mandrel mounted on the taper hole inside a drill spindle. The mandrel axis is an extension of the drill spindle taper hole axis and the protruding portion of the mandrel surface is perfectly cylindrical. Measurements are taken with the sensor placed at two positions P and Q as shown in the figure. The readings are recorded as R_x = maximum deflection minus minimum deflection, corresponding to sensor position at X, over one rotation. If $R_P = R_Q > 0$, which one the following would be consistent with the observation?

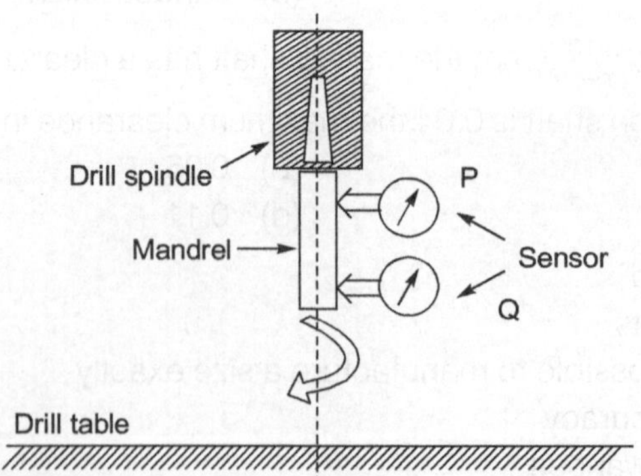

(a) The drill spindle rotational axis is coincident with the drill spindle taper hole axis.
(b) The drill spindle rotational axis intersects the drill spindle taper hole axis at point P.
(c) The drill spindle rotational axis is parallel to the drill spindle taper hole axis.
(d) The drill spindle rotational axis intersects the drill spindle taper hole axis at point Q.

25. A shaft has a dimension, $\phi 35^{-0.009}_{-0.025}$. The respective values of fundamental deviation and tolerance are
(a) –0.025, ±0.008 (b) –0.025, 0.016
(c) –0.009, ±0.008 (d) –0.009, 0.016

26. Expressing a dimension as $25.3^{+0.05}_{-0.05}$ mm is the case of
(a) Unilateral tolerance (b) Bilateral tolerance
(c) Limiting dimensions (d) All the above

27. In an interchangeable assembly, shafts of size $25.3^{+0.010}_{-0.010}$ mm mate with holes of size size mm. The maximum possible clearance in the assembly will be
(a) 10 microns (b) 20 microns
(c) 40 microns (d) 50 microns

28. Dimension of the hole is $50^{+0.02}_{-0.00}$ mm and shaft is $50^{-0.00}_{-0.02}$ mm.The minimum clearance is
(a) 0.02 mm (b) 0.00 mm
(c) –0.02 mm (d) 0.01 mm

29. The tolerance specified by the designer for the diameter of a shaft is 20.00±0.025 mm. The shafts produced by three different machines A,B and C have mean diameter of 19.99 mm, 20.00 mm and 20.01 mm respectively, with same standard deviation. What will be the percentage rejection for the shafts produced by machines A, B and C?
(a) Same for the machines A, B and C since the standard deviation is same for the three machines
(b) Least for machine A
(c) Least for machine B
(d) Least for machine C

30. In order to have interference fit, it is essential that the lower limit of the shaft should be
(a) Greater than the upper limit of the hole
(b) Lesser than the upper limit of the hole
(c) Greater than the lower limit of the hole
(d) Lesser than the lower limit of the hole

31. A shaft and hole pair is designated as $50h_7d_8$. This assembly constitutes
(a) Interference fit (b) Transition fit
(c) Clearance fit (d) None of the above

32. Consider the following statements :
1. The amount of interference needed to create a tight joint varies with diameter of the shaft.
2. An interference fit creates no stress state in the shaft
3. The stress state in the hub is similar to a thick walled cylinder with internal pressure.
Which of the statements given above are correct?
(a) 1,2 and 3 (b) 1 and 2only
(c) 2 and 3only (d) 1 and 3 only

33. Which of the following is an interference fit?
(a) Push fit (b) Running fit
(c) Sliding fit (d) Shrink fit

34. Consider the following fits:
1. I.C. engine cylinder and piston
2. Ball bearing outer race and housing
3. Ball bearing outer race and shaft
Which of the above fits are based on the interference system?
(a) 1 and 2 (b) 2 and 3
(c) 1 and 3 (d) 1, 2 and 3

35. In the specification of dimensions and fits,
(a) Allowance is equal to bilateral tolerance
(b) Allowance is equal to unilateral tolerance

(c) Allowance is independent of tolerance
(d) Allowance is equal to the difference between maximum and minimum dimension specified by the tolerance

36. Match the items in List I and II.

List I	List II
A. Interference fit	1. Viscosity index
B. Cyclic loading	2. interference
C. Gear meshing	3. Notch sensitivity
D. Lubricating of bearings	4. Induced compressive stress

Codes:

	A	B	C	D		A	B	C	D
(a)	3	4	1	2	(b)	4	3	2	1
(c)	3	4	2	1	(d)	4	3	1	2

37. Minimum clearance in a fit is the difference between
(a) Maximum hole size and minimum shaft size
(b) Minimum hole size and Maximum shaft size
(c) Maximum hole size and Maximum shaft size
(d) Minimum hole size and minimum shaft size

38. A master gauge is
(a) A new gauge
(b) An international reference standard
(c) A standard gauge for checking accuracy of gauges used on shop floors
(d) A gauge used by experienced technicians

39. The vernier reading should not be taken at its face value before an actual check has been taken for
(a) Zero error
(b) Its calibration
(c) Flatness of measuring jaws
(d) Temperature equalization

40. Standards to be used for reference purposes in laboratories and workshops are termed
(a) Primary standards
(b) Secondary standards
(c) Tertiary standards
(d) Working standards

41. A feeler gauge is used to check the
(a) Pitch of the screw
(b) Surface roughness
(c) Thickness of clearance
(d) Flatness of a surface

42. CLA value and RMS values are used for measurement of
(a) Metal hardness
(b) Sharpness of tool edge
(c) Surface dimensions
(d) Surface roughness

43. A sine bar is specified by
(a) Its total length
(b) The size of the rollers
(c) The centre distance between the two rollers
(d) The distance between rollers and upper surface

44. The M and E-system in metrology are related to measurement of:
(a) Screw threads
(b) Flatness
(c) Angularity
(d) Surface finish

45. A sine bar has a length of 250 mm. Each roller has a diameter of 20 mm. During taper angle measurement of a component, the height from the surface plate to the centre of a roller is 100 mm. The calculated taper angle is

(a) 21.1 (b) 22.8
(c) 23.6 (d) 68.9

46. Surface roughness on a drawing is represented by

(a) Triangle (b) Circles
(c) Squares (d) Rectangles

47. Match the items in List I and II.

	List I		**List II**
A.	Surface profilometer	1.	Calibration
B.	Light Section Microscope	2.	Form tester
C.	Microkater	3.	Film thickness measurement
D.	Interferometer	4.	Centre line average
		5.	Comparator
		6.	Surfaces lay measurement

Code:

	A	B	C	D		A	B	C	D
(a)	4	1	2	3	(b)	4	3	5	1
(c)	4	2	1	3	(d)	3	1	2	4

48. Auto collimator is used to check

(a) Roughness (b) Flatness
(c) Angle (d) Automobile balance

49. An autocollimator is used to

(a) Measure small angular displacements on flat surface
(b) Compare known and unknown dimensions
(c) Measure the flatness error
(d) Measure roundness error between centres

50. Optical square is

(a) Engineer's square having stock blade set 900
(b) A constant deviation prism having the angle of deviation between the incident ray and reflected ray, equal to 900
(c) A constant deviation prism having the angle of deviation between the incident ray and reflected ray, equal to 450
(d) Used to produce interference fringes

51. Cylindrical pins of $25^{+0.020}_{+0.010}$ mm diameter are electroplated in a shop. Thickness of the plating is 30 ± 2.0 micron. Neglecting gage tolerances. The size of the GO gage in mm to inspect the plated components is

(a) 25.042 (b) 25.052
(c) 25.074 (d) 25.084

52. Match the following

	Feature to be inspected		**Instrument**
A.	Pitch and Angle errors of screw thread	1.	Auto Collimator
B.	Flatness error of a surface plate	2.	Optical Interferometer

C. Alignment error of a machine slide way
D. Profile of a cam

3. Dividing head and Dial gauge
4. Spirit Level
5. Sine bar
6. Tool makers Microscope

Code:

	A	B	C	D		A	B	C	D
(a)	6	2	4	6	(b)	5	2	1	6
(c)	6	4	1	3	(d)	1	4	4	2

53. Match the following

List-I	List-II
A. Talysurf	1. T-slots
B. Telescopic	2. Flatness
C. Transfer callipers	3. Internal diameter
D. Autocollimator	4. Roughness

Code:

	A	B	C	D
(a)	4	1	2	3
(b)	4	3	1	2
(c)	4	2	1	3
(d)	3	1	2	4

54. Match the following

List-I	List-II
A. Pitot-tube	1. R.P.M. of a shaft
B. McLeod Gauge	2. Displacement
C. Planimeter	3. Flow velocity
D. LVDT	4. Vacuum
	5. Surface finish
	6. Area

Code:

	A	B	C	D		A	B	C	D
(a)	4	1	2	3	(b)	3	4	6	2
(c)	4	2	1	3	(d)	3	1	2	4

55. The diameter of a recessed ring was measured by using two spherical balls of diameter d_2=60 mm and d_1=40 mm as shown in the figure beloaw. The distance H_2 =35.55 mm and H_1 =20.55 mm. The diameter (D, in mm) of the ring gauge is

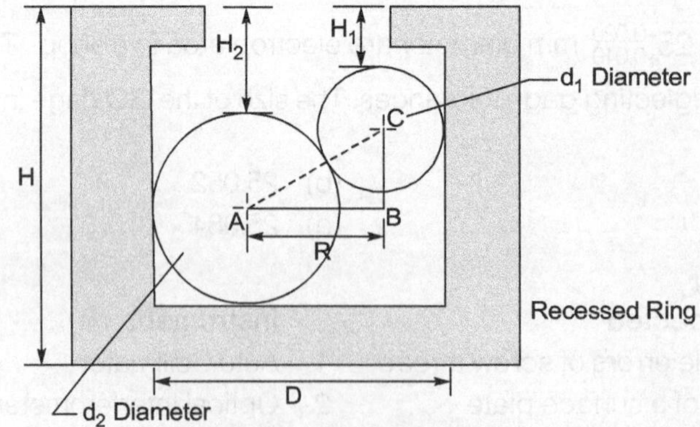

ANSWERS

METROLOGY

1. (c)	**2.** (d)	**3.** (c)	**4.** (b)	**5.** (d)	**6.** (b)	**7.** (b)
8. (d)	**9.** (b)	**10.** (b)	**11.** (a)	**12.** (c)	**13.** (d)	**14.** (c)
15. (d)	**16.** (c)	**17.** (b)	**18.** (a)	**19.** (d)	**20.** (a)	**21.** (b)
22. (a)	**23.** (a)	**24.** (c)	**25.** (d)	**26.** (b)	**27.** (c)	**28.** (b)
29. (c)	**30.** (a)	**31.** (c)	**32.** (d)	**33.** (a)	**34.** (b)	**35.** (c)
36. (b)	**37.** (b)	**38.** (c)	**39.** (a)	**40.** (d)	**41.** (c)	**42.** (d)
43. (c)	**44.** (d)	**45.** (c)	**46.** (a)	**47.** (b)	**48.** (b)	**49.** (c)
50. (c)	**51.** (c)	**52.** (b)	**53.** (b)	**54.** (d)	**55.** (93.3)	

CHAPTER 2

Material Science

All the dead species in the universe are called matter. Material science is a combination of solid state physics and solid state chemistry. The elements are divided into categories like metals, nonmetals, polymer and ceramics.

2.1 METALS

Metal is one which has free electrons. These materials has their characteristic appearance and they are capable of changing shape upon machining and gives good finish. At room temperature they are usually solid and to some extent they are malleable and ductile. Metals are good conductors of heat and electricity e.g. Copper, Silver, Gold etc.

2.2 NON-METALS

The materials in the right portion of the periodic table are called non-metals. These materials are usually dull in appearance, brittle and poor conductor of electricity (except graphite). They does not form alloys but combine chemically to form compounds.

2.3 ORGANIC POLYMER

The long molecules are composed to structural entities called 'mer' which are successively repeated along the chain. All organic polymers are based on C-C bonding, polymers are mostly amorphous and sometimes partly crystalline. The term polymer means many 'mers'. It is in coil form i.e. continuous series of C-C atom. If ethylene (C_2H_4) which is a gas is subjected to some temperature and pressure with some initiator (to break the double bond) it transform into polyethylene.

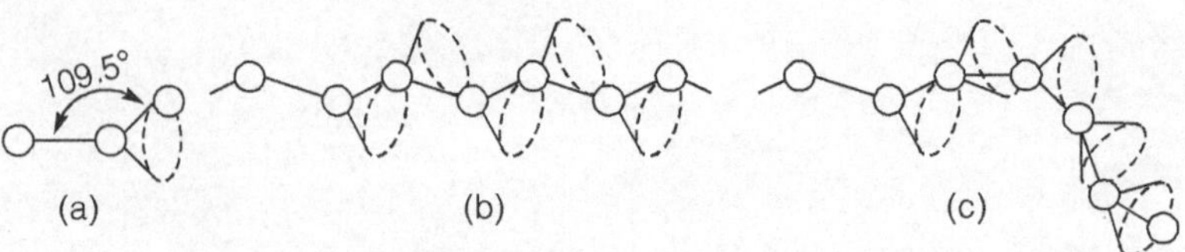

Fig. 2.1 : Schematic representations for how polymer chain shape is influenced by the positioning of carbon atoms

It is also called long chain molecule or macromolecule. Diamond is also a C-C bond and it is the hardest element. It is the phenomenon of conformation which makes the polymers so soft, when a single polymer chain can take different 3D shapes, it is called conformation. It can be seen from Fig. 2.2 that because of bond rotation (conformation) there is considerable increase in strain with slight increase in stress.

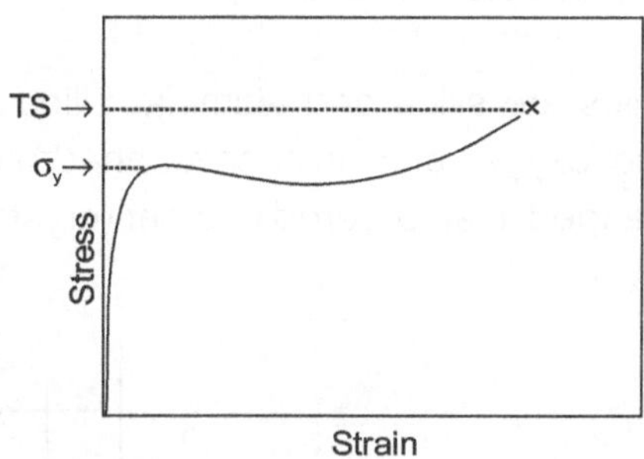

Fig. 2.2 : Schematic stress-strain curve for a plastic polymer

The degree of polymerization(DOP) is referred as the number of mers in a polymer chain.

$$\text{DOP} = \frac{\text{Molecular weight}}{\text{mer weight}}$$

The degree of crystallization(f_C) is referred as

$$f_C = \frac{\text{Volume of crystalline region}}{\text{Total volume of the specimen}}$$

The packing is more dense in crystalline part

Let ρ_C is the density of crystalline part and ρ_a is the density of amorphous part, so the density of polymer sample is

$$\rho V = \rho_a V_a + \rho_c V_c$$

$$\Rightarrow \qquad \rho\left(\frac{V_c}{V_a}+1\right) = \rho_a + \rho_c \frac{V_c}{V_a}$$

$$\rho(1 + f_c) = \rho_a + \rho_c f_c$$

$$\rho = \frac{\rho_a + \rho_c f_c}{1 + f_c}$$

The plastic is defined as the organic polymer that can be easily shaped. It can be further divided into thermoplastic and thermosetting plastics. Thermoplastics softens when heated and becomes hard when cooled. The process can be repeated many times. Thermosetting plastics once shaped cannot be recycled. Upon heating they get burned but does not melt. It is because of cross linking between adjacent molecular chains.

Polyvinyl chloride (PVC), Polypropylene (PP), Polyethelene (PE) and Polystyrene (PS) plastics are known as BIG-4 because 70-80% of market is covered by these plastics only. Other well known thermoplastics are poly tetra fluoro ethylene (PTFE or TEFLON), acrylic etc. The commonly used thermosetting plastics are Epoxy, Polyster, Phenol formaldehyde (Bakelite) etc. Most of these thermosetting plastics are used as matrix in composite materials.

2.4 CERAMICS

Ceramics are burnt in earth and its definition can be given in a negative way i.e. whatever is not a metal, non-metal and organic polymer is called a ceramic. The properties of ceramic are:

1. They are extremely brittle
2. High thermal stability
3. High chemical stability (corrosion resistance)
4. High hardness

The most important ceramics are silica and alumina. Silica (SiO_2) has a 3D structure that is generated when every corner oxygen atom in each tetrahedron is shared by adjacent tetrahedra. If these tetrahedras are arranged in an ordered manner (crystal form) it is called quartz.

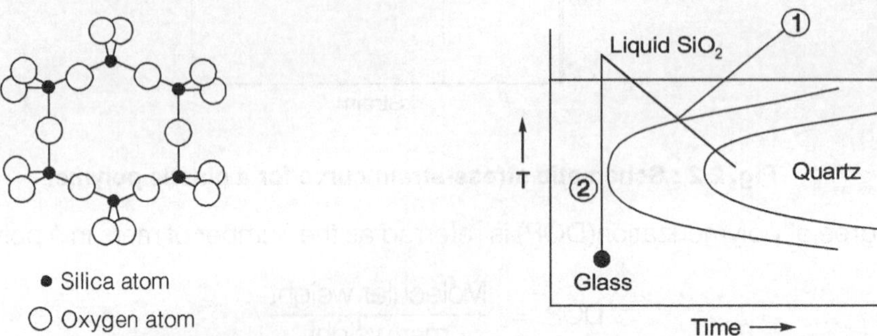

Fig. 2.3 : Structure of Quartz

Fig. 2.4 : TTT Diagram of SiO_2 cooling

Upon Cooling liquid silica at a very slow rate (along line 1 in Fig. 2.4) crystals being formed (Quartz). It is used in making optical flat and also has application in watches. Upon cooling liquid silica at a very fast rate (along line 2) the viscosity of liquid increases as temperature decreases and finally the material becomes solid even though there are no crystals in it (Fig. 2.5). This is called amorphous silica or glass. It is an unstable state and after some time it converts into crystalline form. This is called devitrification of glass. Upon quenching liquid metals at a very fast rate (10^6 °C/sec) the metal will also convert into glass called metallic glass and is used in transformer cores to minimize the eddy current losses. Since the ceramic is extremely brittle, its failure is because of the propogation of crack. **In the larger samples the probability of having larger cracks are more so their strength in tension is less**. So the strength of ceramics is size dependent. Under compression load the cracks do not propagate, hence the ceramica posses high compression strength.

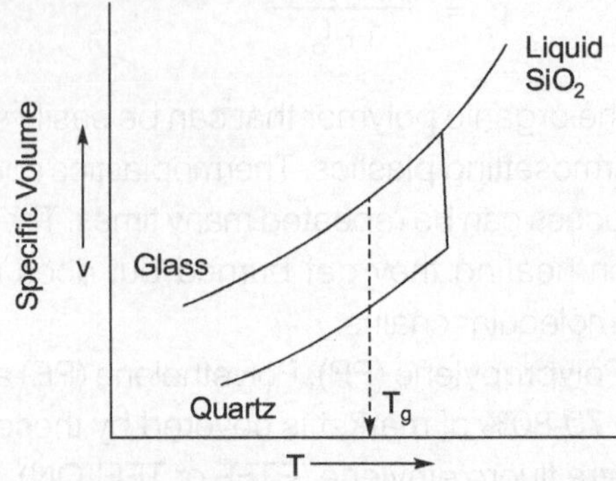

Fig. 2.5 : Specific volume versus-temperature behaviour of crystalline and non crystalline materials

Upon quenching liquid silica the viscosity of liquid increases and there is a temperature at which it is difficult to understand whether it is liquid or solid . This is called glass transition temperature (T_g). This temperature is unique and it depends on the cooling rate. Faster the cooling rate T_g will be higher. $[SiO_4]^{-4}$ is a building block for glass and quartz. If the atoms are arranged in a periodic hexagonal structure (Fig. 2.3), it is called quartz. If the atoms are randomly oriented (Fig. 2.6), it is called glass or fused silica. Since $[SiO_4]^{-4}$ building block is having a long chain, hence the operating temperature of pure silica glass is very high. These glasses are used in high temperature application e.g. furnace window etc. To break this long chain, SiO_2 is mixed with Na_2O and CaO which acts as a catalyst.

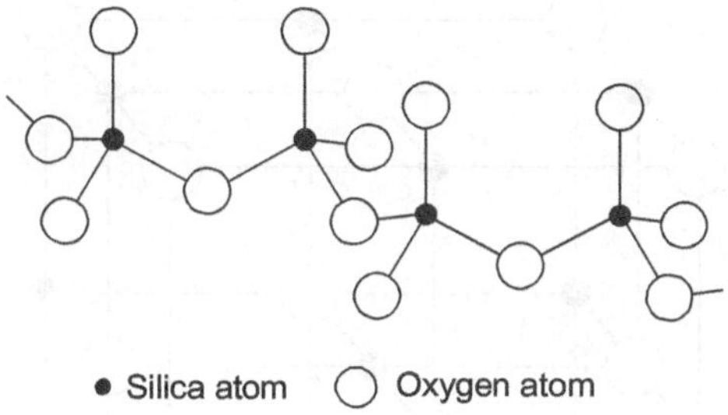

• Silica atom ○ Oxygen atom

Fig. 2.6 : Structure of Glass

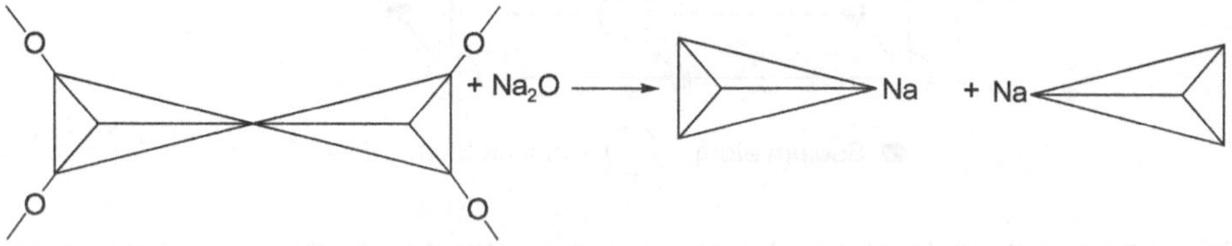

O O + Na_2O ⟶ Na + Na O O

Fig. 2.7 : Soda lime glass

Na_2O acts as a network modifier and the length of silica tetrahedra chain decreases which decreases in the operating temperature. This glass is called soda lime glass. Glass fails by brittle fracture. To increase it's strength, the propogation of crack is made difficult by introducing compressive stresses in the outer layer by the following methods **:**

1. **Thermal tempering or air quenching :** This process involves heating of glass upto softening temperature and then quenching in cool air medium. The outer layer of the glass, quickly becomes rigid but the core remains still hot. When core comes to room temperature its specific volume decreases and it tries to pull the material from the surface towards the core. This introduces compressive stresses in the outer layer.
2. **Ion exchange :** In this process, the glass is dipped into a KNO_3 bath at suitably high temperature. The Na^+ ions are replaced out by K^+ ions and since K^+ are bigger in size the outer layer tries to expand but the core resists the motion of outer layer. This induces compressive stresses in the outer layer. The stresses generated in this method are much higher **(approximately 400 MPa)** but the thickness of compressive layer is much smaller than that of tempering.

   The tempered glass is also called as toughened glass and upon fracture it gives large number of small pieces with blunt edges. Ceramic or glass if loaded at a certain load, it may take that load for some time but it fails at the same load after some time. This phenomenon of ceramics is called **static fatigue**.

2.5 CRYSTAL STRUCTURES OF CERAMICS

2.5.1 AX-Type Crystal Structures of Ceramics

If there are equal numbers of cations and anions in structure, then they are often referred to as AX compounds, where A denotes the cation and X the anion. Perhaps the most common AX crystal structure is the **sodium chloride (NaCl)**, or rock salt, type. The coordination number for both cations and anions is 6. A unit cell for this crystal structure (Fig. 2.8) is generated from an FCC arrangement of anions with one cation situated at the cube center and one at the center of each of the 12 cube edges.

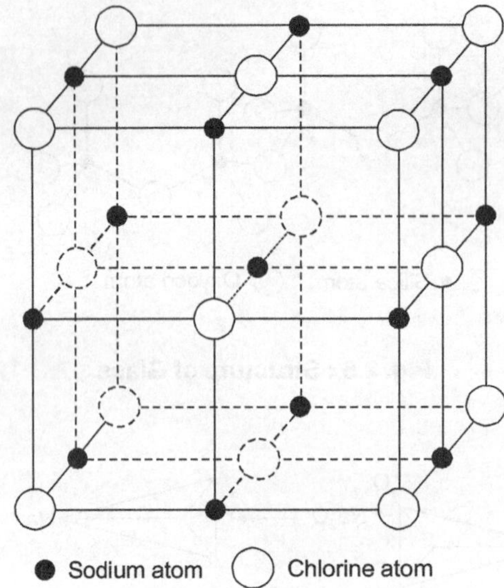

Fig. 2.8 : A unit cell for the rock salt, or sodium chloride (NaCl), crystal structure

2.5.2 Cesium Chloride Structure

Fig. 2.9 shows a unit cell for the cesium chloride (CsCl) crystal structure the coordination number is 8 for both ion types. The anions are located at each of the corners of a cube, whereas the cube center is a single cation. Interchange of anions with cations, and vice versa, produces the same crystal structure. This is not a BCC crystal structure because two different kinds of ions are involved.

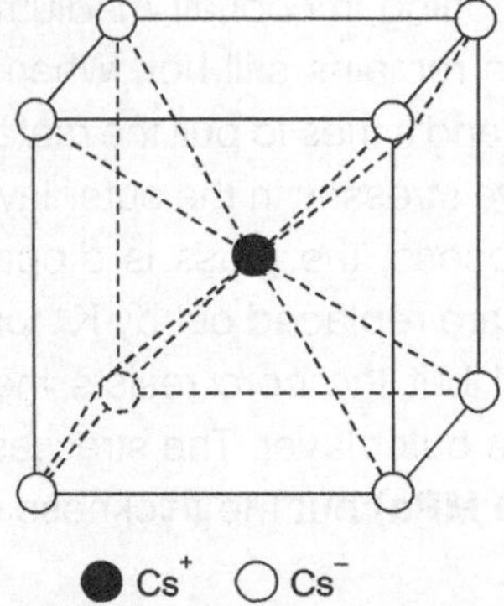

Fig. 2.9 : A unit cell for the cesium chloride (CsCl) crystal structure

2.5.3 Zinc Blende Structure

The coordination number is 4 and all ions are tetrahedrally coordinated.

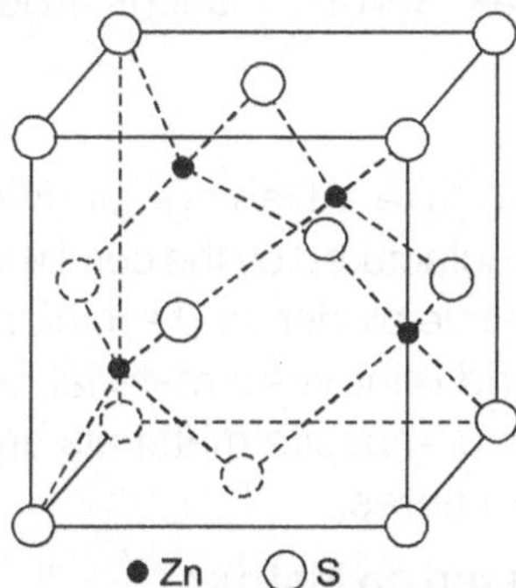

Fig. 2.10 : A unit cell for the zinc blende (ZnS) crystal structure

2.5.4 A_mX_p - Type Crystal Structures

If the charges on the cations and anions are not the same, compound ions exist with the chemical formula A_mX_p and such structure is found in fluorite (CaF_2). Calcium ions are positioned at the centers of cubes, with fluorine ions at the corners.

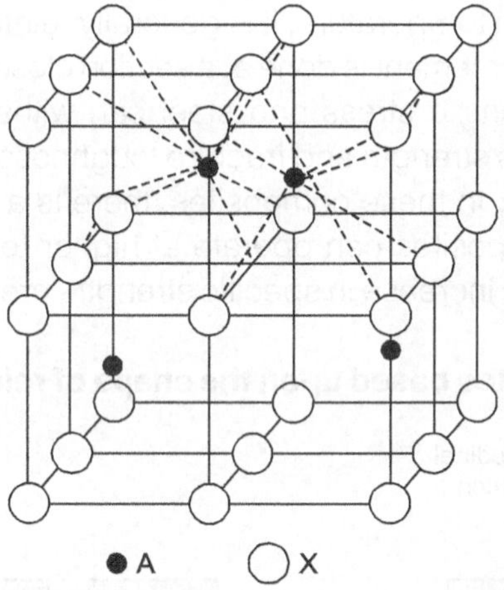

Fig. 2.11 : A unit cell for the fluorite (A_mX_p) crystal structure

2.5.5 $A_mB_nX_p$ - Type Crystal Structures

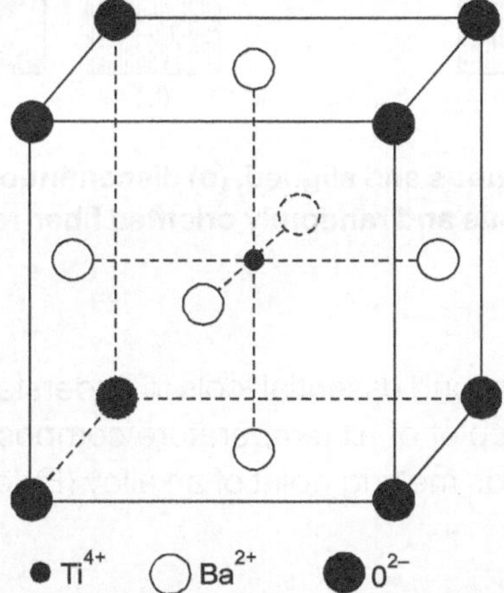

Fig. 2.12 : A unit cell for the $A_mB_nX_p$ type crystal structure

It is also possible for ceramic compounds to have more than one type of cation; for two types of cations (represented by A and B), their chemical formula may be designated as $A_mB_nX_p$. Barium titanate ($BaTiO_3$), having both Ba^{2+} and Ti^{4+} cations, falls into this classification.

2.6 COMPOSITE MATERIALS

Composite materials are made to take advantage of different properties of metal, polymer and ceramics. These materials are manufactured by the combination of 2 or more materials. In composite materials there is a matrix (compatible binder) and a reinforcing material. The strength of composite is generally in between matrix and reinforced material but sometimes it is even higher than the pure matrix and pure fibres. The composite materials are generally classified according to the type of matrix or the orientation of fibres.

Classification of composite Based on Matrix :

Polymer matrix composite : When an organic polymer e.g. epoxy, phenolic resin etc. is used as matrix material, the composite is called polymer composite. In this category, the two most important composites are **glass fiber reinforced plastic (GFRP)** and **carbon fiber reinforced plastic (CFRP)**. GFRP is having high strength but low stiffness and it is used in sports equipments, filament rocket motor cases, pressure vessels, aircraft components etc.

Ceramic matrix composite : A typical example of this composite is RCC column. As discussed before, that ceramics are very weak in tension and bending but have excellent compressive strength and stability at high temperature, it is generally reinforced with steel to increase its load carrying capacity. The reinforcement is done at a section close to the bottom because this surface (bottom) is subjected to tensile stress and steel can withstand that. By increasing the fibre content in the composite, its strength and fracture toughness increases.

Metal matrix composites : In these composites, there is a matrix of any ductile material e.g., Al, Mg, Ti etc., These composites can operate at higher temperature than other composites. The main properties include increase in specific strength, creep resistance, dimensional stability etc.

Classification of Composites based upon the shape of reinforcement :

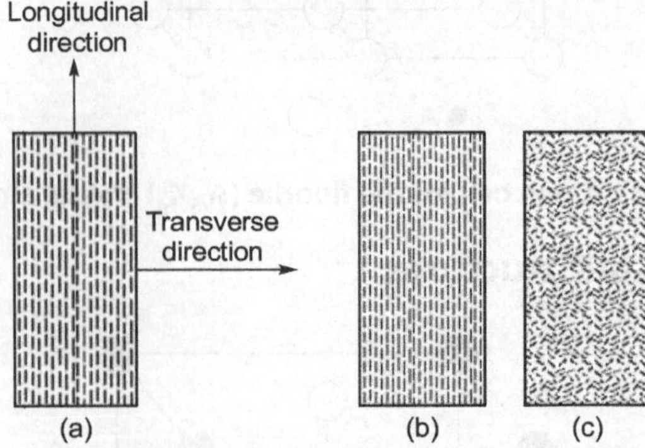

Fig. 2.13 : (a) Continuous and aligned, (b) discontinuous and aligned, and (c) discontinuous and randomly oriented fiber-reinforced composites.

2.7 PHASE DIAGRAMS

Phase diagrams are important and essential tools to understand compositional concentrations of two or more elements. It is a plot of all temperature composition space showing the stability of various phases. In other words melting point of an alloy (Eutectic point) can be understood from phase diagram.

Terminology used in phase diagram :

System : The substances that are isolated and unaffected by their surrounding are called as system.

State : It is the physical condition of the system and is governed by the quantities such as pressure, temperature.

Property : The quantity that govern the state of system is know as property or Characteristic of a system is known as property.Intensive properties are independent of mass. While extensive properties are mass dependent.

Components : Elements that constitute a system.

Phases : Physically distinct, chemically homogeneous and mechanically separable substances.

Equilibrium : This is the state of a system at any specified condition when the system possesses minimum free energy.

Degree of freedom : The number of independent variables required to describe the state of a system are called as the degree of freedom.

Constraint : A system that has zero degree of freedom is said to be constraint.

Gibb's Phase Rule

$$F = C - P + 2$$

where,

F = degree of freedom

C = no. of component

P = no. of phases

The degree of freedom decreases when number of phases increases. A two component system cannot have more than 4 phases. **Based on number of components phase diagrams classified as :**

2.7.1 Unary phase diagram

It is mainly used to show the phase diagram of carbon or any other pure material. Graphite at 3000 K and 125 bar pressure converts to diamond. There is very limited practical utilities of such diagrams plotted between temperature and pressure axis like water.

2.7.2 Binary phase diagram (Two components)

These diagrams are further divided into 3 categories.

2.7.2.1 Isomorphous system:

The materials which are completely soluble in the liquid as well as solid state.

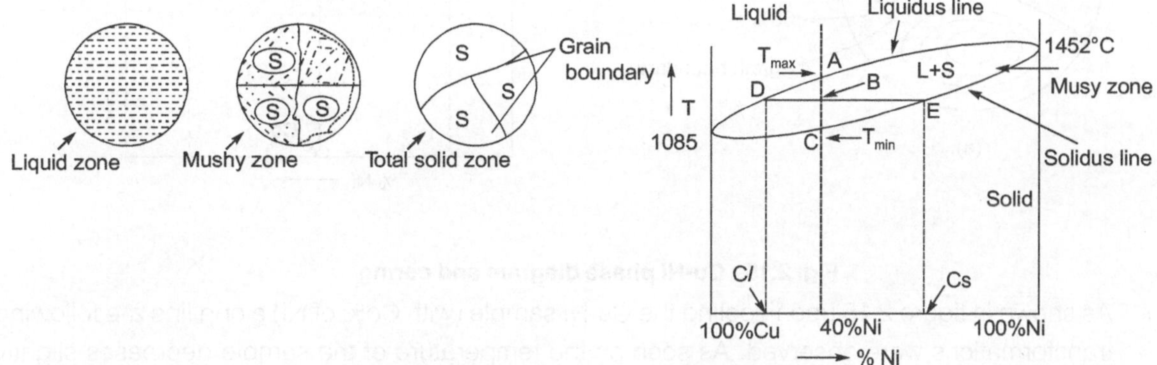

Fig. 2.14 : Cu – Ni phase diagram

In the above phase diagram the region, where a mixture of liquid and solid indicated is called the mushy zone.The line which separates the liquid phase with that of mushy zone is called liquidus. The line which separate the mushy zone with solid phase is called solidus. On such phase diagram solidification takes place over a range of temperature. It can be seen in the diagram that by decreasing the temperature from points 'A' to point 'C' mass fraction of solid increases. Let m_s and m_l be the mass fraction of solid and liquid phase at point B. C_l and C_s are the percentage of Ni in the liquid and solid phase.

$$m_l C_l + m_s C_s = C_o$$

and

$$m_s + m_l = 1$$

$$m_l C_l + (1 - m_l) C_s = C_o$$

We get

$$\Rightarrow \quad m_l = \frac{C_s - C_o}{C_s - C_l}$$

$$m_s = \frac{C_o - C_l}{C_s - C_l}$$

This is called lever rule.

Binary phase diagrams are plotted at a particular pressure, so purposefully we are restricting one degree of freedom, so in case of binary phase diagram Gibbs phase rule converts into

$$F - 1 + P = C + 2 - 1$$

$$F^* + P = C + 1$$

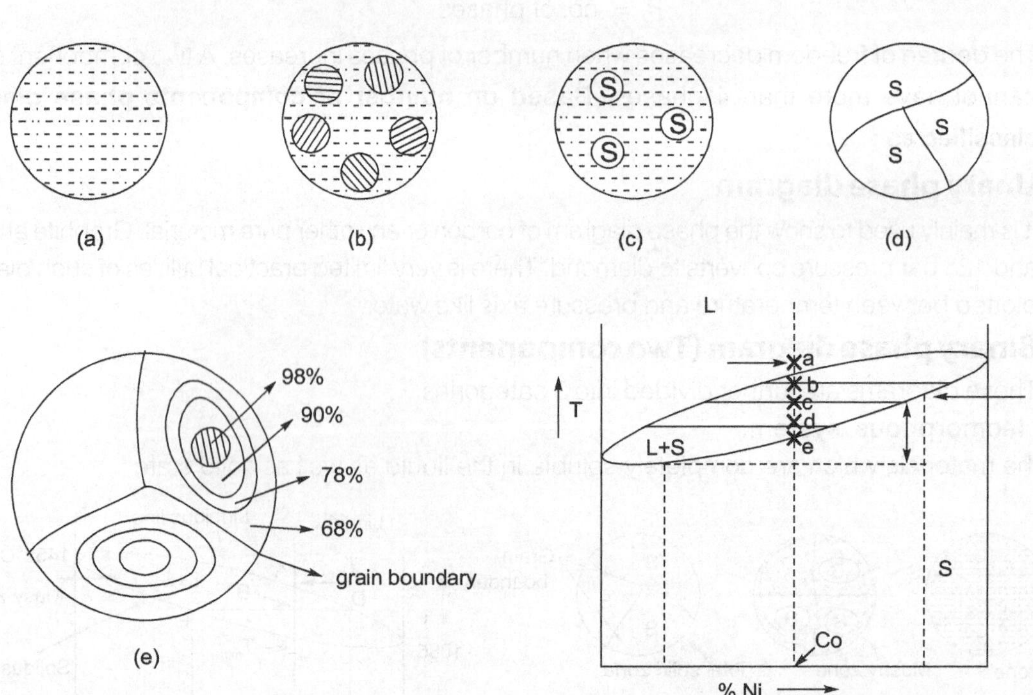

Fig. 2.15 : Cu-Ni phase diagram and coring

As shown in figure 2.15 upon cooling the Cu-Ni sample (with Co% of Ni) along line a-e following transformations were observed. As soon as the temperature of the sample decreases slightly below the liquidus line to point 'b', nucleation will start at number of places. Upon decreasing the temperature to 'c' or 'd' mass fraction of solid phase will keep on increasing. It can be seen from the phase diagram (within the mushy zone) that by decreasing the temperature the % of Ni within the solid phase keeps on decreasing. If the cooling is slow (or) equilibrium cooling Ni will diffuse from the centre to the out ward direction making the solid phase homogeneous. This happens by the process of diffusion and for every 20°C increase in temperature the diffusion gets doubled.

Upon cooling the same sample rapidly since there is no sufficient time for the diffusion to take place, the centre of the solid phase will have higher percentage of Ni and in the outward direction % of Ni will keep on decreasing (as shown in Fig. 2.15(e). It is observed that at the grain boundary % of Ni is low which will have lower melting point. Upon hot working the material, grain boundary will liquify producing micro cracks with in the material. This phenomenon is called coring or miscibility gaps. It can be avoided by heat treatment.

2.7.2.2 Binary phase diagram of a Type-II

The material which are completely soluble in the liquid state but partially soluble in the solid state (Partial Eutectic phase diagrams) :

The binary phase diagram of type-II is called partial Eutectic phase diagram (in Latin the word, Eu means nice, and tectic means melting). One of the important diagram in this category is that of Pb-Sn.

α phase is the solid solution of tin in lead and β phase is the solid solution of lead in tin. There appears a point at which liquid converts into two different solids. This reaction is called eutectic reaction. This reaction is invariant in nature and appears at particular temperature and composition.

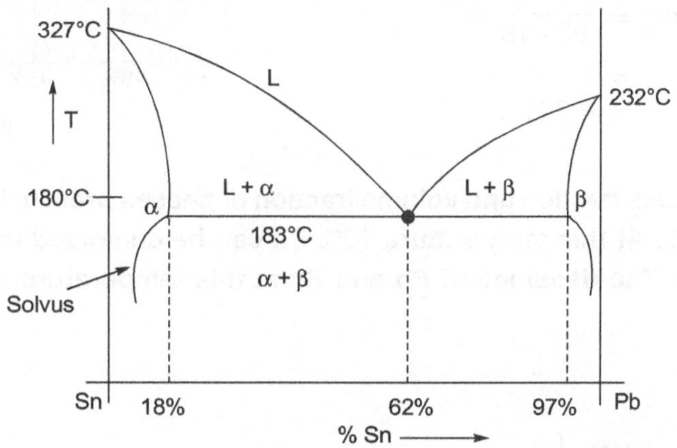

Fig. 2.16 : Pb-Sn phase diagram

$$\underset{\text{Liquid}}{L} \underset{\text{Heating (62\% Sn and 183°C)}}{\overset{\text{Cooling}}{\rightleftharpoons}} \underset{\text{(Solid 1)}}{\alpha} + \underset{\text{(Solid 2)}}{\beta}$$

At eutectic point degrees of freedom are zero. So by changing parameter, the eutectic point disappears. So from Gibb's phase rule,

$$F + P = C + 1$$
$$O + P = 2 + 1$$
$$P = 3$$

So there are three phases which exists i.e. α, β, L at eutectic point.

Example : 2.1

Determine the mass fraction of the phases present at 184°C in a sample of lead and tin with 45% Sn in it.

Solution :

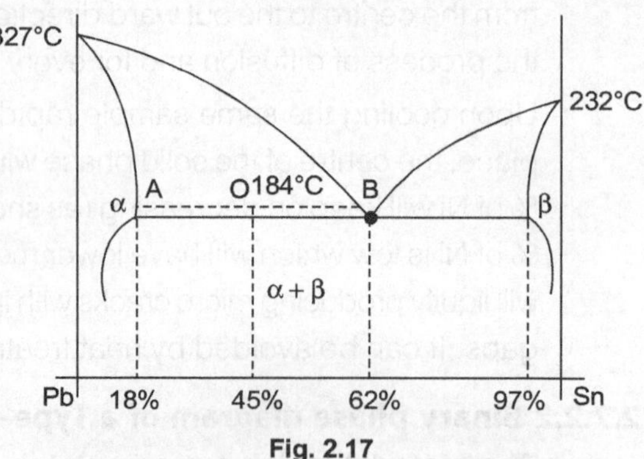

Fig. 2.17

$$m_\alpha = \frac{62-45}{62-18}$$

$$= \frac{17}{44} = 0.38$$

$$m_l = \frac{45-18}{62-18}$$

$$= 1 - m_\alpha = 0.62$$

Example : 2.2

Determine the mass fraction of the phases present at 182°C in the sample of Sn and Pb with 45% Sn in it.

Solution :

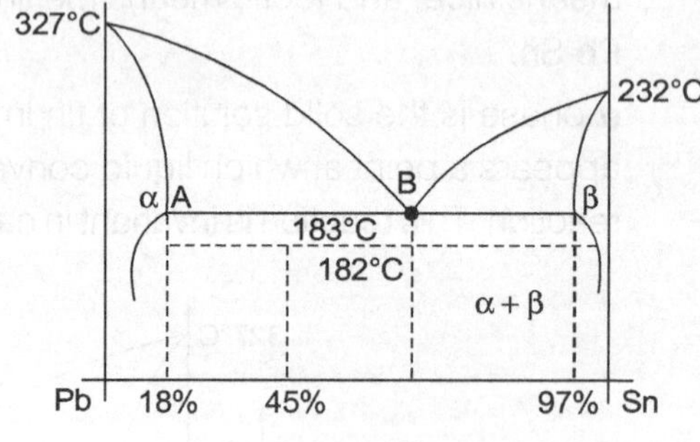

Fig. 2.18

$$m_\alpha = \frac{97-45}{97-18}$$

$$= 0.6582$$

$$m_\beta = \frac{45-18}{97-18}$$

$$= 1 - m_\alpha$$

$$= 0.3418.$$

Example : 2.3

Determine the mass fraction and volume fraction of phases present in an alloy at Pb-Sn with 40% Sn at 100°C. At this temperature 10% Sn can be dissolved in Pb and 2% Pb can be dissolved in Sn. The densities of Pb and Sn at this temperature are 11.23 and 7.24 g/cc respectively.

Solution :

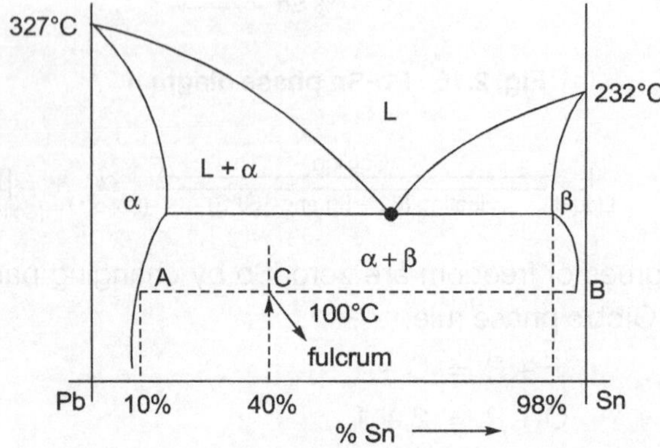

Fig. 2.19

From the above figure

$$m_\alpha = \frac{98-40}{98-10} = 0.659$$

$$m_\beta = 0.340$$

Before calculating the volume fraction, the density of α and β has to be calculated

$$\frac{100}{\rho_\alpha} = \frac{10}{\rho_{sn}} + \frac{90}{\rho_{pb}} \qquad \text{So, } \frac{100}{\rho_\alpha} = \frac{10}{7.24} + \frac{90}{11.23}$$

and

$$\frac{100}{\rho_\beta} = \frac{98}{\rho_{sn}} + \frac{2}{\rho_{pb}} \qquad \text{So, } \frac{100}{\rho_\beta} = \frac{98}{7.24} + \frac{2}{11.23}$$

So $\rho_\alpha = 10.64$ g/cc

and $\rho_\beta = 7.29$ g/cc

volume fraction of α

$$V_\alpha = \frac{\dfrac{m_\alpha}{\rho_a}}{\dfrac{m_\alpha}{\rho_\alpha} + \dfrac{m_\beta}{\rho_\beta}}$$

$$= \frac{\dfrac{0.659}{10.64}}{\dfrac{0.659}{10.64} + \dfrac{0.34}{7.29}} = \frac{0.0019}{0.0619 + 0.0466} = 0.57(57\%)$$

$$V_\beta = 1 - V_\alpha$$

$$= 0.43\,(43\%)$$

2.7.2.3 Binary Phase Diagram of Type-III (Eutectic phase diagram)

The materials which are completely soluble in the liquid state and completely insoluble in the solid state. Alloys of Bi follows these type of diagrams.

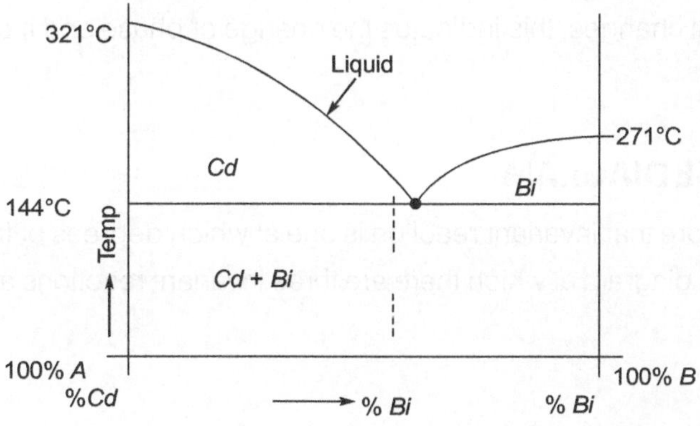

Fig. 2.20

2.8 EXPERIMENTAL METHODS OF CONSTRUCTING PHASE DIAGRAMS

2.8.1 Thermal Analysis Allotropy method

One of the indication of phase change is that there will be latent heat transactions at a particular temperature and a horizontal line will appear over temperature verses time graph(cooling curve)

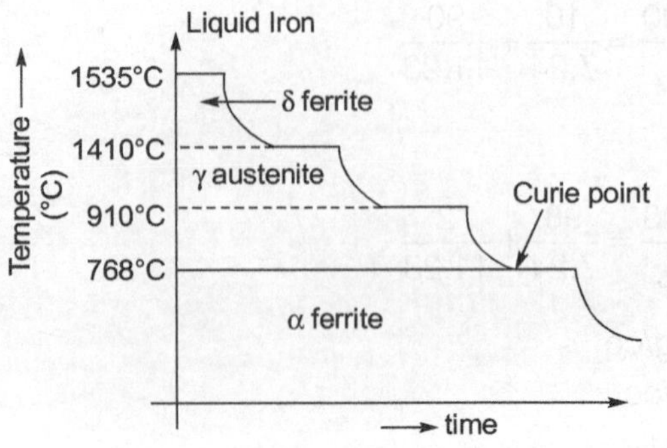

Fig. 2.21 : Allotropy of iron

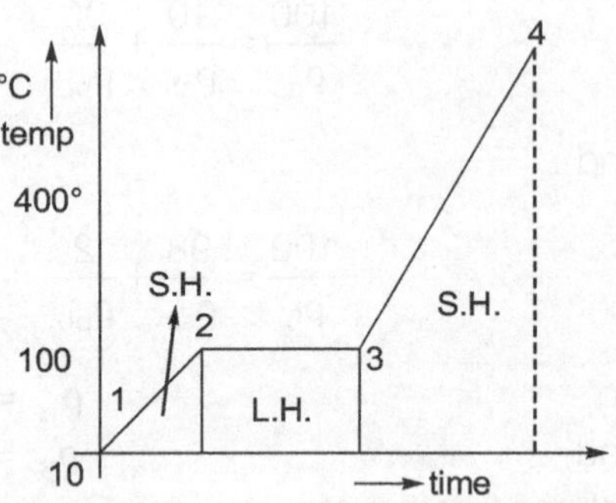

Fig. 2.22 : Phase change of water

For example, water converts into another phase steam by consuming latent heat at 100°C. (Fig. 2.22). Similarly there are four horizontal lines on temperature verses time graph of iron as shown is figure Fig. 2.21 indicating phase change. But and it was observed that at 768°C. There is no change in the micro structure (or) phase and only magnetic properties are disappearing. This temperature is called curie point. In paramagnetic materials, electrons are unpaired and due to this alloy exhibit colour (Ex: copper alloys). In diamagnetic materials electrons are paired and such alloys are colourless. Iron is such a material which some time exhibit paramagnetic and some time diamagnetic. This characteristic is called **"Ferro Magnetism"**.

2.8.2 Metallographic Method

Retained microstructuresnfor different temperatures and compositions of an alloy system are observed using microscopes to construct the phase diagrams.

2.8.3 X-ray Diffraction Method

Every material is a building block of some unit cell. If the type of building block changes (or) the size of the unit cell changes, this indicates the change of phase and it can be identified by x-ray diffraction.

2.9 Fe-Fe_3C PHASE DIAGRAM

As discussed before that invariant reaction is one at which degrees of freedom are zero. fe-fe_3C diagram is such a diagram at which there are three invariant reactions appear.

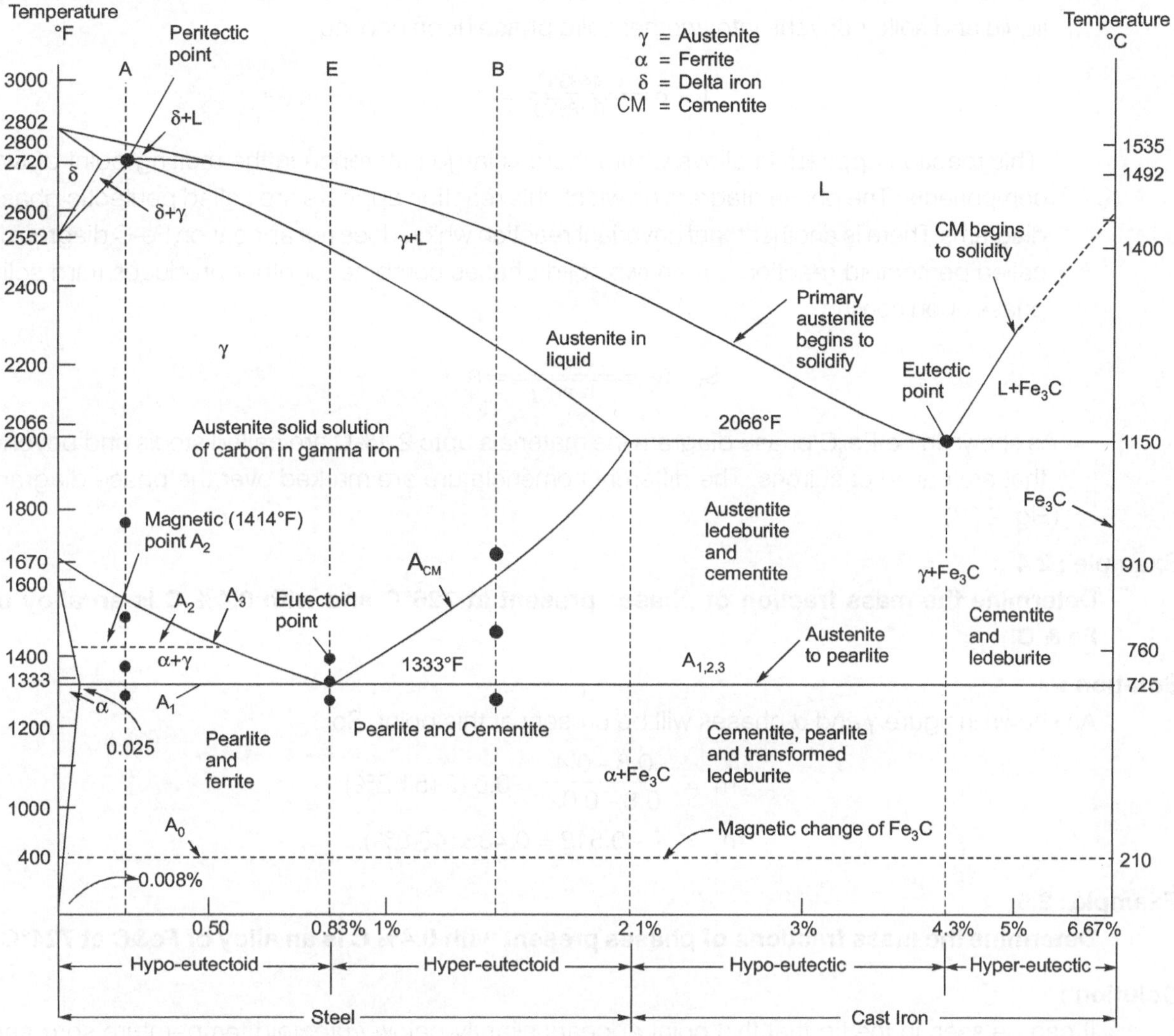

Fig. 2.23 : Fe-Fe_3C phase diagram

1. **Eutectic reaction : Pearlite**

$$L \xrightleftharpoons[1150°C]{4.3\%\,C} \gamma + Fe_3C$$

2. **Eutectoid reaction :** It is a reaction when a solid phase converts into two different solid phases.

$$\underset{\text{(austenite)}}{\gamma} \xrightleftharpoons[0.8\%\,C]{725°C} \underset{\text{(ferrite)}}{\alpha} + \underset{\text{(cementite)}}{Fe_3C}$$

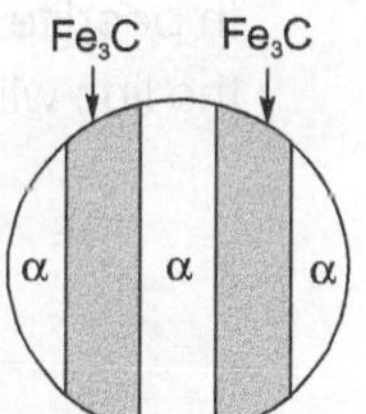

Fig. 2.24 : Pearlite

725°C and 0.83% C are called eutectoid temperature and eutectoid composition respectively and at eutectoid point 3 phases will co-exist in equilibrium simultaneously. Carbon diffuses through the lattice structure and produces alternate plate of α and Fe_3C. This structure is called pearlite and it is produced by diffusion. So pearlite is not ahaving orthorhombic structure. phase, it is a phase mixture of α and Fe_3C. α is having BCC structure and Fe_3C is having orthorhombic structure.

3. **Peritectic reaction :** This reaction appears at 1493°C and at 0.18%C where a mixture of liquid and solid converts into another solid phase upon cooling.

$$L+\delta \underset{0.18\%C}{\overset{1493^\circ C}{\rightleftharpoons}} \gamma$$

This reaction appears in alloys where there is large difference in the melting point of two components. The phase diagram on which this reaction appears are called peritectic phase diagrams. There is another fourth invariant reaction which does not appear on Fe-C diagram is called peritectoid reaction. When two solid phases combine together produces third solid phase upon cooling.

$$S_1 + S_2 \underset{heating}{\overset{cooling}{\rightleftharpoons}} S_3$$

As shown in $Fe\text{-}Fe_3C$ phase diagram the materials upto 2.1%C are called steels and beyond that are called cast irons. The different nomenclature are marked over the phase diagram (Fig. 2.23).

Example : 2.4

Determine the mass fraction of phases present at 726°C and with 0.4% C is an alloy of Fe & C.

Solution :

As shown in figure γ and α phases will be present at this point. So

$$m = \frac{0.8-0.4}{0.8-0.02} = 0.512\ (51.2\%)$$

$$m_\gamma = 1-0.512 = 0.488\ (48.8\%).$$

Example : 2.5

Determine the mass fractions of phases present with 0.4% C is an alloy of Fe&C at 724°C.

Solution :

It can be seen in the fig that this point appears slightly below eutectoid temperature so α and Fe_3C phases will be present.

$$m_\alpha = \frac{6.67-0.4}{6.67-0.02} = 0.942\ (94.2\%)$$

$$m_{Fe_3C} = 1 - 0.942 = 0.058 = (5.8\%)$$

Example : 2.6

Determine the mass fraction of phases present in pearlite.

Solution :

In pearlite there is eutectoid composition, so α and Fe_3C phases will be present. The fulcrum of the line will be at the eutectoid point. So

$$m_\alpha = \frac{6.67-0.8}{6.67-0.02} = 0.882\ (88.2\%)$$

$$m_{Fe_3C} = 1-0.882$$

$$= 0.118\ (11.8\%)$$

2.10 DEVELOPMENT OF MICROSTRUCTURE IN FE-C SYSTEM

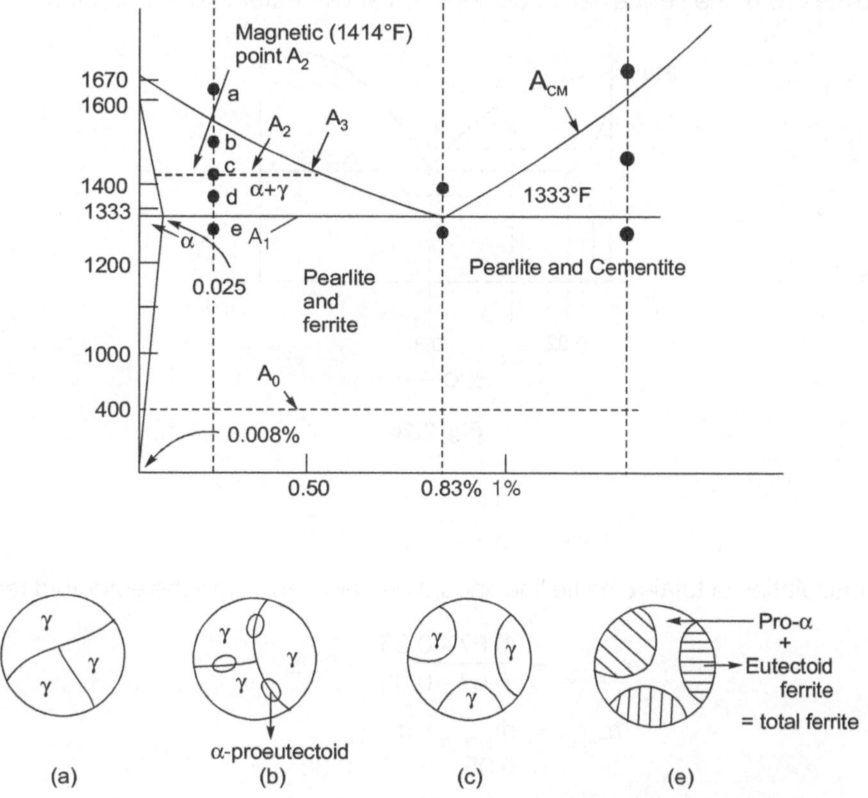

Fig. 2.25 : Development of Microstructure for Hypoeutectoid Steels

If % of Carbon in steel is equal to the eutectoid composition i.e. 0.83%, upon decreasing the temperature of the sample slightly below the eutectoid point, Austenite will convert into ferrite and cementite i.e. 0.83% carbon will bifurcate into 0.02% ferrite and 6.67% cementite. This takes place by the process of diffusion. Carbon jumps from one interstitial site to another making alternate plate like structure of ferrite and cementite, this is called pearlite. If the steel sample is having composition in the hypoeutectoid region the development of micro structure along line a-d will be as follows. At point 'a' the entire structure will be austenite within the grain boundaries. Upon slightly decreasing the temperature to point 'b' ferrite will appear within boundaries. The ferrite which appears before the eutectoid temperature is called pro-eutectoid ferrite.

Upon decreasing the temperature from 'b' to 'c' following observations can be made

1. Mass fraction of pro-eutectoid ferrite increases and at point 'c' there will be canals of pro-eutectoid ferrite and the grain boundaries.
2. Percentage of carbon in austenite increases and it approaches towards the eutectoid composition.

   Upon slightly decreasing the temperature below eutectoid temperature austenite with in grain boundaries will convert into pearlite. Only eutectoid composition produces pearlite. It can be understood from Fe-Fe_3C diagram that in the hypereutectoid steel region pro-eutectoid cementite (Pro-Fe_3C) will appear over the grain boundaries.

Example : 2.7

Determine the mass fraction of pro-α, eutectoid-α and total α in Fe-C alloy with 0.35% C.

Solution :

To calculate pro-α, the tie line has to be selected above eutectoid temperature

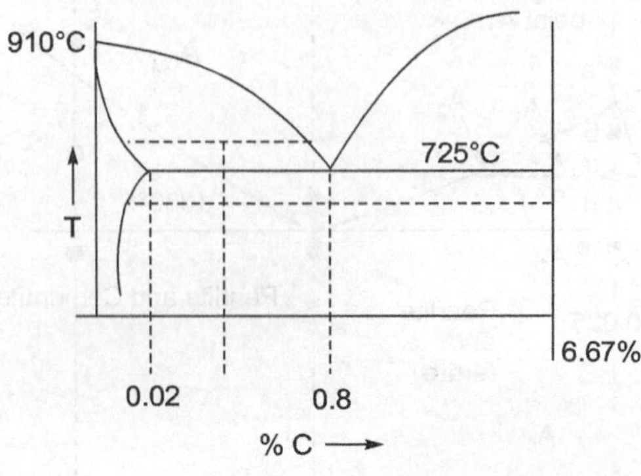

Fig. 2.26

$$m_{pro\text{-}\alpha} = \frac{0.8-0.35}{0.8-0.02} = 0.57$$

For the calculation of total-α the tie line should be selected below the eutectoid temperature

$$m_{total\text{-}\alpha} = \frac{6.67-0.35}{6.67-0.02} = 0.95$$

So

$$m_{eut\text{-}\alpha} = m_{total\text{-}\alpha} - m_{pro\text{-}\alpha} = 0.95 - 0.57 = 0.38$$

Eutectoid-α is present in the microstructure of pearlite.

2.11 FAMILY OF CAST IRONS

1. **Gray Cast Iron :** In pure iron carbon system, if percentage of 'C' in iron increases beyond 6.67% a large amount of carbon will appear in free form (graphite flakes). This is called gray cast iron and it is used to make machine beds (to absorb small vibrations of the machine), piston rings etc.
2. **White Cast Iron :** In which carbon is in the combined form as iron carbide Fe_3C
3. **Chilled Cast Iron :** The cast iron of such composition by which it will normally freeze as gray but forced to appear as white by rapid cooling is called chilled cast iron.
4. **Spheroidal Cast Iron (or) Nodular Cast Iron :** These are ductile in nature. These cast irons are produced Fe-c alloy with 'C' content is same as in Gray cast iron in the presence of Mg or Ce, C comes out in the form of spheroids. These spheroids are ductile in nature.

Properties of Cast Iron	Properties of Steel
Hard (Strong)	Ductile
Good fluidity and easy to cast materials	Higher toughness
Extremely brittle	High wear resistance
Low melting point	Poor castability due to shrinkage during
Good machinability defined	phase change

2.12 EFFECT OF ALLOYING ELEMENT ON IRON

1. **Carbon :** By increasing the % of carbon, steels will slowly convert into cast irons, ductility decreases and brittleness increases as represented in stress and strain diagram.

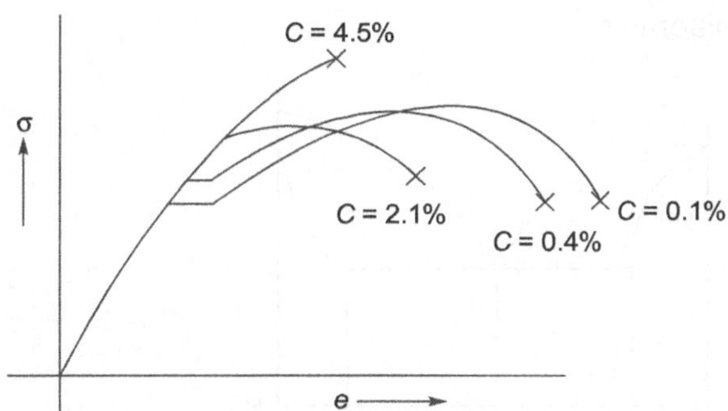

Fig. 2.27 : Stress-strain curve for different percentages of C

2. **Silicon :** The carbon equivalent can be expressed by

$$\text{C equivalent} = \%\,\text{C} + (1/3)\%\,(\text{Si} + \text{P})$$

By Addition of silicon eutectic point shifts towards left and graphite formation takes place at a much lower percentage of carbon. At high temperature graphite will be in the red hot condition and due to buoyancy red hot graphite will jump over liquid iron and sparkles. This phenomenon is called kish. Apart from graphatising agent silicon also increases there hardness. Addition of silicon, helps in removing oxygen from steel castings. Such steels are called killed steels.

3. **Sulphur :** Every solid and liquid matter in the nature will have some impurity of sulphur. Sulphur after reacting with iron produces iron sulphite (FeS). FeS gets accumulated on the grain boundaries and is having lower melting point. Upon hot working the material FeS melts out producing number of cracks on the grain boundaries. As a result of that material will fail as brittle fracture. This phenomenon is called "Hot Shortness". To avoid this phenomenon 'Mn' is added into the material. "Mn" captures sulphur before it can react with iron. Also a large number of "MnS" pockets will be present in the materials. Which improves the machinability because MnS is having low shear strength. If "Mn" is around 12%, the strength of material appears to be very high called "Hadfield Steel", which is used to make bull-dozer rolls.

COMPOSITION				
Element	**Gray iron**	**White iron (malleable iron)**	**High strength gray iron %**	**Nodular iron**
Carbon	2.5 - 4.0	1.8 - 3.6	2.8 - 3.3	3.0 - 4.0
Silicon	1.0 - 3.0	0.5 - 1.9	1.4 - 2.0	1.8 - 2.8
Manganese	0.40 - 1.0	0.25 - 0.80	0.5 - 0.8	0.15 - 0.90
Sulphur	0.05 - 0.25	0.06 - 0.20	0.12 max	0.03 max
Phosphorus	0.05 - 1.0	0.06 - 0.18	0.15 max	0.10 max

2.13 RELATIONSHIP BETWEEN THE PHASE DIAGRAMS AND PROPERTIES OF ALLOYS

When the two materials are completely insoluble solvus lines merges into the temperature axis. As the solid solubility increases solvus lines approaches towards the eutectic point and if there is complete solid solubility both the solvus lines will merge into the eutectic point to produce isomorphus phase diagrams.

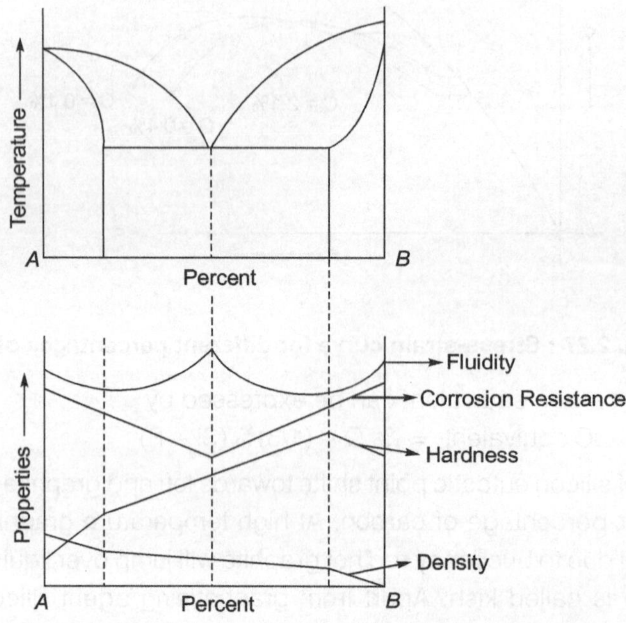

Fig. 2.28 : Properties of Material for eutectic system

1. Fluidity is maximum for pure materials and eutectics. It is directly proportional to the extent of mushy zone. Larger is the mushy zone lower will be the fluidity.
2. Corrosion resistance is very poor for eutectic composition and upon deviating from the eutectic composition, corrosion resistance increases due to the presence of pro-phase, boundaries are covered by the pro-phase.

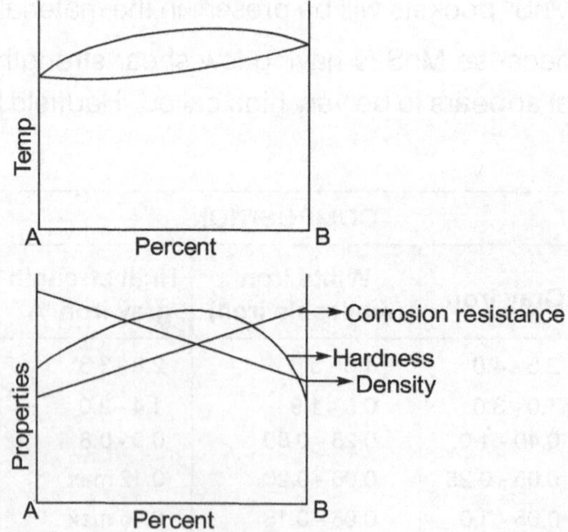

Fig. 2.29 : Properties of alloy for isomorphous system

3. Hardness (or) strength of pure material is very poor and the max strength will be at a point of max solid solubility of one material into another because alloys are always stronger than the pure metal.
4. Generally Density is higher for low melting point materials.

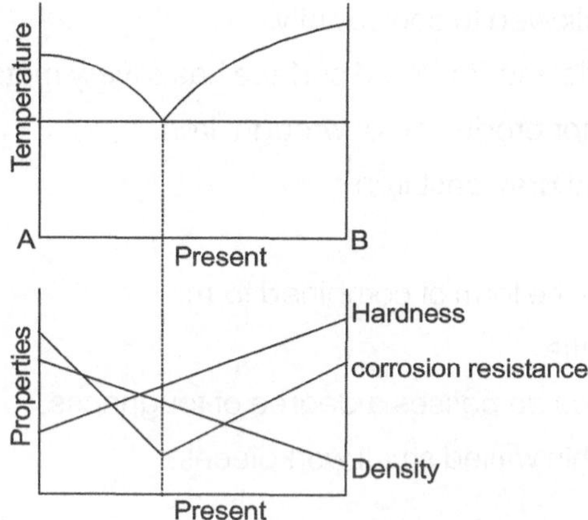

Fig. 2.30 : Properties of alloy for completely insoluble system

2.14 HIGH STRENGTH LOW ALLOY STEEL (HSLA)

The major alloying element in HSLA are copper, vanadium, nickel and molybdenum. In high strength low alloy steels copper is used as an anticorrosive agent. It improves the corrosion resistance 3 times to that of chromium. Vanadium increases hardness and promotes fine grain structure which leads to stronger the material. Nickel strengthens by solid solution and increases the tendency to retain austenite, with higher % of carbon. Molybdenum resists course grain formation and increases the hot hardness. HSLA possesses higher strength than plane carbon steels. It's strength can be increased by heat treatment process. These materials are highly machinable, formable and more resistant to corrosion. In HSLA, % of carbon is in the low carbon steel range, so yield point phenomenon appears during deformation. This phenomenon produces stretcher strains in which material gets accumulated in the different regions which leads to fracture. So, before using the material yield point phenomenon should be avoided.

POINTS TO PONDER

1. Iron ore are Magnetite (Fe_3O_4) and real hematite (Fe_3O_3).

2. Blast furnace comprise a throat, stack, body, bosh and hearth.

3.

Material	% of carbon	Material	% of carbon
Plain carbon steel	0.05 - 2.0%	White cast iron	1.75 - 2.30%
Low C steel	0 - 0.3%	Malleable cast iron	2.20 - 3.60%
Medium C steel	0.3 - 0.7%	Gray cast iron	2.50 - 3.75%
High C steel	Beyond 0.7%	Pig iron	3.00 - 4.00%
Wrought iron	0.02 - 0.03%		

4. Gray cast iron

- This is obtained by allowing molten metal to cool and solidify slowly.
- It contains greater part of carbon in graphite flake form.
- It is brittle and its strength is greater in compression than in tension.

- Used for making parts of intricate shape.
- Free graphite structure seems to acts as lubricants.

5. **White cast iron**
   - Contains carbon exclusively in form of cementite.
   - Obtained when allowed to cool rapidly.
   - Is very hard, brittle and fractured surface has silvery metallic appearance.
   - It is widely used for producing of wrought iron.
   - It does not rust as gray cast iron.
6. **Malleable cast iron**
   - Having carbon in the form of combined form.
   - It is hard and brittle
   - Malleable castings do passes a degree of toughness.
   - Mainly used for thin walled small workpieces.
7. **Ductile iron**
   - Spheroidal graphite iron, also called nodular cast iron.
   - Graphite is in the form of spheroids not in form of flakes.
   - High fluidity and ductility.
   - Used in hydraulic cylinders, valves pipe fitting cylinder head for compressors and diesel engines, etc.
8. **MeChanite cast iron**
   - Cast iron treated with calcium silicide which acts as a graphatiser and produces a fine graphite structure giving cast iron of excellent mechanical properties.
   - It has high strength toughness, ductility and easy machinability.
9. **Wrought iron**
   - It has iron as much as 99%.
   - Noncorrosive slag constituent causes wrought.
   - It cannot be melted while heating it becomes soft enough to take any shape.
   - It can withstand sudden and excessive shock loads.
   - It can neither be hardened nor termpered like steels.
10. **Iron** / **% of carbon**

Iron	% of carbon
Black Iron	2.2 - 2.8%
White Iron	3.2 - 3.6%

11. **Iron Type** / **hardness** / **Application**

Iron Type	hardness	Application
Malleable cast iron	120	Valve bodies, hinges, machine castings etc.
Gray cast iron	130	Ingots, moulds, pistons, machine castings, automobile cylinders etc.
Nodular cast iron	180	Parts subjected to bending and vibration and where strength and ductility are important.
White	400	Rolls, dies wearing plates, stamping shoes, etc.

12. Effects of elements on cast iron

Element	% age	Effect
Carbon	~ 6.67%	Graphite (Cooled slowly) tough iron, easy to machine cementite (rapidly cool) high tensile strength, difficulty to machine element.
Silicon	2.5%	Promotes the formation of free graphite which makes the iron soft and machinable.
Sulphur	0.1%	Makes hard and brittle.
Manganese	0.75%	Harden the cast iron by encouraging the formation of carbide.
Phosphorus	1.0%	Aids fusibility and fluidity but induces brittleness.
Nickel	0.25 - 2.0%	Acts as a graphitiser, to resist corrosion and low expansivity
Chromium	0.15 - 0.9%	Acts as a carbide stabilizer, increases strength, hardness and wear-resistant.
Molybdenum	0.25 - 1.5%	To improve tensile strength, hardness to casting and also improves toughness, fatigue strength machinability, hardenability.
Copper	0.02 - 2.5%	Promotes formation of graphites.
Vanadium	0.01 - 0.5%	Increases hardness strength and machinability.

13. Furnace TypeTonne of steel produced in 24 hours

Furnace Type	Tonne of steel produced in 24 hours
Bessemer converter	1200
Open hearth furnace	800
Electric Furnace	30 - 80

14. Killed steels

- Steels containing more than 0.25% carbon are called killed steels and denoted by symbol 'K'.
- Steel is sufficiently deoxidised, no evolution of gas and freedom from blow-holes (i.e. soundness).

15. Semi-Killed steels

- Steel containing carbon ranging from 0.15 - 0.25% are called semi-killed steels.
- It aims to produce metal free from surface blow holes.
- No symbol is used to represent this type of steel
- Plates and structural products are normally made from this steels.

16. Rimed Steels

- It aims to produce a clean surface but there are blow holes in the core of part.
- Steels are partially deoxidised.
- 'R' stands for rimmed steels

17. Effects of elements on steel

Element	%	Effect
Carbon	0.3%	Increases strength, elasticity, hardness but lowers the ductility and impact strength.
Silicon	0.05 - 0.3%	To prevent steels from porous, acts as a good deoxidiser makes tougher and harder.

Manganese	0.3 - 1.0%	Serves as deoxidising and purifying agent it combines with sulphur to decrease the harmful effect of element on steel.
Sulphur	0.1%	Red-shortness i.e. increases brittleness, increase tensile strength, yield point, fatigue limit and decreases corrosion resistance.
Phosphorus	0.2%	Imparts coldshortness. It increases tensile strength but at same time reduces impact strength and ductility.

18. Free cutting steels

- Subjected to light loads
- high machinability and high quality surface finish because of high sulphur and phosphorous content.
- Sulphur increase the surface finish by reducing the friction.
- Phosphorous reduces the brittleness.

19. Fine grained steels have less tendency to crack during heat treatment but have better toughness.

20. Effect of alloying elements in alloy steels

Alloying Element	Effects
Nickel	Toughness, corrosion resistance and deep hardening
Managanese	Deoxidizes, contribute to strength and hardness
Chromium	Toughness, corrosion resistance and hardenability
Silicon	Deoxidizer increases susceptibility of steel to decarburization and graphitization
Molybdenum	Hardenability, increases tensile and creep strength
Vanadium	Deoxidizes and promotes fine grained structure
Copper	Increases resistance to corrosion and strengthening agent
Aluminium	Deoxidizer
Boron	Increases hardenability

21. Different applications of steels

Steels	Application
Dead Mild	Chains, rivets, seam welded pipe, tin plates
Mild steel	Screws, gears, shafts, drop forgings, free cutting steels
Medium carbon	Connecting rods, axles, shafts, crank shafts, die block etc.
High carbon	loco-tyres, rails, wire ropes, drop-hammer dies, cold chisels, punches, hammers etc.

23. Special alloy steel	**Application**
Magnet steel	Magnetos, loud speakers, electrical machine and instruments 15-40% Co, 0.4-10%W
Heat Resisting steels	Working at high temperature. For furnace parts, annealing boxes Cr, Ni and W, 0.3% C
Shock Resisting steels	In leaf and coil springs and where in fatigue stresses 2.25% W, 1.5% Cr. 0.25% V 0.5 C
Stainless steels	(a) Austenitic stainless steels ❖ high ductility (18/8 stainless steel)

- ❖ are non magnetic (15-20% Cr, 7-10% Ni)
- ❖ Titanium and niobium added to stabilise the carbon

(b) Martensitic stainless steels
- ❖ These steels are hardenable by heat treatment 10-14%Cr

(c) Ferritic stainless steel
- ❖ These can not be hardened by heat treatment
- ❖ Stainless steels have high strength and toughness as well as good resistance to corrosion
- ❖ These are used in all type of surgical and dental instruments house hold appliances utensils etc.
- ❖ When carbon contents is as low as 0.09% the product is called stainless iron.

Maraging steels — Iron based 18 Ni-8 Co
High fracture toughness
5 Mo + small amount of Al and Ti + 0.03% C.

High Speed steels — (a) 18-4-1 HSS 18% W, 4% Cr 1%V
(b) Cobalt HSS 20% W, 4% Cr, 2% V, 12% Co
Cobalt is added to increase hot hardness and wear resistance
(c) Vanadium HSS 0.7 C + more than 1% V excellent abrasive resistant
(d) Molybdenum HSS 6% Mo, 6% W, 4% Cr, 2% V excellent toughness and cutting ability.

24.

Alloying element	**Effect on HSS**
Chromium	Reduces scaling
Vanadium	Abrasion resistance
Tungston	Hot hardness
Molybdenum	Hardenability
Cobalt	Hot hardness and wear resistant

2.15 DEFECTS IN THE MATERIALS

2.15.1 Point defects

1. **Vacancy Defect :** It appears due to the missing of atoms from the lattice.

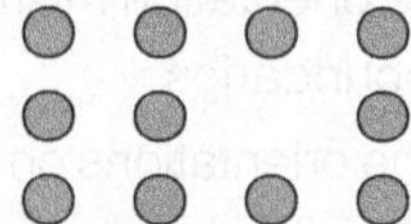

Fig. 2.31 : Vacancy Defect

2. **Interstitial defect :**

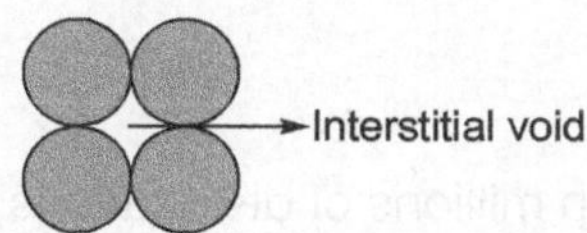

Fig. 2.32 : Interstitial void

When the foreign atom occupies the interstitial site that defect is called interstitial.

3. **Substitutional defect :** If the regular atom is replaced by another foreign atom this defect is called substitutional defect.

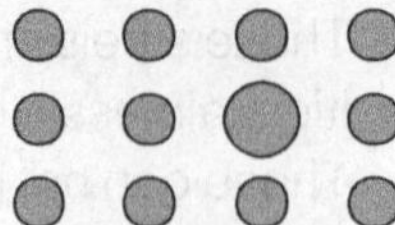

Fig. 2.33 : Substitutional Defect

4. **Frenkel defect :** When the atom in the lattice point goes and occupies the interstitial void of other atom then it is called frenkel defect.

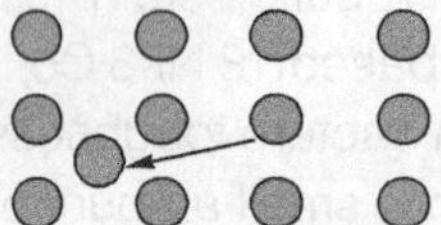

Fig. 2.34 : Frenkel Defect

5. **Schottky defect :** In the combination of cation and anion if there is a vacancy defect it is called schottky defect.

2.15.2 Surface Defects

1. **Grain boundary defect :** The bond length is more at the grain boundary due to orientation mismatch and can easily be broken. So, atmospheric oxygen reacts with the atom at grain boundary and corrodes. So finer the grain structure lower will be the corrosion resistance.

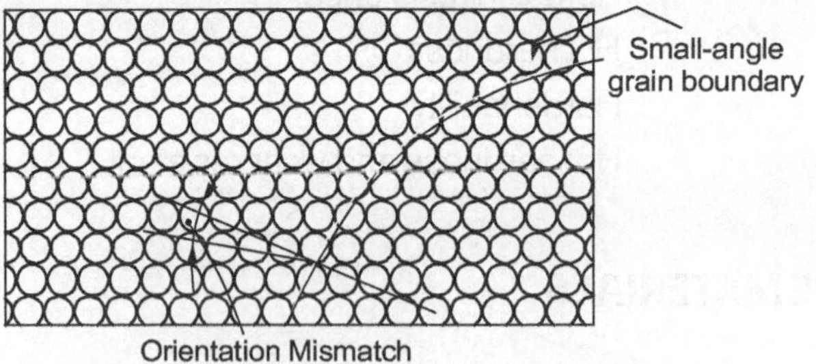

Fig. 2.35 : Grain boundary defect

2. **Tilt boundary defect :** When the orientation mismatch at the grain boundaries is 0.5° to 1°, grain boundaries are called tilt boundaries.
3. **Twin boundary defect :** When the orientations on one side are mirror image of opposite side, such grain boundary defects are called twin boundary defects.
4. **Stacking faults :** When the grain boundaries on both the sides are parallel to each other with slight disturbance of bonds in a very small region, such grain boundaries are called stacking faults.

2.15.3 Line Defects

1. **Edge dislocation :** Although millions of dislocations are already there is the material but to understand the concept let us assume that there is a perfect material and one face of the material is fixed i.e. restricted to move. On the opposite face a uniform pressure is applied on 50% area. Once this pressure exceeds beyond a certain value, slipping of atoms takes place. This unit amount of plastic deformation which appears in the direction of applied load

is called burger vector. There will be a boundary 'AB' as shown in Fig. 2.36 between slipped and unslipped regions called edge dislocation. In this case burger vector is perpendicular to the dislocation line.

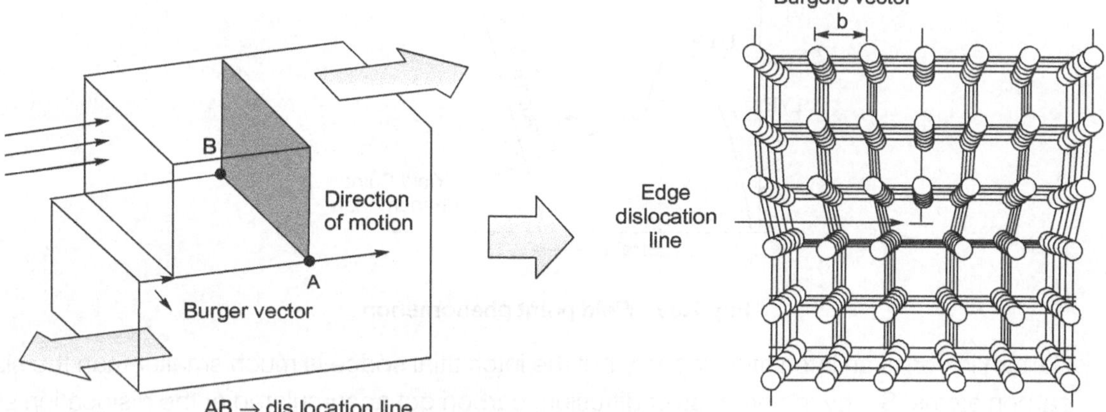

Fig. 2.36 : Edge dislocation

(a) There appears an extra half plane at the dislocation line.
(b) The interstitial void at the dislocation line is larger in size.

Upon applying load on any material, dislocations move in the direction of applied load. If the dislocations comes out of the surface it is said that plastic deformation has begun in the material. The representation of dislocation motion is shown in Fig. 2.37. If extra half plane is above the dislocation line it is called positive edge dislocations represented by ⊥ and if half plane is below the dislocation line it is called negative edge dislocation represented by symbol T.

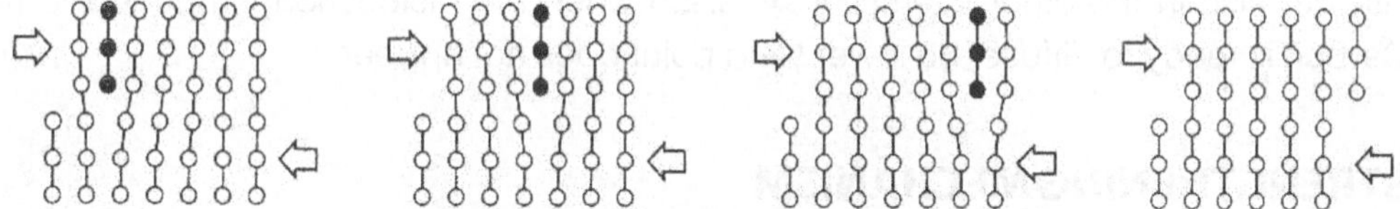

Fig. 2.37 : Dislocation Motion

2. **Screw Dislocation :** It is formed by a shear stress that is applied to produce the distortion shown in figure 2.38. The screw dislocation derive its name from the spiral or helical path or ramp that is traced around the dislocation line by the atomic planes of atoms. The burger vector in this case is parallel to the dislocation line.

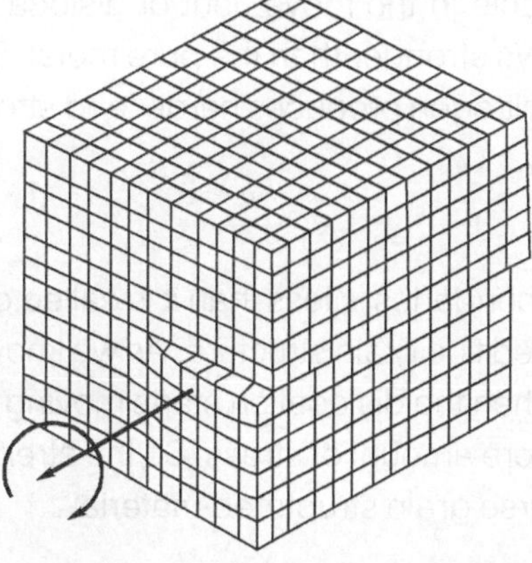

Fig. 2.38 : Screw Dislocation

2.16 YIELD POINT PHENOMENON

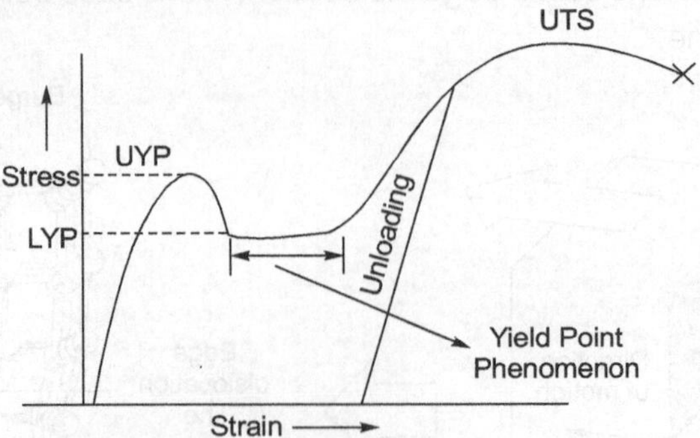

Fig. 2.39 : Yield point phenomenon

Carbon in iron is an interstitial impurity but the interstitial space is much smaller than the size of carbon atoms. So, by the process of diffusion, carbon get accumulated at the dislocation site, it is called cottrell atmosphere. This produces lattice strains. So, extra stresses are required to break this cottrell atmosphere and jump the dislocation to the new site. That is why upper yield point appears. Once the dislocation jumps to the new position, since there is no cottrell atmosphere, lesser stresses are required to keep the dislocation moving so lower yield point and yield point phenomenon appears in the low carbon steel material. Upon unloading the specimen from the work hardening region (as shown in Fig. 2.39) and reloaded again, the yield point phenomenon will not appear because of absence of Carbon at the dislocation site. But after a certain period of time when the sample is reloaded the yield point phenomenon will reappear because Carbon or N will again diffuse at the dislocation site. This time period is called strain aging time. This phenomenon does not appear in medium and high Carbon steels because Carbon is not only present at dislocation site but also in the other interstitial sites. So when the dislocation jumps to the new position Carbon is ready to diffuse. So lower yield point does not appear.

2.17 STRENGTHENING MECHANISM

2.17.1 Alloying

If the foreign element occupies the interstitial site it produces compressive strain field in the host atoms. It is because interstitial void is always smaller than the size of impurity atom. If the impurity atom occupies the substitutional site and if its size is smaller, it produces tensile strain field. If the impurity size is bigger than the host atoms it produces compressive strain field. These strain fields create an obstacle in the movement of dislocation. This increases the strength of material. So, alloys are always stronger than the pure metal. Both carbon and nitrogen occupies the interstitial site in iron but nitrogen produces harder and stronger materials because it produces strain field of larger intensity.

2.17.2 Grain refinement

When the number of grain boundaries is less than it is called coarse grain structure and when the number is more then it is called fine grain structure. As we know the region of orientation mismatch are called grain boundary when the dislocation starts moving at the grain boundary it changes its directions which requires more amount of stress. So the strength of fine grain structure materials will be more than that of coarse grain structured material

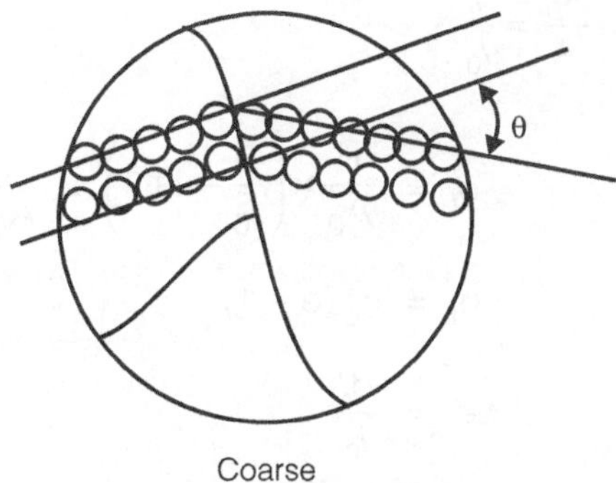

Fig. 2.40 : Orientation mismatch of grains

As soon as the dislocation approaches the grain boundary it has to change its orientation because atoms are arranged in a different direction this requires extra stress. So, grain boundaries creates an obstacle in the movement of dislocation. So, finer the grain better will be the strength of material.

2.17.3 Work (strain) Hardening

Upon cold working (temperature less then 0.4 temperature where temperature is melting point in K) the number of dislocations in the material increases. How much dislocation multiplication takes place, is the property of material and it is expressed by work hardening exponent (n) in the flow curve equations $\sigma_f = K\epsilon^n$, where k is strength coefficient, σ_f is true stress and ϵ is true strain. Higher is the value of n, more dislocations will be created upon cold working the material. These dislocations piles up at the grain boundary and creates a dislocation forest. Since dislocation-dislocation interactions are repulsive in nature, this dislocations forest create a back stress in the movement of dislocation. This increases the strength of material upon unloading the specimen from the region of work hardening and reloading in opposite direction, back stresses now support the movement of dislocation. As a result of that yield point in opposite direction appears prematurity. This phenomenon is called Bauschinger effect.

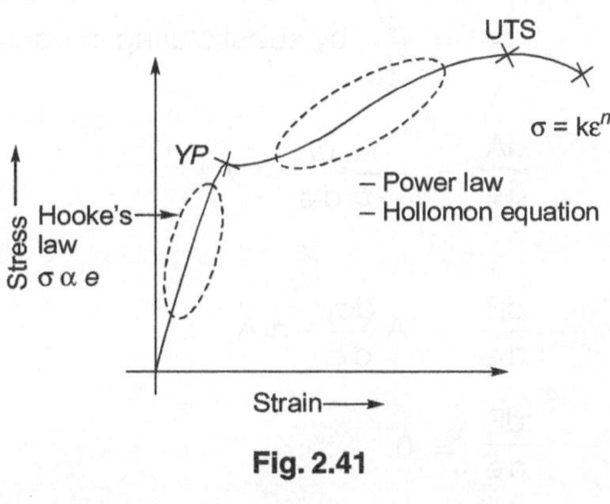

Fig. 2.41

$$\sigma_0 = \frac{P}{A_0}$$

where σ_0 is nominal stress, A_0 is the nominal area and P is the applied load.

$$e = \frac{dl}{l_0}$$

So, $$\frac{A_0}{A} = \frac{l}{l_0}$$

$$\sigma_f = \frac{P}{A_0}\times\left(\frac{l}{l_0}-1+1\right) = \frac{P}{A_0}\times(e+1)$$

$$\sigma_f = \sigma_0(e+1)$$

$$d\epsilon = \frac{dl}{l_0}$$

where l is instantaneous lengths, upon integrating it

$$\epsilon = \int_{l_0}^{l}\frac{dl}{l} = l\,n\left(\frac{l}{l_0}\right)$$

So $$\epsilon = ln\,(e+1)$$

Also $$\sigma_f = \frac{P}{A}$$

$$\Rightarrow \quad P = \sigma_f A \quad \text{...(1)}$$

$$\frac{dP}{d\epsilon} = \frac{d}{d\epsilon}(\sigma_f A) = \frac{d\sigma_f}{d\epsilon}A + \sigma_f\frac{dA}{d\epsilon} \quad \text{...(2)}$$

Volume $$V = Al$$

Differentiating this equation we get

$$0 = \frac{dA}{d\epsilon}l + A\frac{dl}{d\epsilon} \quad \text{...(3)}$$

Also we know that

$$d\epsilon = \frac{dl}{l}, \text{ by substistuting in equation (3)}$$

$$\frac{dA}{d\epsilon} = -\frac{A}{l}\frac{dl}{d\epsilon} = -A$$

From equation (2)

$$\frac{dP}{d\epsilon} = A\frac{d\sigma_f}{d\epsilon} - \sigma_f A$$

at UTS $$\frac{dP}{d\epsilon} = 0$$

So, $$\sigma_f = \frac{d\sigma_f}{d\epsilon} \quad \text{...(4)}$$

This is true only at UTS point

From holloman equation

$$\sigma_f = K\epsilon^n$$

$$\frac{d\sigma_f}{d\epsilon} = Kn\epsilon^{n-1} = K\frac{n\epsilon^n}{\epsilon} = \frac{n}{\epsilon}\sigma_f \qquad ...(5)$$

Comparing equation (5) and (4) we can see that at UTS **[n = ∈]**.

Example : 2.8

The ultimate tensile strength of a material is 400 MPa and elongation up to maximum load is 35%. If the material obeys power law then what is the true stress-strain relationship.

Solution :

Nominal UTS

$$\sigma_0 = 400 \text{ MPa}$$

$$e = 0.35$$

So $\epsilon = ln(1 + e) = ln(1.35) = 0.3$

$$\sigma_f = \sigma_0(1 + e) = 400(1.35) = 540 \text{ MPa}$$

at UTS $n = \epsilon = 0.3$

$$\sigma_f = K\epsilon^n$$

$$540 = K(0.3)^{0.3}$$

$$K = 775$$

So power law equation is $\sigma_f = 775\,\epsilon^{0.3}$.

2.18 TYPE OF BOND

Solid state exhibits a crystal structure which is having a definite geometry except amorphous state like glass. In crystal structures there are four types of bonding.

Ionic Bond : Strong electrostatic attraction between cation and anion is called ionic bond. These bonds are permanent and atom doesn't drift throughout the lattice structure.

Covalent Bond : In this type of bond there is a sharing of one or more electrons from the adjacent atoms.

Metalic bonds : Metallic materials have one, two (or) three valency electrons. These electrons are not bound to any particular atom in the solid and drift through out the entire metal.

Vander Walls Forces : This is the bond arises in neutral atoms like inert gases. When atoms are brought close to each other there is a separation of +ve and –ve charges and a weak attractive force develops.

2.19 TIME TEMPERATURE TRANSFORMATION (TTT) DIAGRAM

Prof. Bain heated some steel samples having an eutectoid composition at a temperature at which austenite is stable. These samples are subsequently quenched to different temperatures below 725°C. The first sample was quenched to 700°C and it was observed that for a substantial period of time there was no change in the micro structure of 'Austenite'. Although austenite cannot be stable below 725°C. This period is called incubation period. Upon decreasing the quenching temperature incubation period decreases and at 550°C it is only 'one sec'. Upon decreasing the quenching temperature further, again there is increase in the incubation period.

s → starting point

f → final state (or) point

Fig. 2.42 : Construction of TTT diagram

Upon projecting the % decomposition V_S time graph, to temperature vs time graphs 2 'C' curves appears this diagram is called **Time-Temperature-Transformation curve (or) TTT Curve or C-Curve or S-curve or Bains Curve.**

1. **Driving Force :** $\Delta G = G_{final} - G_{initial}$ If DG is +ve, it means the final phase is stable. If its value is zero it means the phase is neutral and if it is –ve, it means the final phase is unstable. As there is a decrease in temperature the (DG) driving force value increases and so incubation period decrease from 725°C to 550°C.

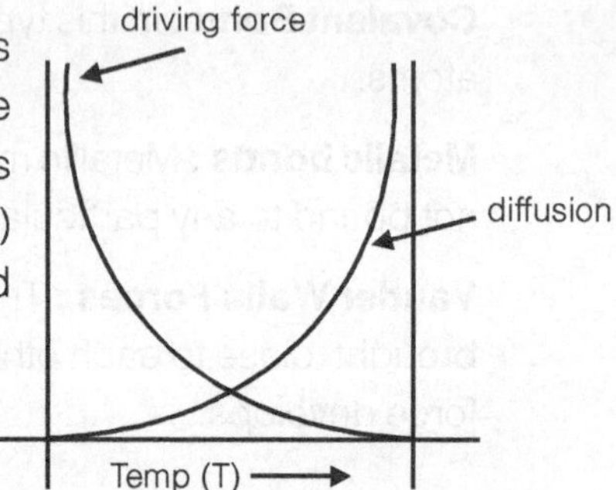

Fig. 2.43 : Driving force and diffusion variation with temperature

2. **Atomic Mobility (Diffusion) :** It decreases with decrease in temperatures slightly below 725°C, the driving force is very low that is why incubation period is more. Below 550°C although the value of driving force is very high but due to lower temperature, diffusion is very low. This increases the incubation period.

   In the TTT-diagram there appears two lines other than C-curves at 220°C and 100°C horizontal to time axis.

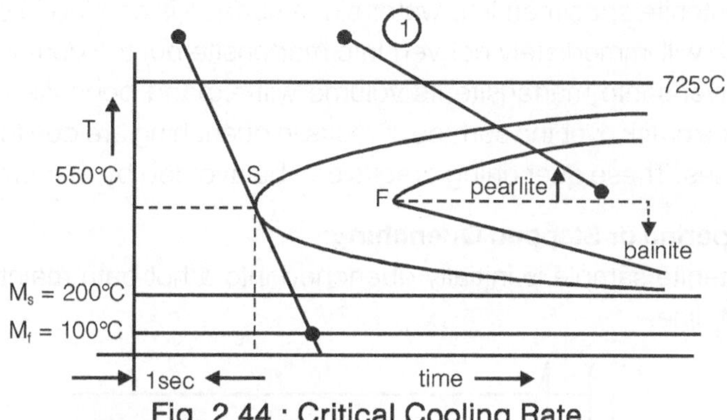

Fig. 2.44 : Critical Cooling Rate

220°C-M_s → Start of martensite % of decomposition

100°C-M_f → Finish of martensite

If the cooling rate is such that, it just touches the nose of TTT diagram is called critical cooling rate (CCR). Any cooling rate greater than (or) equal to CCR will not produce pearlite and is the microstructure carbon will freeze at its location and forms a super saturated solution of carbon in α iron with body centered tetragonal(BCT) structure. This phase of iron is called 'martensite', which is the hardest phase of iron. All the lines on TTT diagram are that of decomposition of austenite into some other micro structure. Once the austenite converts into some other microstructure it never reconverts again. The slow cooling process produce coarse structure and fast cooling process produce fine structure. It can be observed that bainite cannot be produced by continuous cooling. To produce bainite, the sample has to be quenched below the nose of TTT diagram but above martensite start line (220°C). The sample is then maintained at this temperature for substantial period of time till entire austenite converts into bainite. The process is called Austempering.

The advantages of austempering are:

1. Brittle martensite does not formed.
2. Quenching cracks between core and surface do not develop.
3. Ductility is increased.
4. Impact strength and toughness are increased.

Only small sections are suitable for austempering because big sections can not be cooled rapidly to avoid formation of pearlite. Steel sections less than 12 mm thick are suitable for austempering.

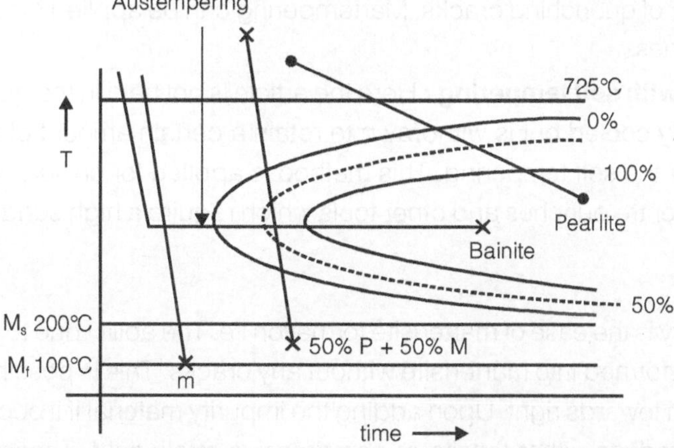

Fig. 2.45 : Different cooling rate results to different microstructure

Upon quenching austenite specimen into water since surface is coming in contact with the quenched medium. The surface will immediately convert into martensite but the core is still austenite. After some time when core converts into martensite its volume will expand because martensite is having lower density. This produces crack over the surface. Cracks in quenching are due to density difference and not due to thermal stresses. These quenching cracks can be avoided by martempering :

(1) Martempering or Stepped Quenching:

(a) The austenite sample is initially quenched into a hot bath maintained below the nose but above M_s line.

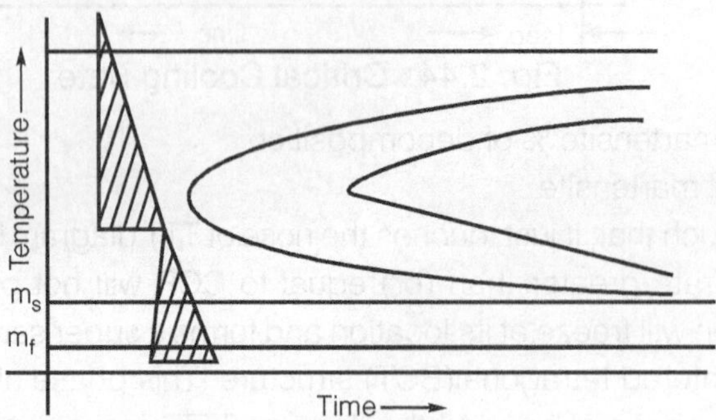

Fig. 2.46

(b) Quenching in two medium or Martempering : Articles are first quenched in water to a temperature of 300-400°C and then quickly transferred to a less intensive medium like oil or air where they are held until they are completely cooled. The purpose of transfer to the second medium is to reduce internal stresses associated with the austenite to martensite transformation. This method is widely used in heat treatment of Carbon-steel tools like taps, disc, milling cuttersetc. A larger holding time will cause austenite decomposition. Austenite is transformed into martensite during the subsequent period of cooling to the room temperature. This treatment gives a structure of martensite and retained austenite.

Martempering has the following advantages over conventional quenching.

(i) Less volume changes occurs due to the presence of a large amount of retained austenite and the possibility of self tempering of the austenite.

(ii) Less distortion since the transformation occurs almost simultaneously in the entire volume.

(iii) Less danger of quenching cracks. Martempering can be applied for thin sections from 5 to 8 mm is thickness.

2. Hardening with self tempering : Here the article is not held in the quenching medium until it is completely cooled but is withdrawn to retain a certain amount of heating the core which accounts for the self tempering. This method is applied for chisels, sludge hammers, hand hammers, centre punches and other tools which require a high surface hardness and tough core.

3. Alloying

Hardenability is the ease of martensite formation i.e. The ability due to which austenite sample can be transformed into martensite without any cracks. This is possible by shifting the entire TTT diagram towards right. Upon adding the impurity material if it occupies the substitutional site it will produce either tensile or compressive strain field. These strain field creates an obstacle in the diffusion process of Carbon atoms. So, for the equilibrium to be happen not only carbon has to jump from one interstitial site to another but also the substitutional impurity has to jump to some vacancy site. This process takes time and as a results of that entire TTT-diagram shifts towards right. Hence martensite can be produce by air quenching.

2.20 HEAT TREATMENT

Heat treatment is an important operation in the manufacturing process of many machine parts and tools which imparts high mechanical properties to steels. Heat treatment can be classified into 5 main groups. The first group comprises first order annealing which is employed to relieve internal stresses, reduce the hardness and to increase the ductility of the strain hardened metal, it is also known as stress relieving.

The second group or full annealing which involved phase recrystallization consists in heating alloys above the temperatures require for phase transformation. This is followed by slow cooling. Full annealing substantially changes the physical and mechanical properties and may refine grain structure.

In the third group, hardening alloys are heated above the phase transformation temperature and then rapidly cooled to certain at room temperature either a phase stable at high temperature or various stages of decomposition or transformation of this phase. Thus hardening causes the formation of unstable structures.

The fourth group is tempering it involves the reheating of hardened steels to a temperature below that required for phase transformation so as to bring it nearer to equilibrium state.

The fifth group chemical heat treatment involves the saturation of the surface of steel with a certain element by the diffusion of this element, from the surrounding medium at a high temperature it is also called case hardening.

2.20.1 Annealing

Annealing is the heating of steel to austenite temperature and then cooling slowly in the furnace. The purpose of annealing is

1. To reduce hardness
2. To improve machinability
3. To increase or to restore ductility
4. To relieve internal stresses
5. To reduce or eliminate structural in homogeneity
6. To refine grain size
7. To prepare steel for subsequent heat treatment

Slow cooling results in the formation of spheroidal carbide and (coarse) lamilar pearlite. These products are very soft. The cooling rate during annealing varies from depending upon the alloying element in the steel and lower rate of cooling is used for alloy steels as compared to plain C-steels. Annealing results in the formation of ferrite, spheroidal cementite and coarse pearlite. All theses phases and micro-constituents are relatively soft and therefore this is known as softening treatment and produces relatively lower hardness values while ductility increases. There are 4 types of annealing process.

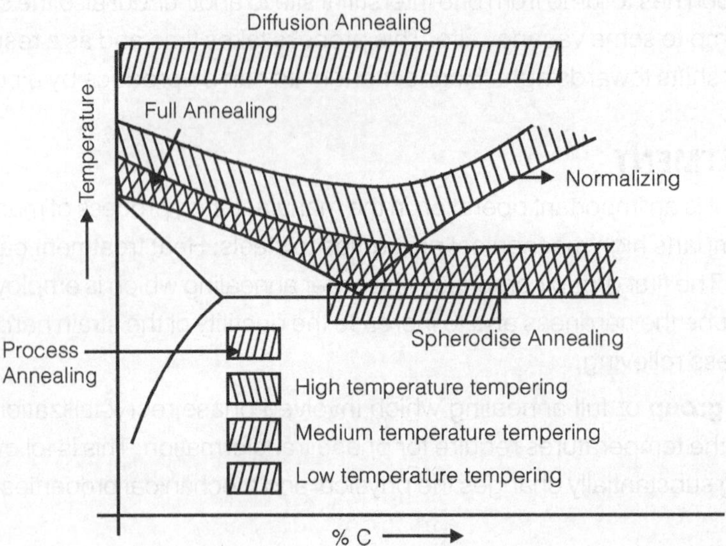

Fig. 2.47 : Regions of temperature for different heat treatment processes

1. **Full Annealing :** Primary objective of this process is to reduce hardness and increase ductility. The process involves.
   (a) Heating the steel to about 50 to 75°C above the upper critical temperature for hypoeutectoid steels and above the lower critical temperature for hyper eutectoid steel.
   (b) Holding it at this temperature for a sufficient time depending upon the thickness of object the holding time is 3-4 min/mm of thickness of the largest sections.
   (c) Slowly cooling in the furnace. The rate of cooling varies from 30°C-200°C per hour depending upon the composition and stability of austenite, when slow cooling in done austenite is decompose into coarse pearlite and ferrite structures in hypoeutectoid steels and in eutectoid steels is converted into pearlite. Hyper-eutectoid steels may undergo full annealing and after cooling austenite decomposes into pearlite and cementite.
2. **Process Annealing :** It is usually carried out to remove the effects of cold working and to softer to make it suitable for further plastic deformation as in the case of sheet and wire industries. It is the re-crystallization of cold worked steel by heating below the lower critical temperature. The exact temperature depends upon the extent of cold working, grain size, composition and holding time. During re-crystallization the deformed grains are reoriented hence they increase plasticity and remove internal stresses. It should be noted that this process produces no change in microstructure and the process is primarily used for low carbon steels.
3. **Spheroidise Annealing : This process is applied to high carbon steels** which are difficult to machine. It causes the formation of all carbides in the steel in the form of very small globules or spheroids like spheres. The process consists of heating the steel near the lower critical temperature (730-770°C), holding at this temperature and then cooling slowly to 600°C. The rate of cooling in the furnace is 25-30°C/hr. Varieties of heat treatment can be used to produce a spheroidized structure but all of them are relatively lengthy and costly.
4. **Diffusion Annealing :** Diffusion annealing or homogenizing is applied to alloy steel ingots and heavy complex castings for eliminating the chemical in homogeneity within the separate crystals by diffusion. Homogenizing is carried out at temperature 1000-1200°C. The optimum temperature being 1150°C at which diffusion proceeds quite easily. The steel is heated to 1150°C as the highest rate available for the furnace. Holding time is kept at minimum, followed

by cooling with the furnace for 6-8 hours to 800-850°C and then further cooling in air. After Homogenizing castings undergo full annealing to refine their structures. Higher temperatures are selected to enable diffusion more and more.

2.20.2 Normalizing

It is a final heat treatment process which is given to a product, which are subjected to relatively high stresses. The process consist of heating steel to a temperature 40-50°C above the line which austenite is stable, holding at that temperature for a short period and subsequently cooling in air at room temperature. **This is known as air quenching,** Normalizing produces micro-structures consisting of ferrite and pearlite for hypo-eutectic steel and pearlite and cementite for hypereutectoid steels. This alloy steel structure consist of ferrite and sorbite (medium pearlite). Normalizing raises the yield point ultimate tensile strength and impact strength in steels. Normalized steels are harder and stronger but less ductile than annealed steels with the same composition.

Comparison between Normalizing and Annealing :

1. Normalizing requires a heating range which is about 40°C above that of annealing.
2. Mechanical properties of steels are better than those produced by annealing.
3. Heat treatment process is of short duration due to increased rate of cooling of the metal in air.
4. If mechanical properties is not the main aim of the heat treatment, better machinability and removal of internal stresses is possible in annealing than those obtained by normalizing.

2.20.3 Hardening

Hardening is a process in which steel is heated to austenite temperature held at this temperature and then quenched in water oil or molten salt baths. Hypoeutectoid steels are heated from 30-50°C above the Upper Critical Temperature (UCT), while hypereutectoid steels are heated above the lower critical temperature. Cooling at a rate higher then the critical value enables the austenite to supercooled to martensite point. Due to rapid cooling carbon freezes at its location and the microstructure appears like collidal solution of cementite is ferrite. This microstructure is called martensite which is very hard and brittle. Hardened steel is very brittle and can not be used for practical purposes. After hardening steel must be tempered to reduce brittleness relieve the internal stresses caused by hardening and to obtain the desired mechanical properties. Hardening followed by tempering improves the mechanical properties of steels and in case of tool steels, it increases the hardness and wear resistance and toughness at the same time. In case of structural steel, it improves strength, ductility and toughness.

The process is carried out in three stages:

1. Heating the object to a temperature above the critical point.
2. Holding the object at this temperature for a definite period.
3. Quenching in a suitable medium.

The hardening process is based on the metallurgical reaction of eutectoid. This reaction is dependent upon.

(a) Adequate C-content to produce hardening.
(b) Austenite decomposition
(c) Heating rate and time.
(d) Quenching medium.
(e) Quenching rate.
(f) Size of the part.
(g) Surface condition.

(a) **Adequate C-content :** In order to produce hard structures like martensite, atleast 0.5% C must be present. Carbon increases the hardness and upto 1% of carbon wear resistance surface is obtained. These are some alloys in which carbon remain dissolved in the austenite and become very hard during decomposition of austenite to martensite.

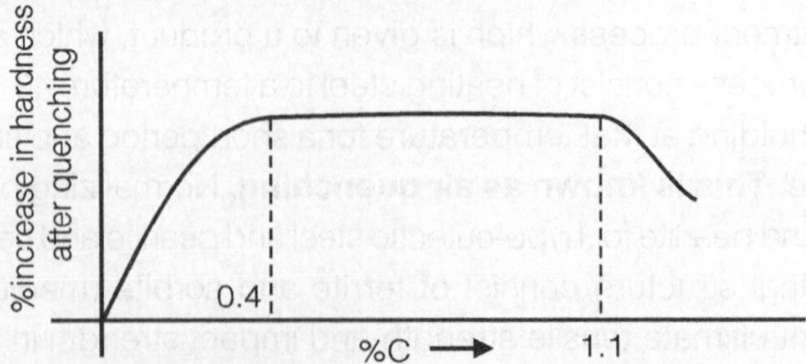

Fig. 2.48 : Hardness Vs Percentage C

It can be seen from the figure that only medium carbon steels can be heat treated. The reason is that only eutectoid composition leads to martensite formation. The pro-eutectoid phase does in low and high carbon steelsnot contribute to the formation of martensite.

(b) **Austenite Decomposition** : The TTT diagram relates the decomposition of austenite with respect to time and temperature conditions. This diagram explains the decomposition of austenite to various structural components like pearlite, banite and martensite.

(c) **Heating rate and heating time :** The heating rate and time depends on the composition of steel, its structure, residual stresses, form and size of the part to be hardened. If the heating rate is too high there will be a temperature gradient between the surface and the core, the best way is to first heat to the required temperature and hold it at that temperature. The heating time for carbon tool steels and medium alloy structural steel is 25-50% more than for C-structural steels. For high alloy structural steels and tool steels it should be 50-100% higher. Heating in salt baths may be conducted more uniformly and rapidly than heating in box furnaces.

(d) **Quenching Medium :** The quenching must provide for a cooling rate above the critical value to prevent austenite decomposition into pearlite and intermediate regions. On the martensitic transformation temperature range, cooling should be slower to avoid high internal stresses, distortion and cracking. The most widely employed quenching media are water aqueous solutions, oil, air and molten salt.

Water and aqueous solution are most widely used as quenching media in hardening carbon and certain low alloy steels which have a high cooling rate (critical). Some features of water quenching are :

(i) It tends to form a vapour blanket which reduces the cooling rate.

(ii) It gives sharp reduction of cooling capacity at higher temperatures.

(iii) It gives a high cooling rate in the temperature range of martensite formation which leads to high structural stresses causing deformation distortion and even quenching cracks.

An addition of NaCl, alkalies, soda and H_2SO_4 to water substantially increases its cooling capacity and provides more uniform cooling. The maximum cooling rate is obtained when 10-15% salt is added. These solutions provide fastest cooling rate.

Mineral oils are suitable for quenching alloy steels in which austenite is highly stable and the critical cooling rate is therefore low. As compared to quenching liquid oil has the following advantages :

(i) The cooling rate in the martensitic range is comparatively low. This prevents quenching defects.

(ii) The quenching capacity of oil is subject to relatively small change in the temperature range which reduces quenching stress.

(iii) It provides uniform cooling to the sample and there baths are used in case of alloy steels.

Disadvantage includes the comparatively low cooling rate in the range of pearlite and intermediate transformations, high inflammability of oil and its tendency to thicken in the course of time which reduces its quenching capacity.

A water and air mixture apply at a pressure of 3 atm is sometimes use of quenching in hardening heavy articles. Various cooling rates can be obtained by changing the amount of water in the mixture.

(e) Quenching rate : It depends upon the quenching medium. The higher the quenching rate the more is the temperature gradient between the core and the surface.

(f) Size and Rate : Long articles of cylindrical and other cross-section should be quenched with their main axis perpendicular to the bath surface. Thin and flat article, should be immersed at an edge. The direction of movement during cooling should coincide with the direction of immersion. Heavy articles should be held stationary and the liquid should be agitated.

(g) Surface Condition : When a part of a surface is quenched in water only its surface is in contact and the heat is dissipated from the part to the water through surface only. Cooling rate depends upon the surface area of the part.

2.20.4 Tempering

The martensite which is formed during quenching is too brittle and hence cannot be used in many cases. The residual stresses are also developed during martensite formation hence hardening should be followed by tempering. Tempering consist in heating the hardened steel to a temperature below the lower critical temperature holding it for sometime and then cooling slowly. It is the final operation in heat treatment.

Some advantages of tempering are :

1. Residual stresses are relived.
2. Ductility is improved.
3. Toughness is increased.

The higher the tempering temperature, more will be the recrystallization phenomenon this will reduce hardness and increase toughness. The work is cooled slowly after tempering. The cooling rate considerably affect the residual stresses. The slower the cooling the lesser will be the stresses. Rapid cooling in water develops new thermal stresses.

Tempering is classified according to the tempering temperature, because it has much effect on properties of steel.

1. **High Temperature Tempering :** It is employed at 500-650°C. The resulting structure consist of sorbite which gives good strength and toughness. Residual stresses are completely relieved if holding temperature is 100-120°C the holding time can be increased to give desired properties to steel.
2. **Medium Temperature Tempering :** This type of tempering is employed at 350-500°C the resulting steel structure consist of troostite. **This process increases the endurance limit**

and elastic limit. After tempering the work is cooled in water then by increasing the endurance limit in case of springs. This process is used for spring steel and die steels.

3. **Low Temperature Tempering :** The temperature employs at 250°C and the holding time is 1-3 hrs. This process reduces internal stresses, increases strength and toughness and provides high wear resistance. This method is employed in manufacture of measuring tools and cutting tools. This is also employed to the components which are surface hardened by carburizing, nitriding or carbo-nitriding.

2.20.5 Case Hardening

Hard surfaces are produced on relatively soft cores by case hardening. A hard wear resistance surface is called the case and relatively soft and tough inside is called the core. The parts which are subjected to wear and impact are hardened by surface treatment. The hard surface has good wear resistance and the soft core has good toughness. Since mild steels can not be hardened by quenching, so their strength is increased by case hardening. Following are case hardening techniques :

2.20.5.1 Carburizing

It is applied to low carbon steel up to 0.18% carbon. In carburizing the carbon content of the surface layer is increased to obtain a hard layer on the work piece surface after heat treatments. The surface layer is enriched with carbon upto 1%. Carburizing is followed by hardening and low temperature tempering. Low C-steel is heated upto 870°C in the atmosphere which contain carbon.

$$Fe + 2CO \rightarrow FeC + CO_2$$

FeC is austenite with dissolve carbon. At 870°C the maximum amount of carbon can be dissolved in austenite. Carburizing is performed in the following ways.

1. **Pack Carburizing :** The parts are surrounded by carburizing mixture and packed in a steel box which is tightly covered and sealed with clay on all sides to prevent entry or exit of gases. The container is heated to proper temperature for the required amount of time. The temperature is generally 950-1000°C and time is 6-8 hrs. higher temperatures are selected to enable diffusion more and more. During this time the carbon diffuses into low Carbon -steel during this time the carburising mixture consist of **50% good Charcoal 20% $BaCO_3$, 5% $CaCO_3$ and 5-12% Na_2CO_3.**

   Barium carbonate acts as energizer it increases the action of carbon on low C-steel. The carburizing mixture is in the form of coarse particles so that sufficient air is trapped, visual the container to form Co.

   $$2\,CO \rightarrow CO_2 + C \text{ (atomic state).}$$

   This atomic state carbon diffuses into austenite. The addition of carbonates activates the carburizer to form the atmosphere of CO.

   $$CaCO_3 + C \rightarrow CaO + 2CO$$

   When heating period is over the boxes are cool down to 450-500°C and then opened. The carburized parts are taken out and quenched if required. This method is a batch production method.

   Disadvantages :

   (a) High labour cost for packing and unpacking.
   (b) It is difficult to quench directly from carburising temperature.
   (c) Much time is needed in heating and cooling.

2. **Gas Carburising :** The work is treated in a medium of gases containing CO and hydrocarbon gases such as CH_4. Propane butane etc. and natural gas are also used which gives very high

quality of carburized case. These gases are mixed with definite amount of air which provides O_2 and forms CO from hydrocarbon gases. Work pieces are heated in an atmosphere of regulating carburizing gases. The temperature, maintained generally at 930-950°C and the holding time is 3-12 hrs depending upon the thickness of the case. The chief reactions of carburising are the dissociation of methane and CO

$$2CO \rightarrow CO_2 + C \text{ (Atomic)}$$
$$CH_4 \rightarrow 2H_2 + C \text{ (Atomic)}$$

This atomic stage of Carbon diffuses into austenite this process may be performed in continuous furnaces. The work pieces are placed in the furnace such that the gases will have easy access to the surface to be carburized. The work pieces are then cool down to room temperature and further heat treatment can be done afterwards. This process has many advantages over pack carburizing and hence it is extensively used the process can be make fully automatic.

Advantages :

(a) High quality carburized case can be produced.
(b) Heat consumption is less.
(c) Time consumption is also less.
(d) Process is clean.
(e) Production cost is low.
(f) Closer quality limits can be achieved.
(g) There is more flexibility of operation.

3. **Liquid Carburising :** In this process molten state bath containing 20% NaCN is used. Cyanide provides C and N. The temperature is maintained at 950°C. The process consist of heating the work pieces into molten bath at 950°C. At this high temperature C and small amount of N are diffused in the surface of work pieces the case is hardened. The only disadvantage is that the cost of carburizing salt is very high.

2.20.5.2 Nitriding

In nitriding the N content of the surface is increased, this is done by heating the steel in the atmosphere of NH_3 gas. The parts to be nitrided are placed in an air tight container. NH_3 is passed continuously over the work pieces at a temperature of 500 to 650°C, NH_3 gas dissociates as follows

$$2NH_3 \rightarrow 2N + 3H_2$$

This atomic state of nitrogen diffuses into steel surfaces and react with alloys of steel making a case containing the alloy nitrides. Thus the presence of alloying elements such as Al, Cr, Mo etc. is necessary, in order to obtain a case of maximum hardness Medium C-steel containing 3% of alloying elements produces hardest case on the steel surface.

The most common size of nitriding case is 0.3 to 5 mm thick, it requires 30-40 hrs at 500-520°C. The time can be reduced by using a double state process in which the work is first heated at 500-520°C and then at 550-600°C for the nitriding process. **After nitriding the work is cooled in the furnace in the spring of ammonia.**

Advantages :

(a) It increases the hardness of surface layer which is harder as compared to carburising.
(b) It increases wear resistance, endurance limit and resistance to corrosion.
(c) No heat treatment is required after nitriding.

Uses : Nitriding is used during manufacture of gears components of M/c tools cylinders of powerful engines, cylinder liners, gauges, cams, values etc.

Disadvantages :

(a) The case produced is brittle.

(b) The process is costly compared to carburising.

2.20.5.3 Cyaniding

In cyaniding a thin case of high hardness and wear resistance is produced on C-alloy steels. Work is immersed in molten salt bath containing NaCN which is heated to 820-860°C. This is usually followed by water quenching. The cyanide bath consists of 20-30% NaCN, 25-50% NaCl and 25-50% Na_2CO_3. The time required for cyaniding depends on the depth of case required. For 0.15- 0.5mm, 30-90 minutes are required.

$$2NaCN + O_2 \rightarrow 2\,NaCNO\ (Air)$$

$$2NaCNO + O \rightarrow NaCO_3 + CO + 2N\ (Atom)$$

$$2CO + O \rightarrow CO_2 + C\ (Atomic)$$

The atomic C diffused into steel the work can be directly quenched as soon as it is taken out of bath then low temperature tempering is done at 200°C. The case obtained by cyaniding is of high wear resistant and endurances limit compared to carburising. It should be noted that hardness at case varies in the following sequence

Nitriding > Cyaniding > Carburizing

Cyaniding is used to produce light cases on small parts like small shafts, worms, gears, nuts, springs, pins etc.

2.20.5.4 High temperature or Deep cyaniding

A deeper or thick case of 0.5-2 mm is obtained by heating the work piece in a salt bath at 930-360°C for 1.5-6 hrs depending upon the depth of case required. The bath consist of 82% $BaCl_2$, 8% NaCN and 10% NaCl. The reaction are :

$$BaCl_2 + 2NaCN \rightarrow 2NaCl + Ba(CN)_2$$

$$Ba(CN)_2 \rightarrow BaCN_2 + C\ (Atomic)$$

$$BaCN_2 + \underset{(Air)}{O_2} \rightarrow BaO + CO + 2N\,(Atomic)$$

This atomic C and N diffuse into steel after deep cyaniding. The work is cooled in air to refine the grains. The work is again hardened and quenched, this is followed by low temperature tempering at 200°C.

Advantages :

1. It requires less time than carburising.
2. There is less distortion of the work piece
3. Resistance to corrosion and wear is high.

Disadvantages :

1. High cost.
2. Cyaniding bath is toxic hence the worker needs protection.

2.20.5.5 Carbo-nitriding (Gas Cyaniding)

A mixture of ammonia and hydrocarbon gas is used. The work is heated to 850°C in the mixture for 2-10 hrs. This is followed by quenching then tempering is employed at 180°C. Martempering can also be employed instead of low temperature tempering. During the process both Carbon and Nitogen diffuses simultaneously with Carbon diffusing at a higher rate than nitrogen. The alloying elements reacts with carbon and nitrogen and from carbides and nitrides. Lower the stability of austenite and troostite is formed in the case. This reduces the endurances limit, ductility and toughness of the steel.

Advantages : Compared to gas carburising it has the following advantages

1. Carbo-nitriding is performed at lower temperature (850 - 870°C) as compared to Carburizing (930 - 950°C)
2. Higher resistance to corrosion and wear.
3. Operation is clean and there is no soot on workpiece. When compared with liquid cyaniding there are following advantages :
   (a) Medium and large size work can be processed.
   (b) Nitrogen and Carbon contents in the case can be controlled.
   (c) Toxic salts are not required.
   (d) The work can be machined.

The above case hardening techniques meant for low Carbon steels but the following case hardening methods are meant for steels have % in the medium Carbon steel range.

2.20.5.6 Induction Hardening

This process is employed to increase hardness, wear resistance and endurances limit of the surface of the work piece. The heat treatment is given to the surface only by supplying excessive heat to the surface followed by drastic quenching, the surface is heated to the austenite range and then quenched immediately to form martensite. The structure of core remains un changed because it is not effected by heat, the components should contain 0.4 - 0.5% carbon or sufficient alloying elements as chromium Ni, or Mo. The work is placed in Helical coils called inductor the coil consist of several turns of water cooled cu tubing. Alternate current is passed through the inductor and thus alternate magnetic field sets up. The field induces eddy currents on the surface layers and heat is generated. The current density is not uniform through out the cross-section of the work piece. Approximate the 90% of the heat is generated in the work layer of thickness (x) represented by

$$x = 5000\sqrt{\frac{\rho}{\mu f}}$$

Where f is the frequency, μ is the magnetic permeability and ρ is the electrical resistivity of the metal of the work piece. The heating rate in the range of transformation temperatures is upto 300°C/sec for hypoeutectoid steels and subsequently the heating time is very small (2 - 50 sec's). The high frequency current is supplied by motor generator and vacuum tube generator. When it is required to increase the strength of the core the work should be normalized before induction hardening. The quenching temperature depends upon the rate of heating if the rate is 500°C/sec it is 1000°C and if it is 250°C/sec it is 900°C. Induction hardening can be performed in three ways :

1. The whole work surface is heated at one time and quenched. This procedure is used for hardening surfaces of small components like shafts and sharp of tools.
2. The sections of work are heated and quenched consequently. This procedure is used for hardening generals of crank shaft teeth of gears and cams of cam shaft.
3. The work in made to travel with respect to stationary inductor or vice versa and spray quenching is done. This procedure is used for hardening long shafts and axles. Generally induction heating is followed by low temperature tempering at about 180°C.

Advantages :

1. Time required is very small hence the process is very quick and productivity is high.

2. Scale is not formed hence machining time is saved moreover material is not lost due to scale formation.
3. The process can be automated.
4. Depth of hardness can be controlled easily.
5. Distortion is reduced.
6. Due to high speed grain growth decarburising do not occur.
7. Both external and internal surfaces can be hardened.
8. It is an efficient mass production method. Almost all components can be uniformly hardened i.e. the quality of all the components is uniform.
9. Where ever required the hardening can be localised and there is no need to protect the remaining surfaces.

Disadvantages : Each type of work piece requires different fixtures for its holding, each time one piece can be heat treated where as different work pieces can be treated at one time in carburizing and nitriding.

Uses : Induction hardening is mostly used in industry e.g., for hardening the surfaces of cam shaft, crank shaft, gears, automobile components, cold rolling mills, splined shaft, crane wheels, spindles, Brake drums etc.

2.20.5.7 Flame Hardening

The process consists of heating the surface of **medium carbon steel by high temperature** gas flame at 2400°C-3300°C and immediately cooling in air or in water. Heat may be supplied by oxyacetylene torch. The fuel used by the flame may be oxyacetylene, natural gas or kerosene. The flame rapidly imparts large amount of heat to the surface. The heat is supplied so quickly to the surface and for so short time the core remains unaffected.

As soon as the desired temperature is achieved water is immediately sprayed which cools the surface. By proper control of heating and cooling operation the core is not affected by the treatment. The thickness of hardened layer is 2-4 mm and its structure is martensite. The structure of underline layers is troostomartensite.

Advantages :

(a) There is practically no distortion of the work piece because only small sections of the work piece are heated.
(b) As heating rate is high the work surface remains clean.
(c) The process can easily be automated.
(d) The process is more efficient for large work piece as compared to induction heating. It is very economical for large work.

Disadvantages :

(a) Very thin sections may get distorted extensively.
(b) Over heating may cause crack.

Uses : It is used to increase wear resistance and surface hardness of piston pins, large gears, hand tools shafts, cams, mill rolls etc.

As it can be seen from the TTT diagram of high carbon steel that martensite finish line appears approximately at 25°C. Hardenability of hypo and hyper eutectoid steels is lower than that of eutectoid steels. Because, upon quenching the material into a bath maintained at room temperature, martansite, pro-eutectoid cementite and retained austenite will be present in the sample. Therefore to convert the sample into 100% martensite, steel sample is quenched into a bath of "liquid N_2". This process is called cryogenic treatment of materials.

2.21 TTT DIAGRAMS FOR HYPO AND HYPER-EUTECTOID STEELS

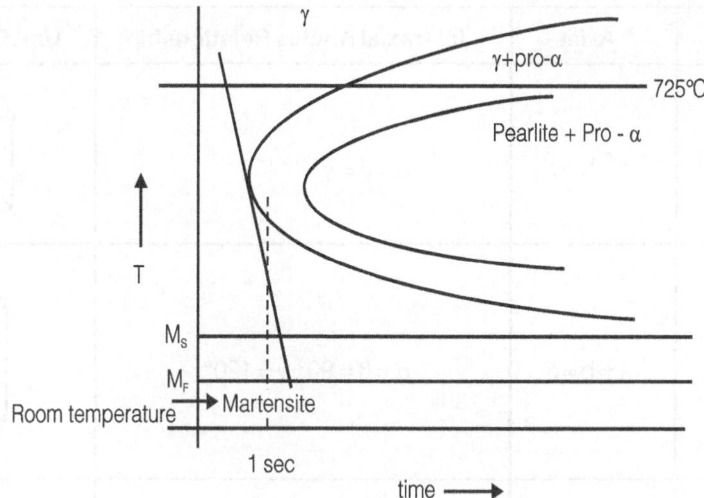

Fig. 2.49 : TTT diagram of Hypo-eutectoid steel

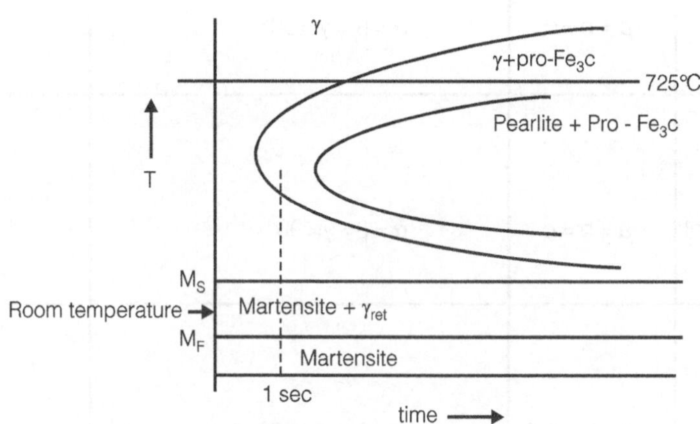

Fig. 2.50 : TTT Diagram for Hyper-eutectoid steel

As it can be seen in the diagram that due to the presence of proeutectoid phase, start line shifts towards left and nose of TTT diagram appear prior to 1 sec. So hardenability of both hypo and hyper eutectoid steels are inferior to eutectoid steels. In the microstructure along with normal pearlite phase there will be proeutectoid phase. It can be observed from the TTT diagram of hypereutectoid steel that martensite finish line appears below room temperature. So retained austenite will appear while quenching the sample to room temperature. To produce uniform phase martensite, such sample are quenched in liquid nitrogen and the process is called cryogenic treatment of materials.

2.22 CRYSTAL SYSTEMS

Generally all the metals or alloys are building blocks of small repeated entity called '**Unit cell**'. The unit cell posses axial lenght (a, b, c) and axial angles (α, β, γ). By assigning different values of axial length and angles seven different types of crystal systems are identified. These are also called premitive latices, and a unit cell of premitive lattices contains atoms, only at its corners.

Lattice Parameter Relationships and Figures Showing Unit Cell Geometries for the Seven Crystal Systems:

Crystal system	Axial	Interaxial Angles Relationship	Unit Cell Geometry
Cubic	$a = b = c$	$\alpha = \beta = \gamma = 90°$	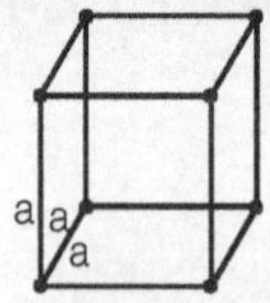
Hexagonal	$a = b \neq c$	$\alpha = \beta = 90°, \gamma = 120°$	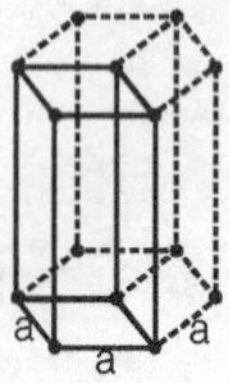
Tetragonal	$a = b \neq c$	$\alpha = \beta = \gamma = 90°$	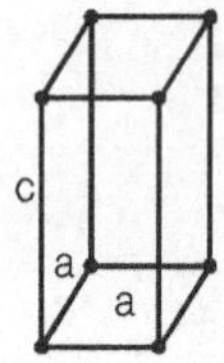
Rhombohedral	$a = b = c$	$\alpha = \beta = \gamma \neq 90°$	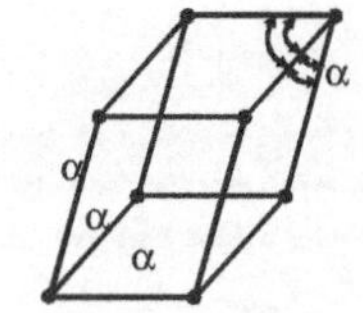
Orthorhombic	$a \neq b \neq c$	$\alpha = \beta = \gamma = 90°$	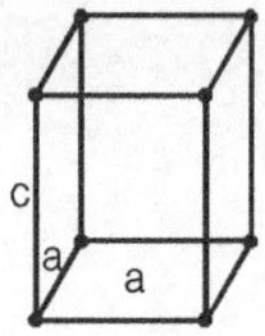
Monoclinic	$a \neq b \neq c$	$\alpha = \gamma = 90° \neq \beta$	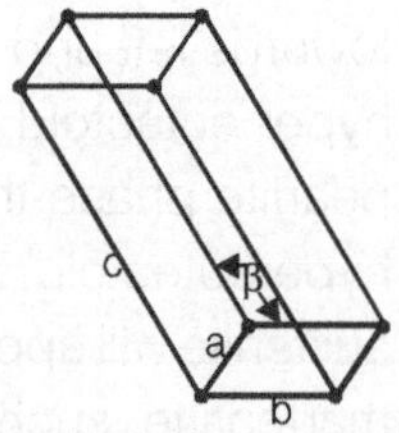
Triclinic	$a \neq b \neq c$	$\alpha \neq \beta \neq \gamma \neq 90°$	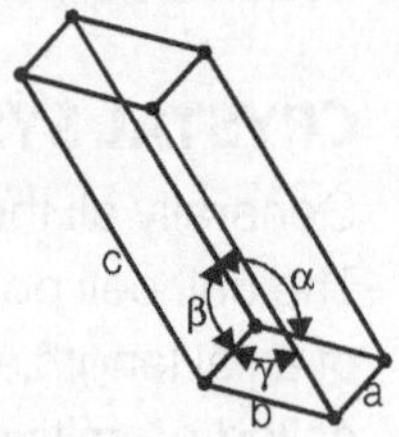

There are 14 possible arrangements of atoms in space called Bravai's lattices. The Bravai's lattices are belong to the 7 crystal systems.

Atomic packing factor (APF) is defined as the volume of unit cell covered by the atoms. The coordination number is defined as the no. of atoms that are in the physical proximity of any atom. Let us elaborate on cubic crystal structure.

2.22.1 Simple Cubic structure:

There is one atom at each corner and this atom is shared by 8 unit cells. So number of atoms per unit cell

$$n = \frac{1}{8} \times 8 = 1$$

size of unit cell (lattice constant) = a

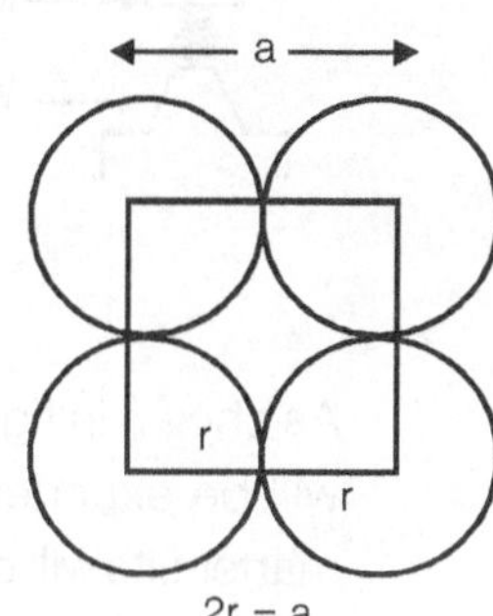

Fig. 2.52

$$APF = \frac{1 \times \frac{4}{3} \times \pi r^3}{a^3} \quad \text{also } a = 2r$$

So APF = 0.52

The coordination number of simple cubic is = 6.

2.22.2 Body Centred Cubic (BCC) structure:

Along with one atom at each corner there is atom at the centre of each unit cell which is physical contact with all the corner atoms.

$$n = \frac{1}{8} \times 8 + 1 = 2$$

$$APF = \frac{2 \times \frac{4}{3} \times \pi r^3}{a^3}$$

$$\Rightarrow \quad 4r = a\sqrt{3}$$

So APF = 0.68.

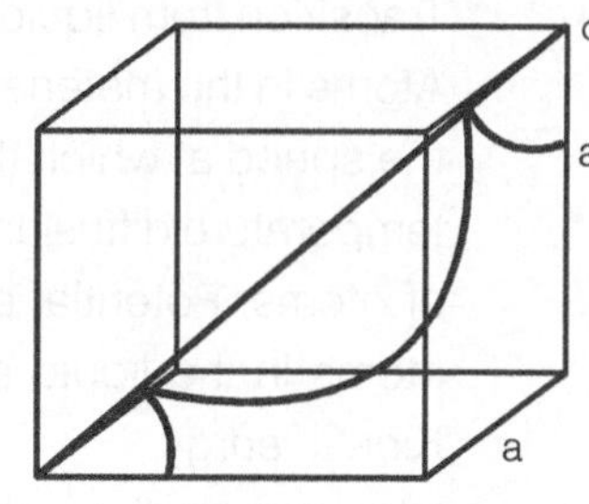

Fig. 2.53

The coordination number of bcc is 8.

2.22.3 Face Centred Cubic (FCC) structure:

Along with one atom at each corner there is an atom on each face.

$$n = 8 \times \frac{1}{8} + 6 \times \frac{1}{2} = 4$$

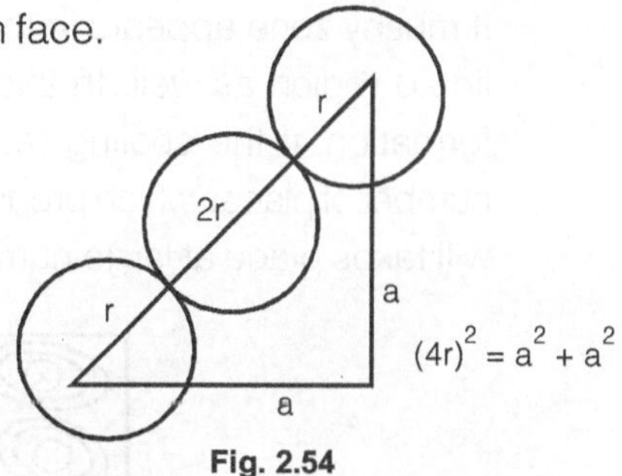

Fig. 2.54

$$APF = \frac{2 \times \frac{4}{3} \times \pi r^3}{a^3}$$

$$\Rightarrow \quad 4r = a\sqrt{2}$$

APF = 0.74

The coordination number of FCC is 12.

2.23 HARDENABILITY CURVES JOMINY END QUENCH TEST

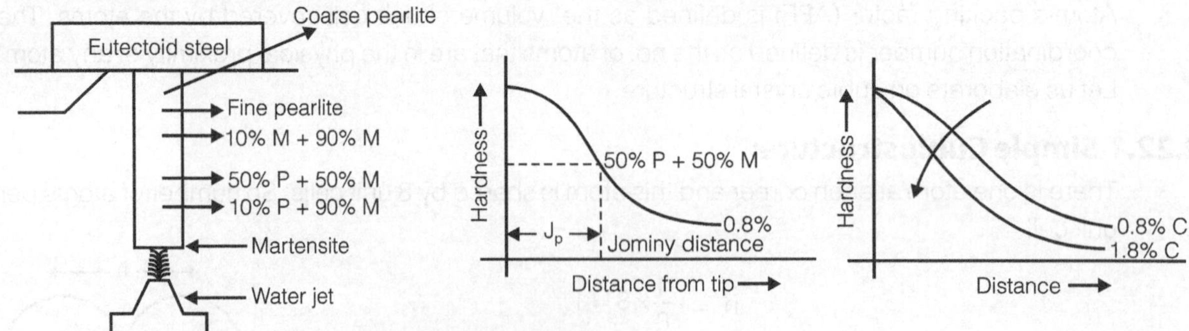

Fig. 2.51 : Jominy end Quench test

As shown in figure (2.51), austenite sample is quenched on one side and different cooling rates will be experienced by the sample from the tip to the other end. Gradually the percentage of martensite will decrease and percentage of pearlite will increase. Upon plotting the graph between hardness and distance from the tip the resulting curve is called hardenability curve. On this curve, the distance upto which there is a mixture of 50% pearlite and 50% martensite is called Jominy distance.

2.24 MECHANISM OF CRYSTALLIZATION

Transition from liquid to the solid state takes place by nuclei formation and growth mechanism. Atoms in the material have both kinetic energy and potential energy. Kinetic energy is related to the speed at which the atom moves and it is a function of temperature. It is because at a higher temperature diffusion phenomenon will be predominant which is caused by higher Kinetic energy of atoms. Potential energy is related to the distance between the atoms. Potential energy of atoms in the liquid as well as solid phase is same but there is a significant difference in the Kinetic enrgy.

If there is an alloy with eutectic composition (or) there is a pure matel, mechanism of crystallization is called skin forming. Since the temperture at the boundary of the liquid metal is low, solid will nucleate. This solidification front will move towards the centre and there will be systematic crystal growth from the surface towards the centre.

If mushy zone appears in the alloy nucleation will takes place not only on the surface but in the liquid region as well. In this case there will be multi directonal crystal growth called dendrite formation. If the cooling rate of the liquid metal is slow nucleation will takes place at smaller number of places which produces coarse grain structure. If the cooling rates are very high, nucleation will takes place at large number of places. This produces fine grain structure.

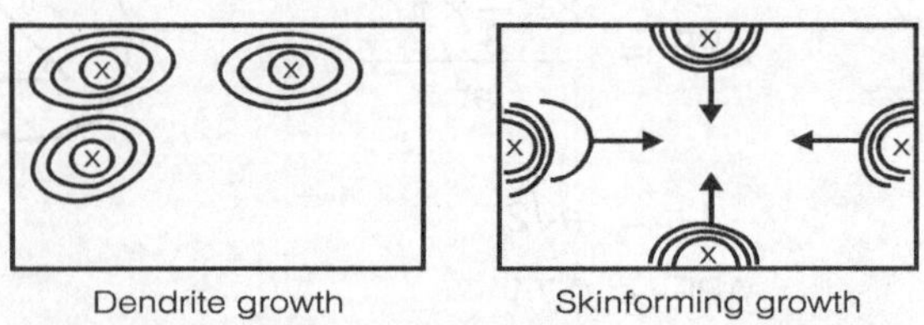

Fig. 2.55 : Dandritic growth and skin forming

2.25 METHODS FOR MEASUREMENT OF GRAIN SIZE

1. **Comparison Method :** The micro structure (100x) of samples obtained from optical microscope is compared with the standard grade grain size(ASTM E112-63 charts). Each chart contain a grain size number(n). When a match is found between a ASTM chart and the micro structure of sample the grain size number is selected. By substituting the n in the following formula, number of grains per in^2 is obtained.

$$N = 2^{n-1}$$

N = Number of grains/ $inch^2$

n = Grain size number

2. **Intercept (Heyn) Method :** The number of grains intersected by one or more straight lines. Grains touched by the ends are counted as 1/2 grains only such measurements are done in all the three directions. In the similar way dislocation density is also defined i.e. number of dislocation per unit length. But in the intercept method the grain diameter is defined as length of line divided by average number of grains.
3. **Planimetric Method :** (Jefferies Method) A circle (or) square of area 5000 mm^2 is drawn over the specimen. Grains falling completely inside the area are counted as full and intersected grains are counted as 1/2. Then the number of grains per $millimeter^2$ is determined by multiplying equivalent number of whole grains by the corresponding magnification factor(f) given in table.

Magnification Used	Magnification factor (f)
1	0.002
25	0.125
50	0.5
75	1.125
100	2
200	8
300	18
500	50
1000	200

2.26 CRYSTAL DIRECTIONS

Crystal directions are represented by bracket [ ] and whatever is written in the bracket can be considered as 3D co-ordinate point. After identifying the point it has to be joined with origin to draw the crystal direction. Bar on any axis indicate negative value. Some of the directions are drawn as under :

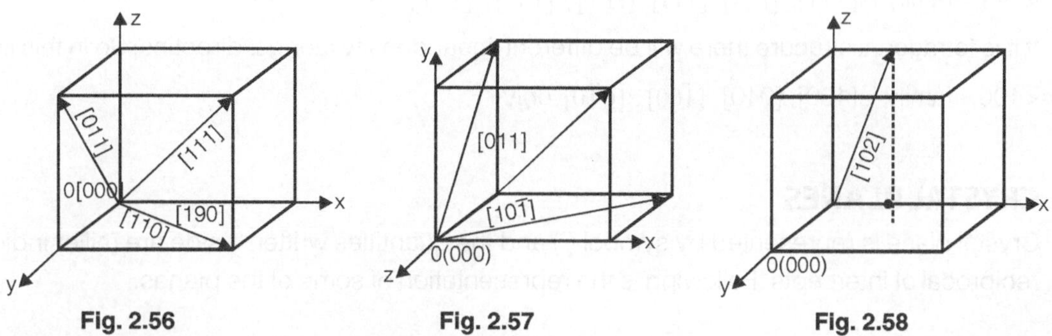

Fig. 2.56 Fig. 2.57 Fig. 2.58

The maximum number appeared is the bracket can be considered as the size of unit cell. The term linear density is defined as the number of atoms appeared on any direction to the length of direction vector. Here the atom is designated by its diameter.

Example : 2.9

Linear density of sample cubic structure along direction [100].

Solution :

Since one complete atom appears along [100].

$$\text{Linear density} = \frac{1}{a}.$$

Fig. 2.59

Example : 2.10

Linear density of FCC along [110].

Solution :

Since two atoms will appear along the directions and the length of direction is $a\sqrt{2}$.

Solution :

So $$\text{Linear density} = \frac{2}{a\sqrt{2}}$$

$$= \frac{\sqrt{2}}{a}.$$

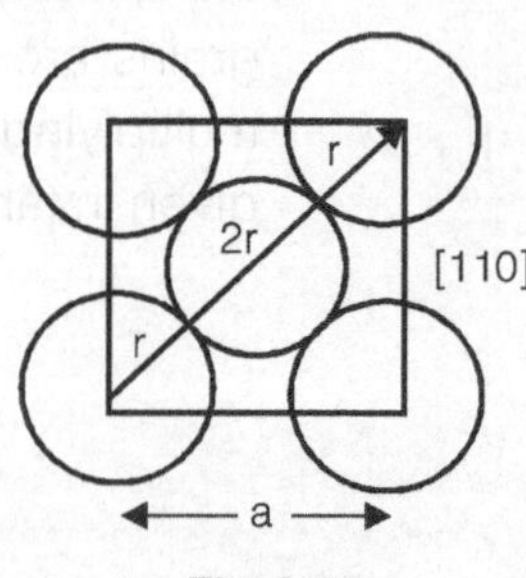

Fig. 2.60

Example : 2.11

Linear density of bcc along [111].

Solution :

Since two atoms appear along this direction. So

$$\text{Linear density} = \frac{2}{a\sqrt{3}}.$$

Family of directions in the unit cell defined as all those possible directions in the unit cell which has same linear density are called family and are represented by symbol < >.

<100> family is[100], [010], [001], $[\bar{1}00]$, $[0\bar{1}0]$, $[00\bar{1}]$

<110> family is[110], $[\bar{1}10]$, $[1\bar{1}0]$, $[\bar{1}\bar{1}0]$, [101], $[\bar{1}01]$, $[\bar{1}0\bar{1}]$, [011], $[0\bar{1}1]$, $[01\bar{1}]$, $[0\bar{1}\bar{1}]$

<111> family is[111], $[1\bar{1}1]$, $[\bar{1}11]$, $[11\bar{1}]$, $[\bar{1}\bar{1}1]$, $[\bar{1}1\bar{1}]$, $[1\bar{1}\bar{1}]$, $[\bar{1}\bar{1}\bar{1}]$

If it is tetragonal strucure there will be different linear density along z directins. So in this limit cell

<100> family is[100], [010], $[\bar{1}00]$, $[0\bar{1}0]$ only

2.27 CRYSTAL PLANES

Crystal plane is represented by symbol ( ) and the quantities written inside are miller indices i.e. reciprocal of intercepts. Following is the representation of some of the planes.

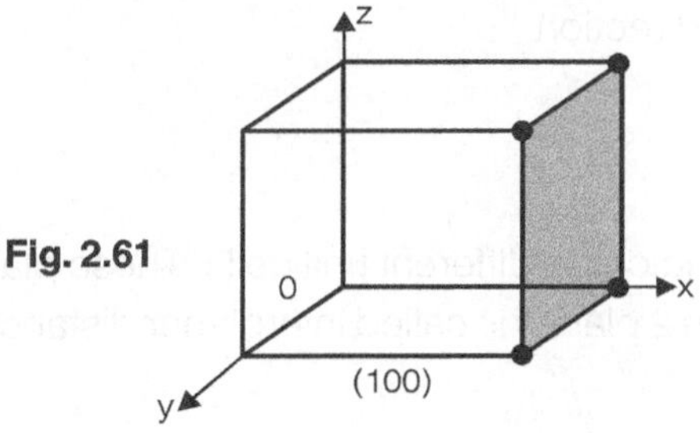

Fig. 2.61

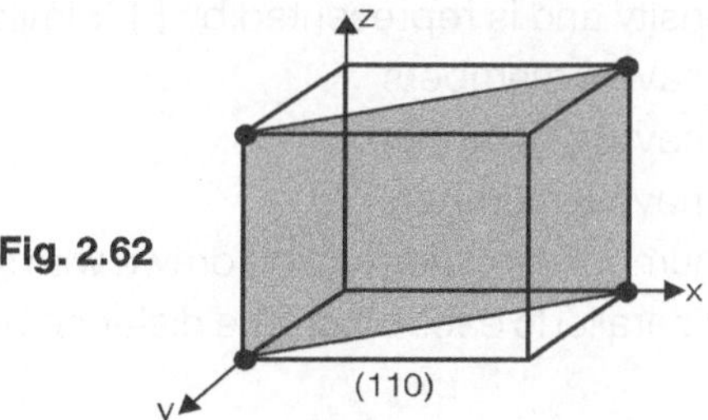

Fig. 2.62

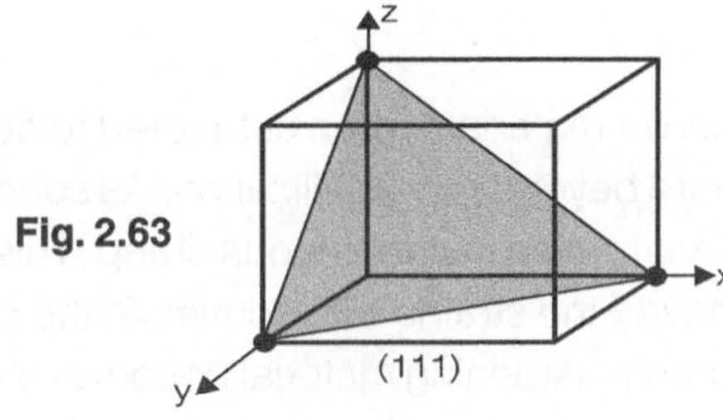

Fig. 2.63

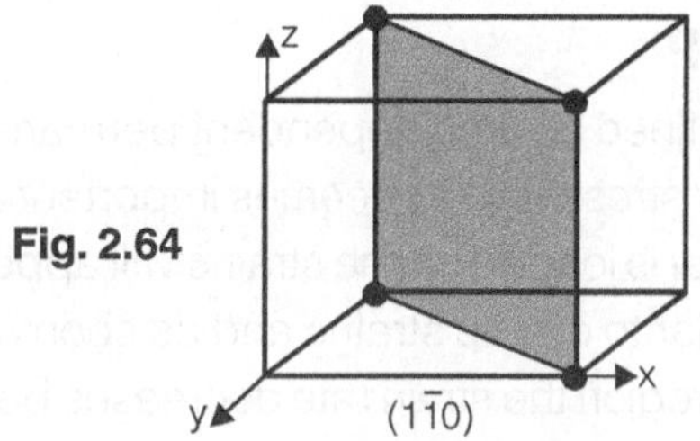

Fig. 2.64

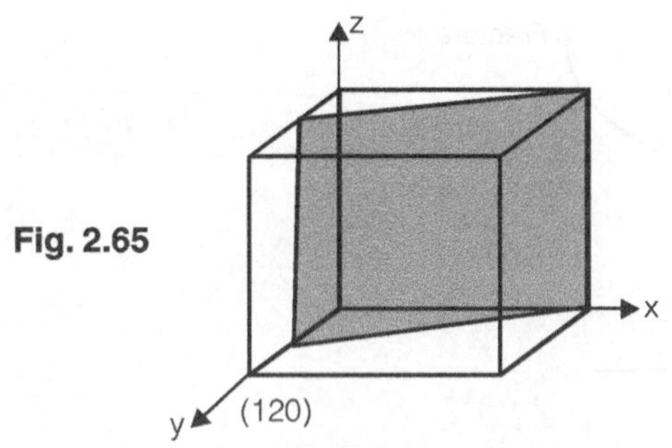

Fig. 2.65

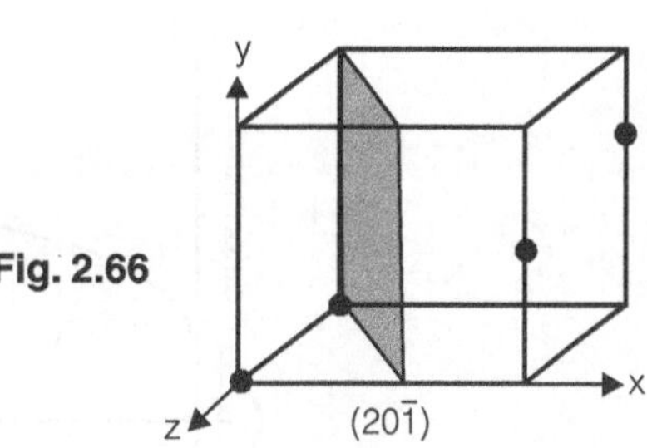

Fig. 2.66

$$\text{Planar density} = \frac{\text{No. of atom on a plane}}{\text{Area of plane}}$$

NOTE: number of atoms here are defined by cross-section area of the atom.

Example : 2.12

Planar density of (100) plane of simple cubic structure is $\frac{1}{a^2}$.

Example : 2.13

Planar density of (100) plane of fcc structure is $\frac{2}{a^2}$.

Example : 2.14

Planar density of (110) plane of bcc structure is $\frac{2}{a^2\sqrt{2}} = \frac{\sqrt{2}}{a^2}$.

Family of plane is defined as all those possible planes inside the unit cell which has same planar density and is represented by { }. Similar to crystal direction

{ } have 6 members

{ } have 12 members

{ } have 8 members.

A number of planes can be drawn with same miller indices in different unit cells. These planes will be parallel to each other. The distance between these 2 planes is called interplanar distance (d_{hkl}).

$$d_{hkl} = \frac{a}{\sqrt{h^2 + k^2 + l^2}}$$

2.28 CREEP

It is defined as time dependent permanent deformation of materials when subjected to constant load or stress and it becomes important at a temperature beyond recrystallization. As soon as the material is loaded, some strains will appear in the material called instantaneous creep. This creep is similar to elastic strains and as soon as load is removed the strains will recover. In the primary creep region the strain rate decreases because due to work hardening material becomes stronger. In the secondary creep region a balance is achieved between recrystallization and work hardening. So almost constant strain rates are observed.

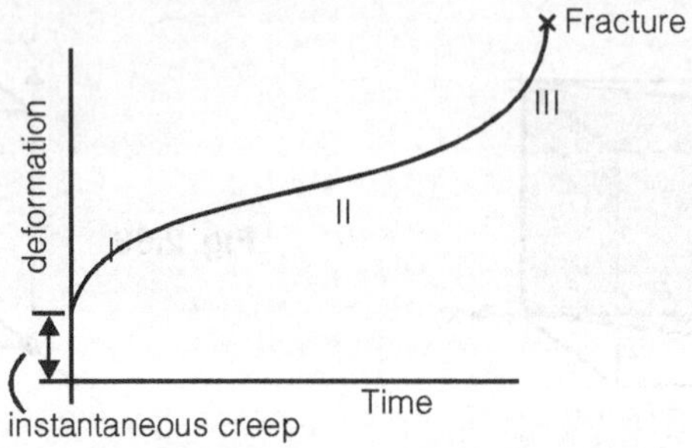

Fig. 2.67 : Creep Curve

But there will be a limit upto which work hardening can take place and in the tertiary creep regular recrystallization phenomenon takes over giving rise to sudden increase in strains and finally material breaks. By increasing the temperature the creep curve shifts as shown in figure because higher is the temperature, more predominant recrystallization phenomenon will be discussed as before that grain boundary material is at a higher energy level, so at elevated temperature there is a tendency in the material to be low. So finer the grain structure lower will be the creep resistant.

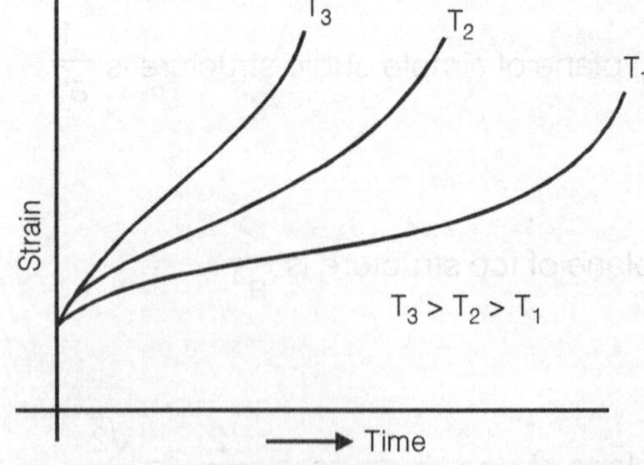

Fig. 2.68 : Variation of creep curve with temperature

2.29 TENSILE TEST

Tensile tests are being carried out to find out the mechanical properties of material. Samples are prepared generally according to ASTM standards and loaded in a UTM machine. Once the stress goes beyond the yield strength there will be decrease in crossection area hence increase is stress. This increase in stress is compensated by increase in strength by work hardening. A unique neck will appear (anywhere in the parallel length) at a point of plastic instability where the increase in strength due to work hardening fails to compensate for the decrease in crossectional area. Gradually the sample fails like cup and cone fractures.

2.30 HARDNESS TESTS

Following is the comparision of different tests to quantity hardness

Test	Indenter	Measurement	Load (P)	Formula
Brinel hardness	Ball of 10mm Hardened Steel	Area of indentation	500 Kg for soft material to 3000 kg for hard material	$BHM = \frac{2P}{\pi D [D - \sqrt{D^2 - d^2}]}$ D = dia of ball d = dia of indentation
Vickers hardness	Diamond pyramid of 136°	Area of indentation	5 to 120 Kg	$VH = \frac{1.854 P}{d_1^2}$ d_1 = diagonal size of indentation
Rockwell hardness	Ball of 1.58 mm diameter for soft materials and cone of 120° with spherical tip of 0.2 mm radius for hard material	Depth of indentation	60 - 150 Kg	$B\ scale = \frac{130 - depth}{0.002}$ $C\ scale = \frac{100 - depth}{0.002}$

2.31 IMPACT TEST

	Charpy	Izod
Specimen	55 mm, 0.25 mm, 10, 2 mm, 45°	45°, 0.25 mm, 28 mm, 75 mm, 10
Placed like	Simple supported beam	Cantilever beam
Strain rate	$10^3\ s^{-1}$	$10^3\ s^{-1}$
Measurement	Energy absorbed in fracturing specimen generally is in joules	Energy absorbed in fracturing specimen generally is in joules

2.32 ALUMINUM ALLOYS

The addition of small quantities of various elements in aluminum greatly improves the mechanical properties. In aeroplanes and other structures the extreme lightness of the alloy enables much greater volume to be used for a given weight. Aluminium-alloys has coefficient at linear expansion twice to that of cast irons or steels and allowance has to be made when aluminium has to be casted. The corrosion resistance of Aluminium and its alloys is quite high.

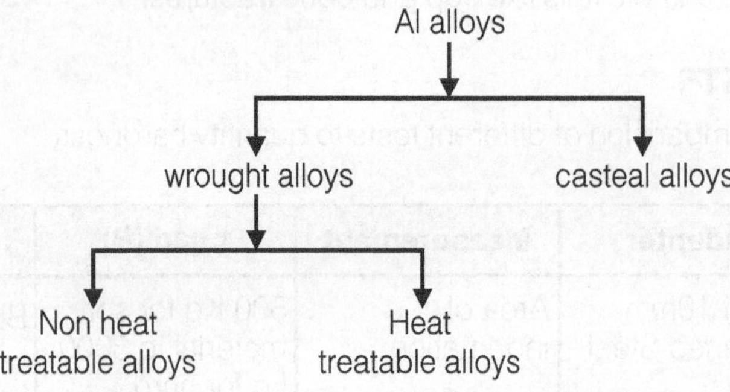

2.32.1 Non heat treatable Al alloys

These alloys included Al-Mn (1.3% Mn) and Al-Mg (2.5-5% Mg). Al-Mn alloys have structure comprising the solid solution of Mn in Al and forms the compound $MnAl_6$. This alloy is strengthened by work hardening. Al-Mg alloys are the binary alloys at Al and Mg but most of the Al-Mg alloys are single phase in structure. All these alloys have comparatively low strength and high ductility and good weldability

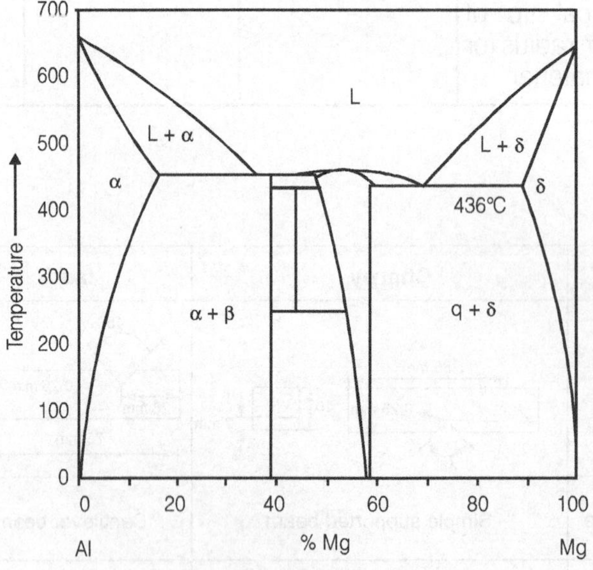

Fig. 2.69 : Al-Mg Phase Diagram

2.32.2 Heat treatable Aluminium alloys (Duralomin)

Any alloy of Al which produces phase diagram similar to Al-Cu with % Cu less then 5.5% can be heat treated. Cu is added in Al less than 5.5%. This alloy is heated to a temperture between T_{min}-T_{max} so that a uniform phase α appears in the sample. The sample is then quenched to room temperature. As a result of that Cu freezes in the structure of Al.

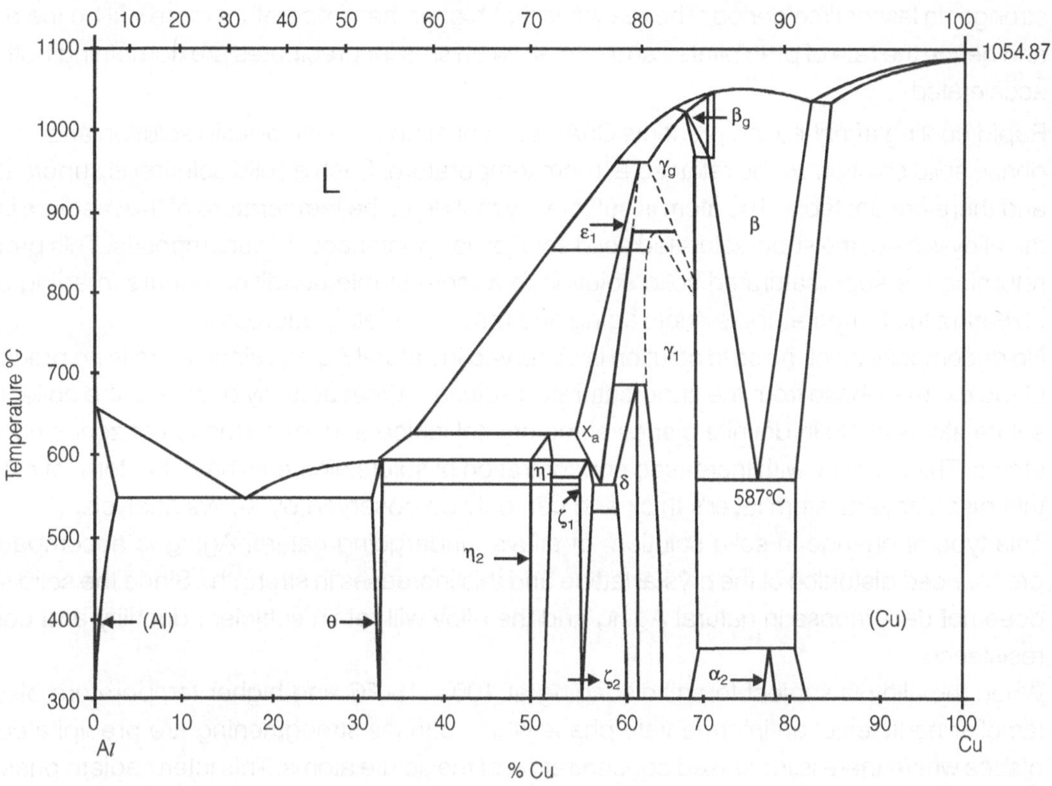

Fig. 2.70 : Al-Cu phase diagram

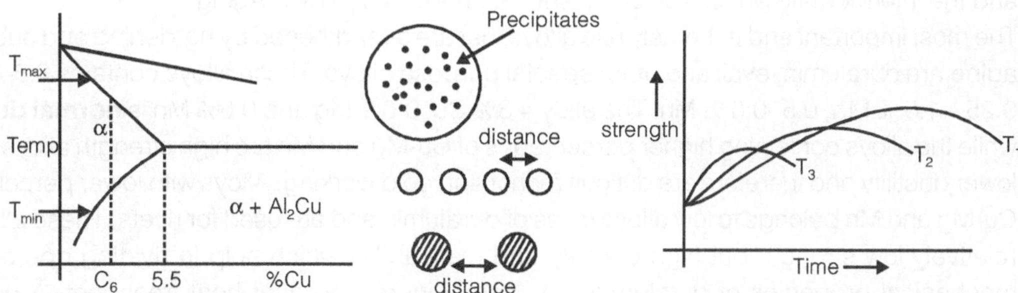

Fig. 2.71 : Portion of Al-Cu phase diagram

Fig. 2.72 : Age hardening

As the time passes precipitate start forming in the sample. These precipitate creates an obstacle in the movement of dislocation. This increases the strength. This phenomenon is called age (precipitation) hardening. As the time passes more and more precipitates nucleate. This decreases the average distance between the precipitates and have dislocation will meet more number of obstacles. So the strength of material keeps on increasing. Since Cu in Al is limited quantity so nucleation stops after some time this refers to the peak strength. Later on smaller precipitates are comparing together and forming bigger precipitates, this increases the distance between precipitates and hence strenght of sample will decrease. This period on the curve is called overaging. Aging performed at room temperature is called natural aging and when it is performed at elevated temperature is called artificial aging. It can be seen from the figure (2.72) that as temperature increases the peak strength of material decreases and simultaneouly material becomes stronger in lesser time period. The reason is that higher the temperature more will be the diffusion and hence the rate of precipitates and rate at which smaller precipates are combining both will be accelerated.

Rapid cooling after heating presents $CuAl_2$ precipitating from the α-solid solution and this single phase solid solution will be retained at room temperature. Such a solid solution is supersaturated and therefore unstable. The atom is sufficiently mobile at the temperature of the medium in which the alloy is held, the supersaturated solid solution is spontaneously decomposes. This process of returning the supersaturated solid solution to a more stable condition occurs in aging as well. Lowering the temperature restricts aging and may completely supress it.

No decomposition of the solid solution is observed in natural Aging , since there is no precipitation of the excess phase from the supersaturated solution. What actually occurs is the collections of solute atoms of Cu in definite planes of the crystal lattice and the formation of zones rich in Cu-atoms. These zones with increased concentration of solute elements have the form of extremely thin disc's several atom layer's thick and can only be observed by X-rays analysis.

This type of change in solid solution of alloys, undergoing natural Aging is accompanied by pronounced distortion of the crystal lattice and this increases in strength,. Since the solid solution does not decompose in natural Aging and the alloy will retain sufficient ductility and corrosion resistance.

When the alloy is subject to artificial aging at 100 – 150°C and higher temperature dispersed lamellar particles of an intermediate phase which causes strengthening are precipitated at the places where there is increased concentration of the solute atoms. This intermediate phase does not appear in composition from the stable $CuAl_2$ phase but has a distorted tetragonal lattice. If the alloy is held for a long period of time at the above temperature or if the temperature is raised to 250°C or higher coagulation of the particles formed at the centres of decomposition is observed and the intermediate phase becomes stable. This is called over Aging.

The most important and extremely use alloys that are strengthened by hardening and subsequent aging are duralumin, avail and other special purpose alloys. These alloys contains 2.6-4.9% Cu, 0.25 – 1.5 % Mg, 0.5 -0.8 % Mn. The alloy 4.3% Cu, 0.6% Mg and 0.6% Mn is **normal duralumin** while the alloys containing higher percentages of Cu-Mg and Mn are high strength alloys but have lower ductility and therefore are difficult for hot and cold working. Alloys with lower percentages of Cu-Mg and Mn belongs to low alloys class of duralumin and are used for rivets. These alloys have relatively low strength but high ductility and malleability which help in riveting operation. The mechanical properties of duralumin are sharply increased after heat treatment. A necessary component of all duralumin alloy is Mn which impoves corrosion resistance.

Fe to some extent and Si are impurities of duralumin alloys. Fe is a harmful impurity and by forming the compound (Mn, Fe)Al_6, Fe considerably reduces the strength and ductility. Another Fe compound comprising both Cu and Al reduces the strengthening effect of aging. Therefore Fe-content should never exceed 0.5 – 0.6%. To a certain extent silicon eliminates the harmful effects of Fe by fixing it in the α-phase which can more easily be broken up while working. If the Fe content is low Si has a bad effect on natural aging.

Duralumin is available at sheet, strip, rod, shapes and wires. Sheet duralumin is usually cladded to protect it against corrosion. Cladding consists in coating with pure Al (99.5%) cladded Al is obtained by placing the duralumin ingot between two sheets of pure Al and rolling this pack or sandwich together this pack is hot rolled and reduce to a thickness of 6 mm and then cold rolled with intermediate process annealing operations.

Duralumin and similar other alloys may be comparatively easily hot and cold rolled, forged, formed and have good machinability. Intermediate annealing operations have 350°C - 370°C are required in cold rolling.

The heat treatment for strengthening duralumin consist of :

1. Heating followed by water quenching
2. Natural aging
3. Recovering i.e. heating for a short time (20-40° s) for 250-270°C after natural aging to restore the properties.
4. Strain hardening or work hardening

Avial : It contains 0.4-0.9% Mg, 0.5-1.2% Si and 0.2-0.6% Cu and it is heat treated. The heat treatment consist of heating to 530-540°C quenching in water and subsequent natural aging or artificial aging at 150-160°C for 12-16 hrs. Aging must be conducted directly after quenching otherwise the strengthening effect is reduced. Various semifinished products such as sheets, pipes, shapes etc are made form avail.

High strength Al-alloys : It contains 6% Zn 2.3% Mg 1.7% Cu and 0.4% Mn. After hardening by heating to 465 – 475°C and water quenching and subsequent artificial aging for 24 hrs at 120- 125°C. This alloy develops high strength and ductility. It has low machinability which may be improved by hardening and aging. It is comparatively low corrosion resistant.

Al-forging alloys : These alloys have different compositions ranging from 1.8-4.8% Cu, 0.4-1.8 % Mg, 0.4-1% Mn, 1-2.3% Mi, 0.5 - 1.6% Fe and 0.5-1.2 % Si. These alloys are used to manufacture machine parts as pistons, screw blades, crank cases, rings corers, discs etc. Forging of these alloys are subjected to heat treatment which consists in heating to 510-520°C, quenching in water and aging at 150-160°C for 12-18 hrs. Most Al alloys are forged at temperature 470° Cand 430°C

2.32.3 Al – base Casted Alloys

The following alloys are generally used :

1. **Alloys based on Al-Si system** called **silumin alloys** which contain 8 – 13% Si, 0.17 – 3% Mg and 0.25 – 0.5% Mn. A typical silumin is the eutectic alloy which is distinguished for its high castability (high fluidity and low shrinkage), good corrosion resistance, high dultility and low specific gravity (2.6 gram/cm^3). The alloy is non heat treatable and is strengthened by adding about 0.01% Na to melt just before casting. Silumin is modified by adding 2% of a mixture of 2/3 NaF and 1/3 NaCl . Other modifying mixture like 60% Na, 25% NaCl, 15% $NaAlF_6$, degasifies the alloy and removes oxide.

   A disadvantage of silumin as a casting alloy is the porosity of the casting and it can be avoided by casting under pressure. The addition of Mg and Cu enables silumin alloys to be heat treated. Such alloys posses excellent casting properties but has a tendency to foam and porosity. The alloy has strength by modifying and subsequent heat treatment. The alloy can be easily machined and has a high corrosion resistance. It is used for large castings which are used to operate under heavy duty condition.

2. **Alloys bas4ed on the Al-Mg System** which contain 9.5 – 11.5% Mg and have comparatively

poor castability. Special foundry procedures like application of special fluxes addition to the molding sand and degasification of the melt are required in casting. Due to its high mechanical properties good corrosion resistance and low weight, It is employed for castings of relatively simple form subject to dynamic loads in operation and expose to corrosive environment. This alloy is used only after heat treatment.

3. **Alloys based on Al-Cu system** which contains either 4-5% Cu or 7-14% Cu. The former posseses comparatively low castability but develops good mechanical properties after heat treatment. The latter is used without heat treatment. It has much better casting properties and a high strength which is retain up to 250-300°C but is less ductile.
4. **Alloys based on Al-Si-Cu system** which contains either 4.5-5.5% Si, 1-1.5% Cu, 0.35-0.6% Mg or 4.5-6% Cu and 2-3% Cu. These alloys have good foundary characteristics. The latter of the two is used for sand mold permanent mold and die castings. Dense air tight castings are obtained which may be welded if required. The former heat treatable alloy is extensively used becaue its casting properties are good and it gives excellent service for large heavily loaded machine parts and parts used at elevated temperatures.
5. **Zn – silumin alloys** which contains 6-9% Si and 10-14% Zn have excellent castability and can be modified to improve mechanical properties. These are used for castings of complex form requiring high mechanical properties. A disadvantage is its increased specific weight (2.93 gm/cc).

2.33 BRASSES

The alloys of Copper and Zinc are generally called brasses. The equilibrium diagram of Copper and Zinc is shown in figure.

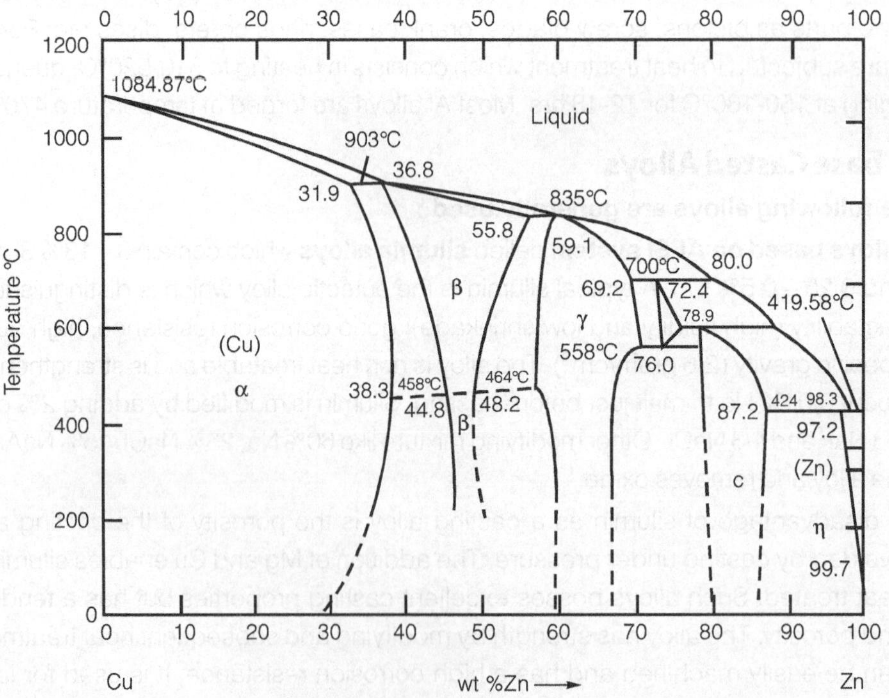

Fig. 2.73 : Cu-Zn phase diagram

As shown in Fig. (2.73) Cu-containing up to 39% Zn has a single phase structure consisting of crystals of the solid solution of Zn in Cu(α – phase). A new structural component i.e. β phase appears in alloys containing from 39-50% Zn. β-phase is an electron compound of BCC structure. The β-phase has a disordered distribution of atoms at high temperature. Upon lowering the temperature to 453-470°C, the Cu and Zn atoms in the solid solution acquire an ordered structure. Which is designated as β'. It is clear from Fig. (2.73) that the β phase has a wide range of homogeneity at high temperatures. It decomposes as the temperature is lowered and crystals of the excess α and γ phases begin to precipitate from the solution. The γ-phase formed in high Zn alloys is a very brittle compound, has a complex cubic lattice and therefore brasses intended for room temperature service consist completely of the α-phase (α brasses or of the mixture ($\alpha + \beta$ brasses)).

Brass acquires maximum relative elongation at a Zn content from 30-32% which approximately coincides with the boundary between the α-phase and ($\alpha + \beta'$) phase regions. The α phase is the most ductile structure where as the β' phase appears in the structure as hard phase. The relative elongation is reduced and reaches very low values when the zinc content corresponds to the point at which the β' phase structure is formed. Upon increasing the zinc content 45 to 47% a tensile strength first increase and then rapidly decreases.

The α-brasses have good workability at room temperature. They are less ductile at temperature from 300-700°C. Brasses are strain hardened by cold working i.e. rolling drawing etc. Their ductility may be restored by annealing at 500-700°C. In ductility the ($\alpha + \beta'$) brasses are inferior to α-brasses. For hot working the ($\alpha + \beta'$) brasses are heated to the range of homogenous β solution which posseses high ductility.

When stored for a long period of time or when exposed to a corrosion reactive agent Brass disintegrates and the failure of the metal is know as seasonal cracking. This may be eliminated by annealing the strain hardened brass at 200-300°C for secural hours. The considerable toughness makes the brass more difficult to machine. A small amount to lead is often added to improve machinability .Such alloys are called leaded or free cutting brasses. Brasses are used in the sheet or strip form for products made by deep drawing for radiators curling rings connection pipes, tubings etc. Some types are used as rolled or extruded bar stock for making machine parts such as bushings, nuts, rings, ecps etc.

In addition to plain brasses i.e. Cu and Zn alloys of **muntz metal, (Cu-Zn-Pb),** special brasses find wide applications in industries. There are complex alloys that differ from ordinary brasses in structure and properties. This is due to the addition of various alloying elements like Sn, Si, Mn, Al and Fe. These are used for castings and tubings for ship building and general engineering casting for heavily loaded components like large worms, large critical castings in ship buildings, load carrying components for aircraft automobiles and ship building, bearings, bushings etc.

2.34 BRONZES

Alloys of Cu with all other elements except Zn are called bronzes. Zn may be a constituent but only together with other components and in relatively small amount. Cu_6sn_5

2.34.1 Tin-bronzes

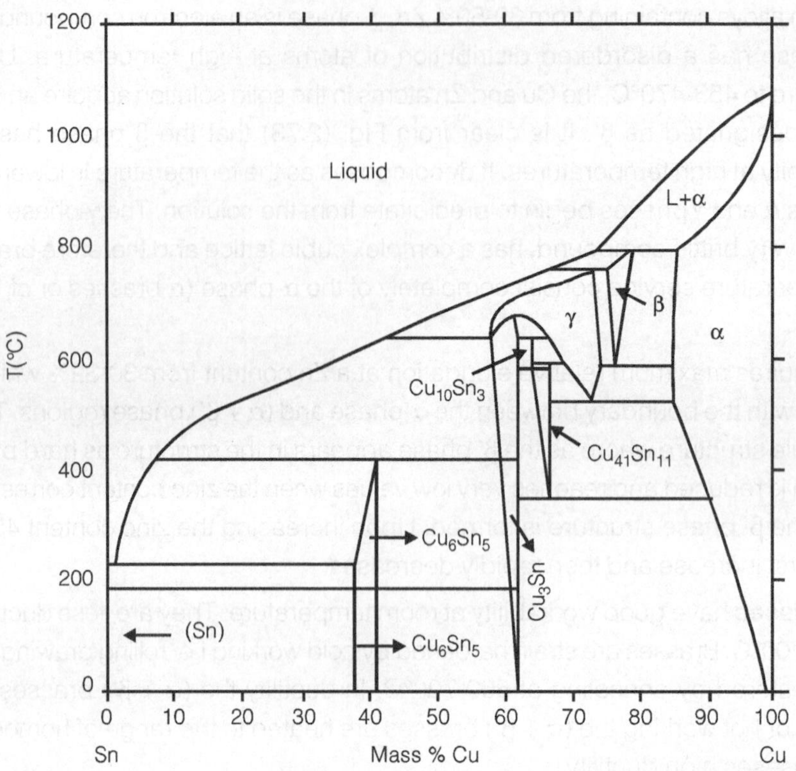

Fig. 2.74 : Cu-Sn phase diagram

The Cu-Sn equilibrium diagram is shown in figure (2.74) where the α-phase is a solid solution of Sn in Cu. The diagram shows that there are several peritectic transformation and two eutectoid transformation in the solid state at 587°C. The β-phase under goes a eutectoid decomposition and form a mixture of α and γ phases while at 520°C the γ-phase crystals de-compose into the α-phase and $Cu_{41}Sn_{11}$.

At a temperature of about 350°C the electron compound $Cu_{41}Sn_{11}$ decomposes into the α-phase and the compound Cu_3Sn. This transformation occurs only in a state of complete equilibrium under actual cooling conditions the structure of bronze consist of the phase α and Cu_3Sn. Only the right hand side of the diagram is a practical importance which corresponds to the formation of the α-solid solution of Sn in Cu. The tensile strength of bronze increases until the tin content reaches 20%. It drops sharply upon further increase in tin content due to the presence of a considerable quantity of eutectoid which contains the brittle compound $Cu_{31}Sn_8$. The relative ductility increases upto a Sn content of 5-6% and then it drops sharply as the eutectoid begins to appear in the structure consequently only bronzes containing upto 5-6% Sn are workable.

Sn-bronzes have good capability and are widely used in the foundry. A special feature of bronzes is anti friction property. Zn may be added to Sn bronzes to allow the Sn-content to be reduced which in turn reduces the cost and improve castability.

Machinability of bronze may be improve by adding lead (Pb). Phosphorus acts as a de-oxidizer when added to bronze, as well as an element for strengthening the alloy. An addition of about 1% P improves castability and increases the mechanical and antifriction property. Wrought tin bronzes undergo heat treatment consisting of annealing at 600-650°C as an intermediate operation in cold working or as a final operation to provide the required properties. In the semifinished products like sheet and strip, a large quantity of expressive Sn are replaced by cheaper alloys. Low Sn bronzes to which Zn, Pb, Ni and other element have been added and is used for fittings and for antifriction components.

2.34.2 Tin Free Bronzes

As Sn is very expensive, various grades of Sn free bronzes have found extensive applications in industry along with tin bronzes. Most tin free bronzes have higher mechanical properties than tin bronzes, better corrosion resistance excellent antifriction properties etc. Their machinability and castability are however interior to those of tin bronzes.

Al bronzes are extensively used. The effect of Al-addition on the machinability properties of Cu is quite substantial. Bronzes containing upto 6-8% Al can be successfully worked either hot or cold. The mechanical properties of these bronzes are considerably improved by stress hardening. Bronzes containing 10% Al can only be worked at high temperatures but are easily casted and are much used as foundry materials. All Al bronzes have good corrosion resistance. An addition of Fe to these materials refine their structure and substantially. Thereby improves their mechanical properties. A better combination of mechanical properties is obtained by normalizing or hardening which consist in heating to 650°C followed by water quenching. The high mechanical properties developed at room temperature by the addition of Ni are retained at elevated temperature as well. A desirable addition of bronze in Mn, improves both mechanical and fabricating properties.

2.34.3 Si-Bronzes

In certain application e.g. for spring type components tin bronzes may be replaced by cheaper Si-bronzes alloyed in Mn which also have high mechanical properties.

2.34.4 Be-Bronzes

They develop exceptionally high mechanical properties after heat treatment. After hardening by heating to 800°C and water quenching these alloys acquire high tensile strength and relative elongation. A subsequent tempering or age hardening at 350°C for 1-2 hrs raises the tensile strength to almost 3 times, though the relative elongation falls to quite low values. Be-bronze is an excellent material for springs due to its high elasticity and resistance to fatigue and corrosion. At the same time it has a high hardness and wear resistance. A disadvantage is its high cost.

2.34.5 Leaded Bronzes

It is applied for the bearing lines for machinary operating at high speeds. In the solid state Pb and Cu are almost insoluble in each other. The structure of these bronzes consist of crystals of Cu and soft Pb. The excellent antifriction property are due to this structure. Leaded bronzes give satisfactory service as bearings subject to high specific pressures and speeds since their fatigue limits reach very high values and their resistance to impact loads exceeds that of other bearing materials. A disadvantage of leaded bronzes is their tendency to segregate. To avoid this special casting procedures such as additions of special elements are required.

A Cu base alloy with nickel called cupronickel is used in industry as a corrosion resistant material at elevated temperatures and can also be used in surgical equipments, kitchen utensils coins and various ornaments. Cupro-nickel with 30% Ni is widely used for navel condenser tubes due to its good resistance against corrosion by sea water. A Cu alloy containing 40% Ni and 1.5% Mn called constantar is widely used for rheostats, thermocouples and heating devices operating at moderately high temperatures.

2.35 INTRODUCTION TO TITANIUM ALLOYS

Sheet metal forming is a general term, for a large group, that includes a wide variety of manufacturing processes. Sheet metal forming processes are characteristic in that the metal being processed is plastically deformed in order to shape it into a desired geometry. They are commonly used in the industry because it eradicates costly operations such as welding, machining and manufactures parts with reduced weight and good mechanical properties with high production rates. Specifically, in automotive and aerospace industries, sheet metal forming processes are extensively used. Sheet metal forming processes became a backbone of modern manufacturing industry. Moreover, many sheet metal components are produced using a deep drawing and stretching process. It is one of the common and popular process in assessment of formability of sheet metal.

Sheet metal forming technologies are constantly challenged by the improvements in the automotive industry in the last several decades. Due to increasing customer expectations, safety requirements and market competitions, there is a strong need for products, which can be manufactured more successfully, more economically and rapidly to satisfy ever-increasing market needs. On the other hand, in recent years, environmental and safety concerns have forced the industry to choose lighter-yet-safer materials for aerospace and automotive production. Therefore, 'alternative' materials are being studied to replace conventional steel materials to reduce weight in many parts. For this purpose, various light weight alloys such as titanium alloy, magnesium alloy, aluminum alloys sheet metal forming gain a special attentions.

Nowadays, titanium alloy are extensively used in aerospace and automotive sectors due to its attractive combination of characteristics in terms of high mechanical properties, elevated corrosion resistance and low density. Presently, it has been extensively used in aerospace, automotive sectors for many sheet metal applications. Despite the obvious advantages of the high strength to weight ratio and corrosion resistance of titanium alloys, they have a distinguishable downside in that their formability is considerably lower than traditional steel alloys at room temperature conditions. It is usually caused by the high alloy percentages that are required for high strength.

Generally, titanium alloys are considered more difficult to form and often have less predictable forming characteristics than other metallic alloys such as steel and aluminum. The literature states that titanium alloys can be formed in room temperature or at lower elevated temperatures to some extent. A high degree of spring back is common in cold and hot forming processes due to the high yield stress in combination with a low elastic modulus. Further, the hexagonal crystal structure (HCP) of the α-phase has anisotropic characteristics which implying consequences to the elastic properties.

Considering this, titanium components are commonly formed at elevated temperatures. Formability of titanium alloys can be greatly improved by warm forming. Since elevated temperature results in decreased flow stress and increased ductility in the sheet, it allows deeper drawing and more stretching to form products. Therefore, by increasing the forming temperature an increased formability is obtained in which the spring back and the scatter in yield stress can be reduced. However, such techniques require heat resistant forming tools and the commonly slow forming velocities with subsequent holding times implying long exposure times at high temperatures and additional costs. As the temperature exposure increases a higher degree of contamination occurs. Particularly for Ti-6Al-4V alloy, Hot forming should generally not be performed at temperatures higher than ~400°C without a protective atmosphere to avoid deterioration of the mechanical properties. When oxygen enrichment of titanium occurs, the material becomes more brittle. Ti-6Al-4V alloy is sensitive to the strain rate and a higher formability is generally obtained when forming titanium alloys at lower strain rates. Titanium alloys are also sensitive to the Bauschinger effect, which is reported to be most pronounced in room temperature.

2.36 CORROSION DEFINITION

Corrosion is defined as deterioration or destruction of material due to its interaction with environment or in other words it is reverse of metallurgy. In metallurgy, metals are extracted from their ores (oxides or hydrides of the metals) and refined and then converts into slabs, tubes or sheets etc. The material then starts transforming into again oxides or hydrides due to interaction of the environment, therefore, in the material world nothing is permanent.

Practically, all the environments are corrosive in nature and all the materials including ceramics, plastics, rubber and other non-metals also corrode with varying corrosion rates depending on environment. For example, 18-8 stainless steel deteriorates rapidly, within hours, by polythionic acid. The rail tracks even though looks always rusty but their performance is many years. The Delhi Pillar in India was made almost 2000 years back bust still now, looks new. In essence, corrosion of a material depends entirely on the environment in which it is situated. In general, the organic materials are more corrosive resistive than the inorganic materials. For example, in petroleum industries, the corrosion of components is more due to Sulfur, NaCl, H_2SO_4 and HCl than the oil, naphtha or gasoline.

Corrosion rate generally expressed in terms of loss of weight of material but it does not give idea about penetration depth. The most prominent expression for corrosion rate is *corrosion penetration rate (CPR)* and it may be expressed as 'mils per year (mpy)'or millimetres per year (mm/yr).

$$\text{Corrosion penetration rate (CPR)} = \frac{KW}{\rho A t} \quad \text{...(i)}$$

Where, 'W' is loss of weight, ρ is density, A is exposed specimen area, t is exposure time. K is a constant which depends on unit system with which CPR is expressed, see Table.

Note : 1 mil = 0.001 inch.

Table : *CPR Unit systems*

CPR unit systems	K	W	ρ	A	t
Mils per year	534	mg	g/cm^3	inch^2	hr
Millimetres per year	87.6	mg	g/cm^3	cm^2	hr

The corrosion rate, r, in terms of degradation of surface area without touching the penetration depth (mol/m^2-s) is also expressed by current (i) per unit surface area of material corroding.

$$r = \frac{i}{nF} \quad \text{...(ii)}$$

Where, i is current density, n is number of electrons involved in ionization of each metal atoms and F is 96,500 C/mol.

2.36.1 Electrochemical aspects in corrosion

Electrochemical reactions in corrosion

For the purpose of explanation, let us say, Zn is dipped in HCl, then,

$$Zn + 2\,HCl \longrightarrow ZnCl + H_2 \quad \text{...(iii)}$$

Noting that, the chlorine is not involved in reaction (chlorine has not been converted into chlorine ions), it can be written as

$$Zn + 2H^+ \longrightarrow Zn^{2+} + H_2 \quad \text{...(iv)}$$

Furthermore, Eq. (iv) is actually combination of two reactions, firstly, the Zn has given up $2e$ and dissolved in electrolyte as Zn^{2+}, secondly, the same number of electrons are absorbed by H^+ions and then it is reduced to H_2 gas (see fig.2.75). These can be written as

Oxidation (anodic) reaction:

$$Zn \longrightarrow Zn^{2+} + 2e \text{ in general } (M \rightarrow M^{n+} + ne) \quad \text{...(v)}$$

Reduction (cathodic) reaction:

$$2H^+ + 2e \longrightarrow H_2 \quad \text{...(vi)}$$

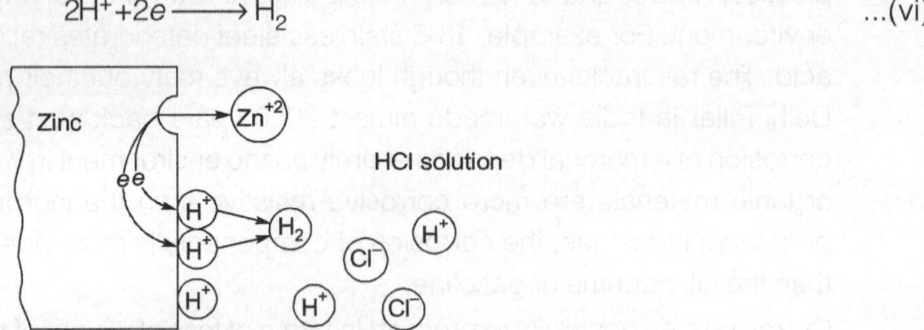

Fig. 2.75 : Electrochemical reactions involved in corrosion of Zn in HCl

In summary, the corrosion involve electrochemical reactions or in other words the reactions which can be divided into two, called, oxidation (anodic) in which electrons are released and reduction (cathodic) in which electrons are absorbed. *It is important to note that during the metallic corrosion both the reactions take place and the oxidation rate and reduction rate are equal (in terms of electrons received or released) at any point of time.*

In general (presence of air or some impurity), the corrosion may involve one anodic and more than one cathodic reaction therefore some of the important *cathodic reactions* are listed here

1. Hydrogen evolution:

$$2H^+ + 2e \longrightarrow H_2$$

2. Oxygen reduction (acidic solution):

$$O_2 + 4H^+ + 4e \longrightarrow 2H_2O \text{ (for acidic solution)}$$

$$O_2 + 2H_2O + 4e \longrightarrow 4OH^- \text{ (for basic or neutral solutions)}$$

3. Metal ion reduction:

$$M^{3+} + e \longrightarrow M^{2+}$$

4. Metal deposition:

$$M^+ + e \longrightarrow M$$

If, one of the reaction rate, either cathodic or anodic, is reduced, the entire corrosion rate will be reduced, because they are inter dependent on each other. Corrosion inhibitors (organic compounds: high molecular weight amines) are used to decrease the corrosion rate because they interrupt one of the reactions or form scale on the metals. Reducing the flow of electrons in the system (material and electrolyte) is another way of controlling the corrosion, but, the electrical conductivity of the material cannot be changed. But, fortunately, the electrical resistance of electrolyte can be increased so that corrosion will be reduced. For example, pure water results in higher corrosive resistance than impure water.

2.36.2 Polarization

The polarization term, in this particular context, refers to a key environmental factor which controls the entire corrosion rate. A complete corrosion rate, if corrosive species (in the fig1, Hydrogen) available on the surface of material, is depended on *one of the slowest steps* out of (a) electron transfer (b) reduction of the species to atoms (c) formation of molecules (d) formation of gas bobbles from the molecules. It is called activation polarization, where *slowest step* controls the corrosion rate. If the electrolyte is a dilute (i.e., in the case of dilute acids), then, a small number of reducible species (in Fig. 1, Hydrogen) available near the surface, and therefore, they are consumed in the reactions. Now, there is deficiency of the species, therefore, it is require to diffuse the required species form concentrated zone to near the material to continue the reactions. Therefore, the entire corrosion rate of the material is depended on the diffusion of reducible species from the concentration zone so it is called concentration polarization. The concentration polarization can easily be neglected because, if there is a small agitation in the electrolyte the corrosion rate increases, whereas in the case of the activation polarization, the corrosion rate does not be affected by the agitation. In a simple way we can say that, the activation polarization is related to surface of the material whereas the concentration polarization is related to the electrolyte.

2.36.3 Passivity

The loss of chemical activity of metals or alloy under certain environmental conditions is called *passivity*. Mostly, platinum and gold, in some cases the structural metals such as iron, nickel, silicon, chromium, titanium and their alloys, in very rare cases, the other materials such as cadmium, zinc, tin, uranium, and thorium exhibit passivity and these are called active-passive metals. In case of ordinary metals, the corrosion rate is increased with the increase in oxidizing power by adding oxygen or iron, but in the case of active-passive materials, though initially the corrosion rate is increased, it is drastically reduced in a certain range of potential (Fig. 2.76).

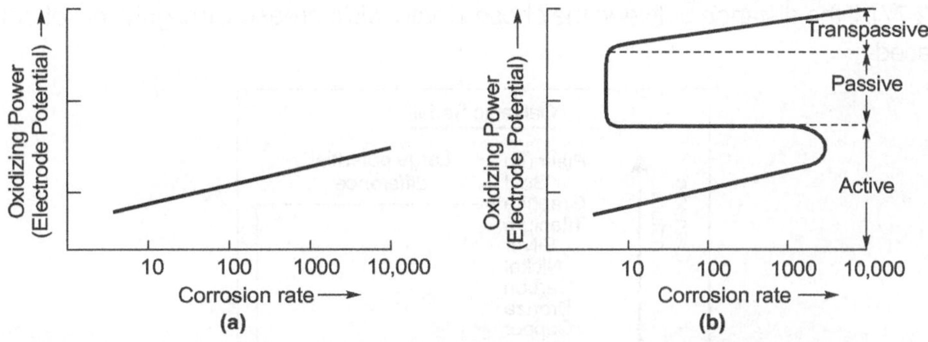

Fig. 2.76 : (a) corrosion rate in ordinary metals (b) Corrosion rate in active-passive metals

It is because of formation of a stable surface film (generally metallic oxides) over the surface in the particular oxidizing power range. Then, the film prevents oxygen to penetration into material thereby protecting the material from further degradation. Upon increasing the oxidizing power, the film will be broken, and there is again increase in corrosion rate.

2.36.4 Eight Forms of Corrosion

Although corrosion is classified in number of ways, based on temperature such as low temperature corrosion, high temperature corrosion, and also like dry corrosion and wet corrosion but most of the time based on the *appearance* of the corrosive products and surface of the material it has classified in to eight forms.

Uniform attack

It is nothing but the entire surface of the material is attacked uniformly by the electrochemical reactions and therefore the material becomes thinner and thinner. For example, Zn in dilute HCl, Steel iron roof, steel tanks etc. Though *it is the greatest loss of material, by time, in weight basis, but it is much safer than the other types of corrosions because it can be easily detected and it is easy to predict life span of the material to replace the materials.* The other types of corrosion are localized attack and difficult to detect and therefore there is risk of premature failure of machines, tool etc.., So, these are explained in detail in the following sections.

Galvanic or Two-Metal Corrosion

Naturally, when two dissimilar metals are dipped in an electrolyte (corrosive media) and they are in contact (or electrically connected), due to emf difference, electrons flow from anode to cathode through electrolyte (corrosive environment). *As a result, corrosion resistant of less corrosion resistant material (anode) decreases and corrosion resistant of the more corrosion resistant material (cathode) increases comparatively with the corrosion resistances of them when they are not in contact. This* is called *Galvanic Effect. In* essence, if two materials are in contact, one will be more corroded and other one will be more protected comparatively with their natural corrosion (when they are not in contact).

The galvanic effect can be estimated from emf series and galvanic series of the materials. The 'standard emf series' gives emf values of metals, but, it does not give any information regarding that of alloys, therefore, 'galvanic *series'* is generally preferred to estimate the corrosive nature. The 'galvanic series' is a list of metals and alloys arranged from their cathodic nature to anodic nature (without the emf values). Therefore, if we select two materials from the list, which are in close to each other in the list, there will be less galvanic effect. On the other hand, if the selected materials are apart from each other, then the galvanic effect will be more. In summary, in the list (Fig. 2.77), if the distance between the chosen materials increases, the galvanic effect will also be increased.

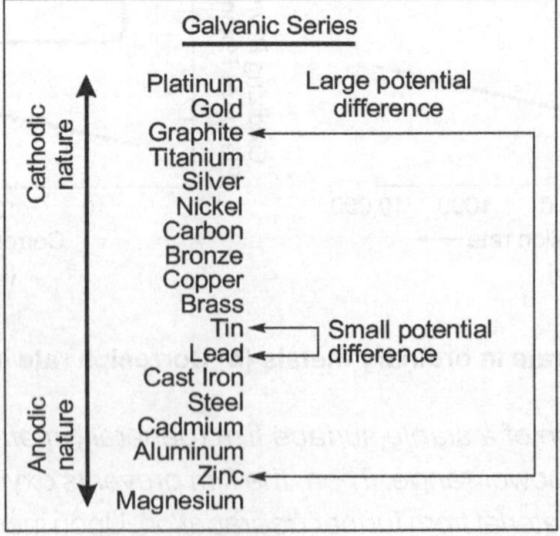

Fig. 2.77 : Application of Galvanic series

Area effect: When the two materials are in contact in such way that lesser area of anode is exposed to environment compared to that of cathode then the anode corrodes rigorously because there will be more electron density at anode surface.

In one hand, the galvanic effect principle is effectively used for protecting required metal by other anodic sacrificial metal coating, on the other hand, there is a risk of catastrophic failure if the combination of the two metals is reversed. For example (see fig. 2.78), if steel is coated by zinc, steel is cathode and zinc is anode according to their relative position in the galvanic series. Then, even if there is a hole in the zinc coating, the exposed steel will be protected due to the galvanic effect. Conversely, if zinc is coated by steel then there is a rigorous localized degradation of anodic zinc due to high electron density.

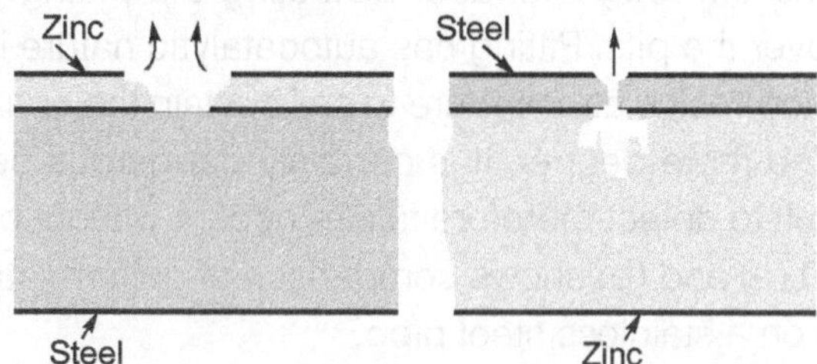

Fig. 2.78 : Application and misapplication of galvanic effect of two metals Zinc and Steel

Crevice Corrosion

It is highly localized corrosion generally appears in shielded areas like gasket surfaces, lap joints, under bolts and rivet heads where small amount of stagnant solutions (like dust), surface deposits, are settled, then, these are the means to create an acidic atmosphere to automatically propagate the corrosion into thickness. Furthermore, the resulted corrosive productives are prone to improve the acidic nature more and more. It is also called as deposit or gasket carrsion (See fig. 2.79).

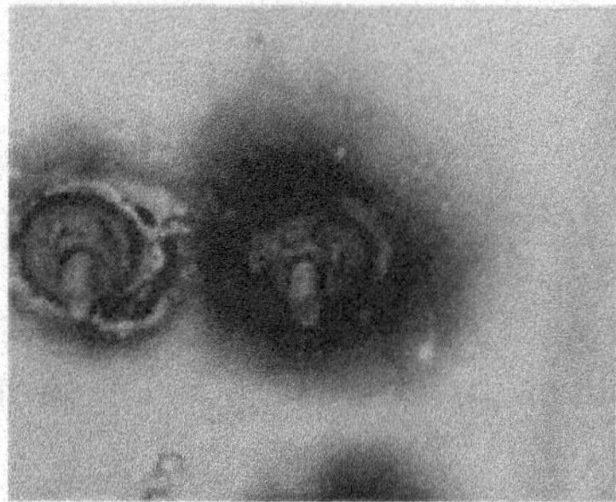

Fig. 2.79 : Crevice corrosion under the nuts due to the gaskets

Prevention:

1. Use but joint instead of lap-joints.
2. Clean thoroughly to remove stagnant.
3. Avoid sharp corners while designing.
4. Use non-absorbable solid gaskets such as Teflon

Filliform Corrosion: Filliform corrosion is a special type of crevice corrosion where the corrosion take place under a thin film. It appears on steel, magnesium, and aluminium surfaces covered by tin, silver, gold, phosphate, enamel, and lacquer coatings and it also observed at the paper-aluminium interface. It moves in between film and metal surfaces like a worm with blue-green head and red-brown tail. It is not a dangerous kind but the appearance of the surface look very dirty, mainly, on the beverage cans and other food packing therefore the market sales value goes down.

There is no method to prevent the filliform corrosion. It is better to store the coated materials in less humid atmosphere, but still, it is not for long time.

Pitting

It is extremely localized corrosion, results in small holes grouped very close, sometimes, the very close small holes looks like rough surface. Detecting the pitting is also difficult because the corrosive products cover the pits. Pitting has autocatalytic nature in other words, the corrosive products create more corrosive atmosphere to self sustain the corrosive reactions and then the pits becomes more and more deeper. It is certainly dangerous because it is localized and in some cases it is difficult to detect therefore failure occurs without prior indications with extreme suddenness. Fig. 2.80 (a) and (b) shows comparison of uniform attack and pitting and fig. 5(c) displays sever pitting on a stainless steel pipe.

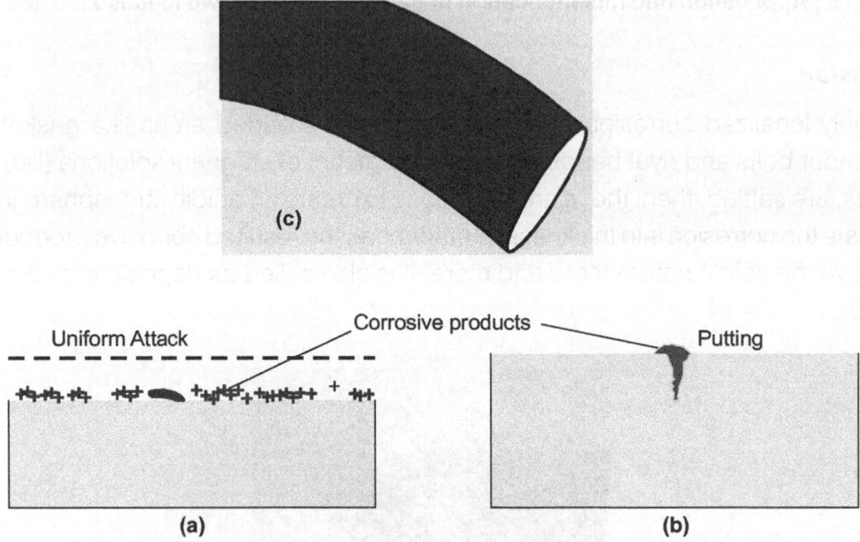

Fig. 2.80 : (a) uniform attack (b) Pitting (c) a stainless steel attacked by pitting

Prevention:

1. Don't use the materials which show pitting nature in building big plants or equipments.
2. Some material show high pitting resistance than others such as stainless steel -316 has high pitting resistance than that of stainless steel-304 due to addition of 2% molybdenum.
3. In general the pitting resistance: Stainless steel 304 < Stainless steel 316 < Hastelloy F < Nione or Durimet 20 < Hastelloy C or Chlorimet 3 < Tianium.

Intergranular Corrosion

In materials, deviation in the proportion of one of the alloying element causes intergranular corrosion and it generally happens at grain boundary areas (active regions) at high temperature due to segregation or enrichment of the alloying element. For example, Cr(>10%) improves corrosion resistant in stainless steels, but if it is heated to sensitizing temperature (950 -1450°F) (susceptible to corrosion), it will be separated out from the solid solution as $Cr_{23}C_6$ compound at grain boundary area. Further, the depletion of Cr at grain boundary area results in less corrosion resistance therefore the material at grain boundary area is corroded and eventually failure occurs catastrophically in brittle manner as shown in fig 2.81.

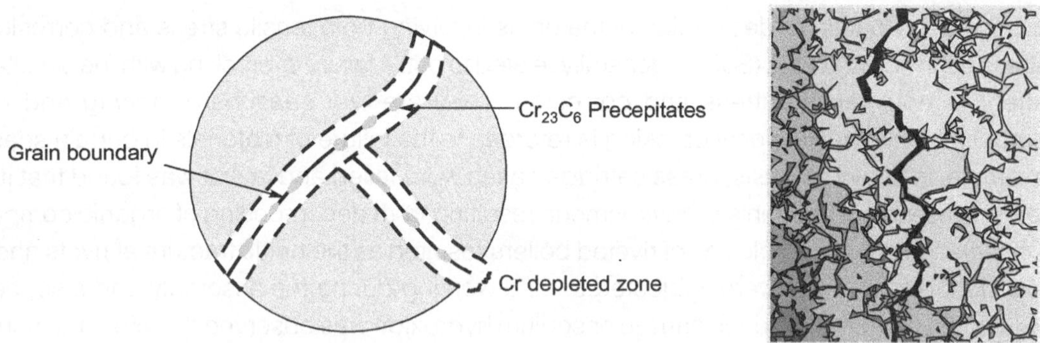

Fig. 2.81 : Inter granular corrosion at Cr depleted zones

Selective Leaching

It is nothing but loss of one chemical element from an alloy. For example, dezincification, where the Zn element is depleted from yellow brass (30% Zn & 70% Cu) consequently, it will be turned into red. It is because, Zn is very active metal compared to the Cu (noble metal) therefore Zn will be oxidized. Initially, there is a reduction of H_2O in to H_2 as and hydroxide ions without oxygen, but later on, the rate increases with environmental oxygen. Similarly, in some alloy systems, aluminium, cobalt, iron, chromium and other elements are removed and they are called as dealuminumification, decobaltification, grafitization etc.

Prevention:

1. Reducing the aggressiveness of the environment, that is, reducing oxygen content, but it is very economical method therefore less susceptible alloy is selected (like red brass with 15% Zn than the yellow brass). Moreover, in the case of sever selective leaching environments, the cupronickels are used.
2. Adding "inhibitors". For example, 1% Sn is added to the yellow brass to improve the corrosion resistance and the resulted alloy is called *Admirality Metal.*

Erosion Corrosion

It is nothing but mechanical wear or abrasion of material when it is in relative motion with corrosive fluids. This is due to development of surface films (passivity) and then damaged by the fluids due to high velocity and abrasive nature of fluids. For example Aluminium, lead, and stainless steels are susceptible to this by hot gases, aqueous solutions, organic systems and liquid metals etc.

Prevention or control:

1. *Changing design:* Improving thickness of pipes and increased diameter to make sure velocity should be less and laminar flow.

2. *Alteration of environment:* Settling and filtering the fluids and very important is to make sure temperature should not increase.
3. *Cathodic protection:* Even though it works very well but is not popular in industries to prevent erosion corrosion. One steel plant uses steel plate to cover heat exchanger head to protect inner ends of tube from hot seawater.

Stress corrosion

It is the degradation or destruction of materials involving both tensile stress and corrosion. The stress corrosion cracking (SSC) is generally related to brittle failure or cracking with the simultaneous effect of both tensile stress and corrosion. They are two *seasonal cracking* and *caustic embrittlement.* The seasonal cracking is referees to the failure of materials in certain season for example, in heavy rain falls, brass catridge cases were failed and later it was found that it is due to the presence of ammonia in environment, resulting from decomposing of organic compounds. Observation of many explosion of riveted boilers resulted as the brittle fracture at rivets and it was found that those rivets were subjected to cold working during the assembly and also, near the failure, the corrosive product, caustic or sodium hydroxide was observed therefore the name was given to this failure as caustic embrittlement.

Though nobody knows exact mechanism of the corrosion, still, it is proposed that, a pit, on the surface of material act as stress riser, then once crack starts, there is a formation surface films(with less strength) on newly explosive material and then once the sufficient stress is raised at the notch, the film brakes and in this way the crack continues.

Prevention:

1. Lowering the stress
2. Eliminating the critical environmental species
3. Changing the alloy if possible
4. Applying cathodic protection with an external power supply
5. *Adding inhibitors:* Phosphates are generally useful inhibitors to reduce stress corrosion cracking
6. Coatings
7. Shot-peening: for example, 410 stainless steel show high stress corrosion cracking resistance by peening with glass beads at room temperature.

2.36.5 Corrosion Prevention

Materials selections

Most important method in preventing corrosion is selection of materials, most of the people belief that stainless steel is stainless, but it is not. In fact stainless steel is not one alloy, stainless steels refer to 30 different alloys containing 11.5 -30% Cr to 0-22% nickel. In some environments such as chloride-containing mediums and stressed structures these are more susceptible than ordinary steel. Localized corrosions such as pitting, crevice attack more on stainless steel than ordinary ones. Therefore, stainless steels are not possible to use in all circumstances. The general combination of alloys and medium is given below for better resistance

1. Stainless steels - nitric acid
2. Nickel and nickel alloys – caustic

3. Monel – hydrofluoric acid
4. Hastelloys (Chlorimets) – hot HCl
5. Tin – distilled water
6. Lead- dilute sulphuric acid
7. Al – nonstaining atmospheric exposure
8. Titalnium-hot strong oxidizing solutions
9. Tantalum-ultimate resistance
10. Steel- concentrated sulphuric acid

Alteration of Environment

Altering the corrosive environment involves either changing the medium or reducing the aggressiveness of the medium by adding inhibitors. The changes in medium are often considered as lowering temperature, decreasing velocity, removing oxygen or oxidizers and changing concentration. The inhibitors are retarding catalysts and often these are added to corrosive medium to reduce aggressiveness, in fact, nobody knows how inhibitors works and they are being find by trial and error method. Though the working mechanism is not know, they are classified as

1. *Adsorption inhibitors*, organic compounds, which adsorb metal surface and decrease the metal dissolution and reduction reactions (example: organic amines)
2. *Hydrogen-evolution poisons* are arsenic and antimony ions which reduce hydrogen evolution reaction therefore these are very effective in acidic solutions
3. *Scavengers* are those which remove corrosive reagents from solution. For instance, sodium sulphite and hydrazine remove dissolved oxygen from water solutions.
4. *Oxidizers* such as chromate, nitrate and ferric salts activate passive region in active-passive transitions.
5. *Vapor-phase inhibitors* are like adsorption type but these have high vapour pressure and these are placed in the vicinity of the components without in contact and due to sublimation and condensation it occupies the environment near to the component and then protects the component.

Design

The most important design rule for combating disasters related to corrosion is wall thickness. Some allowance must be provide for corrosion damage while manufacturing tanks, pipe lines etc. Some of the important design rules are given below:

1. As riveted joints are sites for crevice corrosion, welding is preferred more than riveting tanks.
2. Tank bottoms must be designed in some inclination towards hoes for easy draining otherwise the left out liquids results in sever corrosion. For example, although concentrated sulphuric acid does not corrode a steel but if it is left out in small quantity in steel tank, it absorb moisture from atmosphere and become diluted and as a result it corrode the steel severely.
3. Excessive mechanical stresses or residual stresses are avoided to reduce stress corrosion cracking.
4. Eliminate electrical contact between dissimilar metals to prevent galvanic corrosion by using similar materials or using insulate materials.
5. Avoid sharp bends in piping systems when high velocities are abrasion materials are involved to prevent erosion corrosion.

Cathodic Protection

In essence, the electrochemical reactions involved in corrosion are metal dissolution and hydrogen evolution as shown below:

$$M \longrightarrow M^{n+} + ne \quad \text{...(vii)}$$

$$2H^{+} + 2e \longrightarrow H_2 \quad \text{...(viii)}$$

Therefore, if the electrons are released from metal, into electrolyte, the metal corrodes rigorously. Conversely, as long as electrons are supplied to the metal it is protected. Hence, this principle, *supplying electrons to metal through electrolyte is called cathodic protection.* It can be accomplished in two ways (1) by external power supply (2) by galvanic effect.

The accomplishment of cathodic protection of a tank in soil by external DC supply with an inert anode is depicted in the given fig. Negative terminal of the supply is connected to the buried tank whereas that of positive is connected to an anode which is surrounded by backfill consisting of coke breeze, gypsum, or bentonite to improve electric contact between soil and anode. Then, the electrons travel through soil (electrolyte) to tank therefore tank is protected.

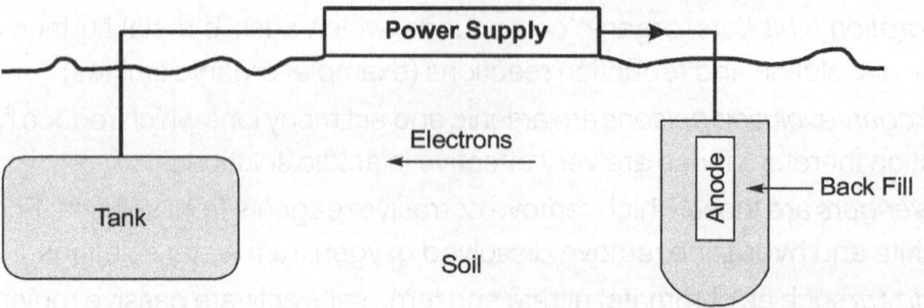

Fig. 2.82 : Cathodic protection by an external power supply

The cathodic protection by galvanic effect of steel and zinc is illustrated in the fig. The galvanic effect involves the increase of the rate of corrosion of anode, zinc, and protection of steel component.

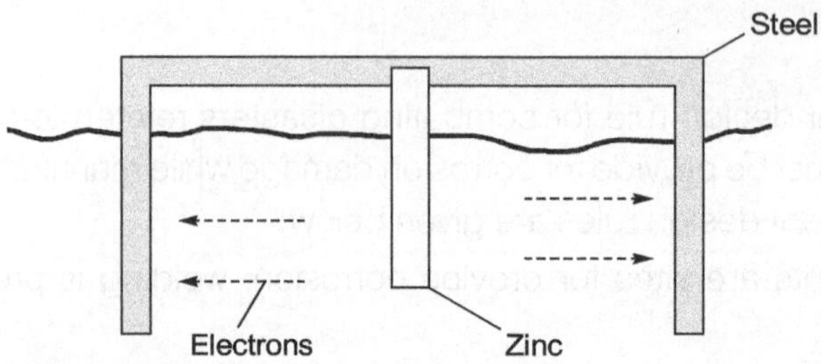

Fig. 2.83 : Cathodic protection by galvanic effect

Anodic Protection

According to the electrochemical equations (vi) and (vii), one can conclude that anodic current (leaving electrons from metal into electrolyte) degrade the material but it is not fact for all the materials and also it depends on the kinetics of the reaction. In the case of active-passive transition metals such as nickel, iron, chromium, titanium and their alloys, if carefully controlled anodic current is supplied corrosion rate decrease drastically by forming surface films on them. It is called anodic protection which is entirely depended on formation of surface films which protect the material from further degradation. An interesting application is the use of 316 stainless steel tubes to cool concentrated sulphuric acid by protecting them by anodic protection. To maintain required emf difference an electrical instrument *potentiostatis* used.

Coatings

Apart from the use of sacrificial coatings such as zinc, the metallic and other inorganic coatings serve very active barriers. The coatings broadly classified as metallic or inorganic and organic. Metallic coatings can be applied very easily and they are quite deformable, further, there is a risk of localized corrosion due two metal contact whereas the inorganic coatings are to be applied by diffusion or chemical conversion and these are brittle. Besides, the organic coatings are very easy to apply and protect more than the metallic coatings on a tonnage basis.

Metal coats-examples: automobile bumpers and trim, silverware, galvanized steel, tin cans.

Ceramic coats-examples: Bathtubs, glassed steel tanks

Diffusion coatings: anodized aluminium, chromized steel.

Organic coatings: Paints, Varnishes, lacquers.

2.37 COPPER

Table : Important properties and applications of copper

Sl. No.	Properties	Applications
1.	High electirical conductivity	The electric tough-pitch copper (ETP) or oxygen free high conductivity copper(OFHC) has 99.9% Cu and it is used for electric wires
2.	High thermal conductivity	Radiator elements, Domestic boilers, cooking utensils
3.	Corrosive resistance	Arsenic copper (0.3% Arsenic) used in condenser and heat exchangers
4.	Machinability	Free cutting copper (0.6% Tellurium) used for bolts, studs, weldtips and contact pins

Note: The oxygen free copper has high strength. The copper, initially, reacts with the oxygen and forms copper oxide. The copper oxide and copper forms a eutectic mixture and during work-hardening, the eutectic mixture is broken and is distributed like small hard particles in the copper matrix, therefore its strength increases.

Copper alloys:

I. Brass-alloys of copper and zinc
   - *A.* α–brasses: alloys containing up to 36% Zn
     1. Red brasses (5-20% Zn)
     2. Yellow brasses (20-36% Zn)
   - *B.* $\alpha+\beta'$-brasses: (54-62% Cu)

II. Bronzes- alloys containing primarily up to 12% of alloying element
   - A. Tin bronzes
   - B. Aluminium bronzes
   - C. Beryllium bronzes
   - D. Silicon bronzes

III. Cupronickels- alloys of copper and nickel

IV. Nickel silvers- alloys of Cu, Nickel and Zn

(I) ***Brasses***

The alloy system Cu-Zn contains mainly and phases at room temperature and it is shown in phase diagram (fig 1). The maximum solubility of Zn in solid solution is 36% and it has FCC structure. The phase is an electron compound with BCC crystal structure and indicates the ordered phase of the in such a way that Cu is at corner and Zn is at centre of unit cell.

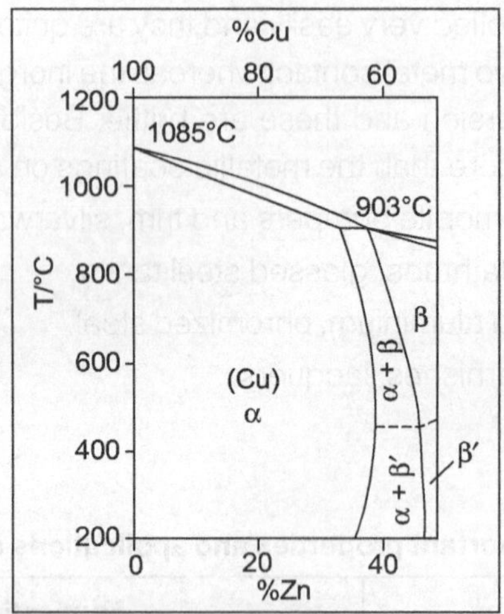

Fig. 2.84 : A portion of Cu-Zn alloy stem shown as phase diagram

A. **α-brasses:** These contain upto 36% Zn. These have relatively good working properties. The colour of the -brasses is changed with the proportion of Zn, in such a way that, the increase of Zn percentage will changes the colour from red to yellow. Therefore, these are divided as red brasses which contain maximum up to 20% of Zn, and yellow brasses, which contain 20-36% of Zn. Comparatively the red brasses are more corrosive resistance than that of yellow brasses, because, more copper content available in the red brasses. Surprisingly, the increase of Zn in Cu increases not only ductility of the alloy but also its strength, therefore, yellow brasses are more stronger than the red ones. Inspite of the yellow brasses possess good strength, they are susceptible to seasonal or stress corrosion cracking. Therefore, they are annealed after cold-working to avoid the seasonal cracking. The yellow brasses are also susceptible to pitting and dezincification which leads to leakage of pipes, and to avoid it, small amounts of Sn or antimony is added to the alloy. Some of the important -brasses are given in table 2 and 3.

Table : Important red brasses and their applications

Red Brasses		
Sl. No.	**Alloy**	**Application**
1.	Gliding metal (95%Cu-5%Zn) contains high strength	Coins, medals, tokens, base for articles to be gold plated
2.	Commercial Bronze (90% Cu-10% Zn) contains excellent cold and hot working properties	Costume jewellery, forgings and rivets
3.	Red brass (85%Cu-15%Zn)	Electrical conduit, screw shells, sockets, hardware, condenser and heat exchanger tubes
4.	Low brass (80%Cu-20%Zn)	Ornamental work, musical instruments, deep drawn articles.

Table :Important yellow brasses and their applications

Yellow Brasses		
Sl. No.	**Alloy**	**Application**
1.	Catridge brass (70%Cu-30%Zn)	Radiators, flash light shells, eyelets, fasteners, rivets etc.,
2.	Yellow brass (65%Cu-35%Zn)	Radiators, flash light shells, eyelets, fasteners, rivets etc.,
3.	Leaded brass (adition of 0.5 to 3 % of lead improves machinability)	Screw-machine parts, keys, lock parts, watch parts, etc.
4.	Admiralty Metal (71% Cu-28% Zn-1%Sn), the addition of Sn improves corrosion resistance	Condenser and heat exchanger tubes, steam power plant equipment.
5.	Aluminium brass (76% Cu-22% Zn-2% Al), the addition of Al increases the self healing property	Marine and land power stations.

B. **α+β′-brasses:** The alloys contain 54-62% of Cu. These have an ordered electron compound (β′) along with a solid solution. The being a compound, exhibit very brittle nature at room temperature, therefore, it is very difficult to deform-brasses at room temperature. In the phase diagram, it can be observed that most of these alloys can be heated to a temperature where it becomes single solid solution. Therefore these are very easy to deform at high temperatures. Most widely used brass is Muntz metal. Some of the important brass and their applications are given in table 4.

Table : **α+β′- brasses and their important properties and applications**

Sl. No.	α + β′ brasses	Properties	Application
1.	Muntz metal (60% Cu-40% Zn)	Excellent hot working properties	Ship-sheathig, condenser heads and in architectural work
2.	Free cutting brass (61.5 Cu-35.5 Zn-3 Pb)	Best machinability with good mechanical and corrosion resistance	Gears, hard ware and high speed screw machine parts
3.	Forging brass (60 Cu-38 Zn-2 Pb)	Best hot working properties	Hot forgings, hard ware, grilles, hinges, lock bodies
4.	Architectural bronze (57 Cu-40 Zn-3Pb)	Excellent forging and high machinability	Decorative moldings, grilles, hinges, lock bodies
5.	Naval brass (Tobin bronze) (60 Cu-39.25 Zn-0.75 Sn)	Resistance to salt water corrosion	Condenser plates, propeller shafts, welding rods, piston rods
6.	Naval brass (Tobin bronze) (60 Cu-39.25 Zn-0.75 Sn)	High strength combined with excellent wear resistance	Clutch disks, extruded shapes, forgings, pump rods, welding rods.

II. Bronzes

Originally, the copper-tin alloys are called bronzes, but later on, the alloys of copper and any other elements except Zn are called as bronzes. Most probably, the important element is up to 12%. The bronzes are more stronger than that of brass. Some times, the term 'bronze' is incorrectly applied to special type of brasses such as Tobin bronze, Manganese bronze, Architectural bronze. The bronzes are mainly alloys of copper and tin, silicon, aluminium and beryllium and small amounts of zinc, phosphorous, lead and nickel. Some of them are given table 5.

Table : Bronzes and their important properties and applications

Sl. No.	Bronze	Properties	Application
1.	Phosphorous bronze (93.7 Cu-6 Sn-0.3 P)	High fatigue strength	Springs, bearings, gears
2.	Aluminium bronzes (90-95) Cu-(5-10) Al	High strength, resistance to corrosion	Marin engineering, flanges, seats, moulding dies for plastic
3.	Gun metal (88 Cu-10 Sn-2Zn)	Easy to cast, high strength, resistance to sea water corrosion	Casting guns, boiler fittings, bolts, nuts and for many parts in naval constructions
4.	Bell metal (80 Cu-20 Sn)	Hard and resistance to surface water	Making bells, gongs and utensils
5.	Silicon bronze (97.5 Cu-2.5 Si)	High strength, corrosion resistance	Marine applications and high strength fastners

III. Cupronickels

These are alloys of Cu and Ni. The Cu and Ni both are completely soluble one in another at all proportions and at any temperature, therefore, they have only single phase alloys, hence, they are not heat treatable. So, their strength can only be increased by cold working. These have high erosion corrosion resistance to the high velocity of sea water therefore, these are mostly used in distiller, condenser, heat exchanger pipes.

IV. Nickel-silvers

These are alloys of Cu (50-70), Ni (5-30) and Zn (5-40) but not silver.

The nickel-silvers that contain more than 60% of Cu are single phase alloys therefore they are ductile and workable at room temperature. The nickel gives pleasing silver- blue-white colour and corrosion resistance. These act as excellent base material for electrodeposition of silver, Cr, Ni elements. These are used to manufacture screws, rivets, table flatware, zippers, name plates and costume jewelry.

The nickel-silvers that contain 50-60% of Cu are two phase alloys and are readily hot workable and have modulus of elasticity, therefore, these are used to make springs, contacts in telephone equipment, hardware, resistance wires, and also in surgical and dental equipments.

2.38 NANO-MATERIALS

Introduction

The materials, which have at least one external dimension or any internal structure (like grain size or any repeated small structures in materials) is less than 100 nm, are called *nano materials. (One* nano meter is approximately equal to one lack times smaller than human hair). In another words, these are any materials (devices, things, fluids etc.) which contain an internal morphology is less than 100 nm. The internal morphology may being the form of sphere, clusters, fibres, tubes, films, or particles etc.

Classification

The nano materials are classified according to the internal morphology, as zero dimensional (sphere, clusters), one dimensional (tubes, fibres or rods), two dimensional (films) and three dimensional (particles) nano material.

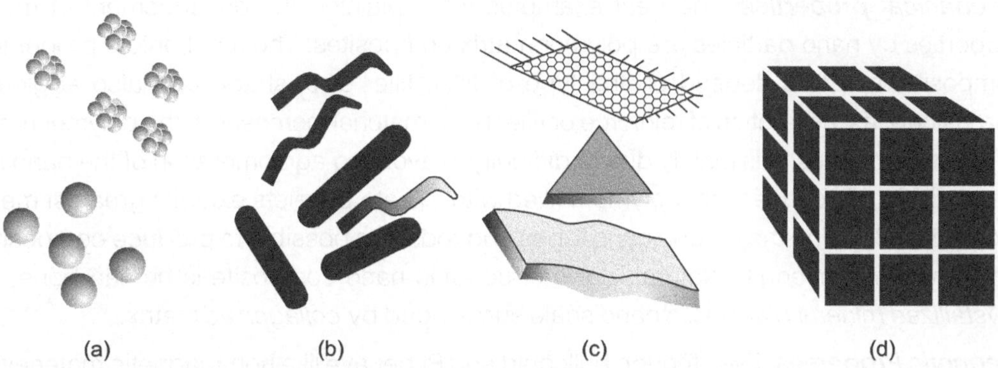

Fig. 2.85 : (a) zero dimensional, (b) one dimensional, (c) two dimensional and (d) three dimensional

Properties of Nano materials

The nano materials have been attracted world explosively due to their *different properties* than that of the base bulk materials. For example, nanophase ceramics attracted the attention of manufacturers due to their higher ductility at elevated temperature compared to the conventional coarse-grain ceramic materials. The reasons, that the nano materials have different properties, are mainly due to their higher specific surface area (surface area to volume ratio) and new quantum effects. Due to the small dimension of the nano materials, they have high fraction of surface atoms, surface energy and very less imperfections. Since they have highly surface dependent properties, it reflects on their bulk materials. For example, Chemical sensors, made from nanoparticles and nano wires have enhanced sensitivity and selectivity compared to conventional sensors. Due to the less distance requirement in the movement of impurities and imperfections by the application of thermal annealing, they have fewer imperfections, which alter their mechanical properties.

Optical Properties: Optical properties mainly depend on size and shape of the nano particles. As the size of particle shape of the particle changes in a nano-fluid, its colour changes as shown in the figure 2.86.

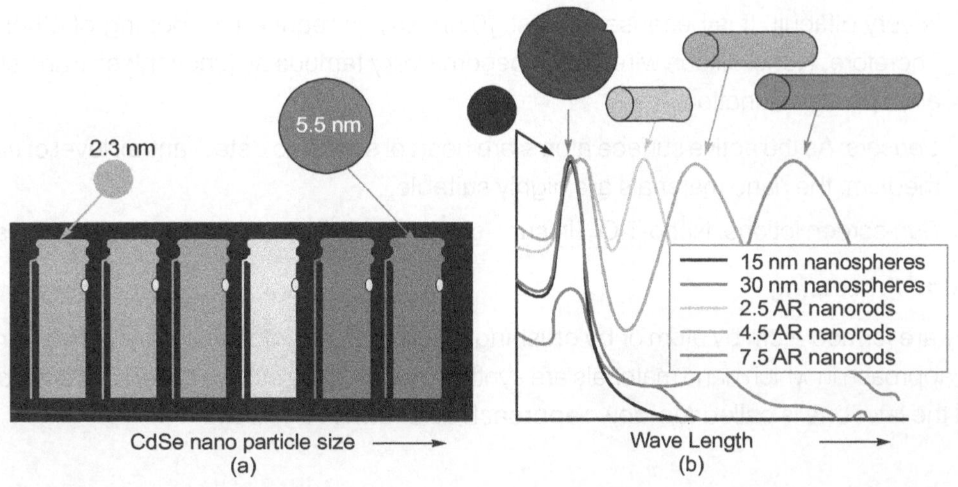

Fig. 2.86 : (a) Fluorescence emission of (CdSe) ZnS quantum dots of various sizes (b) absorption spectra of various sizes and shapes of gold nanoparticles (Chem. Soc.Rev., 2006, 35, 209–217).

Mechanical properties: The best examples for explaining the enhancement of mechanical properties by nano particles are polymer matrix composites. The mechanical properties of the composite materials depend on the type of filler, filler size, shape and also agglomeration. Surprisingly, as the fraction of reinforce or filler nano material increases in the polymer matrix, their strength decreases. It is mainly due to difficulty in avoiding agglomeration of the nano reinforce materials. In this connection, polymers filled with silicate platelets exhibits greatest mechanical properties. By using the carbon nano tubes and rods, it is possible to produce composite fibres, which has high strength. Natural, polymer-ceramic nano composite is human bone, contains *crystallized mineral platelets* in nano scale surrounded by *collagen* as matrix.

Magnetic Properties: Even though bulk gold and Pt behave like non-magnetic materials. But the bulk gold made from nano particles less than 2 nm shows ferromagnetic property, it is due to the change of d band width, by localized carriers or chemical bonding is induced in clusters

Application of Nano materials

1. *Fuel cells*: The performance of fuel cells has been increased by changing the physical structure of electrodes by using the nanomaterials. It is because of increased contact surface of electrodes with electrolyte, catalyst.
2. *Nano-catalyst*: Bulk aluminium is generally used as utensils and other engineering structural applications but, nano-aluminium is used as solid-fuel in rocket propulsion. It is because of the availability of higher active surface area.
3. *High-Definition TV*: Using nano-crystalline zinc selenide, nano zinc sulphide, nano cadmium sulphides the pixel size can be reduced and hence monitor's resolution increases drastically. Because resolution of monitor or TV depends on the size of pixel.
4. *Automobile Catalytic Converters:* As nano materials are very active due to their more amount of grain boundary area, they can easily react with toxic gasses, (CO, NO_2) exits from automobiles.
5. *Computer Chips:* Even though computer devices are greatly decrease in their size and are enhanced in their performance, maintaining high quality, especially, in the case of transistors is very difficult. It is because at least 10 nm size is required for doping of other elements. Therefore, Nano-Silicon wires have become very famous as junctionless transistor, without any doping or junction.
6. *Sensors:* As the active surface atoms are heart of sensor to detect antrax level of surrounding medium, the nano materials are highly suitable.
7. *Sun*-screen lotions: Nano-TiO_2, in sun –screen lotions, protects skin form UV-rays.

Synthesis and Processing

They are formed atom by atom or by crushing a bulk material into small nano sized particles. The first approach in which nano materials are synthesized atom by atom is called *bottom-up* approach and the later one is called *top-down* approach as shown in figure 2.87.

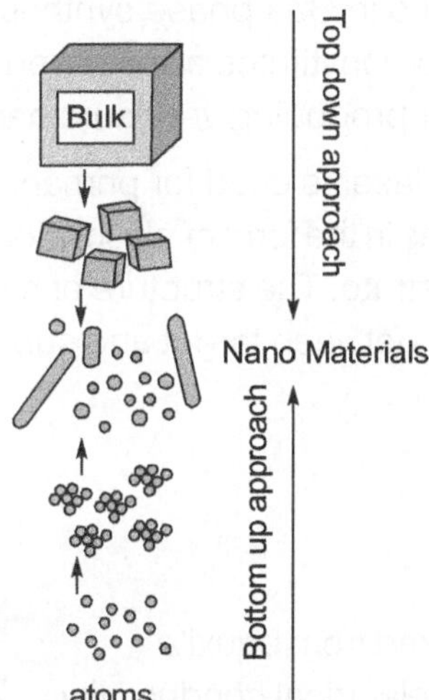

Fig. 2.87 : Preparation methods of Nano materials

Methods to produce Nano structures

Mechanical Grinding: As shown in figure 2.88, bulk materials/ large particles are crushed between hard balls in mills. It is very much suitable for making uniform alloy powders. The major problem involved in this method is contamination of ball material/ cylinder material.

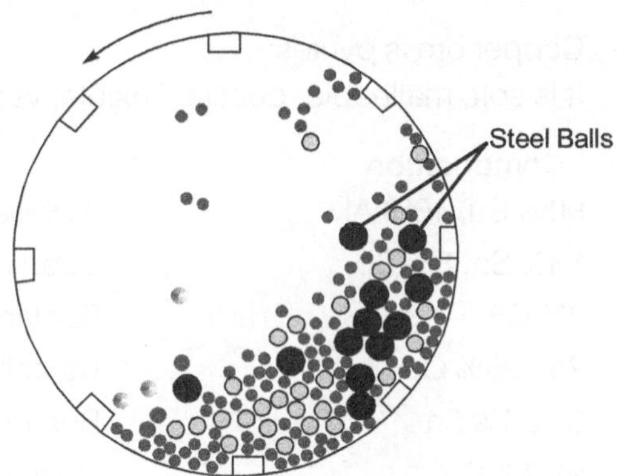

Fig. 2.88 : Ball mill

Sol-gel method: It involves hydrolysis and condensation of alkoxide-based precursors (metal alkoxides) to form metal oxide nano materials. It is a very popular method among chemists. It involves the following reactions.

$$MOR + H_2O \longrightarrow MOH + ROH \text{ (hydrolysis)}$$

$$MOH + ROM \longrightarrow M\text{-}O\text{-}M + ROH \text{ (condensation)}$$

Manin advantage by this method is that it give uniform size particles but it is very slow process and production rate is very less.

Gas phase synthesis: Size, shape and chemical composition are precisely controlled by this synthesis. Gas phase synthesis has different techniques such as Thermal evaporation, chemical vapour deposition (CVD), flame assisted ultrasonic spray pyrolysis, Gas condensation processing (GPC), chemical vapour condensation (CVC). In all these techniques, generally, materials is vaporized, clustered and condensed in to nano materials.

Sputtering: In this method, unlike in Gas phase synthesis, the material (target) is subjected to sputtering and then the resulted constitutes are allowed to cluster to form nano materials. This method is very much suitable for producing very pure nano materials.

Laser Ablation: In this method, laser is used for primary excitation source of the target material from which the resulted constitutes in the form of plasma focuses towards substrate. The constitutes are then condensed on the substrate. The structure of nano materials are controlled by ambient gas flow, laser density, distance between target and substrate, substrate surface temperature.

POINTS TO PONDER

Nonferrous Metals and Alloys

1. **Aluminium**
   — Prepared from bauxite
   — Good electrical conductivity
   — High resistance to corrosin and is non-toxicity

Al. Alloys	**Composition**	**Application**
Duralium	3.5 - 4.5% Cu, 0.4 - 0.7%	Forging stamping bars, tubes, rivets, sheets
y-alloy	Mn & Mg + Al 3.5 - 4.5% Cu, 1.2 - 1.7% Mg	Piston and components of aeroengines

2. **Copper**
   — Copper ore is pyrites
   It is soft, malleable, ductile, flexible, very tough and strong.

Copper Alloys	**Composition**	**Application**
Al bronze	90% Cu, 10% Al	Jewellary decorative purpose
Babbit Metal	88% Sn, 8% Sb	Used in bearing metals that
	4% Cu	Subjected to high pressure and load
Tin Bronze	75 - 95% Cu,	Utensile,s bearings
	5 - 25% Sn	Bushes, sheet rod, wires, articles
Phosphorous	93.7% Cu,	Heavy loads, worm
	6% Sn	Wheels, gears
Bronze	0.3% P	Nuts for machine screws, springs
Brass	50% Cu,	Hydraulic fitting,
	50% Zn	Pumps, lining, utensils, bearing, bushes etc.

3. **Lead**
   — Extracted from lead ore "galena".
   — It is soft malleasble and cuctile metal

Soft solder	Lead 37% + Tin 63%
Medium solder	Lead 50% + Tin 50%
Electrician Solder	Lead 58% + Tin 42%
Plumbers Solder	Lead 70% + Tin 30%

4. **Bearing Metals**

Bearing Metals	**Composition**	**Application**
Copper-base	80% Cu	Harder and stronger
BM	10% Sn, 10% Pb	

Babbit BM	88% Sn	High pressure
	8% Sb, 4% Cu	Load
Lead-Base	85% Pb	Light load
BM	10% Sb, 5% Sn	Pressure
Cadmium-base	at elevated temperature	
BM		

5.Copper - Tin alloys

Alloys	Composition	Application
Bronze		Boller, Tanks
Phosphorous-		Stoves where high
Bronze		
Silicon Bronze	96% Cu	Strength and good
	3% Si, 1% Mg Zn	Corrosion resistance are required
Gun Metal	88% Cu	Bioler fitting
	10% Sn, 2% Zn	Bushes, beam glands etc. (Zn for fluidity)
Bell Metal	80% Cu	Bells, gongs, utensil
	20% Sn	etc.
Mazanese	Cu, Zn	Bushes, Plungers
Bronze	Pb, M	Feed pump rods etc.

6. **Zinc** — The ore is zinc - blende (zince - sulphide) and calamine.
— Low cost, corrosion resistance and alloying properties.

Alloys	Composition	Application
Brasses	50% Cu	Hydraulic fitting
	50% Zn	Pump lining utensils, bearing, bushes
Muntz Metal	60% Cu	Electrical equipment, fuses, bolts, rods,
	40% Zn	tubes, etc.
Admiralty	70% Cu	Condenser
brass	29% Zn, 1% Sn	Tubes

7. **Nickel Alloys**

Alloys	Composition	Application
Monel Metal	60%Ni, 38%Cu + Al & Mn	Rods, sheets, wire, welded tubing
German Silver	25 - 50% Cu, 10 - 35% Ni,	Utensil, electrical works
	25 - 35% Zn	
Inconel	75 - 80% NI, 10 - 15% Cr	Resistance wire for electrical appliances

8. **Other important alloys**

i. Timken alloy - 10% W, 10% Co, 59% Ni, 9% Cr, 0.25% Fe

ii. Beryllium Bronze - Be 1.5 - 2%, Sn 8%, + Rest Cu

iii. Elektron - 12 - 13% Al, 0.03% Mn, 2% Zn, + Rest Mg

iv. Stellite - 40 - 50% Co, 25 - 30% Cr, 15 - 30% W, 2 - 4%

v. CementedCarbide - 10% TaC, 15% TiC, + Co

vi. UCON - 50% Nb, 30% Ti, 20% W

9.

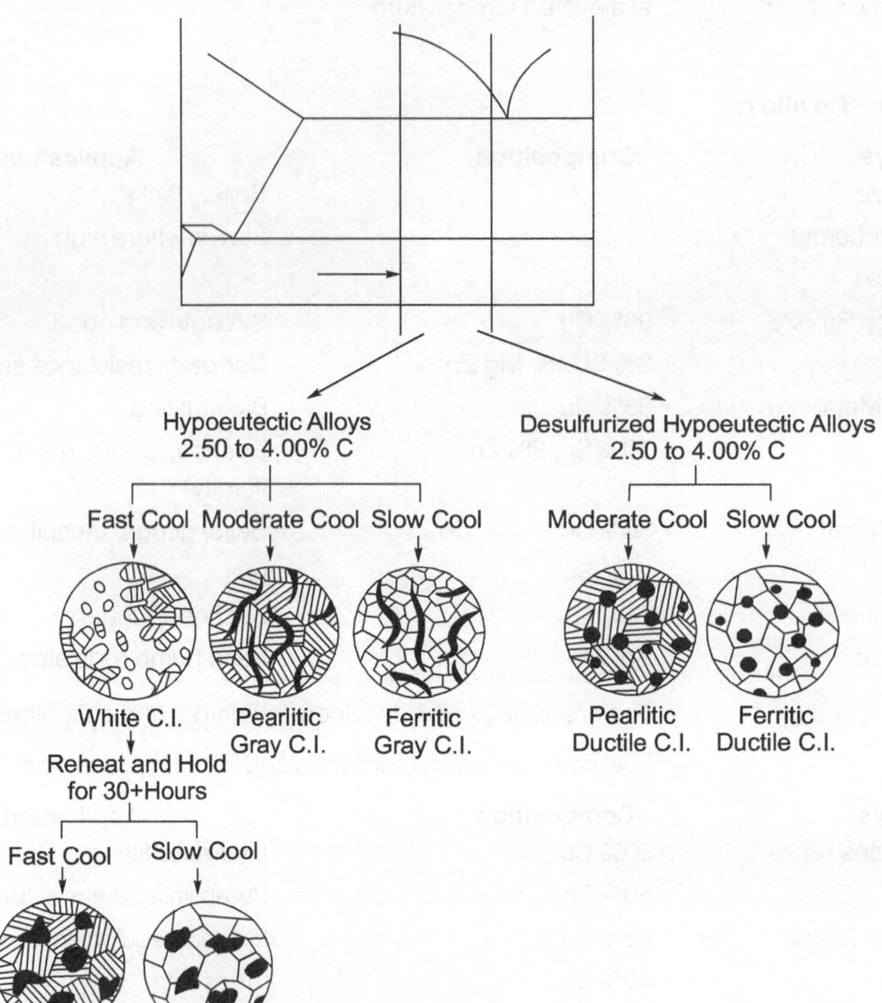

■■■■

PRACTICE QUESTIONS

1. Which one of the following materials can be subjected to an age hardening process ?
   (a) HSS (b) Aluminium
   (c) Pure iron (d) Stellite

2. Consider the following statements:
   Addition of silicon to cast iron
   1. Promotes graphite module formation
   2. Promotes graphite flake formation
   3. Increases the fluidity of the molten metal
   4. Improves the ductility of cast iron

   Of these statements
   (a) 1 and 4 are correct (b) 2 and 3 are correct
   (c) 1 and 3 are correct (d) 3 and 4 are correct

3. Match **List-I** (Heat treatment) with **List-II** (Effect on the properties) and select the correct answer using the codes given below the lists:

	List-I		List-II
A.	Annealing	1.	Refined grain structure
B.	Nitriding	2.	Improves the hardness of the whole mass
C.	Martempering	3.	Increases surface hardness
D.	Normalising	4.	Improves ductility

Codes:

	A	B	C	D
(a)	4	3	2	1
(b)	1	3	4	2
(c)	4	2	1	3
(d)	2	1	3	4

4. Eutectoid reaction occurs at
   (a) 600°C (b) 723°C
   (c) 114°C (d) 1493°C

5. Match **List-I** (Name of Material) with **List-II** (% Carbon Range) and select the correct answer using the codes give below the lists:

	List-I		List-II
A.	Hypo-eutectoid steel	1.	4.3 – 6.67
B.	Hyper-eutectoid steel	2.	2.0 – 4.3
C.	Hypo-eutectic cast iron	3.	0.8 – 2.0
D.	Hyper-eutectic cast iron	4.	0.008–0.8

Codes:

	A	B	C	D
(a)	4	3	2	1
(b)	1	3	4	2
(c)	4	1	2	3
(d)	1	2	3	4

6. Which one of the following sets of constituents is expected in equilibrium cooling of a hypereutectoid steel from austenitic state ?
(a) Ferrite and pearlite (b) Cementite and pearlite
(c) Ferrite and bainite (d) Cementite and martensite

7. Match **List-I** (Alloy) with **List-II** (Use) and select the correct answer using the codes give below the lists:

	List-I		**List-II**
A.	Low carbon steel	1.	Bearing
B.	Hadfield manganese steel	2.	Thermocouple
C.	Constantan	3.	Wire nails
D.	Babbitt alloy	4.	Bulldozer blades

Codes:

	A	**B**	**C**	**D**
(a)	1	2	3	4
(b)	3	4	1	2
(c)	3	2	1	4
(d)	3	4	2	1

8. The structure of a polymer is shown in the given figure. This polymer finds special application in

$$\left(\begin{array}{cc} F & F \\ | & | \\ -C- & C- \\ | & | \\ F & F \end{array} \right)$$

(a) packaging (b) adhesives
(c) bearings (d) fertilizer

9. Addition of magnesium to cast iron increases its.
(a) Hardness (b) Ductility and strength in tension
(c) Corrosion resistance (d) Creep strength.

10. 18/8 stainless steel contains
(a) 18% stainless, 8% chromium (b) 18% chromium, 8% nickel
(c) 18% tungsten, 8% nickel (d) 18% tungsten, 8% chromium

11. Tin base white metals are used where the bearings are subjected to
(a) large surface wear (b) elevated temperature
(c) light load and pressure (d) high pressure and load,

12. Alloy steel which is work hardenable and which is used to make the blades of bulldozers, bucket wheel excavators and other earth moving equipment contain iron, carbon and
(a) chromium (b) silicon
(c) manganese (d) magnesium

13. Consider the following statements:
Thermosetting plastic are
1. Formed by addition polymerization
2. Formed by condensation polymerization

3. Softened on heating and hardened on cooling for any number of times
4. Modulated by heating and cooling

Of these statements

(a) 1 and 3 are correct
(b) 2 and 4 are correct
(c) 1 and 4 are correct
(d) 2 and 4 are correct

14. Match **List-I** and **List-II** and select the correct answer

List-I		**List-II**	
A.	Polyethylene	1.	Adhesive
B.	Polyurethane	2.	Film
C.	Cyanoacrylate	3.	Wire
D.	Nylon	4.	Foam

Codes:

	A	**B**	**C**	**D**
(a)	2	4	3	1
(b)	4	2	3	1
(c)	2	4	1	3
(d)	4	2	1	3

15. Consider the following statements Fibre Reinforced plastic are
1. Made of thermosetting resins and glass fibre
2. Made of thermoplastic resins and glass fibre.
3. Anisotropic
4. Isotropic

Of these statements which are correct?

(a) 1 and 4
(b) 1 and 3
(c) 2 and 3
(d) 2 and 4

16. During tensile-testing of a specimen using Universal Testing Machine, the parameters actually measured include

(a) True stress and true strain
(b) Poisson's ratio and Young's modulus
(c) Engineering stress and engineering strain
(d) load and elongation

17. Consider the following statements When a metal or alloy is cold worked
1. It is worked below room temperature
2. It is worked below recrystallization temperature
3. Its hardness and strength increases
4. Its hardness increases but strength does not increase

Of these statements

(a) 1 and 3 are correct
(b) 1 and 3 are correct
(c) 2 and 3 are correct
(d) 2 and 4 are correct

18. Which of the following pairs regarding the effect of alloying elements in steel are correctly matched?
1. Molybdenum : Forms abrasion resisting particles
2. Phosphorous : Improves machinability in free cutting steels

3. Cobalt : Contributes to red hardness by hardening ferrite
4. Silicon Reduces oxidation resistance the correct answer using the codes given below :

(a) 2, 3 and 4 (b) 1, 3 and 4
(c) 1, 2 and 3 (d) 1, 2 and 4

19. Machines tool guideways are usually hardness by

(a) vacuum hardening (b) martempering
(c) induction hardening (d) flame hardening

20. Guideways of lathe beds are hardened by

(a) carburizing (b) cyaniding
(c) nitriding (d) flame hardening

21. A given steel test specimen is studied under metallurgical microscope. Magnification used is 100 X. In that different phases are observed. One of them is Fe_3C.
The observed phase Fe_3C is also known as

(a) ferrite (b) cementite
(c) austenite (d) martensite

22. Match **List-I** with **List-II** and select the correct answer using the codes given below the Lists:

	List-I		**List-II**
A.	Toughness	1.	Moment area method
B.	Endurance strength	2.	Hardness
C.	Resistance to abrasion	3.	Energy absorbed before fracture in a tension test
D.	Deflection in beam	4.	Fatigue loading

Codes:

	A	B	C	D
(a)	4	3	1	2
(b)	4	3	2	1
(c)	3	4	2	1
(d)	3	4	1	2

23. In metals subjected to cold working, strain hardening effect is due to

(a) slip mechanism (b) twining mechanism
(c) dislocation mechanism (d) fracture mechanism

24. Match **List-I** with **List-II** and select the correct answer using the codes given below the Lists:

	List-I		**List-II**
A.	Neoprene	1.	Electric switches
B.	Bakelite	2.	Adhesive
C.	Foamed poly-urethane	3.	Thermal insulator
D.	Araldite	4.	Oil seal

Codes:

	A	B	C	D
(a)	4	1	2	3
(b)	1	4	2	3
(c)	4	1	3	2
(d)	1	4	3	2

25. Which one of the following materials is used for car tyres as a standard material?
(a) Styrene-butadiene rubber (SBR)
(b) Butyl rubber
(c) Nitrile rubber
(d) Any one of the above depending upon the need

26. Which one of the following refractory materials is recommended for steel furnaces containing CaO slag?
(a) Alumina (b) Silica
(c) Magnesia (d) Fireclay

27. Which of the following properties of a solid are dependent on crystal imperfections?
1. Yield stress 2. Melting point
3. Semiconductivity 4. Ductility
Select the correct answer using the codes given below:
(a) 1 and 3 (b) 1, 3 and 4
(c) 2, 3 and 4 (d) 2 and 4

28. Which of the following components can be manufactured by powder metallurgy methods?
1. Carbide tool tips 2. Bearings
3. Filters 4. Brake linings
Select the correct answer using the codes given below:
(a) 1, 3 and 4 (b) 2 and 3
(c) 1, 2 and 4 (d) 1, 2, 3 and 4

29. Assertion (A) : The notch sensitivity of cast iron component is zero.
Reason (R) : Cast iron does not have yield point.
(a) Both A and R are individually true and R is the correct explanation of A.
(b) Both A and R are individually true but R is not the correct explanation of A.
(c) A is true but R is false.
(d) A is false but R is true.

30. Assertion (A) : Austenitic stainless steel contains 18% chromium and 8% nickel. Since it retains its austenitic structure at room temperature, it is called austenitic stainless steel.
Reason (R) : Chromium present in the steel improves its corrosion resistance by forming a thin film of chromium oxide on the surface.
(a) Both A and R are individually true and R is the correct explanation of A.
(b) Both A and R are individually true but R is not the correct explanation of A.
(c) A is true but R is false.
(d) A is false but R is true.

31. Match **List-I** (Alloying element in steel) with **List-II** (Property conferred on steel by the element) and select the correct answer using the codes given below the Lists:

List-I	**List-II**
A. Nickel	1. Corrosion resistance
B. Chromium	2. Magnetic permeability
C. Tungsten	3. Heat resistance
D. Silicon	4. Hardenability

Codes:

	A	B	C	D
(a)	4	1	3	2
(b)	4	1	2	3
(c)	1	4	3	2
(d)	1	4	2	3

32. Two cooling curves A and B for eutectoid iron-carbon alloy are superimposed on a continuous cooling transformation diagram as shown in the given figure. Fine pearlite microstructure is represented by the points labelled

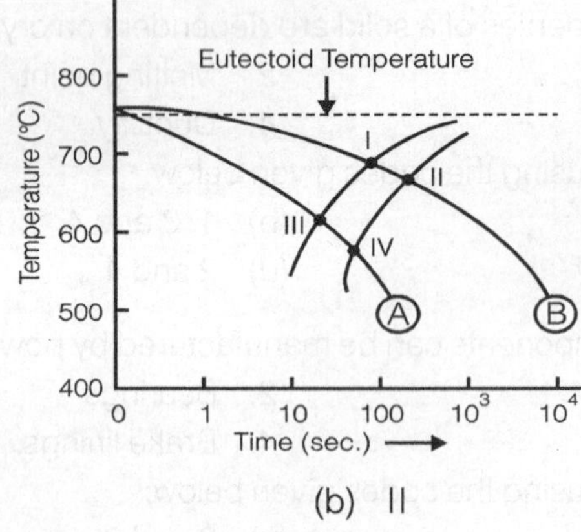

(a) I and III (b) II
(c) IV (d) I

33. Match **List-I** (Alloys) with **List-II** (Applications) and select the correct answer using the codes given below the Lists:

List-I	**List-II**
A. Chromel	1. Journal bearing
B. Babbit alloy	2. Milling cutter
C. Nimonic alloy	3. Thermocouple wire
D. High speed steel	4. Gas turbine blades

Codes:

	A	B	C	D
(a)	3	1	4	2
(b)	3	4	1	2
(c)	2	4	1	3
(d)	2	1	4	3

34. A unit cell of a crystal is shown in given figure. The miller indices of the direction (arrow) shown in the figure is

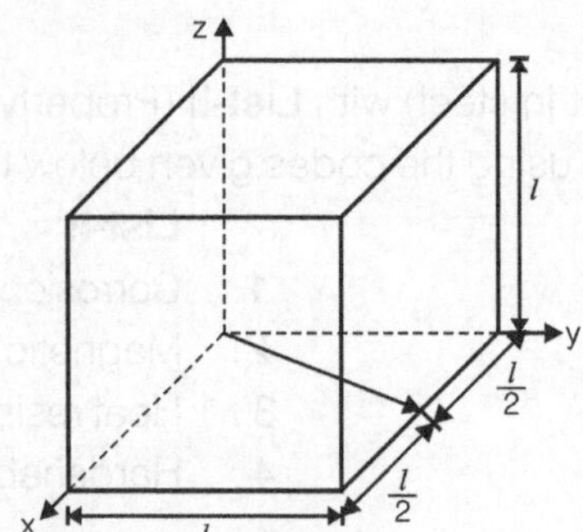

(a) [0 1 2] (b) [0 2 1]
(c) [2 1 0] (d) [2 0 1]

35. Match **List-I** (Material) with **List-II** (Structure) and select the correct answer using the codes given below the Lists:

List-I		**List-II**	
A.	Charcoal	1.	F.C.C.
B.	Graphite	2.	H.C.P.
C.	Chromium	3.	Amorphous
D.	Copper	4.	B.C.C.

Codes:

	A	B	C	D
(a)	3	2	1	4
(b)	3	2	4	1
(c)	2	3	4	1
(d)	2	3	1	4

Direction: The following items consist of two statements, one labelled the 'Assertion (A)' and the other labelled the 'Reason (R)'. You are to examine these two statements carefully and decide if the Assertion (A) and the Reason (R) are individually true and if so, whether the Reason is a correct explanation of the Assertion. Select the your answer to these items using the codes given below and mark your answer sheet accordingly.

Codes:
(a) Both A and R are individually true and R is the correct explanation of A.
(b) Both A and R are individually true but R is not the correct explanation of A.
(c) A is true but R is false.
(d) A is false but R is true.

36. **Assertion (A):** Refining the grain size of polycrystal line material renders it harder and stronger.
Reason (R): Grain boundaries provide easy paths to dislocation motion.

37. **Assertion (A):** Plastic deformation in metals and alloys is a permanent deformation under load. This property is useful in obtaining products by cold rolling.
Reason (R): Plastic or permanent deformation in metal or alloy is caused by movement of dislocations.

38. **Assertion (A):** Carbon would form an interstitial solid solution with iron.
Reason (R): The atomic radius of iron is smaller than that of carbon.

39. **Assertion (A):** Cast iron is generally hard, brittle and wear resistant.
Reason (R): Cast iron contains more than 20% carbon and as such the percentage cementite in it is higher.

40. Consider the following statements:
The strength of the fibre reinforced plastic product
1. Depends upon the strength of the fibre alone
2. Depends upon the fibre and plastic
3. Is isotropic
4. Is anisotropic

Which of these statements are correct?

(a) 1 and 3 (b) 1 and 4

(c) 2 and 3 (d) 2 and 4

41. Match **List-I** (Crystal structure) with **List-II** (Atomic packing factor) and select the correct answer using the codes given below the lists:

List-I		**List-II**	
A.	Simple cubic	1.	74%
B.	Body-Centered cubic	2.	74%
C.	Face-Centered cubic	3.	52%
D.	Hexagonal close packed	4.	68%

Codes:

	A	B	C	D
(a)	3	4	2	1
(b)	4	3	2	1
(c)	3	4	1	2
(d)	4	3	1	2

42. Consider the following pairs of plastics and their distinct characteristics:

1. Acrylics...Very good transparency to light
2. Polycarbonate...Poor impact resistance
3. PTFE...Low coefficient of friction
4. Polypropylene...Excellent fatigue strength

Which of these statements are correct?

(a) 2 and 3 (b) 1 and 3

(c) 1 and 4 (d) 2 and 4

43. Gibb's phase rule is given by

(F = number of degree of freedom

C = number of components

P = number of phases)

(a) F = C + P (b) F = C + P – 2

(c) F = C – P – 2 (d) F = C – P + 2

44. The set of Miller indices of the plane shown in the given figure is

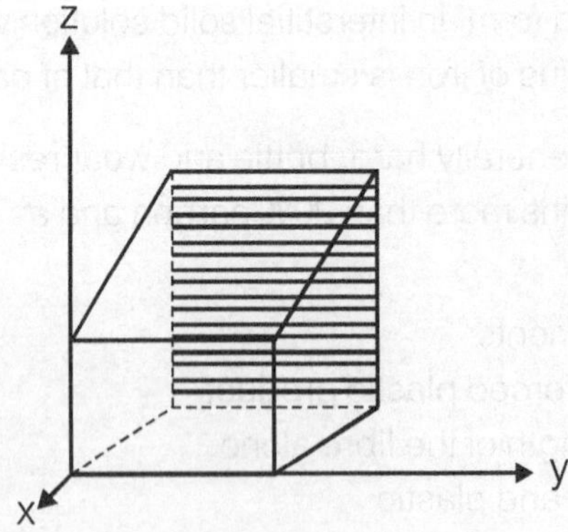

(a) (1 0 0) (b) $[0\bar{1}0]$

(c) (1 0 1) (d) (1 1 0)

45. Match **List-I** (Material) with **List-II** (Applications) and select the correct answer using the codes given below the Lists:

List-I		**List-II**	
A.	Tungsten carbide	1.	Abrasive wheels
B.	Silicon nitride	2.	Heating elements
C.	Aluminium oxide	3.	Pipes for conveying liquid metals
D.	Silicon carbide	4.	Drawing dies

Codes:

	A	B	C	D
(a)	3	4	1	2
(b)	4	3	2	1
(c)	3	4	2	1
(d)	4	3	1	2

46. Heating the hypoeutectoid steels to 30°C above the upper critical temperature line, soaking at the temperature and then cooling slowly to room temperature to form a pearlite and ferrite structure, is known as
(a) Hardening
(b) Normalising
(c) Tempering
(d) Annealing

47. In a eutectic system, two elements are completely
(a) Insoluble in solid and liquid state
(b) Soluble liquid state
(c) Soluble in solid state
(d) Insoluble in liquid state

48. **Assertion (A):** Specimens for impact testing are never notched.
Reason (R): A notch introduces triaxial tensile stresses which cause brittle fracture.
(a) Both A and R are individually true and R is the correct explanation of A.
(b) Both A and R are individually true but R is not the correct explanation of A.
(c) A is true but R is false.
(d) A is false but R is true.

49. **Assertion (A):** Carbon forms interstitial solid solution when added to iron.
Reason (R): The atomic radius of carbon atom is much smaller than that of iron.
(a) Both A and R are individually true and R is the correct explanation of A.
(b) Both A and R are individually true but R is not the correct explanation of A.
(c) A is true but R is false.
(d) A is false but R is true.

50. Which one of the following features improves the fatigue strength of a metallic material?
(a) Increasing the temperature
(b) Scratching the surface
(c) Overstressing
(d) Understressing

51. Percentages of various alloying elements present in different steel materials are given below:
1. 18% W; 4% Cr; 1% V; 5% Co; 0.7% C
2. 8% Mo; 4% Cr; 2% V; 6% W; 0.7% C
3. 27% Cr; 3% Ni; 5% Mo; 0.25% C
4. 18% Cr; 8% Ni; 0.15 C

Which of these relate to that of high speed steel?
(a) 1 and 3
(b) 1 and 2
(c) 2 and 3
(d) 2 and 4

52. Pearlite consists of

(a) 6.67% C and 93.33 ferrite (b) 12% Fe and 87% cementite
(c) 13% C and 87% ferrite (d) 13% cementite and 87% ferrite

53. Addition of vanadium to steel results in improvement of

(a) heat-treatability by quenching
(b) hardenability
(c) fatigue strength
(d) resistance to oxidation at elevated temperature

54. Atomic packing factor (APF) in the case of copper crystal is

(a) 0.52 (b) 0.68
(c) 0.74 (d) 1.633

55. During peritectic solidification, one liquid

(a) combines with one solid to form a second new solid
(b) Solidifies into two different solids
(c) forms one solid
(d) forms one solid and another liquid

56. Assertion (A): Addition polymerization is a primary summation of individual molecules into long chains.

Reason (R): In addition polymerization, the reaction produces a small molecule as by-product.

(a) Both A and R are individually true and R is the correct explanation of A.
(b) Both A and R are individually true but R is not the correct explanation of A.
(c) A is true but R is false.
(d) A is false but R is true.

57. Assertion (A): Normalized steel will have lower hardness than annealed steel.

Reason (R): The pearlite of normalized steel is finer and has lower intermolecular space.

(a) Both A and R are individually true and R is the correct explanation of A.
(b) Both A and R are individually true but R is not the correct explanation of A.
(c) A is true but R is false.
(d) A is false but R is true.

58. Martensite is a supersaturated solution of carbon in

(a) alpha iron (b) beta iron
(c) gamma iron (d) delta iron

59. The correct sequence of creep deformation in a creep curve in order of their elongation is

(a) steady state, transient, accelerated (b) transient, steady state, accelerated
(c) transient, accelerated, steady state (d) accelerated, steady state, transient

60. Nodular grey cast iron is obtained from the grey cast iron by adding a small amount of

(a) Manganese (b) Phosphorus
(c) Magnesium (d) Chromium

61. Which one of the following pairs of axis lengths (a, b, c) and interaxial angles (α, β, γ) represents the tetragonal crystal system?

(a) $a = b = c$; $\alpha = \beta = \gamma = 90°$ (b) $a = b \neq c$; $\alpha = \beta = \gamma = 90°$
(c) $a \neq b \neq c$; $\alpha = \beta = \gamma = 90°$ (d) $a = b = c$; $\alpha = \beta = \gamma \neq 90°$

62. Consider the following quenching media:

1. Oil
2. Water
3. Water + NaOH
4. Brine

The correct sequence of these media in order of increasing hardness of steel undergoing heat treatment is

(a) 1, 3, 2, 4 (b) 2, 1, 3, 4
(c) 1, 2, 3, 4 (d) 4, 3, 2, 1

63. The alloy steel designated as 40 Cr 18 Ni 2 by Bureau of Indian Standards contains

(a) 0.4% C, 18% Cr and 2% Ni (b) 4.0% C, 1.8% Cr and 0.2% Ni
(c) 0.4% C, 1.8% Cr and 2% Ni (d) 0.4$ C, 1.8% Cr and 0.2% Ni

64. 'Tempering' of quenched martensitic steel is necessary to improve the

(a) hardness of the metal (b) surface texture of the metal
(c) corrosion resistance of the metal (d) ductility of the metal

65. The molecular weight of vinyl chloride is 62.5. Thus the molecular weight of a polyvinyl chloride with a degree of polymerization of 20000 is

(a) $\frac{20000}{62.5}$ (b) $\frac{62.5}{20000}$
(c) 62.5×20000 (d) 20000

66. Match **List-I** (Name of the Element) with **List-II** (Crystal Structure) and select the correct answer using the codes given below the lists:

List-I	**List-II**
A. Fluorspar	**1.** Body-centered cubic
B. Alpha-Iron	**2.** Hexagonal close packed
C. Silver	**3.** Simple cubic
D. Zinc	**4.** Face-centered cubic

Codes:

	A	**B**	**C**	**D**
(a)	3	2	4	1
(b)	4	1	3	2
(c)	4	2	3	1
(d)	3	1	4	2

67. Which of the following factors govern solubility of two nonferrous metals both in liquid state as well as in solid state?

1. Crystal structure
2. Relative size factor
3. Chemical-affinity factor
4. Relative valence factor

Select the correct answer using the codes given below:

(a) 1, 2 and 3 (b) 2, 3 and 4
(c) 1 and 4 (d) 1, 2, 3 and 4

68. Which of the following are fabricated using engineering plastics?

1. Surface plate
2. Gears

3. Guide ways for machine tools
4. Foundry patterns

Select the correct answer using the codes given below:

(a) 1, 2 and 3 (b) 1
(c) 2, 3 and 4 (d) 1, 2, 3 and 4

69. Consider the following statements:
Polytetrafluoroethene is
1. A thermoplastic material.
2. Having high friction coefficient.
3. A thermosetting material.
4. Having high friction coefficient.
5. An electric insulator.
6. Non sticking to surfaces.

Which of the above statements are correct?

(a) 1, 2 and 5 (b) 2, 3 and 6
(c) 3, 4 and 5 (d) 3, 2 and 5

70. Which of the following fibre materials are used for reinforcement in composite materials:
1. Glass
2. Boron carbide
3. Graphite

Select the correct answer using the codes given below:

(a) 1 and 2 (b) 1 and 3
(c) 2 and 3 (d) 1, 2 and 3

71. Match **List-I** (Phase diagram) with **List-II** (Characteristic) and select the correct answer using the codes given below the lists:

List-I	**List-II**
A. Isomorphous system	1. One liquid decomposes into another liquid and solid
B. Eutectic system	2. One liquid and another solid combine to form a new solid
C. Peritectic system	3. Two metals are completely soluble in liquid state and completely insoluble in solid state
D. Monotectic system	4. Two metals, soluble in solid and liquid state

Codes:

	A	B	C	D
(a)	2	3	4	1
(b)	4	1	2	3
(c)	2	1	4	3
(d)	4	3	2	1

72. Match **List-I** (Material) with **List-II** (Application) and select the correct answer using the codes given below the Lists:

List-I	**List-II**
A. Ceramics	1. Construction of chemical plants
B. Refractories	2. Columns and pillars
C. Stones	3. Lining of furnaces
D. High silica glass	4. Tiles

Codes:

	A	B	C	D
(a)	4	3	2	1
(b)	2	1	4	3
(c)	4	1	2	3
(d)	2	3	4	1

73. T.T.T. diagram indicates time and temperature transformation of
(a) Cementite (b) Pearlite
(c) Ferrite (d) Austenite

74. The correct composition of austenitic stainless steel used for domestic utensils is
(a) 0.08% C, 18% Cr, 8% Ni, 2% Mn, 1% Si
(b) 0.08% C, 24% Cr, 12% Ni, 2% Mn, 1% Si
(c) 0.15% C, 12% Cr, 0.5% Ni, 1% Mn, 1% Si
(d) 0.30% C, 12% Cr, 0.4% Ni, 1% Mn, 1% Si

75. Which one of the following is true?
(a) Structure of metallic consists of atoms having valence of 5, 6 or 7
(b) Ceramic materials have long range electron matrix bond
(c) Polymers are composed of long chain of repeating molecules
(d) Ceramics are weaker than metals because of weak electrostatic bond

76. In the grain-size determination using standard charts, the relation between the given size number n and the average number of grains 'N' per square inch at a magnification of 100 X is
(a) $N = 2^n$ (b) $N = 2^{n-1}$
(c) $N = 2^{n+1}$ (d) $N = 2^n + 1$

77. Chemicals attack atoms within grain boundaries preferentially because they have
(a) lower energy than those in the grains
(b) higher energy than those in the grains
(c) higher number of atoms than in the grains
(d) lower number of atoms than in the grains

78. As per Gibb's phase rule, if number of components is equal to 2 then the number of phases will be
(a) ≤ 2 (b) ≤ 3
(c) ≤ 4 (d) ≤ 5

79. **Assertion (A):** The hardness test is a slow, expensive method of assessing the mechanical properties of materials.
Reason (R): The hardness is a function of yield stress and the work hardening rate of material.
(a) Both A and R are individually true and R is the correct explanation of A.
(b) Both A and R are individually true but R is not the correct explanation of A.
(c) A is true but R is false.
(d) A is false but R is true.

80. Monel metal is an alloy of
(a) Iron and carbon (b) Copper and zinc
(c) Aluminium and copper (d) Copper and nickel

81. Primary object of full annealing is to
   (a) Increase toughness and yield point
   (b) Reduce ductility and resilience
   (c) Remove foreign impurities and improve surface finish
   (d) Increase ductility and machinability

82. Globular form of cementite in the structure of steel is obtained through
   (a) Normalising (b) Malleabilising
   (c) Spheroidising (d) Carbonizing

83. Gunmetal, which is used in journal bearings, contains
   (a) 88% Cu, 10% Sn, 2% Zn (b) 80% Cu, 10% Zn, 10% Al
   (c) 85% Cu, 5% Mg, 10% Al (d) 85% Cu, 5% Sn, 10% Pb

84. The correct sequence of elements of 18-4-1 HSS tool is
   (a) W, Cr, V (b) Mo, Cr, V
   (c) Cr, Ni, C (d) Cu, Zn, Sn

85. Teflon is a
   (a) Thermosetting fluorocarbon polymer
   (b) Thermoplastic fluorocarbon polymer
   (c) Inorganic compound of fluorine and carbon
   (d) Laminated phenolic material

86. In the case of rubber, vulcanization refers to the process of producing a
   (a) Linear polymer (b) Branched polymer
   (c) Cross-linked polymer (d) Network polymer

87. Polyesters can be defined as the condensation products of
   (a) Dicarboxylic acids with dihydroxy alcohols
   (b) Biphenol-A and epichloro-hydrin
   (c) Phenol and formaldehyde
   (d) Benzene and toluene

88. The coordination number for FCC crystal structure is
   (a) 4 (b) 8
   (c) 12 (d) 16

89. Match **List-I** (Crystal Structure) with **List-II** (Example) and select the correct answer using the codes given below the Lists:

	List-I		**List-II**
A.	Simple Cubic	1.	Zinc
B.	Body-centered Cubic	2.	Copper
C.	Face-centered Cubic	3.	Alpha iron at room-temperature
D.	Hexagonal Close Packed	4.	Manganese

Codes:

	A	B	C	D
(a)	4	3	1	2
(b)	4	3	2	1
(c)	3	4	2	1
(d)	3	4	1	2

90. Which one of the following pairs is not correctly matched?
(a) Point defect in crystal lattice : Self interstitial
(b) Linear defect in crystal lattice : Grain boundary
(c) Planar defect in crystal lattice : External surface
(d) Volume defect in crystal lattice : Other phases

91. A screw dislocation
1. Lies parallel to its Burger's vector
2. Lies perpendicular to its Burger's vector
3. Moves in a perpendicular direction to the Burger's vector
4. Moves in an inclined direction to the Burger's vector

Select the correct answer using the codes given below:
(a) 1 and 4 (b) 1 and 3
(c) 2 and 3 (d) 2 and 4

92. According to Gibbs' phase rule, the number of degrees of freedom of an eutectic point in a binary system is
(a) 1 (b) 2
(c) 0 (d) 3

93. **Assertion (A):** Unlike in the case of ionic bonds, the coordination numbers for covalently bonded atoms are not controlled by the radii ratio.
Reason (R): A covalent bond has a specific direction of bonding in space.(a) Both A and R are individually true and R is the correct explanation of A.
(b) Both A and R are individually true but R is not the correct explanation of A.
(c) A is true but R is false.
(d) A is false but R is true.

94. **Assertion (A) :** The machinability of steels improves by adding sulphur to obtain so called 'Free Machining Steels'.
Reason (R) : Sulphur in steel forms manganese sulphide inclusion which helps to produce thin ribbon like continuous chip.
(a) Both A and R are individually true and R is the correct explanation of A.
(b) Both A and R are individually true but R is not the correct explanation of A.
(c) A is true but R is false.
(d) A is false but R is true.

95. Match **List-I** (Name of treatment) with **List-II** (Media used) and select the correct answer using the codes given below the lists:

List-I	**List-II**
A. Pack carburizing	1. Ammonia gas
B. Gas carburizing	2. Sodium cyanide
C. Cyaniding	3. Carburising compound
D. Nitriding	3. Carburising compound
	4. Ethane

Codes:

	A	B	C	D
(a)	3	4	2	1
(b)	2	1	3	4
(c)	3	1	2	4
(d)	2	4	3	1

96. Consider the following pairs

	Heat treatment		Effects on medium carbon steel
1.	Normalising	:	Grain refinement
2.	Full annealing	:	Uniform grain structure
3.	Martempering	:	Decreased ductility
4.	Spheroidizing	:	Maximum softness

Which of the pairs given above are correctly matched?

(a) 1 and 2 (b) 2 and 3
(c) 3 and 4 (d) 1, 2, 3 and 4

97. Match **List-I** (Material) with **List-II** (Typical use) and select the correct answer using the codes given below the Lists:

List-I	List-II
A. Branched polyethylene	1. Bottles
B. Polyester	2. Textile fibres
C. Polyvinyl chloride	3. Films for packaging
D. Linear Polyethylene	4. Transparent film

Codes:

	A	B	C	D
(a)	3	2	1	4
(b)	2	3	4	1
(c)	2	3	1	4
(d)	3	2	4	1

98. Match **List-I** (Steel type) with **List-II** (Product) and select the correct answer using the codes given below the lists:

List-I	List-II
A. Mild steel	1. Screws
B. Tool steel	2. Commercial beams
C. Medium carbon steel	3. Crane hooks
D. High carbon steel	4. Blanking dies

Codes:

	A	B	C	D
(a)	2	4	1	3
(b)	3	1	4	2
(c)	2	1	4	3
(d)	3	4	1	2

99. Match **List-I** (Alloy) with **List-II** (Application) and select the correct answer using the codes given below the lists:

	List-I		List-II
A.	Silicon steel	1.	Marine bearings
B.	High carbon steel	2.	Cutting tools
C.	High speed steel	3.	Springs
D.	Monel metal	4.	Transformer laminations

Codes:

	A	B	C	D
(a)	1	2	3	4
(b)	4	3	2	1
(c)	4	2	3	1
(d)	1	3	2	4

100. Match **List-I** (Alloying Element) with **List-II** (Effect on Steel) and select the correct answer using the codes given below the lists:

	List-I		List-II
A.	Vanadium	1.	Increases endurance strength
B.	Molybdenum	2.	Improves creep properties
C.	Silicon	3.	Increases hardness
D.	Chromium	4.	Increases resistance to high temperature oxidation

Codes:

	A	B	C	D
(a)	2	1	3	4
(b)	1	3	2	4
(c)	2	1	4	3
(d)	1	2	4	3

101. A 60 C-plain carbon steel has, approximately,
(a) 75% of pearlite and 25% of ferrite
(b) 25% of pearlite and 75% of ferrite
(c) 75% of cementite and 25% of ferrite
(d) 75% of pearlite and 25% of cementite

102. Austempering is employed to obtain
(a) 100% martensitic structure
(b) 100% bainitic structure
(c) 50% martensitic and 50% bainitic structure
(d) 100% pearlitic structure

103. Match **List-I** (Steel) with **List-II** (Application) and select the correct answer using the codes given below the Lists:

	List-I		List-II
A.	Mild Steel	1.	Ball bearing
B.	Tool Steel	2.	Cold chisels
C.	High Carbon Steel	3.	Shaft and axles
D.	Medium Carbon Steel	4.	Rolled steel sections

Codes:

	A	B	C	D
(a)	2	1	4	3
(b)	4	3	2	1
(c)	2	3	4	1
(d)	4	1	2	3

104. Increase of ferrite phase in steel increases
(a) Strength (b) Hardness
(c) Ductility (d) Brittleness

105. Consider the following statements about FCC and HCP crystal structure:
1. Both have same coordination number and atomic packing fraction.
2. Both represent closely packed crystal structures.
3. Both structures are generated by stacking of close packed planes on top of one another, but only the stacking sequence is different.

Which of the statements given above are correct?
(a) 1 and 2 (b) 2 and 3
(c) 1, 2 and 3 (d) 1 and 3

106. The eutectoid of carbon in iron, above lower critical temperature, when cooled, results in
(a) Ferrite and austenite (b) Ferrite and cementite
(c) Cementite and austenite (d) Ferrite, cementite and austenite

107. Consider the following statements about medium carbon steel:
1. It can be quench-hardened but not case-hardened
2. It cannot be quench-hardened but case-hardening can be done
3. It exhibits distinct yield point under tension test

Which of the statements given above are correct?
(a) 1 and 2 (b) 2 and 3
(c) 1 and 3 (d) 1, 2 and 3

108. Match **List-I** (Alloy) with **List-II** (Major constituent) and select the correct answer using the codes given below the Lists:

List-I	**List-II**
A. Babbit	1. Nickel
B. Invar	2. Tin and lead
C. Gun Metal	3. Aluminium
D. Duralumin	4. Copper

Codes:

	A	B	C	D
(a)	2	4	1	3
(b)	3	1	4	2
(c)	2	1	4	3
(d)	3	4	1	2

109. Which of the following materials is used in the manufacture of extrusion nozzles?
(a) Grey cast iron (b) Malleable cast iron
(c) White cast iron (d) Nodular cast iron

110. Consider the following statements:
1. Cast Iron has poor ability to damp vibrations.
2. Cast Iron has higher compressive strength compared to that of steel.
3. Cast Iron parts are suitable where permanent deformation is preferred over fracture.

Which of the statements given above is/are correct?
(a) 1, 2 and 3 (b) 1 and 3
(c) 3 only (d) 2 only

111. Assertion (A) : Carburizing is done on nonferrous alloys to increase the surface hardness.
Reason (R) : Precipitation hardening of nonferrous alloys involves solution heat treatment followed by precipitation heat treatment.
(a) Both A and R are individually true and R is the correct explanation of A.
(b) Both A and R are individually true but R is not the correct explanation of A.
(c) A is true but R is false.
(d) A is false but R is true.

112. Match **List-I** (Composition) with **List-II** (Application) and select the correct answer using the codes given below the Lists:

	List-I		**List-II**
A.	Commercial bronze (10% Zn)	1.	Radiator
B.	Red brass (15% Zn)	2.	Spring metal
C.	Aluminium brass (22% Zn 2% Al)	3.	Forging and stamping
D.	P-bronze (11% tin, small amount of P)	4.	Power plant and chemical equipment

Codes:

	A	B	C	D
(a)	2	4	1	3
(b)	3	1	4	2
(c)	2	1	4	3
(d)	3	4	1	2

113. Phenol formaldehyde is a/an
(a) thermoplastic polymer
(b) thermoset polymer
(c) elastomer
(d) rubber

114. The pattern known as Widmanstatten structure is encountered in
(a) tempering
(b) normalizing
(c) spheroidizing
(d) annealing

115. Assertion (A) : In Addition Polymerization method, polymer is produced by adding a second monomer to the first, then a third monomer to this dimer and so on.
Reason (R) : There must exist at least one double bond in the monomer for Addition Polymerization reaction.
(a) Both A and R are individually true and R is the correct explanation of A.
(b) Both A and R are individually true but R is not the correct explanation of A.
(c) A is true but R is false.
(d) A is false but R is true.

116. In case of power screws, what is the combination of materials used for the screw and the nut?
(a) Cast iron screw and mild steel nut
(b) Carbon steel screw and phosphor bronze nut
(c) Cast iron screw and cast iron nut
(d) Aluminium screw and alloy steel nut

117. In Zinc Blende structure each atom is surrounded by four atoms of the opposite kind which are located at the corners of which one of the following?
(a) Tetrahedron
(b) Hexahedron
(c) Cube
(d) Orthorhombic

118. Match List-I (Fe-Fe_3C Phase Diagram Characteristic) with List-II (Phase) and select the correct answer using the codes given below the Lists:

List-I

A. Alpha (α) iron
B. Iron carbide having crystal lattice
C. BCC pure allotrope of iron is stable between 1388°C & its melting point at 1535°C

List-II

1. δ iron
2. Eutectic with 3 iron and 1 carbon atom
3. Ferrite stable between 1388°C
4. Cementite

Codes:

	A	B	C
(a)	4	2	3
(b)	3	4	1
(c)	4	2	1
(d)	3	1	2

119. Which one of the following pairs is not correctly matched?

Space Lattice Atomic Radius r		Relation between and Edge Element a
(a) Simple cubic structure	:	$a^2 = 4r^2$
(b) Body-centred cubic structure	:	$3a^2 = 16r^2$
(c) Triclinic	:	$2a^2 = 3r^2$
(d) Face-centred cubic structure	:	$a^2 = 8r^2$

120. What is the planar density of (100) plane in FCC (face-centred cubic) crystal with unit cell side a equal to?

(a) $\frac{1.484}{a^2}$ (b) $\frac{2}{a^2}$

(c) $\frac{1}{a^2}$ (d) $\frac{\sqrt{2}}{a^2}$

121. Match List-I (Element) with List-II (Crystal Structure) and select the correct answer using the codes given below the Lists:

List-I	List-II
A. Alpha Iron	1. Hexagonal closed packed
B. Copper	2. Body-centred cubic
C. Zinc	3. Amorphous
D. Glass	4. Face-centred cubic

Codes:

	A	B	C	D
(a)	2	3	1	4
(b)	1	4	2	3
(c)	2	4	1	3
(d)	1	3	2	4

122. Which one of the following factors is more relevant to represent complete solubility of two metals in each other?

(a) Chemical affinity (b) Valency factor
(c) Crystal structure factor (d) Relative size factor

123. Match **List-I** (Effect of Cooling) with **List-II** (Cooling Medium) and select the correct answer using the codes given below the lists:

	List-I		**List-II**
A.	Martensite	1.	Water quenched
B.	Very fine pearlite	2.	Air cooled
C.	Fine pearlite	3.	Furnace cooled
D.	Coarse pearlite	4.	Oil quenched

Codes:

	A	**B**	**C**	**D**
(a)	1	4	2	3
(b)	2	3	1	4
(c)	2	3	4	1
(d)	1	2	3	4

124. Which one of the following is the process to refine the grains of metal after it has been distorted by hammering or cold working?

(a) Annealing (b) Softening
(c) Recrystallizing (d) Normalizing

125. Tempering is a process of annealing

(a) martensite at low temperatures (b) martensite at higher temperatures
(c) bainite at low temperatures (d) bainite at higher temperatures

126. What is the approximate strain energy expression for a dislocation of unit length, irrespectively of its edge or screw character?

(a) $\frac{G^2b}{2}$ (b) $\frac{Gb^2}{2}$

(c) $\frac{G^2b}{4}$ (d) $\frac{Gb^2}{4}$

127. What is the movement of block of atoms along certain crystallographic plane and directions, termed as?

(a) Glide (b) Twinning
(c) Slip (d) Jog

128. Which one of the following is the correct statement?
Pearlite in iron-carbon system is a

(a) phase consisting of ferrite and cerement at room temperature
(b) mechanical mixture of ferrite and cementite at room temperature
(c) eutectic mixture of ferrite and cementite at room temperature
(d) all the above the three are correct

129. Match **List-I** (Name of the invariant reaction) with **List-II** (Critical load reaction during cooling) and select the correct answer using the code given below the Lists:

	List-I		List-II
A.	Monotectic	1.	LIQUID $\rightarrow$ SOLID$_1$ + SOLID$_2$.
B.	Eutectic	2.	LIQUID$_1$ $\rightarrow$ LIQUID$_2$ + SOLID
C.	Eutectoid	3.	SOLID $\rightarrow$ SOLID$_1$ + SOLID$_2$
D.	Peritectic	4.	LIQUID + SOLID$_1$ $\rightarrow$ SOLID$_2$

Codes:

	A	B	C	D
(a)	3	1	2	4
(b)	2	4	3	1
(c)	3	4	2	1
(d)	2	1	3	4

130. Which of the following factors influence hardness in a plain carbon steel ?

1. Percentage carbon
2. Quenching media
3. Work size

Select the correct answer using the code given below :

(a) 1 and 2 only (b) 2 and 3 only
(c) 1 and 3 only (d) 1, 2 and 3

131. Match **List-I** (Material) with **List-II** (Application) and select the correct answer using the code given below the lists :

	List-I		List-II
A.	Fibre reinforced plastics	1.	Automobile tyres
B.	Acrylics	2.	Aircraft
C.	Phenolics	3.	Lenses
D.	Butadiene rubber	4.	Electric switch cover

Codes:

	A	B	C	D
(a)	1	4	3	2
(b)	2	3	4	1
(c)	1	3	4	2
(d)	2	4	3	1

132. Which one among the following is the most effective strengthening mechanism of nonferrous metal?

(a) Solid solution hardening (b) Strain hardening
(c) Grain size refinement (d) Precipitation hardening

133. Wood is a natural composite consisting of which of the following?

(a) Lignin fibres in collagen matrix (b) Lignin fibres in apatite matrix
(c) Cellulose fibres in apatite matrix (d) Cellulose fibres in lignin matrix

134. Which one of the following is not a ceramic?

(a) Alumina (b) Porcelain
(c) Whisker (d) Pyrosil

135. Which one of the following elements/alloys exhibits season cracking?

(a) Iron (b) Brass
(c) Aluminium (d) Steel

136. Which one of the following is the correct ascending order of packing density for the given crystal structures of metals?
(a) Simple cubic – Face central cubic – Body centrad cubic
(b) Body centered cubic – Simple cubic – Face centred cubic
(c) Simple cubic – Body centred cubic – Face centred cubic
(d) Body centred cubic – Face centred cubic–Simple cubic

137. Vibration damping in machinery is best achieved by means of base structures made of which one of the following material?
(a) Low carbon steel
(b) Nodular iron
(c) Grey cast iron
(d) White cast iron

138. For a Rhombohedral space lattice, which one of the following is correct ?
(a) $\alpha = \beta = \gamma = 90°$
(b) $\alpha = \beta = \gamma \neq 90°$
(c) $\alpha = \gamma = 90° \neq \beta$
(d) $\alpha \neq \beta \neq \gamma \neq 90°$

139. Match **List-I** (Component) with **List-II** (Required Property) and select the correct answer using the code given below the lists :

List-I	List-II
A. Blades of bulldozer	1. High wear resistance and high toughness
B. Gas turbine blades	2. Low Young's modulus and high fatigue strength
C. Drill bit	3. High wear and abrasion resistance
D. Spring of automobiles	4. High creep strength and good corrosion resistance

Codes:

	A	B	C	D
(a)	3	2	1	4
(b)	1	4	3	2
(c)	3	4	1	2
(d)	1	2	3	4

140. Structure of a polymer is:
(a) Long chain
(b) Rhombic
(c) Cubic
(d) Closed pack hexagonal

141. In the atomic hard-sphere model of the crystal structure of Copper, what is the edge length of unit cell?
(a) $2 \times$ Atomic radius
(b) $(4/\sqrt{3}) \times$ Atomic radius
(c) $(2/\sqrt{2}) \times$ Atomic radius
(d) $\sqrt{2} \times$ Atomic radius

142. Which one of the following is correct?
When "devitrification" of inorganic glasses done,
(a) glass transforms from crystalline to noncrystalline
(b) glass transforms into a fully transparent material
(c) glass transforms from monocrystalline state to polycrystalline state
(d) glass is relieved of internal stresses

143. What is a surface imperfection, which separates crystals of different orientations a polycrystalline aggregate, called?

(a) Edge dislocation (b) Stacking fault
(c) Grain boundary (d) Screw dislocation

144. Which one of the following statements is correct in the case of screw dislocations?
($\vec{b}$ = Burger's Vector; $\vec{t}$ = Imaginary Vector)

(a) $\vec{b}$ is perpendicular to $\vec{t}$ (b) $\vec{b}$ is parellel to $\vec{t}$
(c) $\vec{b}$ is inclined to $\vec{t}$ (d) $\vec{b}$ and $\vec{t}$ are non-coplanar and non-intersecting

145. Which one of the following is correct?
Malleability is the property by which a metal or alloy can be plastically deformed by applying?
(a) tensile stress (b) bending stress
(c) shear stress (d) compressive stress

146. Which of the following elements given below determine(s) the maximum attainable, hardness in steel?
1. Chromium
2. Manganese
3. Carbon
4. Molybdenum

Select the correct answer using the code given below:
(a) 1 only (b) 1 and 2
(c) 3 only (d) 2 and 4

147. Coefficient of Expansion is practically nil in a particular alloy. What is this alloy?
(a) Hadfield Manganese Steel (b) Invar
(c) Vitallium (d) Stellite

148. Consider the following statements relating to mechanical properties of ceramics:
1. Tensile strength is theoretically high but is practice quite low.
2. Compressive strength is many times lower than tensile strength
3. Shear strength is high
4. Transverse strength is easy to ascertain.

Which of the statements given above are correct?
(a) 1 and 3 (b) 1 and 4
(c) 2 and 3 (d) 2 and 4

149. Consider the following statements in respect of austenitic stainless steels:
1. Austenitic stainless steels are hardened and strengthened by cold working.
2. Austenitic stainless steels cannot be quenched and tempered.

Which of the statements given above is/are correct?
(a) 1 only (b) 2 only
(c) Both 1 and 2 (d) Neither 1 nor 2

150. **Asseration (A):** Lever Rule can be applied to determine relative amounts of phases present at any temperature.
Reason (R): Lever Rule is restricted to estimate relative phases, only if they are solid phases.
(a) Both A and R are individually true and R is the correct explanation of A.
(b) Both A and R are individually true but R is not the correct explanation of A.

(c) A is true but R is false.
(d) A is false but R is true.

151. **Asseration (A):** Elements are classified into metals and nonmetals on the basis of their atomic weights.
Reason (R): The valence electron structures contribute to the primary bonding between the atoms to form aggregates.
(a) Both A and R are individually true and R is the correct explanation of A.
(b) Both A and R are individually true but R is not the correct explanation of A.
(c) A is true but R is false.
(d) A is false but R is true.

152. Which one of the following is correct?
Babbitts are used for
(a) gears (b) bearings
(c) bolts (d) clutch liners

153. What is the process by which two or more chemically different monomers are polymerized to form a cross link polymer together with a by-product such as water or ammonia, known as?
(a) Addition polymerization (b) Co-polymerization
(c) Linear polymerization (d) Condensation polymerization

154. Which material is used for bushes in the bushed-pin type of flexible coupling?
(a) Gun metal (b) Plastic
(c) Rubber (d) Aluminium

155. Which one of the following crystal systems is valid for gold ?
(a) Orthogonal (b) Cubic
(c) Hexagonal (d) Triclinic

156. Which one of the following is correct for 'Climb' ?
(a) Dislocation moves parallel to the slip plane
(b) Dislocation moves perpendicular to the slip plane
(c) Sliding of one plane of atoms over the other plane
(d) Dislocation moves from a slip plane to another slip plane

157. Which one of the following is correct for "Burger's vector" in screw dislocation ?
(a) Perpendicular to the dislocation line (b) Inclined to the dislocation line
(c) Parallel to the dislocation line (d) Opposite to the dislocation line

158. Which one of the following defects is 'Schottky defect' ?
(a) Vacancy defect (b) Compositional defect
(c) Interstitial defect (d) Surface defect

159. Which one of the following elements is an austenitic stabilizer ?
(a) Chromium (b) Tungsten
(c) Nickel (d) Molybdenum

160. Which one of the following elements is a ferritic stabilizer ?
(a) Nickel (b) Manganese
(c) Copper (d) Chromium

161. The elements which, added to steel, help in chip formation during machining are
(a) sulphur, lead and phosphorous (b) sulphur, lead and cobalt
(c) aluminium, lead and copper (d) aluminium, titanium and copper

162. Which one of the following cast irons consists of carbon in rosette form ?
(a) White cast iron (b) Gray cast iron
(c) Malleable cast iron (d) Nodular cast iron

163. Which one of the following possesses the property of nonsparking character ?
(a) Hadfield's manganese steel (b) Spring steel
(c) Stellite (d) Invar

164. Nano composite materials are highly preferable in design consideration for their
(a) high resistance to crack propagation (b) vibration resistance
(c) impact resistance (d) high resilience

165. Which one of the following mediums is used for the fastest cooling rate of steel quenching ?
(a) Air (b) Oil
(c) Water (d) Brine

166. Which one of the following materials is not a composite ?
(a) Wood (b) Concrete
(c) Plywood (d) Sialon

167. Which of the following composites are 'dispersion-strengthened composites' ?
(a) Particulate composites (b) Laminar composites
(c) Fiber reinforced composites (d) Short-fiber discontinuous composites

168. Why are Babbitt alloys used for bearing material ?
(a) They have excellent embeddability
(b) They are relatively stronger than other bearing materials
(c) They do not lose strength with increase in temperature
(d) They have high fatigue strength

169. The material property which depends only on the basic crystal structure is
(a) fatigue strength (b) work hardening
(c) fracture strength (d) elastic constant

■■■■

ANSWERS

MATERIAL SCIENCE

1. (b)	**2.** (b)	**3.** (a)	**4.** (b)	**5.** (a)	**6.** (b)	**7.** (d)
8. (c)	**9.** (b)	**10.** (b)	**11.** (a)	**12.** (c)	**13.** (b)	**14.** (c)
15. (b)	**16.** (d)	**17.** (c)	**18.** (c)	**19.** (d)	**20.** (d)	**21.** (b)
22. (c)	**23.** (c)	**24.** (c)	**25.** (a)	**26.** (b)	**27.** (b)	**28.** (d)
29. (b)	**30.** (b)	**31.** (a)	**32.** (c)	**33.** (a)	**34.** (c)	**35.** (b)
36. (c)	**37.** (a)	**38.** (c)	**39.** (c)	**40.** (d)	**41.** (a, c)	**42.** (b, c)
43. (d)	**44.** (b)	**45.** (d)	**46.** (d)	**47.** (b)	**48.** (d)	**49.** (a)
50. (d)	**51.** (b)	**52.** (d)	**53.** (b)	**54.** (c)	**55.** (a)	**56.** (c)
57. (d)	**58.** (a)	**59.** (b)	**60.** (c)	**61.** (b)	**62.** (c)	**63.** (a)
64. (d)	**65.** (c)	**66.** (d)	**67.** (d)	**68.** (c)	**69.** (a)	**70** (d)
71. (d)	**72.** (d)	**73.** (d)	**74.** (a)	**75** (c)	**76.** (b)	**77.** (b)
78. (c)	**79.** (b)	**80.** (d)	**81.** (d)	**82.** (c)	**83.** (a)	**84.** (a)
85. (b)	**86.** (c)	**87.** (a)	**88.** (c)	**89.** (b)	**90.** (b)	**91.** (b)
92. (c)	**93.** (c)	**94.** (a)	**95.** (a)	**96.** (d)	**97.** (a)	**98.** (d)
99. (c)	**100.** (d)	**101** (a)	**102.** (b)	**103.** (d)	**104.** (a)	**105.** (d)
106. (b)	**107.** (b)	**108.** (c)	**109.** (c)	**110.** (d)	**111.** (d)	**112.** (b)
113. (b)	**114.** (a)	**115.** (a)	**116.** (b)	**117.** (a)	**118.** (b)	**119.** (c)
120. (b)	**121.** (c)	**122.** (c)	**123.** (a)	**124.** (c)	**125.** (a)	**126.** (b)
127. (c)	**128.** (b)	**129.** (d)	**130.** (a)	**131.** (b)	**132.** (d)	**133.** (d)
134. (d)	**135.** (b)	**136.** (c)	**137.** (c)	**138.** (b)	**139.** (c)	**140.** (a)
141. (c)	**142.** (c)	**143.** (c)	**144.** (b)	**145.** (d)	**146.** (d)	**147.** (b)
148. (a)	**149.** (c)	**150.** (c)	**151.** (d)	**152.** (b)	**153.** (d)	**154.** (c)
155. (b)	**156.** (b)	**157.** (c)	**158.** (a)	**159.** (c)	**160.** (d)	**161.** (a)
162. (d)	**163.** (b)	**164.** (b)	**165.** (d)	**166.** (d)	**167.** (a)	**168.** (a)
169. (c)						

CHAPTER 3

Casting

Casting is the oldest and still most widely used process. A mould cavity is created out of sand or some permanent material and liquid metal is poured into this cavity. The product is taken out after solidification. If the mould or pattern is broken after each cast then it is called expandable mould/pattern. If the same mould/pattern is used for a number of castings, it is called reusable mould/pattern. Pattern is the replica of the part being produced from the pouring temperature of the liquid metal to the room temperature, the material undergoes following shrinkages :

1. Liquid from the pouring temperature to the liquid at the melting point.
2. Shrinkage during the phase change.
3. Solid at the melting point to solid at the room temperature.

First two stages shrinkages are compensated by providing a riser. The riser design should be such that it solidifies after casting so that liquid metal is available to compensate liquid shrinkage. Third stage of shrinkage is compensated by providing allowance on the pattern. Following are the allowances that are provided on the pattern :

1. Shrinkage allowance : to compensate the third stage of shrinkage.
2. Machining allowance: casting produces poor finish and tolerances. so some margin is given to the pattern for machine to produce after solidification.
3. Draft allowance: around 1/2° to 2° taper is provided over the pattern for easy removal from the sand mould. Otherwise straight pattern may destroy the shape of product during the removal.

It should be noted here that while casting gray cast iron due to the appearance at graphite (low density material) the casting expands upon solidification. The percentage volume change can be expressed as

$$\% V = 2 \times (\% \text{ graphite} - 2.8)$$

So negative shrinkage allowances are given to the pattern while casting gray cast iron having more than 2.8% Carbon. Generally the shrinkage is expressed in linear units. In certain cases like large flat plates or domes, distortion allowances are given to the pattern. The commonly used patterns are as follows :

1. **Loose patterns :** These are the single piece pattern but incorporating the allowances, generally these patterns are made up of wood. Gating system is cut from sand manually.
2. **Gated patterns :** The gating system and runner is the integral part of the pattern. This would eliminate the hand cutting of the runners and gates and hence improved productivity.
3. **Match plate pattern :** The cope and drag portions of the pattern are mounted on the opposite side of wood or metal plate contouring to the parting line.

4. **Cope and drag patterns :** The cope and drag halves at the mould are made separately but it requires accurate alignment by guide and locating pins. These types of patterns are used for castings which are heavy and inconvenient for handling as also for continuous production.

3.1 DESIGN OF RISERS

As discussed before the purpose of riser is to provide liquid metal to compensate liquid and shrinkage during phase change, the secondary purpose of riser is to show that whether casting is full or not. So the riser design should be such that it solidify after casting, so that liquid metal is available to compensate shrinkage. The solidification time for both riser and casting can be expressed as :

$$t \propto \left(\frac{V}{A}\right)^2$$

where v is the volume of riser and A is surface area of riser.

Riser should be designed in such a way that it has minimum surface area to begin the design, the riser volume can be taken as three times the liquid shrinkage but this design has to be checked whether riser will solidify before or after casting. The design of risers can be understood by taking following examples.

Example : 3.1

Compare the solidification time of two optimum side risers having same volume, one is cylindrical and other is square parallelepiped.

Solution :

The riser which is placed on the top of casting and one surface which comes in contact with the metal is called top riser. The riser which is placed on the sides both top and bottom faces are exposed to atmosphere is called side riser. In the riser design the area of riser which is exposed to atmosphere is considered.

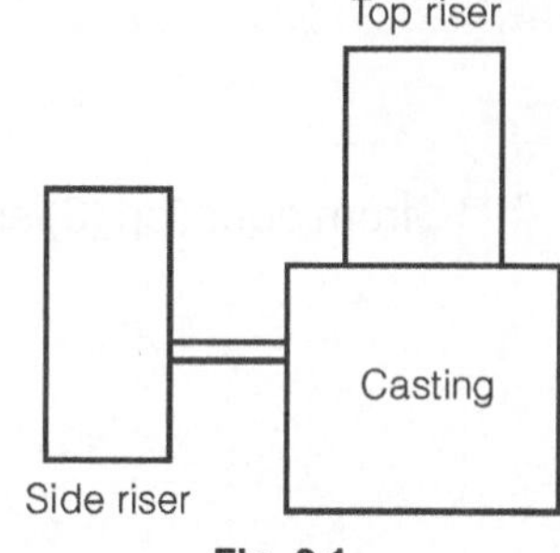

Fig. 3.1

Cylindrical rise :

volume of riser

$$V = \pi r^2 h \quad ...(1)$$

area of riser

$$A = 2\pi r^2 + 2\pi r h \quad ...(2)$$

for minimum surface area of riser

$$A = 2\pi r^2 + 2\pi r\left(\frac{V}{\pi r^2}\right)$$

$$A = 2\pi r^2 + \frac{2V}{r}$$

$$\frac{dA}{dr} = 4\pi r - \frac{2V}{r^2} = 0$$

So

$$r = \left(\frac{V}{2\pi}\right)^{1/3}$$

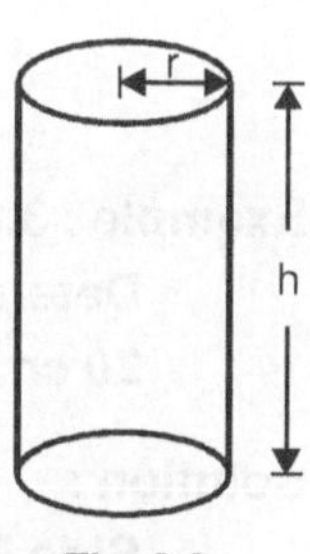

Fig. 3.2

Putting this value in equation (1) we get

$$h = 2r$$

So
$$\left(\frac{V}{A}\right)_{cylinder} = \frac{2\pi r^3}{6\pi r^2} = \frac{r}{3} \quad ...(3)$$

Square Parallelepiped :

$$\text{Volume} = a^2h_1 = V \quad ...(4)$$
$$\text{Area (A)} = 2a^2 + 4ah_1 \quad ...(5)$$

or
$$A = 2a^2 + 4a\left(\frac{V}{a^2}\right)$$

$$A = 2a^2 + \frac{4V}{a}$$

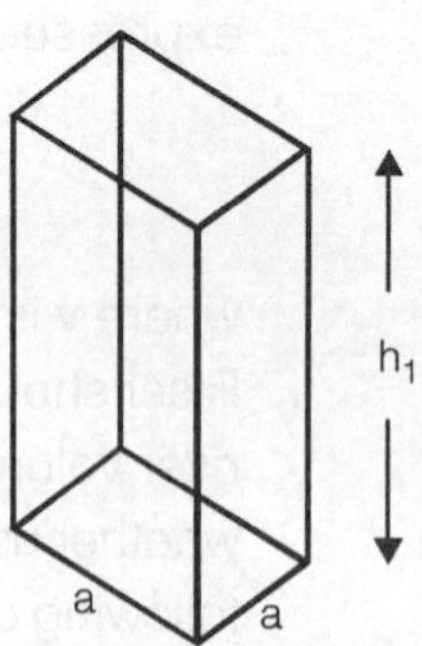

Fig. 3.3

for minimum surface area
$$\frac{dA}{da} = 4a - \frac{4V}{a^2} = 0$$

$$a = (V)^{1/3}$$

Putting this value is equation (4) we get

$$h_1 = (V)^{1/3} = a \quad ...(6)$$

$$\left(\frac{V}{A}\right)_{square} = \frac{a^3}{6a^2} = \frac{a}{6} \quad ...(7)$$

from equation (3) and (7)

$$\frac{t_{cylinder}}{t_{square}} = \frac{\left(\frac{V}{A}\right)^2_{cylinder}}{\left(\frac{V}{A}\right)^2_{square}} = \frac{\left(\frac{r}{3}\right)^2}{\left(\frac{a}{6}\right)^2} = 4\left(\frac{r}{a}\right)^2 \quad ...(8)$$

If the volume at both the risers are equal these

$$2\pi r^3 = a^3$$

so
$$\frac{r}{a} = \left(\frac{1}{2\pi}\right)^{1/3}$$

Putting this value is equation (8) we get

$$\frac{t_{cylinder}}{t_{square}} = 4\left(\frac{1}{2\pi}\right)^{2/3} = 1.17$$

Example : 3.2

Determine the dimensions of a cylindrical side and top riser used for casting brass cube of 20 cm dimension. The volume shrinkage can be taken as 9%.

Solution :

Side Riser : For optimum side riser h = 2r where h is the height and r is the radius. Initially let us take top riser volume to be 3 times the shrinkage.

$$V_r = 3 \times 0.09 \times (20)^3 \text{ cm}^3 = \pi r^2 h$$

since $$\frac{r}{h} = \frac{7 \text{ mm}}{14 \text{ mm}}$$

Confirming these dimensions of riser

$$\left(\frac{V}{A}\right)_{riser} = \frac{\pi r^2 h}{2\pi r^2 + 2\pi rh} = \frac{2\pi r^3}{6\pi r^2} = \frac{r}{3}$$

$$\left(\frac{V}{A}\right)_{riser} = 2.33$$

$$\left(\frac{V}{A}\right)_{casting} = \frac{(20)^3}{6\times(20)^2} = 3.33$$

So $$\left(\frac{V}{A}\right)_{casting} > \left(\frac{V}{A}\right)_{riser}$$

So riser will solidify prior to casting, so these dimensions are not correct since at least riser should solidify along with casting. So

$$\left(\frac{V}{A}\right)_{riser} \geq \left(\frac{V}{A}\right)_{casting}$$

So $$\frac{r}{3} \geq 3.33 \text{ then } r=9.99 \text{ cm} \approx 10 \text{ cm}$$

Therefore h = 2r $$h \geq 20 \text{ cm}$$

Top riser :

$$V = \pi r^2 h$$

$$A = \pi r^2 + 2\pi rh$$

The bottom area of riser coming in contact with casting will not be considered in the design analysis because there is no heat transfer through that area

So $$A = \pi r^2 + 2\pi r\left(\frac{V}{\pi r^2}\right)$$

$$A = \pi r^2 + \frac{2V}{r}$$

for minimum area of riser $$\frac{dA}{dr} = 2\pi r - \frac{2V}{r^2} = 0$$

So $$r = \left(\frac{V}{\pi}\right)^{1/3}$$

$$h = \left(\frac{V}{\pi}\right)^{1/3} = r$$

$$\left(\frac{V}{A}\right)_{riser} = \frac{\pi r^2 h}{2\pi rh + \pi r^2} = \frac{\pi r^3}{3\pi r^2} = \frac{r}{3}$$

$$V_r = 2160 \text{ cm}^3 = \pi r^2 h$$

$$r = h = 8.82 \text{ cm}$$

$$\left(\frac{V}{A}\right)_{riser} = 2.94$$

$$\left(\frac{V}{A}\right)_{casting} = \frac{(20)^3}{5\times 20\times 20} = 4$$

So again this riser dimension is not correct and therefore optimum riser dimension for top riser is

$$\frac{r}{3} \geq 4$$

So $\quad r = h = 12 \text{ cm.}$

3.2 RISER DESIGN FOR STEEL CASTINGS

Steels are difficult to cast because it undergoes large amount of shrinkage during solidification. It also has tendency to absorb atmospheric oxygen. So Silicon is added in the material to remove the trapped oxygen. The sound castings particularly in case of steel can be obtained by either using insulating pad or exothermic materials. As shown in figure 3.4.

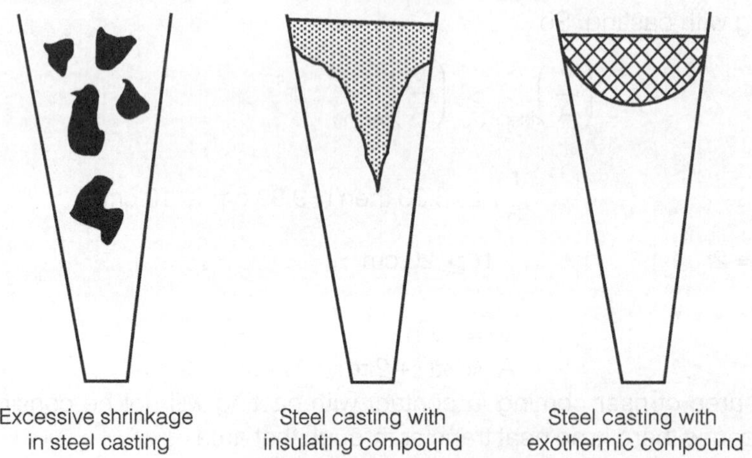

Fig. 3.4 : Steel Casting

When insulting material is kept at the top, the heat transfer decreases and hence the extent of shrinkage at the centre line decreases. But when exothermic compounds are kept at the top of riser, the centre line shrinkage can be completely eliminated. Caine curves are used to calculate the size of riser shown in figure 3.5.

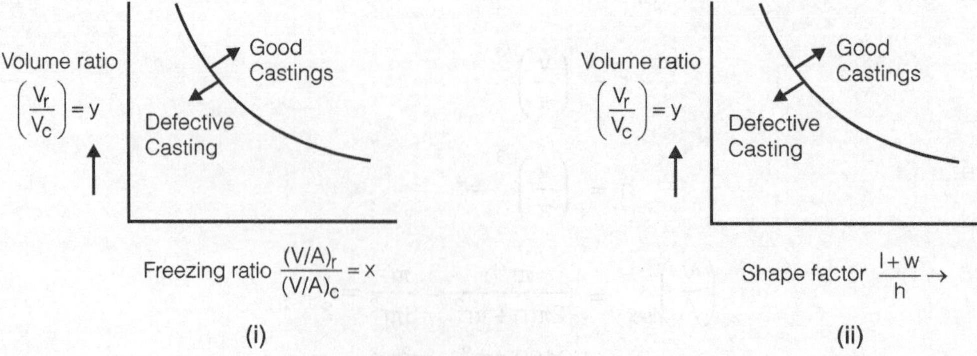

Fig. 3.5 : Caine Curve

The x-axis is freezing ratio $\frac{\left(\frac{V}{A}\right)_{riser}}{\left(\frac{V}{A}\right)_{casting}}$ and y-axis is volume ratio of riser to casting. The equations of their caine curve is

$$X = \frac{a}{y-b}+c$$

where a is freezing constant, b is contraction ratio and c depends upon the medium around the riser. For steel the value of a = 0.1 and b = 0.03.

Example : 3.3

Chvorinov and Caine gave rules for solidification time and freezing ratio for a riser. Using these rules or otherwise find the size of a cylindrical riser of height to diameter ratio as one for a steel casting of size 250 × 250 × 50 mm^3, when the casting is fed horizontally and riser is a side one. Thickness of the casting is 50 mm. For steel a = 0.10, b = 0.03 and c = 1.00. Consider it as a long bar of cross-section 250 × 50.

Solution :

Cylindrical Side riser :

It was proved before that for optimum side riser

$$h = d$$

The parameter's of Caine's curve

$$X = \frac{\left(\frac{V}{A}\right)_{riser}}{\left(\frac{V}{A}\right)_{casting}} = \frac{\dfrac{\frac{\pi}{4}d^2h}{\frac{\pi}{4}d^2\times 2+\pi dh}}{\dfrac{250\times 250\times 50}{2\times(250\times 50)+2\times(250\times 250)+2\times(50\times 250)}}$$

$$X = \frac{\dfrac{\frac{\pi}{4}d^3}{\frac{3\pi}{2}d^2}}{17.8}$$

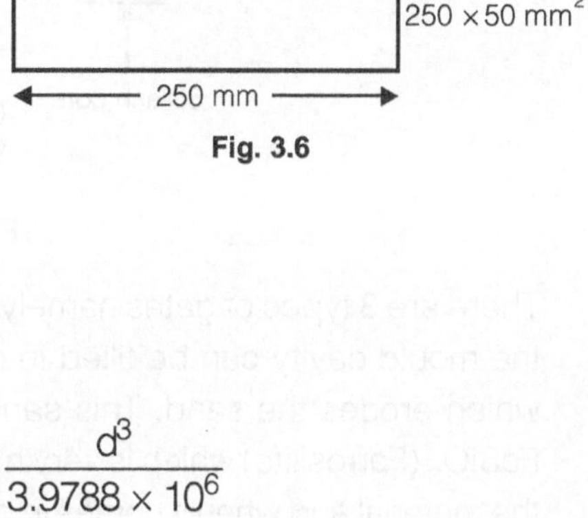

Riser

250 × 50 mm^2

250 mm

Fig. 3.6

$$X = \frac{\frac{d}{6}}{17.8}$$

$$X = \frac{d}{107.1}$$

$$Y = \frac{V_r}{V_c} = \frac{\frac{\pi}{4}d^3}{250\times 250\times 50} = \frac{d^3}{3.9788\times 10^6}$$

Putting these values in Caine's equation;

$$\frac{d}{107.1} = \frac{0.1}{\frac{d^3}{3.9788\times10^6} - 0.03} + 1$$

$$\frac{d}{107.1}\left(\frac{d^3}{3.9788 \times 10^6} - 0.03\right) = \frac{d^3}{3.9788 \times 10^6} + 0.07$$

$$d = 128.4 \text{ mm}$$

3.3 GATING SYSTEM

It refers to all the elements which are used to push the liquid metal into mould cavity. The detailed process is shown in the figure 3.7. Initially the liquid metal is poured into a pouring basin to settle the turbulence of liquid metal and almost with laminar flow liquid enters into the sprue. In the sprue there is a ceramic filter primarily to remove dross (lighter impurities) and also some larger impurities. A ceramic splash core is placed at the bottom of the sprue to avoid the erosion of sand in the mould. Compensation for vena contracta is provided at a place where liquid metal just enters into the horizontal runner otherwise vacuum will be formed in the gating system and atmosphere gases will be dissolved in the liquid metal. Skin bob is placed in the horizontal runner just before liquid metal enters into the mould cavity, so that heavier impurities settles down and lighter impurities float over it and almost pure metal enters into the mould cavity. The cross-section of horizontal runner increases in the forward direction to maintain uniform pressure (atmospheric) in the gating system.

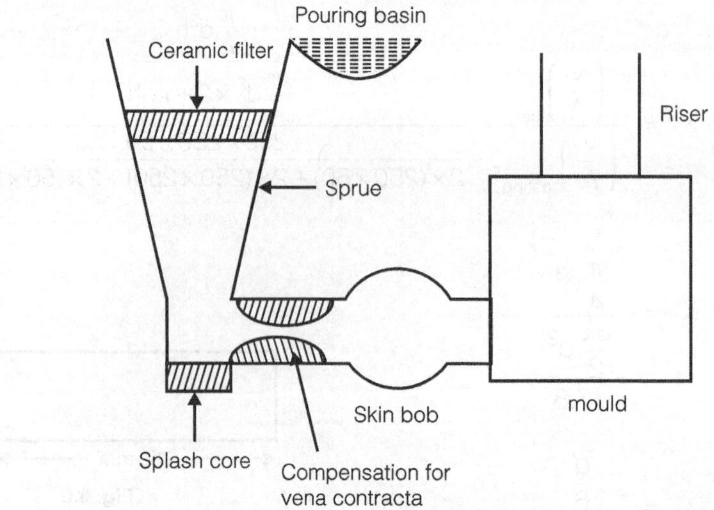

Fig. 3.7 : Gating System

There are 3 types of gates namely top gating, bottom gating and horizontal gating. In top gating the mould cavity can be filled in minimum time but there will be turbulence in the liquid metal which erodes the sand. This sand mixed with the liquid metal and after reaction it produces $FeSiO_3$ (Ferrosilite) which is very hard. These micro faylite products will be distributed throughout the material and when it comes in contact with the cutting tool, there will be sudden increase in the cutting forces and tool will be eroded gradually. In the bottom gating system the flow will be laminar but it will take more time to fill the casting and it is not favourable situation for the alloys having larger mushy zone, so top gates are used in case of large castings or for the alloys having larger mushy zone.

1. Top Gate :

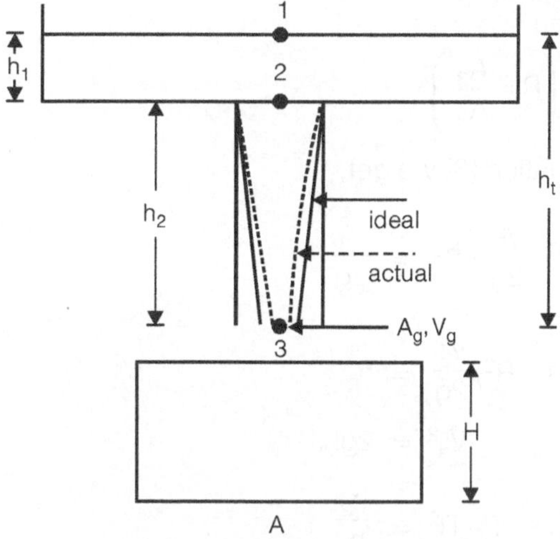

Fig. 3.8 : Design of Sprue

A_g is the cross-section area and v_g is the velocity of liquid metal coming out of the gate.

$$v_g = \sqrt{2gh_t}$$

A and H are the cross-section area and height of mould cavity. t_{f_1} is the total time to fill the mould by top gate then.

$$A_g v_g t_{f_1} = AH$$

$$t_{f_1} = \frac{AH}{A_g\sqrt{2gh_t}} \quad \text{...(1)}$$

Upon applying Bernoullis equation between 2 and 3 we get,

$$\frac{p_2}{\rho g} + \frac{V_2^2}{2g} + h_2 = \frac{p_3}{\rho g} + \frac{V_3^2}{2g} \quad \text{...(2)}$$

Point 3 is open to atmosphere and if the crossection area of the sprue is same then by continuity equation we can write

$$V_2 = V_3 \quad \text{as } p_3 \text{(gauge)} = 0$$

So equation (2) reduces to $\frac{p_2}{\rho g} + h_2 = 0$

$$\frac{p_2}{\rho g} = -h_2$$

It means that at point (2) in the sprue there will be vacuum and it will capture atmosphere into liquid metal through sand voids. This will produce blow holes in the casted part. This phenomenon is called aspiration effect which is not desirable. So to avoid this phenomenon the cross-section area of the sprue is changed to have uniform atmospheric pressure throughout the sprue. Let us again take equation (2) if the pressure throughout sprue is uniform

$$\frac{V_2^2}{2g} + h_2 = \frac{V_3^2}{2g} \quad \text{...(3)}$$

Also $$A_2 v_2 = A_3 v_3$$

$$v_2 = \frac{A_3}{A_2} v_3 = Rv_3$$

Where R is area ratio $\left(R = \frac{A_3}{A_2}\right)$

Putting this value in equation (3) we get,

$$R^2 \frac{v_3^2}{2g} + h_2 = \frac{v_3^2}{2g}$$

$$(1 - R^2)\frac{v_3^2}{2g} = h_2$$

But $$V_3^2 = 2gh_t$$

So $$1 - R^2 = \frac{h_2}{h_t}$$

$$\Rightarrow \quad R^2 = 1 - \frac{h_2}{h_t} = \frac{h_1}{h_t}$$

$$R = \sqrt{\frac{h_1}{h_t}}$$

This should be the area ratio in the sprue to avoid aspiration effect. According to this expression the cross section area should decrease linearly but ideally it should be a curve as shown in figure because friction loss in the sprue has to be incorporated. Also this curve avoids boundary layer separation.

2. Bottom Gate :

By applying Bernoullis equation between (1) and (2) we get

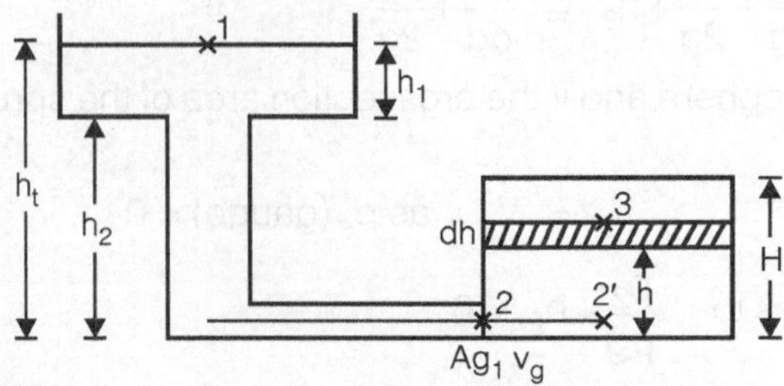

Fig. 3.9 : Bottom Gate

$$\frac{p_1}{\rho g} + \frac{v_1^2}{2g} + h_t = \frac{p_2}{\rho g} + \frac{v_2^2}{2g} + 0 \quad ...(3)$$

v_1 can be neglected because pouring basin area is very large and p_1 is atmospheric pressure. So equation (3) reduces to

$$h_t = \frac{p_2}{\rho g} + \frac{v_2^2}{2g} \quad ...(4)$$

Since pressure on the 2 – 2′ plane will remain same, so by applying Bernoullis equation between this point and (3) we get

$$\frac{p_2}{\rho g} = \frac{p_3}{\rho g} + h$$ (but p_3 is atmospheric pressure and crossection area of casting is same so $v_2 = v_3$)

So
$$\frac{p_2}{\rho g} = h \qquad \text{...(5)}$$

Putting it in equation (4) we get

$$\frac{v_2^2}{2g} = h_t - h$$

$$v_2 = \sqrt{2g(h_t - h)} = v_g \qquad \text{...(6)}$$

Let us consider that in dt_f time the liquid metal comes out of gate increases the level of metal in the cavity by dh. Equating the volume we get

$$A_g v_g dt_f = A\,dh$$

$$dt_f = \frac{A}{A_g}\frac{dh}{\sqrt{2g(h_t - h)}}$$

To calculate total time integrate this exposing from 0 to H.

$$t_{f_2} = \int_0^H \frac{A}{A_g}\frac{dh}{\sqrt{2g(h_t - h)}} = \frac{A}{A_g\sqrt{2g}}\int_0^H \frac{dh}{\sqrt{h_t - h)}}$$

$$= \frac{A}{A_g\sqrt{2g}}\left[\frac{[h_t - h]^{-\frac{1}{2}+1}}{-\frac{1}{2}+1}(-1)\right]_0^H$$

$$= \frac{A}{A_g\sqrt{2g}} \times -2\left[\sqrt{h_t - H} - \sqrt{h_t}\right]$$

$$t_{f_2} = \frac{2A}{A_g\sqrt{2g}}\left[\sqrt{h_t} - \sqrt{h_t - H}\right]$$

Example : 3.4

A casting of size 100 cm × 100 cm × 25 cm was filled by top and bottom gates with manometric height in pouring basic to be 25 cm. Compare the time to fill the casting by different gates. The area of gate is 5 cm².

Solution.

$$\text{Casting area } A = 100 \times 100 = 10^4 \text{ cm}^2$$
$$\text{casting height } H = 25 \text{ cm}$$
$$\text{manometric height } h_t = 25 \text{ cm}$$
$$\text{Gate area } A_g = 5 \text{ cm}^2$$

Top Gate :

$$t_{f_1} = \frac{AH}{A_g\sqrt{2gh_t}} = \frac{10^4 \times 25}{5\times\sqrt{2\times981\times25}} = 225.7 \text{ s}$$

Bottom Gate :

$$t_{f_2} = \frac{2\times10^4}{5\times\sqrt{2\times981}}\left[\sqrt{25}-\sqrt{25-25}\right] = 451.5 \text{ s}$$

It can be noted down that $t_{f_2} = 2t_{f_1}$. This will be the case when manometric height is equal to the height to casting but is all the cases $t_{f_2} > t_{f_1}$.

Example : 3.5

In the above example determine the total time required to full the entire casting and riser.

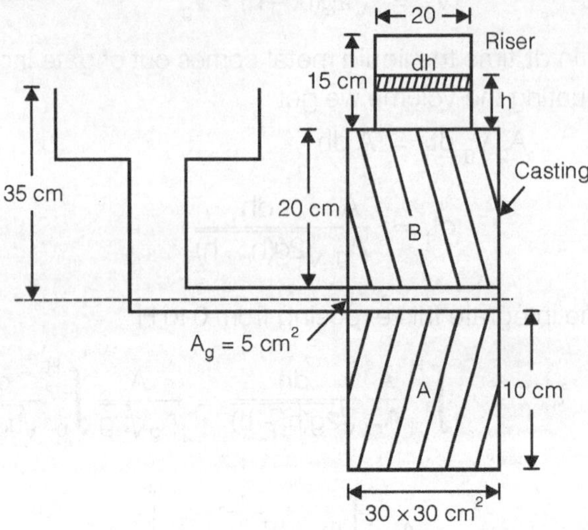

Fig. 3.10

Solution :

It is the horizontal gate. Section A of the mould will be filled like top gate

So $$t_{f_1} = \frac{(30)^2\times10}{5\times\sqrt{2\times981\times35}} = 6.86 \text{ seconds}$$

section B of the mould will be filled like bottom gate.

So $$t_{f_2} = \frac{(30)^2\times2}{5\times\sqrt{2\times981}}\left[\sqrt{35}-\sqrt{35-20}\right] = 16.60 \text{ s}$$

Riser will be filled like bottom gate

$$t_{f_3} = \int_{20}^{35}\frac{\frac{\pi}{4}(20)^2}{5}\frac{dh}{\sqrt{2g(35-h)}}$$

$$t_{f_3} = \int_{20}^{35}\frac{-2\times\frac{\pi}{4}(20)^2}{5\sqrt{2\times981}}\left[\sqrt{h_t-H}\right]_{20}^{35}$$

$$= -2.83\left[\sqrt{35-35} - \sqrt{35-20}\right]$$

$$= 2.83 \times \sqrt{35-20}$$

$$t_{f_3} = 10.98 \text{ s}$$

Total time to full the casting

$$t_{f_1} + t_{f_2} + t_{f_3} = 16.60 + 10.98 + 6.86 = 34.4 \text{ s}$$

3.4 GREEN SAND MOULD

Majority of castings are being produced by sand moulds. Green sand indicates freshly prepared mould or the mould in which there is some moisture. The composition of green sand mould is 70-85% sand, 10-20% clay, 3-6% water and 1-6% additives like cereals, asphalt sea coal etc. Clay is a principal source of strength and plasticity of moulding sand. It is a binder of moulding sand. Following are the properties of mould.

1. **Green Strength :** After mixing with water the sand must have adequate strength for handling.
2. **Dry Strength :** When the liquid metal is poured, moisture immediately vapourizes. The dry mould must have strength to resist erosion.
3. **Permeability :** Gases are formed due to mould metal reaction and also due to vapourization. The mould must be porous to permit the gases to pass off, otherwise casting will have gas holes defect. If the grain size of sand is large, there will be larger void size between particles. This leads to larger permeability.

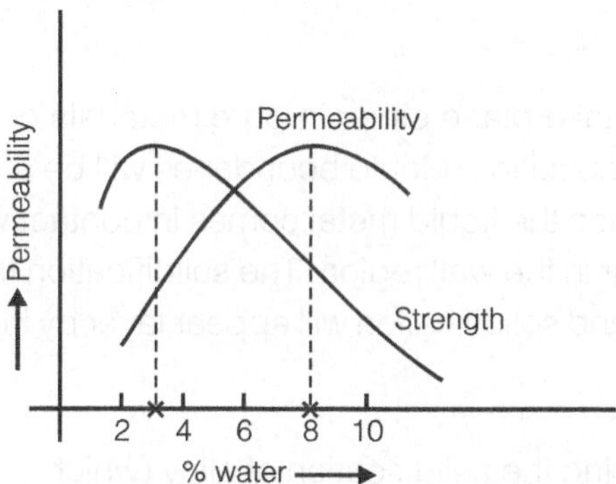

Fig. 3.11 : Permeability and strength Vs moisture content

As shown is figure 3.11 if the water content is less, these voids will be filled with smaller sand particles which results in lower permeability. By increasing the water content, smaller sand particles are also combine together, this increase permeability. Once the percentage of water goes beyond 8%, water start accumulating in these voids and the sand becomes pasty. This decreases permeability and decrease in strength of the mould. Permeability is quantified by Permeability Number (PN) which is defined as

$$PN = \frac{VH}{PAT}$$

PN is the ratio in millimeters at which air will pass through the sand under a standard condition of pressure.

Where

V = volume of air passed through standard cylindrical specimen = 2000 cc
H = height of sand specimen = 5.08 cm
P = Pressure difference across specimen 10 g/cm^2
A = Cross section area of specimen = 20.26 cm^2
T = Time in minute for 2000 cc air to pass through specimens

So, $$PN = \frac{3007.2}{T}$$

4. **Thermal stability :** Heat of the liquid metal may cause rapid expansion and mould surface may crack and destroy its shape.
5. **Refractoriness :** This is the property due to which sand remains solid by increasing the temperature.
6. **Collapsibility :** During the solid contraction of the cast part if mould creates resistance, cracks will appear over it. So the property of sand mould due to which it does not create resistance of the contracting metal is called collapsibility. The additives like cereals, wood flour etc burns and hence creates extra space for the contraction of solid.
7. **Flowability :** It is the ability of sand to flow upon ramming
8. **Loam sand** : As discussed earlier that gray C.I. upon solidification expands, so green sand will have very poor collapsibility. To cast these materials loam sands are used. It is a mixture of sand, silt, clay and excess amount of water. It is 50% sand and 50% silt and clay.

3.5 TYPES OF SOLIDIFICATION

3.5.1 Skin Forming

This type of solidification take place either in pure materials or the alloys having eutectic composition. Mould boundaries will be at the room temperature and when the liquid metal comes in contact with walls, solid will first appear in the wall region. The solidification front will move towards centre and solidification will appear layer by layer.

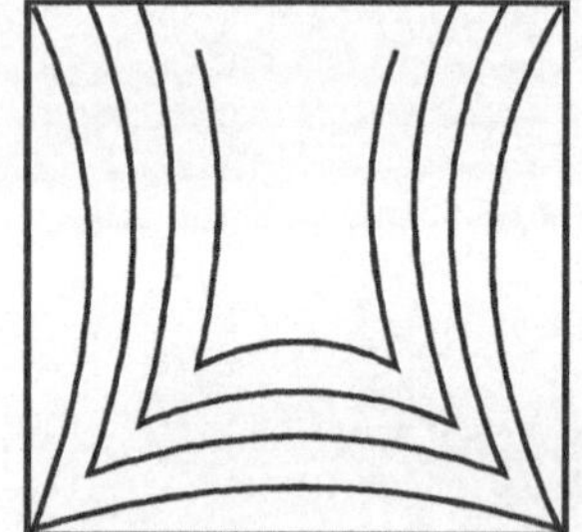

Fig. 3.12 : Skin Forming

3.5.2 Dandritic Growth

If mushy zone appear during the solidification of alloy (which will be there in most of the cases), it results in this type of solidification. Initially a solid will nucleate near the boundary and solidification front will move towards the centre. Solid will also nucleate somewhere in the liquid region and it moves both towards centre and towards the primary dandrite. So in this case the solidification is not uni-directional, the place where primary and secondary dandrites meet there will be micro voids which can not be compensated by riser. As a result of that, casting will be defective.

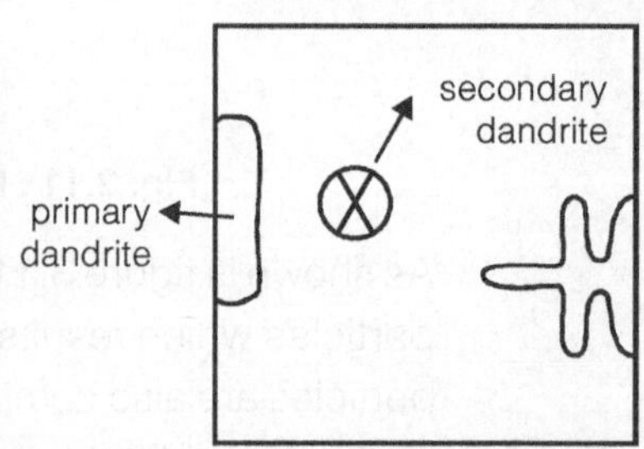

Fig. 3.13 : Dandritic Growth

To provide uni-directional solidification chills are used in such cases. Chills are made of the same material and are kept all around the mould. When the liquid metal is poured, it comes in contact with the chills. Due to rapid heat transfer, a fast moving solidification front moves towards the centre, not giving any chance for the secondary dandrites to form. So directional solidification is achieved by increasing the solidification rate. These chills must be dry to avoid the formation of blow holes. Due to rapidly moving solidification fronts towards centre, there will be some contraction stress associated with it. Especially in case of chills these stress will multiply at the centre. As a result of that cracking appears during the last stage of solidification at the centre. This defect is called hot tear (a metallurgical defect).

If the casting is big or when it is very complex, external chills alone may not completely avoid the formation of secondary dandrites. In such cases internal chills are used. Internal chills and cores are supported by chaplets. Both chills (internal) and chaplets must be pure and are supposed to fuse with the liquid metal, unfused internal chills are highest defect in casting.

Paddings are applied to the patterns to promote directional solidification. Padding is excess metal added to the casting to develop temperature gradients for directional solidification. Directional solidification can also be achieved by decreasing the solidification rate called innoculation. Small amount of another metal (innoculant) is added before pouring the liquid metal into the mould cavity. These metals act like catalyst which retards the formation of secondary dendrites. Examples are, Sodium (Na) is used for Aluminium casting, Magnesium (Mg) is used for cast irons etc.

3.5.3 Pouring Temperature

Section sensitivity is defined as the minimum thickness section attached to the casting that can be filled completely by liquid metal. Higher is the pouring temperature better will be the section sensitivity but the liquid metal will undergoes large amount of shrinkage. As a result of that riser design will become very complex. If the pouring temperature is low there will be defects like cold shut and misruns, appears. When two strems are not able to fuse together, the defect is called cold shut. Misrun appears where all the details of the castings are filled completely.

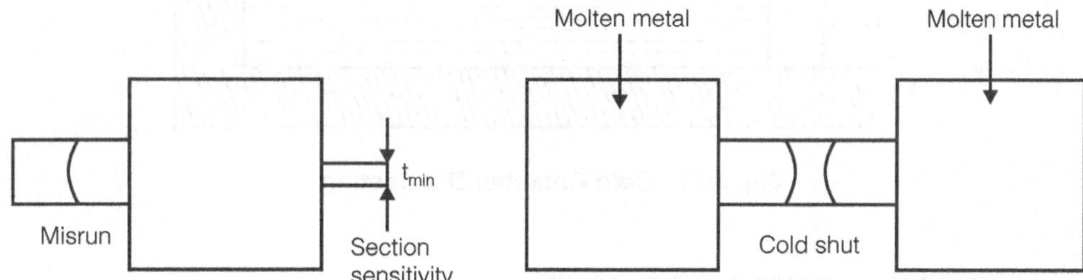

Fig. 3.14 : Cold Shut and Misrun

These defects appear due to poor fluidity. Fluidity is the ability of liquid metal to flow. This can be quantified by spiral test. An spiral mould is being prepared and liquid metal is poured through the centre. How much distance liquid metal covers in the mould before solidification, is the measure of fluidity. If the cross-section area of this cavity is constant, it is called rod test and if the cross-section area is variable, it is called V-test. When the pouring temperature is very high water disintegrates into H_2 and O_2 and as H_2 comes out it leaves large number of fine holes on the surface of casted part. This detect is called pin hole porosity. So the temperature at pouring must be optimum.

3.6 DIE CASTING

In this process the mould is made up of some permanent material like cast iron, die steels, copper and aluminum. Two halves can either be placed horizontally or vertically and when liquid metal is poured under gravity it is called gravity die casting and when liquid metal is injected into this permanent mould, it is called pressure die casting. It is having following advantages

(a) Same mould can be reused again and again. This increases the production rate as high as 150-250 casts per hour.
(b) Dimensional tolerance of the order of 0.001 to 0.003 inch.
(c) Rapid cooling produces high strength.
(d) Better section sensitivity but after some amount of usage, mould contamination takes place i.e. a portion of metal will stick to the mould. This has to be removed by abrasive jet machining (AJM). If atmospheric gases are trapped within the liquid metal, it cannot come out, because these moulds have zero permeability. In gravity die casting there are high chances of atmospheric gases getting trapped.

Pressure die casting :

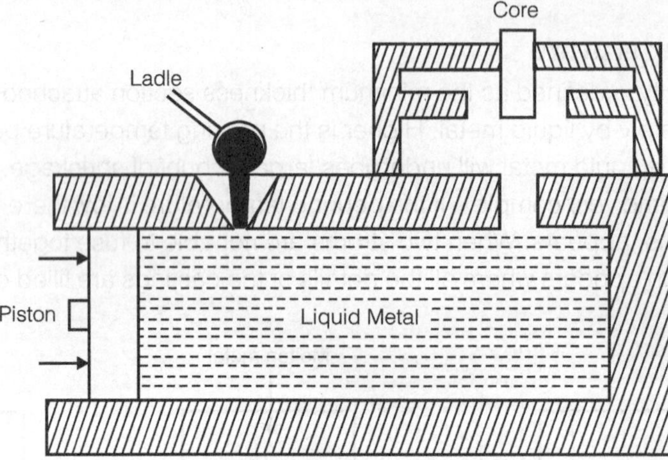

Fig. 3.15 : Cold Chamber Die Casting

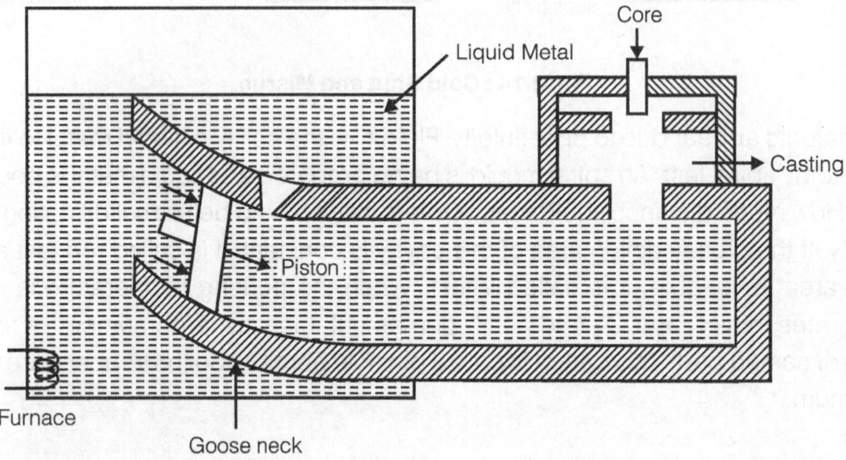

Fig. 3.16 : Hot Chamber Die Casting

In pressure die casting the water cooled dies are lubricated and then damped together. The molten metal is then injected under pressure. This pressure creates turbulence in the liquid metal and as a result of that air will be trapped within the liquid metal. In cold chamber die casting metal is liquified separately and brought to die using crucible and laddle. Since liquid metal comes in contact with atmosphere the chances of air trappment is very high.

In hot chamber die casting process furance is a part of die and liquid metal is not exposed to atmosphere. But since a major portion of the assembly is dipped into the liquid metal, this arrangement is suitable to cast materials which has low melting point. Hot chamber die casting offers fast cycling time (upto 15 cycles/min). Hot chamber die casting can also be used for casting Aluminium alloys because the material has a tendency to pick up some iron due to extended contact with the casting equipment.

3.7 CENTRIFUGAL CASTING

3.7.1 True Centrifugal Casting

In this process a metallic mould (in two parts with flange) is rotated at 3000 rpm using a rotating device. Liquid metal is then poured into it at the centre. The mould is slightly inclined from the horizontal so that the liquid metal covers the entire working length of the mould. Percentage yield in casting is defined as the ratio of useful material to the total liquid material poured into the cavity i.e. useful material after removing risers, gates etc. The percentage yield of true centrifugal casting is around 95-98% because neither riser, nor cores are used is casting and only lighter impurities which settles at the inside surface are removed. Finer grains will appear on outer surface and since cooling rate is decreasing inwards, so grains will become coarser and coarser , such surfaces are called jagged surface. Composite pipe, large ceramic and cast iron pipes are made by this process. Due to permanent material of mould, higher cooling rates will be experienced by the material which produces relatively fine grain structure. So castings will be strong.The process can be either vertical or horizontal, and the need for a center core is completely eliminated. Castings produced by this method will always have a true cylindrical bore or inside diameter regardless of shape or configuration. The bore of the casting will be straight or tapered, depending on the horizontal or the vertical spinning axis used. Castings produced in metal molds by this method have true directional cooling or solidification from the outside of the castingtoward the axis of rotation. This directional solidification results in the production of high-quality defect-free castings without shrinkage, which is the largest single cause of defective sand castings.

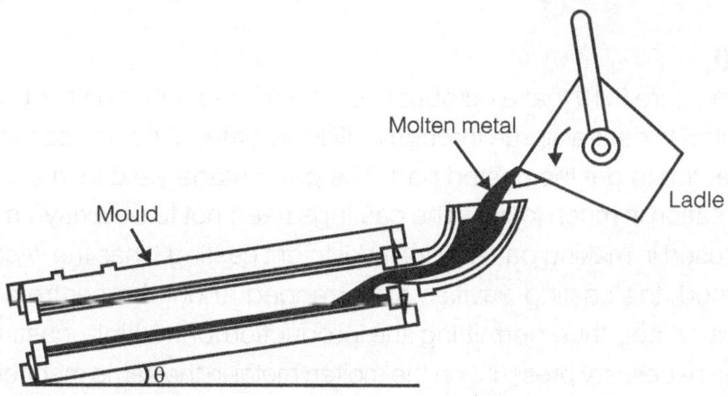

Fig. 3.17 : True Centrifugal Casting

3.7.2 Semicentrifugal Casting

In this process mould is placed on the horizontal plane and it is rotated along the vertical axis. The outer portions of the mould will be filled by purely centrifugal action and as the liquid metal approaches towards centre, the centrifugal component decreases and gravity component increases. The central portion is purely filled by gravity. The speed of rotations and percentage yield is lower then true centrifugal casting. Wheels, pulleys etc. are made by this process.It is used to produce castings with configurations determined entirely by the shape of the mold on all sides, inside and out, by spinning the casting and mold about its own axis. A vertical spinning axis is normally used for this method. Cores may be necessary if the casting is to have hollow sections. Directional solidification is obtained by proper gating, as in static casting. Castings that are difficult to produce statically can often be economically produced by this method, because centrifugal force feeds the molten metal under pressure many times higher than that in static casting. This improves casting yield significantly (85 to 95%), completely fills mold cavities, and results in a high-quality casting free of voids and porosity. Thinner casting sections can be produced with this method than with static casting. Typical castings of this type include gear blanks, pulley sheaves, wheels, impellers, and electric motor rotors.

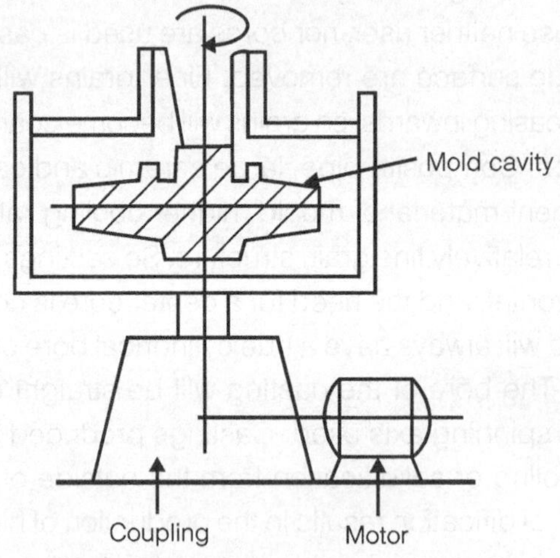

Fig. 3.18 : Semicentrifugal Casting

3.7.3 Centrifuging

As shown in figure 3.19 that a number of castings are placed on the periphery of a drum and are connected to central sprue through individual gates. After the solidification, gating system is disconnected to get the casted part. The percentage yield in this case is only 5-10% and speed of rotation is much lower. The castings need not to be axisymmetric and the process is primarily used in making patterns for investment casting.It has the widest field of application. In this method, the casting cavities are arranged about the centre axis of rotation like the spokes of a wheel, thus permitting the production of multiple castings. Centrifugal force provides the necessary pressure on the molten metal in the same manner as in semi-centrifugal casting. This casting method is typically used to produce valve bodies and bonnets, plugs, yokes, brackets, and a wide variety of various industrial castings.

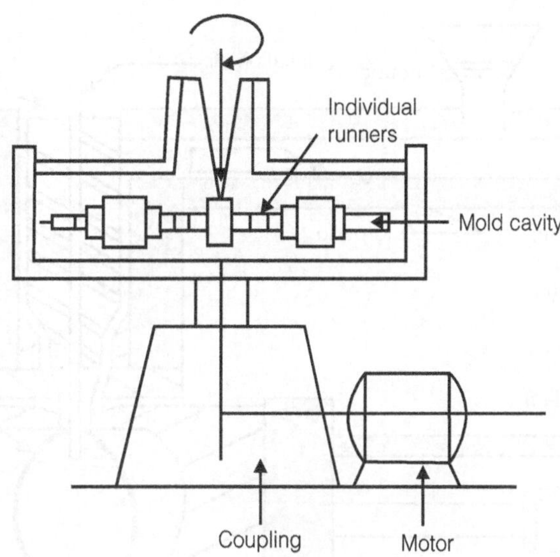

Fig. 3.19 : Centrifuging

3.8 SLUSH CASTING

Gold jewellry, lamp bases, statue and other brass items are made by this process. Liquid metal is poured into the die cavity and since solidification will start from the surface, without completely solidifying the liquid metals, the mould is inverted and excess metal is removed from the cavity.

3.9 SODIUM SILICATE (CO_2) MOULDING

The thumb rule is that harder the mould, better will be the finish and dimensional tolerance and minimum will be the defects such as blow holes. While preparing the mould around 3-4% Na_2SiO_3 (also known as water glass) is mixed with the sand. CO_2 gas is then passed through the mould and following reaction takes place.

$$Na_2SiO_3 + CO_2 \rightarrow Na_2SO_3 + SiO_2 \text{ (colloidal)}$$

This produces very hard mould. Liquid metal is then poured into the cavity and after solidification mould is broken to get the final product out although harder mould is having advantages but it should be noted that the collapsibility to such moulds will be very poor.

3.10 BLOW MOULDING

Glass and plastic bottles, bulb etc are moulded by this process. Glass or plastic in the semi viscous form (called gob) is placed in the die and air is blown into the die. As a result of that material takes the shape of die.Blow moulding is a process used to produce hollow objects from thermoplastic. The basic blow moulding process has two fundamental phases. First, a parison (or a preform) of hot plastic resin in a somewhat tubular shape is created. Second, compressed air is used to expand the hot preform and press it against mould cavities. The pressure is held until the plastic cools. The major applications of the process are Mineral water bottles, Oil bottles, Pharmaceutical bottles, Tablet containers, Mouth wash, bottles, Antiseptic liquid bottles, Chemical containers, Containers for harsh liquid cleansers, Barrier bottles.Pneumatic system extrusion.

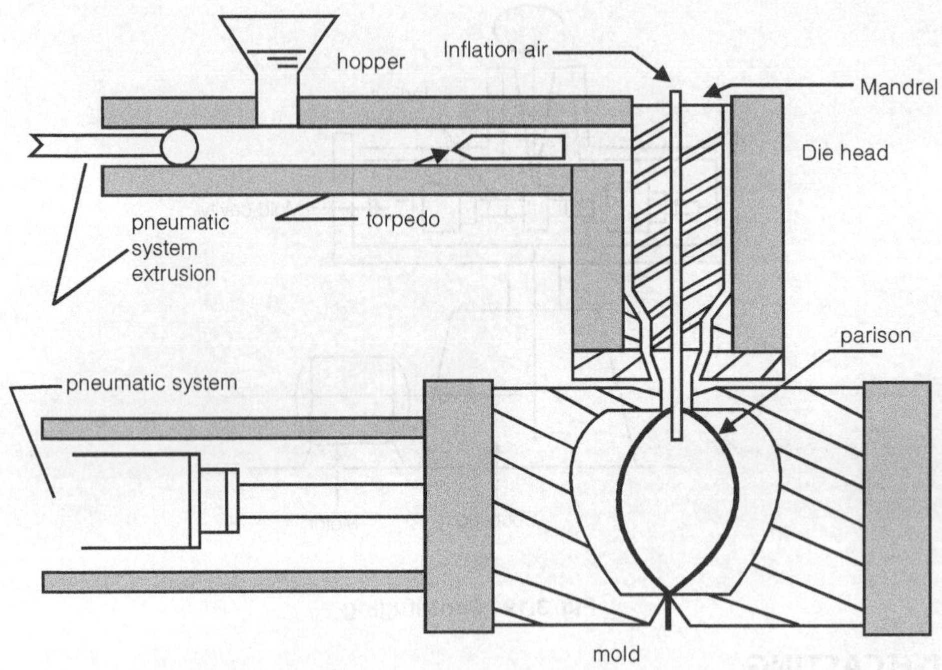

Fig. 3.20 : Blow Moulding

3.11 SHELL MOULDING

Shell moulding produces better surface finish and close dimensional tolerances. The process involves following steps:

1. A pattern made up of some permanent material cast iron is heated to around 230-315°C. It is then clamped over a box containing Silica, Phenolic resins and Alcohol. No water is used in preparing the sand mixture.

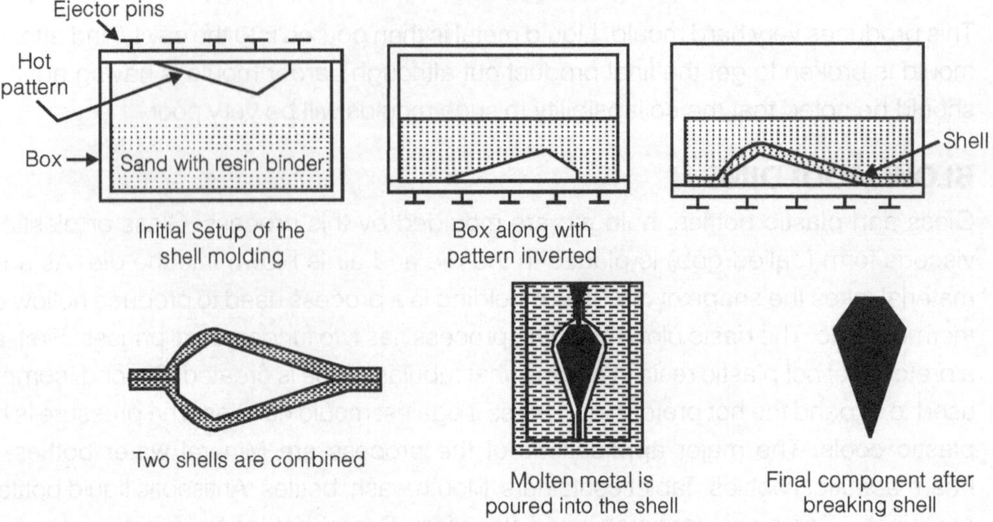

Fig. 3.21 : Shell Moulding

2. The box is then inverted and the sand is allowed to fall over the hot pattern. Depending upon the dwell time a thickness of sand will stick to the metal pattern. The thickness of shell is proportional to the dwell time.
3. The pattern along with shell is backed in an oven at around 300°C. This curing process makes the shell rigid.
4. There are ejector pins over the pattern and by hitting them to ground shell is taken out.
5. Two or more such shells are joined together either by thermo setting plastic or clamped together.
6. Liquid metal is then poured into the shell cavity. This shell provides easy escape of gases which develops due to mould-metal reaction.

3.12 INVESTMENT CASTING

Investment casting produces better tolerances compared to shell moulding. This process is also used to produce intericate shapes which are not symmetrical. One of the application of this process in making jet engine blades. The pattern is made of either wax or mercury. When it is made of wax it is called lost wax method and where the pattern material is mercury it is called mercast process. The process can be understood by following steps.

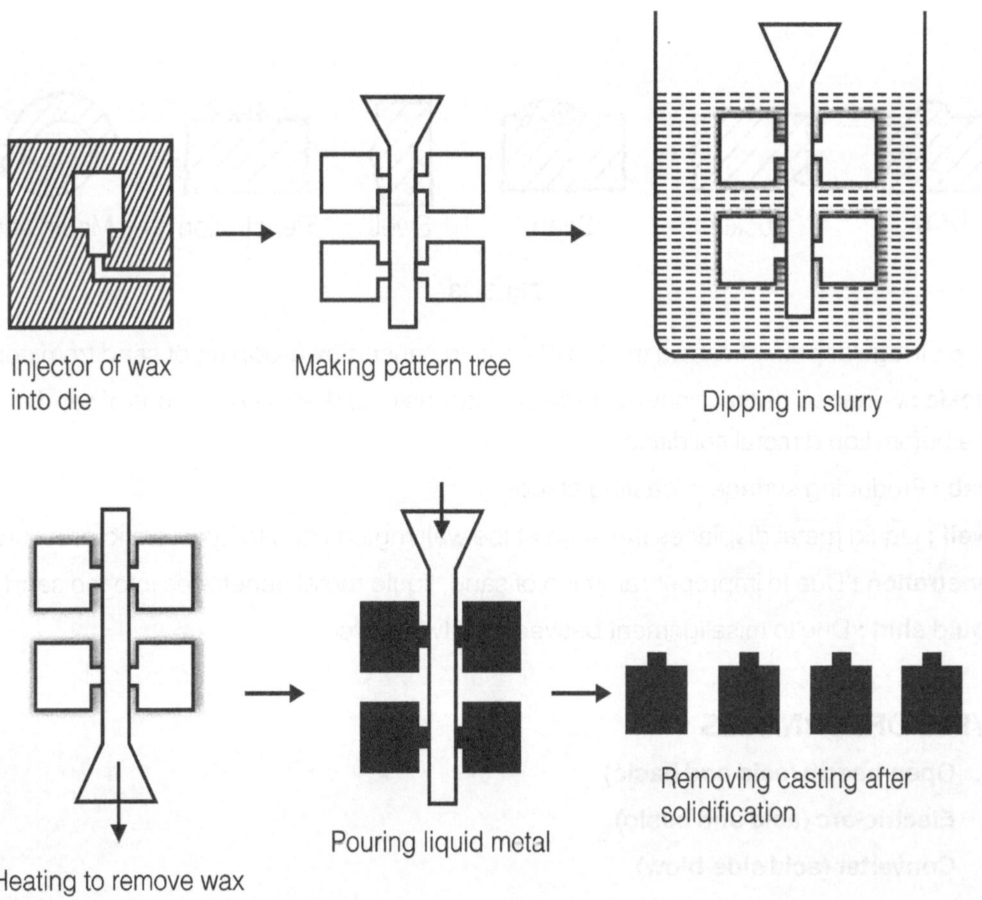

Fig. 3.22 : Investment Casting Procedure

1. Wax is initially injected into a die to make the pattern.
2. A number of these pattern are joined together with a central sprue is the form of a tree.
3. Since workers handle these patterns, it may become dirty. This pattern tree is dipped into alcohol to clean it.
4. Pattern tree is dipped into a slurry which contains water. After removing it from slurry it is dried. The tree is again dipped and then dried. By repeating this procedure for some times a layer of sand (shell) will appear over the pattern.
5. Shell is heated to remove wax from it.
6. The shell is then fired in furnace at 550-1100°C. This ensures complete removal of wax and also gives strength to the shell.
7. Pour the liquid metal when the shell is still hot otherwise shell will catch moisture from the surroundings.
8. After solidification shell is broken to get the final product.

3.13 TYPES OF CASTING DEFECTS

In the previous discussions defects like cold shut misrun, hot tear, shrinkage cavity, sand inclusion, blow holes, pin hole porosity and gas holes are discussed. Following are some other important defects.

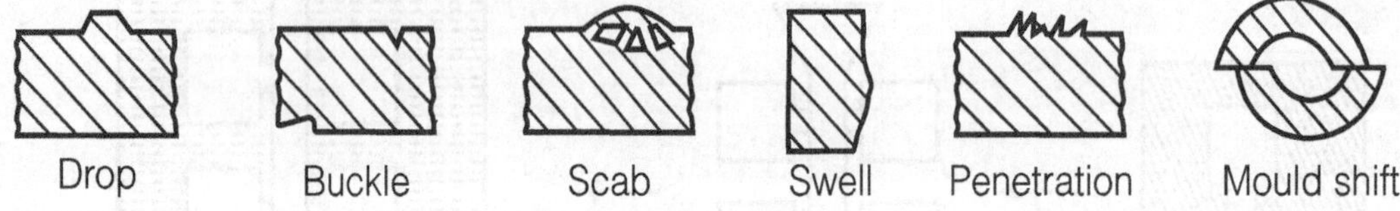

Fig. 3.23

Drop : Irregular projections on the top of coating caused by dropping of sand from cope.

Buckle : v-shaped depressions occurring on flat castings due to expansions of sand at the mould face before liquid metal solidifies.

Scab : Producing surface of casting at roof.

Swell : Liquid metal displaces the sand at the wall regions due to hydrostatic pressure.

Penetration : Due to improper ramming of sand, liquid metal penetrates into the sand.

Mould shift : Due to misalignment between the two halves.

3.14 TYPES OF FURNACES

1. **Open-hearth (acid and basic)**
2. **Electric-arc (acid and basic)**
3. **Converter (acid side-blow)**
4. **Electric induction (acid and basic)**

The distinction between acid and basic practice is in regard to the type of refractories used in the construction and maintenance of the furnace. Furnaces operated by the acid practice are lined with silica base (SiO_2) refractories, and the slags employed in the refining process have a relatively high silica content. Basic furnaces, on the other hand, use a basic refractory such as magnesite or dolomite base and have a high lime (CaO) content in the slag. The choice of furnace and melting practice depends on many variables, including :

1. The plant capacity or tonnage required.
2. The size of the castings
3. The intricacy of the castings
4. The type of steel to be produced, i.e., whether plain or alloyed, high or low carbon, etc.
5. The raw materials available and the prices thereof.
6. Fuel or power costs.
7. The amount of capital to be invested.
8. Previous experience.

Generally, the open-hearth furnace is used for large tonnages and large castings, and the electric furnace for smaller heats or where steels from widely different producers is produced. Special steels or high-alloy steels are often produced in an induction furnace. The converter is used where space is limited and almost continuous pouring is desired.

3.15 METALLURGY OF CAST STEEL

3.15.1 Composition

Cast steel are basically an alloy of iron and carbon. In addition to carbon, which imparts basic properties to steel, the other elements that are normally present in wrought steel are also found in cast steels. There include manganese, silicon, phosphorus and sulfur. These elements usually fall in the following ranges:

Manganese (%)	0.5 - 1.0
Silicon (%)	0.2 - 0.8
Phosphorus (max %)	0.05
Sulfur (max %)	0.06

3.15.2 Melting of Cast Irons (Cupola)

Widespread use of the cupola for gray-iron melting rests upon its unique advantage, which includes :

1. Continuous melting : Foundry production is ircilitated since a ladle of molten iron may be tapped from the furnace at regular intervals. The flow of molten metal and moulds for pouring may be synchronized for quantity production as required by the automotive, agricultural equipment, and similar industries.
2. Low cost of melting : Raw materials and operating costs are lower than any other type of melting furnace producing equivalent tonnage.
3. Chemical-composition control is possible by proper furnace operation with continuous melting.
4. Adcquate temperature control for fluidity in pouring castings can be obtained.

Certain limitations also are characteristic of the cupola furnace. Low carbon percentages in the iron below about 2.80% C are difficult to attain, because of direct contact of molten iron and the carbonaceous fuel.

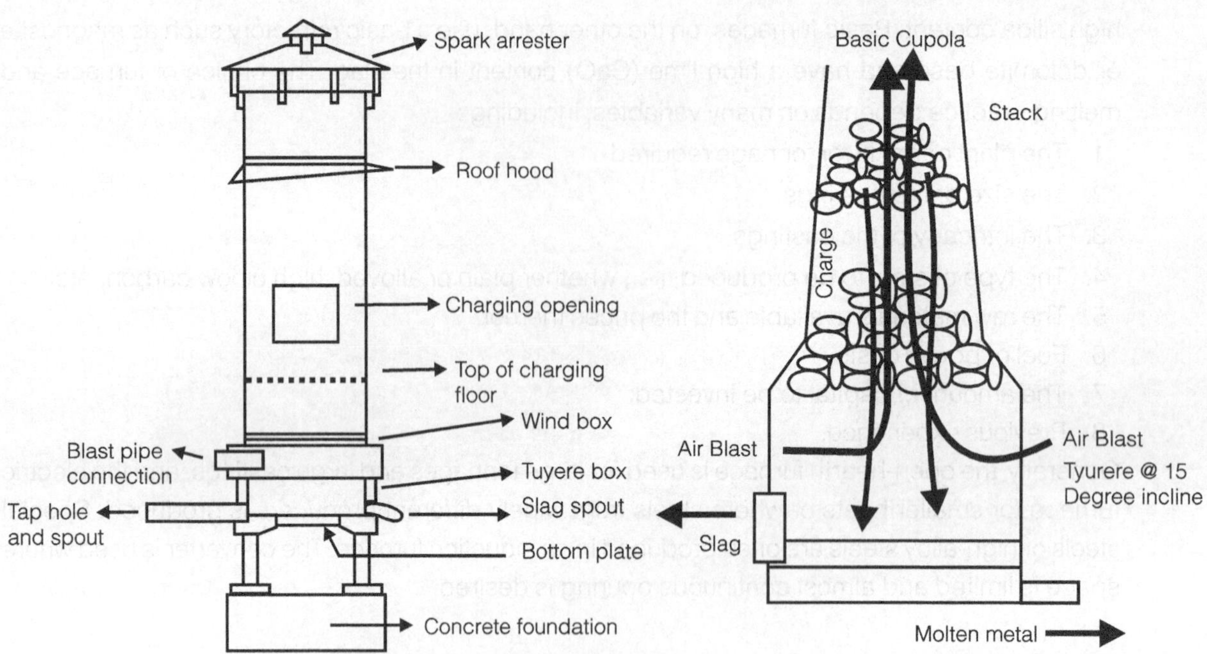

Fig. 3.24 : Cupola

3.15.3 Steps in Operation

A cupola heat includes not only the actual melting operation but all the operations which preceed and follow the period during which iron is being melted. A certain cycle of events occurs each time a heat is made, including the following :

1. Preparation of the refractory lining, bottom, and taphole and slag hole.
2. Lighting and burning in the coke bed.
3. Charging
4. Melting
   (a) Starting the air blast (b) Charging
5. Tapping and slogging
6. Dropping the bottom

Each step above is an important one, and must be properly carried out for a successful heat.

3.16 METHODS OF PURIFYING MATERIAL

1. **Zone refining :** Material is liquified and poured into a mould cavity. Havier impurities will settle down and lighter impurities will float cutting the casted part from top and bottom these impurities are removed. The material is then heated again and then poured into mould. This process is repeated again and again. By this way we approach pure form of material.
2. **Pfann's method :**
   This process is primarily used to purify steels. Material is heated on one side carbon particles will start diffusing on the other side. Heaters are slowly moved towards other side, so carbon will be accumulated on one side. This portion is being cut and the process is then repeated.

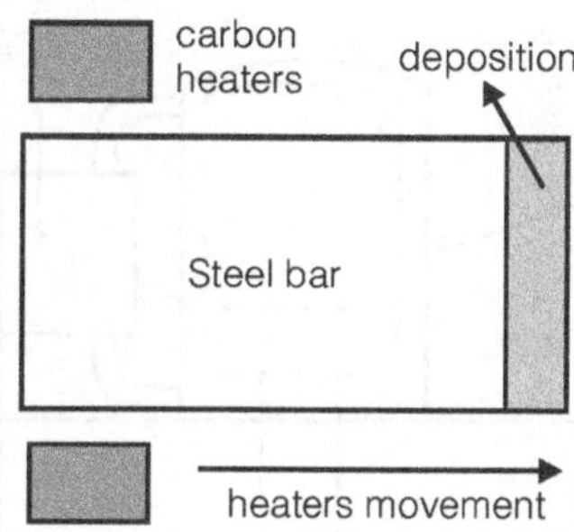

Fig. 3.25 : Pfann's method

3.17 CONTINUOUS CASTING

The casting of ingots is essentially a batch process that produces large sections requiring substantial subsequent processing. Large mechanical equipment that has high construction and operational costs are necessary to break down most ingots. To reduce these costs, continuous casting procedures have been developed for both ferrous and nonferrous metals. Much smaller cross-sectional shapes can be produced directly from the liquid metal with the continuous casting process, thereby eliminating the handling and processing problems associated with large ingots.

Heaters

Continuous casting involves the following sequence of operations:

- Delivery of liquid metal to the casting strand.
- Flow of metal through a distributor (tundish) into the casting mold.
- Formation of the cast section in a water-cooled copper mold.
- Continuous withdrawal of the casting from the mold.
- Further heat removal to solidify the liquid core from the casting by water spraying beyond the mold.
- Cutting to length and removing the cast sections

3.18 INJECTION MOULDING

Injection moulding is the most widely used polymeric fabrication process. It evolved from metal die casting, however, unlike molten metals, polymer melts have a high viscosity and cannot simply be poured into a mould. Instead a large force must be used to inject the polymer into the hollow mould cavity. More melt must also be packed into the mould during solidification to avoid shrinkage in the mould. The injection moulding process is primarily a sequential operation that results in the transformation of plastic pellets into a moulded part. Identical parts are produced through a cyclic process involving the melting of a pellet or powder resin followed by the injection of the polymer melt into the hollow mould cavity under high pressure. The process is used to produce Thinwall containers, Toothbrush handle, Furniture, Luggage shells, Industrial products, Automotive parts, Storage bins, Thermoware and Large appliances.

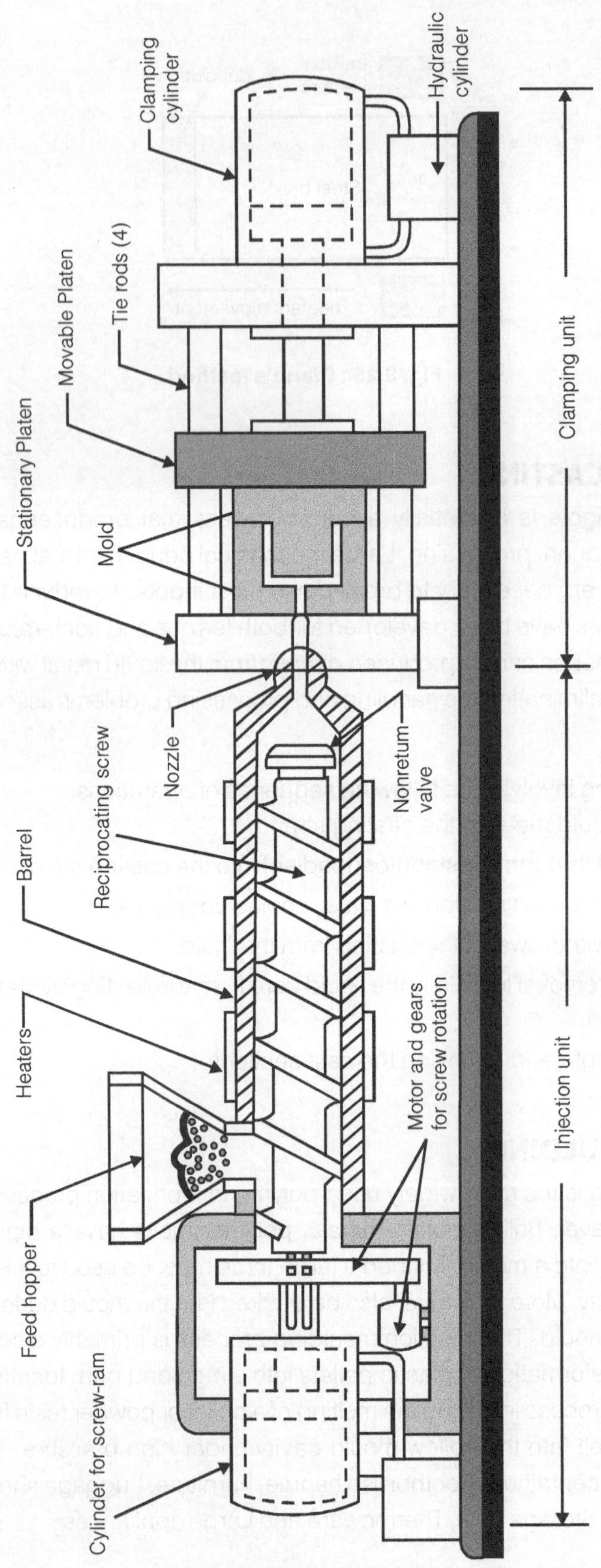

Fig. 3.26 Injection Moulding

Example : 3.6

Design a sprue for avoiding aspiration to deliver liquid iron at a rate of 20 kg/s. Neglecting frictional and orifice effects. Take density of molten iron as 7800 kg/m³. The height of pouring basin is 9 cm and height of sprue as 25 cm.

Solution :

The heights of basin and sprue are 9 cm and 25 cm. So the ratio of areas of sprue at inlet and out let without aspiration effect is

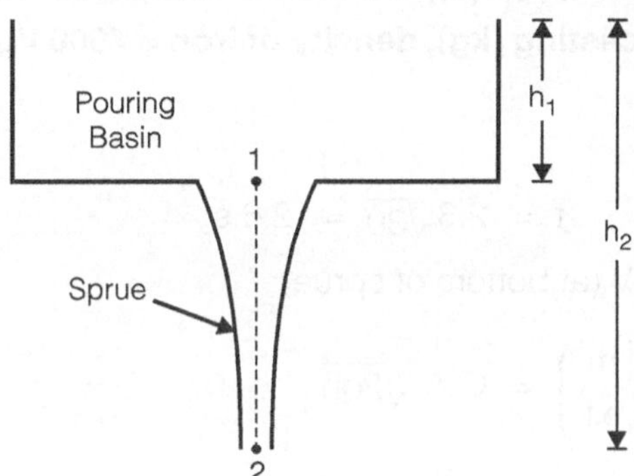

Fig. 3.27

$$\frac{A_1}{A_2} = \sqrt{\frac{h_2}{h_1}}$$

$\because$ $h_1 = 9$ cm, $h_2 = 9 + 25 = 34$ cm

$$\frac{A_1}{A_2} = \sqrt{\frac{34}{9}} = 1.94365$$

Volume flow rate

$$Q = \frac{\dot{m}}{\rho} = \frac{20}{7800} \text{ m}^3/\text{s}$$

Velocity at point '1'

$$V_1 = \sqrt{2gh} = \sqrt{(2 \times 9.81 \times 0.09)}$$

$$= 1.33 \text{ m/s}$$

$$Q = A_1 V_1$$

$$\frac{20}{7800} = A_1 \times 1.33$$

$\therefore$ $$A_1 = 1.93 \times 10^{-3} \text{ m}^2$$

$$A_2 = \frac{A_1}{1.94365} = \frac{1.93 \times 10^{-3}}{1.94365} = 9.93 \times 10^{-4} \text{ m}^2$$

$\therefore$ $$d_1 = \sqrt{\frac{4A_1}{\pi}}$$

$$= 0.0496 \text{ m} = 4.96 \text{ cm}$$

$$d_2 = \sqrt{\frac{4A_2}{\pi}}$$
$$= 0.03555 \text{ m} = 3.555 \text{ cm}$$

Example : 3.7

Calculate the gating dimensions with a gating ratio of 1 : 2 : 3 for a ductile iron casting of section thickness 12 mm weighing 30 kg. Assume a sprue height of 200 mm.

Given : Pouring time, t = $k\sqrt{W}$ (s), k = 2.3 for thickness = 12 mm, flow/discharge factor C_d = 0.9, W = weight of casting (kg), density of iron = 7900 kg/m³.

Solution :

Pouring time of casting

$$t = 2.3\sqrt{30} = 12.6 \text{ s}$$

$\because$ flow rate, $Q = C_d A V$ (at bottom of sprue)

$$\left(\frac{m}{\rho . t}\right) = C_d A_s \sqrt{2gh}$$

$$\frac{30}{7900 \times 12.6} = 0.9 A_s \sqrt{2 \times 9.81 \times 0.2}$$

$\therefore$ sprue base area, $A_s = 1.69 \times 10^{-4} \text{ m}^2$

$\therefore$ diameter of sprue base, $d_s = \sqrt{\frac{4A_s}{\pi}} = 0.01467 \text{ m} = 14.67 \text{ mm}$

Area of runner, $A_r = 2 A_s$
$$= 2 \times 1.69 \times 10^{-4}$$
$$= 3.38 \times 10^{-4}$$

diameter of runner, $d_r = \sqrt{\frac{4A_r}{\pi}}$
$$= 0.02074 \text{ m} = 20.74 \text{ mm}$$

Area of gates, $A_g = 3 A_s = 3 \times 1.69 \times 10^{-4} = 5.07 \times 10^{-4} \text{ m}^2$

diameter of gates, $d_g = \sqrt{\frac{4A_g}{\pi}}$
$$= 0.0254 \text{ m} = 25.407 \text{ mm}$$

Example : 3.8

Determine the riser size for a cylindrical riser for plate like steel casting with 300 mm × 200 mm × 25 mm. Check for pipe height in the riser and make suitable modification in riser design. Take V_r/V_c = 0.26 and percentage shrinkage for steel as 7.

Solution :

$$\because \left(\frac{A}{V}\right)_{Casting} = \frac{2 \times 300 \times 200 + 25 \times 300 \times 2 + 25 \times 200 \times 2}{300 \times 200 \times 25} = \frac{145000}{1500000} = 0.0967$$

Volume of riser, $V_r = 0.26 \times 1500000 = 390000 \text{ mm}^3$

For side placement of riser, d = h

$$V_r = \frac{\pi}{4}.d_r^3 = 390000$$

d_r = 79.2 mm which is far greater than the thickness of casting.

$$\left(\frac{A}{V}\right)_r = \frac{6}{79.6} = 0.07537\ (< 0.0967)$$

Solidification time justify this placement

for top placement $$h = \frac{d}{2}$$

$$V_r = \frac{\pi}{4}.d^2.\frac{d}{2} = 390000$$

d_r = 99.77 mm (less than the cross-section dimensions of casting)

$$\left(\frac{A}{V_r}\right) = \frac{6}{d_r} = 0.06014\ (< 0.0967)$$

This time is also justifying this place. So put the riser in to this position.

Height of riser $$h = \frac{d}{2} = \frac{99.77}{2} = 49.885 \text{ mm}$$

Pipe height: We know that pipe cavity compensate the shrinkage allowance which is around $\frac{1}{6}^{th}$, of riser volume. Let the new modified height of riser is H.

$$\therefore \quad V_s = \frac{7}{100} \times 300 \times 200 \times 25 = 105000 \text{ mm}^3$$

$$\frac{1}{6}.\frac{\pi}{4}.d_r^2 \times H = 105000$$

$$H = \frac{24 \times 105000}{\pi \times (99.77)^2} = 80.6 \text{ mm}$$

Hence the new height of riser should be more than 80.6 mm in order to avoid pipe extension in to the casting.

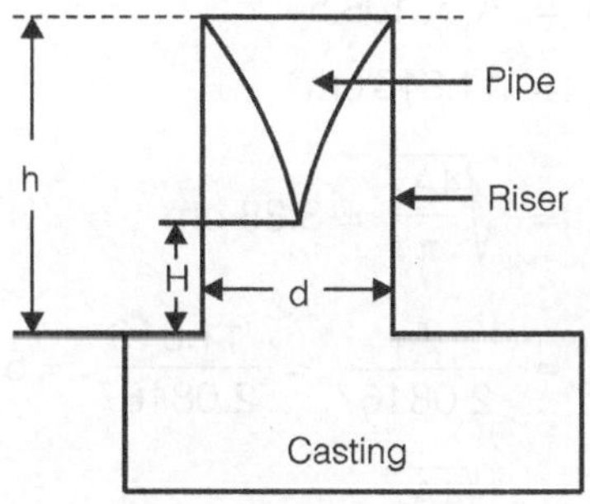

Fig. 3.28

Example : 3.9

Design a down sprue, avoiding aspiration, as shown in fig. 3.27, to deliver liquid cast iron (ρ_m = 7800 kg/m^3) at a rate of 10 kg/s against no head at the base of the sprue Neglect the friction and orifice effects.

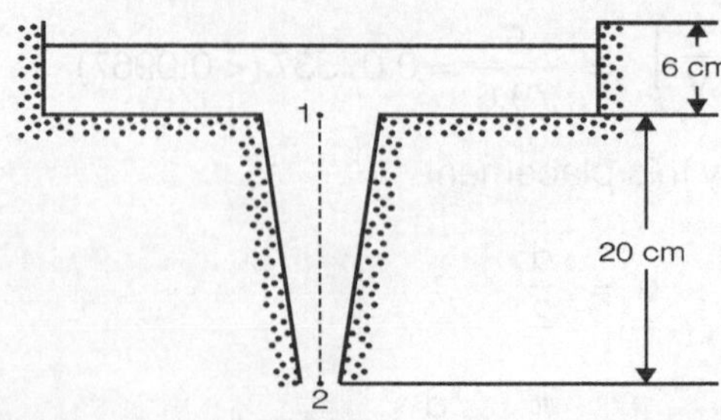

Fig. 3.29

Solution :

In order to avoid aspiration effect

$$\therefore \quad \frac{A_1}{A_2} = \sqrt{\frac{h_2}{h_1}}$$

where A_1 and A_2 are cross-sectional areas at point 1 and 2.

$$h_2 = 20 + 6 = 26 \text{ cm}$$

$$h_1 = 6 \text{ cm}$$

$$\therefore \quad \frac{A_1}{A_2} = \sqrt{\frac{26}{6}} = 2.08167$$

The velocity at 1 $\quad V_1 = \sqrt{2gh_1} \times \sqrt{2 \times 981 \times 6}$

$$V_1 = 108.5 \text{ cm/s}$$

Volume flow rate $\quad \dot{Q} = \frac{\dot{m}}{\rho} = \frac{10}{7800} = \frac{1}{780} \text{ m}^3\text{/s}$

$$Q = \frac{10^6}{780} = 1282.05 \text{ cc/s}$$

$$\therefore \quad Q = A_1 V_1$$

$$1282.05 = A_1 \times 108.5$$

$$A_1 = 11.816 \text{ cm}^2$$

$$d_1 = \sqrt{\frac{4A_1}{\pi}} = 3.88 \text{ cm}$$

$$A_2 = \frac{A_1}{2.08167} = \frac{11.816}{2.08167} = 5.6762 \text{ cm}^2$$

$$d_2 = \sqrt{\frac{4A_2}{\pi}} = 2.69 \text{ cm}$$

POINTS TO PONDER

1. Types of Mould Preparation

I. Bench Moulding:The moulding done on a bench of convenient height to the Moulder is called bench moulding. It is used to produce small moulds which are light and can be easily handled. In this process, loose-piece patterns are normally used. Because of ramming and other manual operations involved this is a slow and laborious method.

II. Floor Moulding: This method is used when moulds are larger and cannot be accommodated on benches. All the work in making the mould is done on the floor of the foundry floor hence calledas floor moulding. It is used for all medium sized and large castings.

III. Pit Moulding: Very large moulds made in a pit or cavity cut in the floor to accommodate very large castings is called pit moulding. The pit acts as a drag. Since pit moulds can resist pressure developed by hot gases therefore it greatly saves pattern expenses.

IV. Machine Moulding: The moulding done by machine is called a machine moulding. Small medium and large moulds may be made with the help of a variety of machines. Machine moulding isusually faster and more uniform than bench moulding. Machine moulding generally requires mounted patterns.

2. Squeezing

More uniform results and higher production rates are obtained by squeezing machines. Hand-operated squeezers were limited to small moulds and are obsolete; air-operated machines permit an increase in the allowable size of moulds as well as in the production rate. These machines are suitable for shallow moulds. Squeezer moulding machines produce greatest sand density at the top of the flask and softest near the parting line of pattern. Air-operated machines are also applied in vertical moulding processes using flaskless moulds. Horizontal impact moulding sends shock waves through the sand to pack the grains tightly.

3. Jolting Machine

In jolt moulding machines the pattern is placed on a platen attached to the top of an air cylinder. After the table is raised, a quick-release port opens, and the piston, platen, and mould drop free against the top of the cylinder or striking pads. The impact packs the sand. The densities produced by this machine are greatest next to the parting line of the pattern and softest near the top of the flask. This procedure can be used for any flask that can be rammed on a moulding machine. As a separate unit, it is used primarily for medium and large work. Where plain jolt machines are used on large work, it is usual to ram the top of the flask manually with an air hammer.

4. Jolt and squeeze machine

Jolt squeeze machines use both the jolt and the squeeze procedures. The platen is mounted on two air cylinders: a small cylinder to jolt and a large one to squeeze the mould. They are widely used for small and medium work, and with match-plate or gated patterns. Pattern-stripping devices can be incorporated with jolt or squeezer machines to permit mechanical removal of the pattern. Pattern removal can also be accomplished by using jolt-rock over-draw or jolt-squeeze-rollover-draw machines.

5. Sand Slinger

The sand slinger is the most widely applicable type of ramming machine. It consists of an impeller mounted on the end of a double-jointed arm which is fed with sand by belt conveyors mounted on the arm. The impeller rotating at high speed gives sufficient velocity to the sand to ram it in the mould by impact. The head may be directed to all parts of the flask manually on the larger machines and may be automatically controlled on smaller units used for the high-speed production of small moulds.

6. Casting Inspection

General inspection procedures for castings are established at the foundry to ensure conformance withcustomer drawings and documents, which are frequently based on various government, technical society, or commercial specifications.

(a) **Methods for Determining Surface Quality.** Cracks and other imperfections at the surface of a casting can bedetected by a number of inspection techniques, including visual inspection, chemical etching, liquid penetrant inspection,eddy current inspection, and magnetic particle inspection (which can also reveal discontinuities situated immediatelybelow the surface). All inspection methods require clean and relatively smooth surfaces for effective results.

(b) **Methods for Detecting Internal Discontinuities.** The principal nondestructive methods used for detecting internaldiscontinuities in castings are radiographic, ultrasonic, and eddy current inspection. Of these methods, radiography is themost highly developed technique for detailed inspection; it can provide a pictorial representation of the form and extent ofmany types of internal discontinuities. Ultrasonic inspection, which is less universally applicable, can give qualitativeindications of many discontinuities. It is especially useful in the inspection of castings of fairly simple design, where thesignal pattern can be most reliably interpreted. Ultrasonic inspection can also be used to determine the shape of graphiteparticles in cast iron. Eddy current and other closely related electromagnetic methods are used to sort castings forvariations in composition, surface hardness, and structure.

(c) **Methods for Dimensional Inspection.** There are a number of techniques used to determine the dimensionalaccuracy of castings. These include manual checks with micro-meters, manual and automatic gages, coordinate-measuringmachines, and three-dimensional automatic inspection stations (machine vision systems).

(i) **Visual Inspection** is a non-destructive testing technique that provides a means of detecting and examining a variety of surface flaws. Visual inspection is also the most widely used method for detecting and examining surface cracks, which are particularly important because of their relationship to structural failure mechanisms. Even when other non-destructive techniques are used to detect surface cracks, visual inspection often provides a useful supplement.

(ii) **Magnetic Particle inspection** is a method of locating surface and subsurface discontinuities in ferromagnetic materials. It depends on the fact that when the material or part under test is magnetized, magnetic discontinuities that lie in a direction generally transverse to the direction of the magnetic field will cause a leakage field to be formed at and above the surface of the part. The presence of this leakage field, and therefore the presence of the discontinuity, is detected by the use of finely divided ferromagnetic particles applied over the surface, with some of the particles being gathered and held by the leakage field. This magnetically held collection of particles forms an outline of the discontinuity and generally indicates its location, size, shape, and extent. Magnetic particles are applied over a surface as dry particles, or as wet particles in a liquid carrier such as water or oil. The principal industrial uses of magnetic particle inspection are final inspection, receiving inspection, in-process inspection and quality control, maintenance and overhaul in the transportation industries, plant and machinery maintenance, and inspection of large components.

(iii) **Dye Penetrantion Test** is a non-destructive method of revealing discontinuities that are open to thesurfaces of solid and essentially nonporous materials. Indications of a wide spectrum of flaw sizes can be found regardless of the configuration of the workpiece and regardless of flaw orientations. Liquid penetrants seep into various types of minute surface openings by capillary action. Because of this, the process is well suited to the detection of all types of surface cracks, laps, porosity, shrinkage areas, laminations, and similar discontinuities. It is extensively used for the inspection of wrought and cast products of both ferrous and nonferrous metals, powder metallurgy parts, ceramics, plastics, and glass objects. In practice, the liquid penetrant process is relatively simple to utilize and control. The major limitation of liquid penetrant inspection is that it can detect only imperfections that are open to thesurface;some other method must be used for detecting subsurface flaws. Another factor that can limit the use of liquid penetrantsis surface roughness or porosity.

(iv) **Pressure test:** For pressurised products, it is important to check the permeability by a pressure test. For low pressure, air is used. For higher pressure a liquid (mostly water) is used and for extreme pressure it can be oil or another liquid. Sometimes the product is filled with liquid without pressure. The liquid is mostly petrol or an acid. The test requires a pressure, applied for a minimum time, indicating a maximum pressure drop during that time. Attention must be paid on the pressure compared to the wall thickness to avoid accidents.

(v) **Ultrasonic testing**is a non-destructive method in which beams of high-frequency sound waves are introduced into materials for the detection of surface and subsurface flaws in the material. The sound waves travel through the material with some attendant loss of energy (attenuation) and are reflected at interfaces. The reflected beam is displayed and then analyzed to define the presence and location of flaws or discontinuities. The problem is that it cannot be performed on every material. The austenitic material is difficult to test as well as the grey (flake) and malleable irons. The other problem is that the operator reads the result on the screen and that there is no proof for later discussion in the form of a paper or file. The advantage is that the equipment is not expensive and the tests can be done on sight, the location of the non-conformance is easy to state (with simple calculations). Standardising should be done before every test, to adapt the readings to the proper material.

(vi) **Radiographic Test:** Because of differences in density and variations in thickness of the part or differencesin absorption characteristics caused by variations in composition, different portions of a test-piece absorb different amount of penetrating radiation. These variations in the absorption of the penetrating radiation can be monitored bydetecting the unabsorbed radiation that passes through the testpiece. Advantage is that it detects internal flaws, significant variations in composition and records permanent raw inspection data.Cracks cannot be detected unless they are essentially parallel to the radiation beam, long testing times and the sample should be accessible from both sides.

7. Fluidity

An ability of metals and alloys to flow through the gating system filling the cavity of the casting mold and conforming its shape. Pouring is a key element in foundry process, the main goal is to get metal to flow into all regions of mould before solidifying. The properties of the melt iin a casting process are very important. The ability of a particular casting melt flow into a mould before freezing is crutial in the consideration of metal casting techniques. This ability is termed as the liquid metal's fluidity.The relative fluidity of a certain metal casting melt can be quantified by spiral test. The geometry of the spiral mould acts to limit the flow of liquid metal through the length of its spiral cavity. The more fluidity possessed by the molten metal, the farther into the spiral it will be able to flow before hardening. The maximum point the metal reaches upon the casting's solidification may be indexed as the relative fluidity of the melt. Following are the factors that determines fluidity:

(i) **Solidification interval** (the difference between liquidus and solidus temperatures of the alloy or mushy zone).The wider solidification interval of an alloy, the lower fluidity of the alloy, as a certain part of the solidifying alloy has an increased viscosity caused by the presence of solid primary crystals, formed in course of cooling the alloy below the liquidus temperature. The fluidity of pure metals is better than that of their alloys. Eutectic alloys, characterized by nil solidification interval, possess good fluidity.

(ii) **Viscosity** of the alloy (the lower viscosity of an alloy, the better its fluidity). As viscosity decreases with temperature, metals and alloys cast at increased temperature fill better the mold cavity.

(iii) **Surface tension** of the alloy. Low surface tension causes better filling of the mold cavity.

8. A note on different Casting Defects

a. Blow holes are round or elongated cavities, usually smooth-walled. They are found on, or close under, the surface of the casting.

b. Gas holes are either caused by insufficient venting of the core, or the core prints are not large enough to allow the gas to escape.

c. Pin holes and porosity are caused by the absorption of gas in the molten metal. The reason for this is incorrect fluxing or de-gassing procedure.

d. Oxide and dross inclusions are caused by the entrapment of surface oxide, or other foreign matter, during pouring of the molten metal.

e. Sand inclusion and drop result from careless ramming or withdrawing of the pattern, leaving loose sand in the mould cavity. This will be included in the casting.

f. Drop is when the upper surface of the mould cracks and pieces of sand fall into the mould cavity. This is caused either by sand of too low a green strength, or by soft ramming

g. Scabs are rough, irregular projections on the surface of the casting containing embedded

h. sand or refractory. There are two types: expansion and erosion scabs.

i. Flashing (also called fin or feather) is the formation of an unwanted run of metal at the parting line, caused by leakage of molten metal into the space between two halves of the mould. This may be caused by poor mould making, or by insufficient weight holding the cope in position.

j. A shrinkage cavity is a void or depression caused by uncontrolled solidification. The causes due to wrong location or inefficient gating and riser system; or poor design of casting or abrupt changes of section thickness; or pouring temperature of metal being too high.

k. Shift is the result of a mismatch of the sections of the casting, usually at the parting line. The most likely cause is carelessness. Core shift is caused by undersized (or oversized) core prints, or chaplets (locating pins) of the wrong size.

l. Swell is the term describing an unwanted positive, usually caused by weak, or inconsistent ramming of the sand, or by the cope lifting because there is not enough weight holding it down. Occasionally a swell is caused by metal flowing into the mould too fast, so building up too much pressure.

m. Warped castings are rare in small foundries. This is more often encountered in large, or flat, castings where insufficient camber allowance has been estimated.

n. A cold shut looks, at first sight, like a crack, but its smooth rounded edges betray its true nature. It is produced when two streams of metal, flowing in different directions, meet without fusing together. This is rare in bronze or brass, but very common in aluminium, a metal which oxidizes rapidly. A layer of oxide, pushed ahead of the melt, prevents the two flows of metal fusing.

o. Rough surface; metal penetration

p. The most common cause is poor quality facing (surface) sand. Sand which has high permeability, large grain size and poor green strength will give a poor surface to the casting. The other cause is soft ramming.

9. Terminology in Casting

1. **As-Cast Condition:** Casting without subsequent heat treatment.
2. **Back Draft:** Reverse taper which would prevent removal of a pattern from a mould or a core from a corebox.
3. **Bar, Flask:** A rib in the cope of a tight flask to help support the sand.
4. **Backing Sand:** The bulk of the sand in the flask. The sand compacted on top of the facing sand that covers the pattern.
5. **Binder:** The bonding agent used as an additive to mould or core sand to impart strength of plasticity in a dry state
6. **Blasting or Blast Cleaning:** A process for cleaning or finishing metal objects by use of an air blast or centrifugal wheel that throws abrasive particles against the surfaces of the work pieces. Small irregular particles of metal are used as the abrasive in grit blasting; sand in sand blasting; and steel balls in shot blasting.
7. **Bleeder:** A defect wherein a casting lacks completeness due to molten metal draining or leaking out of some part of the mould cavity after pouring has stopped.
8. **Burn-On Sand:** Sand adhering to the surface of the casting that is extremely difficult to remove.
9. **Burn-Out:** Firing a mould at a high temperature to remove pattern material residue.
10. **Casting Yield:** The weight of casting or castings divided by the total weight of metal poured into the mould, expressed as a percent.
11. **Cavity:** The portion of a cast which forms the external shape
12. **Charge:** A given weight of metal introduced into the casting furnace.
13. **Cleaning:** Removal of runners, risers, flash, surplus metal and sand from a casting.
14. **Collapsible Core:** A metal insert made in two or more pieces to permit withdrawal from an undercut mould surface.
15. **Cope:** The top half of a horizontally parted mould.
16. **Core:** A sand or metal insert in a mould to shape the interior of the casting or that part of the casting that cannot be shaped by the mould pattern. The portion of the cast which forms the internal shape
17. **Core Assembly:** An assembly made from a number of cores.
18. **Core-box:** The wooden, metal or plastic tool used to produce cores.
19. **Coreprint:** A projection on a pattern that leaves an impression in the mould for supporting the core.
20. **Core Wash:** A liquid suspension of a refractory material applied to cores and dried (intended to improve surface of casting).
21. **Crush:** The displacement of sand at mould joints.
22. **Cupola:** A cylindrical, straight shaft furnace (usually lined with refractories) for melting metal in direct contact with coke by forcing air under pressure through openings near its base.
23. **Cure:** To harden.
24. **Die:** A metal form used as a permanent mould for die casting or for wax pattern in investment casting.

25. **Die Cavity:** The impression in a die into which pattern material is forced.
26. **Ejector Pins:** Movable pins in the pattern die that "push" to remove cast pattern from the dies.
27. **External Undercut:** Any recess or projection on the outside of the die block which prevents its removal from the cavity.
28. **Facing Sand:** The sand used to surround the pattern that produces the surface in contact with the molten metal.
29. **Feeder:** Also called "riser", it is part of the gating system that forms the reservoir of molten metal necessary to compensate for losses due to shrinkage as the metal solidifies.
30. **Flask:** A rigid metal or wood frame used to hold the sand of which a mould is formed and usually consisting of two parts, cope and drag.
31. **Foundry Returns:** Metal (of unknown composition) in the form of gates, sprues, runners, risers and scraped castings returned to the furnace for re-melting.
32. **Gas Porosity:** A condition existing in a casting caused by the trapping of gas in the molten metal or by mould gases evolved during the pouring of the casting.
33. **Inclusions:** Particles of slag, refractory materials, sand or deoxidation products trapped in the casting during pouring solidifications.
34. **Internal Shrinkage:** A void or network of voids within a casting caused by inadequate feeding of that section during solidification.
35. **Lost Foam:** A casting process in which a foam pattern is replaced by molten in a flask filled with loose sand to form a casting.
36. **Master Pattern:** The object from which a die can be made; generally a metal model of the part to be cast with process shrinkage.
37. **Mould Shift:** A casting defect which results when the parts of the mould do not match at the parting line.
38. **Mould Wash:** A slurry of refractory material, such as graphite and silica flour, used in coating the surface of the mould cavity to provide an improved casting surface.
39. **Mould Weight:** A weight that is applied to the top of a mould to keep the mould from separating.
40. **Nitriding:** A process of shallow case hardening in which a ferrous alloy, usually of a special composition, is heated in an atmosphere of ammonia, or in contact with nitrogenous material, to produce surface hardening by formation of nitrites, without quenching.
41. **Nobake Process:** Moulds/cores produced with a resin bonded air-setting sand also known an air-set process because moulds are left to harden under ambient conditions.
42. **Parting Line:** A mark or line produced on the cast, formed at the junction of the parting dies.
43. **Patternmakers Shrinkage:** Contraction allowance made on patterns to compensate for the decrease in dimensions as the solidified casting cools in the mould from freezing temperature of the metal to room temperature. Pattern is made larger by the amount of contraction that is characteristic of the particular metal to be used.
44. **Porosity:** Holes in the produced casting due to Gasses trapped in the mould, the reaction of molten metal with moisture in the molten sand, or the imperfect fusion of chaplets with molten metal.
45. **Rod:** A heavy wire or bar in a sand core used for reinforcing.
46. **Runner:** The portion of the gate assembly that connects the down gate (sprue) with the casting ingate or riser. The term also applies to that part of the pattern which forms the runner.
47. **Runout:** Unintentional escape of molten metal from a mould.
48. **Shrink Hole:** A hole or cavity in a casting resulting from shrinkage and insufficient feed metal, and formed during solidification.

49. **Shrinkage:** Decrease in volume of the metal as it solidifies.
50. **Sprues:** Channels cut into a mould to allow for the entry of metal. Also the name given to the metal rods that assume this shape in the final casting.

10. Distortion/camber Allowances

Certain cast shapes, such as large flat plates and dome or U-shaped castings, sometimes distort when reproduced from straight or perfect patterns. This distortion is caused by the non-uniform contraction stresses during the solidification of irregularly shaped designs. Minor distortions are normally corrected by mechanically pressing or straightening the casting, but if distortions are consistent and prominent, the pattern shape can be intentionally changed to counteract the casting distortions. The "distorted" pattern will then produce a distortion-free casting.

11. Rapping/shake Allowance

Before the withdrawal from the sand mould, the pattern is rapped all around the vertical faces to enlarge the mould cavity slightly, which facilitate its removal. Since it enlarges the final casting made, it is desirable that the original pattern dimension should be reduced to account for this increase. There is no sure way of quantifying this allowance, since it is highly dependent on the foundry personnel practice involved. It is a negative allowance and is to be applied only to those dimensions that are parallel to the parting plane.

12. Comparison of different casting processes

Metal Casting Comparison Chart

	Sand Casting	Permanent Moulding	Die Casting	Ceramic & Investment Casting	Investment Casting
Dimensional Tolerances	± .010" ± .030"	± .010" ± .050"	± .001" ± .015"	± .010" ± .020"	± .003 per inch premium
Relative Cost High Part Quantity	Low	Low	Lowest	Highest	Medium
Relative Cost Low Part Quantity	Lowest	High	Highest	Medium	11 kg
Casting Weight Range	Unlimited	45 kg	35 kg	Ceramic-150 kg Investment 45 kg	11 kg
Minimum Wall Thickness	1/10"	1/8"	1/32"	1/16"	.020"
As Cast Finish (RMS)	150-350	150-200	63-90	60-125	60-125
Draft Angle (Degrees)	5	2–4	1	1/4-1/2	±1/4-1/2
Design Complexity/Detail	Fair-good	Fair	Good	Best	Best
Relative Ease of Design changes in Production	Best	Poor	Poorest	Fair	Good
Castable Alloys	Most Ferrous/ Non-ferrous Metals	Al-base & Cu-base Preferred	Al-base Preferred	Most ferrous/ Non-ferrous Metals	Ferrous, Copper and Copper Based, Cobalt Based, Nickel Based, Aluminium Based

PRACTICE QUESTIONS

1. Consider the following ingredients used in molding.

1. Dry silica sand
2. Clay
3. Phenol formaldehyde
4. Sodium silicate

Those used for shell mould casting include

(a) 1, 2 and 4 (b) 2, 3 and 4
(c) 1 and 3 (d) 1, 2, 3 and 4

2. A sand casting mould assembly is shown in the given figure. The elements marked A and B are respectively.

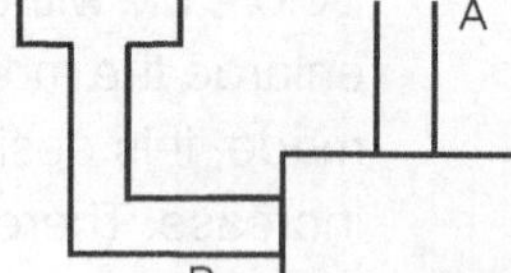

(a) Sprue and riser (b) Ingate and riser
(c) Drag and runner (d) Riser and runner

3. Hardness of green sand mould increases with

(a) Increase in moisture content beyond 6 percent.
(b) Increase in permeability
(c) Decrease in permeability
(d) Increase in both moisture content and permeability.

4. In solidification of metal during casting, compensation during solid contraction is:

(a) Provided by the oversize pattern
(b) Achieved by properly placed risers
(c) Obtained by promoting directions solidifications
(d) Made by providing chills

5. Match **List-I** (Moulding process) with **List-II** (Binding agent) and select the correct answer using the codes given below the lists:

List-I		**List-II**	
A.	Green sand	**1.**	Silicate
B.	Cores sand	**2.**	Organic
C.	Shell moulding	**3.**	Clay
D.	Carbondioxide process	**4.**	Plaster of paris
		5.	Plastic

Codes:

	A	B	C	D
(a)	3	2	5	1
(b)	3	2	4	1
(c)	2	3	5	4
(d)	2	3	4	5

6. In which of the following are metal moulds used?

(a) Green sand mould (b) Dry sand mould
(c) Die casting process (d) Loam moulding

7. The calculation of center line feeding resistance is helpful in

(a) Riser design (b) Choking area calculation
(c) Pouring basin design (d) The sprue design

8. Which one of the following is the correct statements? Gate is provided in moulds to
(a) feed the casting at constant rate (b) give passage to gasses
(c) compensate for shrinkage (d) avoid cavities

9. Volume of a cube of side 'i' and volume of a sphere of radius 'r' are equal. Both the cube and the sphere are solid and of same material. They are being cast. The ratio of the solidification time if the cube to the same of sphere is

(a) $\left(\frac{4\pi}{6}\right)^3\left(\frac{r}{i}\right)^6$ (b) $\left(\frac{4\pi}{6}\right)\left(\frac{r}{i}\right)^2$

(c) $\left(\frac{4\pi}{6}\right)^2\left(\frac{r}{i}\right)^3$ (d) $\left(\frac{4\pi}{6}\right)^2\left(\frac{r}{i}\right)^4$

10. The pattern adopted for those casting where there are some portions which are structurally weak and are likely to break by the force of ramming are called:
(a) Loose piece pattern (b) Follow board pattern
(c) Skeleton pattern (d) Single piece Pattern

11. In a sand casting operation, the total liquid head is maintained constant such that it is equal to the mould height. The time taken to fill the mould with a top gates is t_A. If the same mould is filled with a bottom gate, then the time taken is t_B. Ignore the time required to fill the runner and frictional effects. Assume atmospheric pressure at the top molten metal surfaces. The relation between t_A and t_B is
(a) $t_B = \sqrt{2}\,t_A$ (b) $t_B = 2t_A$
(c) $t_B = \frac{t_A}{\sqrt{2}}$ (d) $t_B = 2\sqrt{2t_A}$

12. Plastic bottles are made using
(a) Blow moulding (b) Injection moulding
(c) Preform moulding (d) Slush casting

13. While cooling, a cubical casting of side 40 mm undergoes 3%, 4% and 5% volume shrinkage during the liquid state, phase transition and solid state, respectively. The volume of metal compensated from the riser is
(a) 2% (b) 7%
(c) 8% (d) 9%

14. If the melting ratio of a couple is 10:1, then the coke requirement for one ton melt will be:
(a) 0.1 ton (b) 10 tons
(c) 1 ton (d) 11 tons

15. Consider the following statements:
Assertion(A): Steel can be melted in hot blast cupola.
Reason(R): In hot blast cupola, the flue gases are used to preheat the air blast to the cupola so that the temperature in the furnace is considerably higher than that in a conventional cupola.
(a) Both A and R are individually true and R is the correct explanation of A.
(b) Both A and R are individually true and R is not the correct explanation of A.
(c) A is true but R is false
(d) A is false but R is true

16. Which of the following are in the requirements of an ideal gating system?

1. The molten metal should enter the mould cavity with as high a velocity as possible.
2. It should facilitate complete filling of the mould cavity
3. It should be able to prevent the absorption of air or gases from the surroundings on the molten metal while flowing through it.

Select the correct answer using the codes given below:

(a) 1, 2 and 3 (b) 1 and 3
(c) 2 and 3 (d) 1 and 3

17. Assertion(A): The rate of flow metal through sprue is NOT a function of the cross-sectional area of sprue, runner and gate.

Reason(R): If respective cross-sectional areas of sprue , runner and gate are in the ratio of 1:2:2, the system is known as unpressurised gating system.

(a) Both A and R are individually true and R is the correct explanation of A.
(b) Both A and R are true and R is not the correct explanation of A.
(c) A is true but R is false
(d) A is false but R is true

18. In gating system design, which one of the following is the correct sequence in which choke area, pouring time, pouring basin and sprue sizes are calculated?

(a) Choke area- pouring time–pouring basin- sprue
(b) Pouring basin-sprue-choke area-pouring time
(c) Choke area- sprue-pouring basin- pouring time
(d) Pouring basin- pouring time-Choke area–sprue.

19. A gating ratio of 1:2:4 is used to design the gating system for magnesium alloy casting. This gating ratio refers to the cross- setion areas of the various gating elements as given below:

1. Down sprue
2. Runner bar
3. Ingates

The correct sequence of the above elements in the ratio 1:2:4 is:

(a) 1, 2 and 3 (b) 1, 3 and 2
(c) 2, 3 and 1 (d) 3, 1 and 2

20. A mould has down sprue whose length is 20 cm and the cross sectional area at the base of the down sprue is 1cm^2. The down sprue feeds a horizontal runner leading into the mould cavity of volume 1000cm^3. The time required to fill the mould cavity will be

(a) 4.05 s (b) 5.05 s
(c) 6.05 s (d) 7.25 s

21. Bottom gating system is some times preferred in casting because

(a) It enables rapid filling of mould cavity (b) It is easier to provide in the mould
(c) It provides cleaner metal (d) It reduces splashing and turbulence

22. A 200mm long down sprue has an area of cross-section of 650 mm^2 where the pouring basin meets the down sprue (i.e. at the beginning of the down sprue). A constant head of molten metal flow rate is 6.5 x 10^{4}mm^3/s. considering the end of down sprue to be open to atmosphere and an acceleration due to gravity of 10mm/s^2, the area of the down sprue in mm^2 at its end (avoiding aspiration effect) should be

(a) 650.0 (b) 350.0
(c) 290.7 (d) 190.0

23. Directional solidification in castings can be improved by using:
(a) Chills and Chaplets
(b) Chills and Padding
(c) Chaplets and Padding
(d) Chills, Chaplets and Padding

24. Match **List-I** (Casting) with **List-II** (Explanation) and select the correct answer using the codes given below the lists:

List-I	List-II
A. Metallic projections	**1.** Consists of rounded or rough internal or exposed cavities including blow holes and pin holes
B. Cavities	**2.** Formed during melting, solidification and moulding
C. Inclusions	**3.** Includes single folds, laps, scars adhering sand layers and oxide scale.
D. Discontinuties	**4.** Include cracks, cold or hot tearing and cold shuts.
	5. Consist of fins, flash or massive projections and rough surfaces.

Codes:

	A	B	C	D
(a)	1	5	3	2
(b)	1	5	2	4
(c)	5	1	2	4
(d)	5	1	3	2

25. With a solidification factor of 0.97×10^6 s/m^2, the solidification time(in seconds)for spherical casting of 200mm diameters is
(a) 539
(b) 1078
(c) 4311
(d) 3233

26. Misrun is a casting defect which occurs due to
(a) very high pouring temperature of the metal
(b) insufficient fluidity of the molten metal
(c) absorption of gases by the liquid metal
(d) improper alignment of the mould flasks

27. Scab is a:
(a) sand casting defect
(b) machining defect
(c) welding effect
(d) forging defect

28. A spherical drop of molten metal of radius 2mm was found to solidify in 10 seconds. A similar drop of radius 4mm would solidify in
(a) 14.14 sec
(b) 20 sec
(c) 18.30 sec
(d) 40 sec

29. According to Chovrinov's equation, the solidification time of a casting is proportional to
(a) v^2
(b) v
(c) 1/v
(d) $1/v^2$
Where, v = volume of casting

30. Consider the following statements
Assertion(A): Aluminium alloys are cast in hot chamber die casting machine
Reason(R): Aluminium alloys require high melting when compared to zinc alloys.

(a) Both A and R are individually true and R is the correct explanation of A.
(b) Both A and R are individually true and R is not the correct explanation of A.
(c) A is true but R is false
(d) A is false but R is true

31. Match **List-I** (Product) with **List-II** (Process of Manufacture) and select the correct answer using the codes given below the listsi

List-I	**List-II**
A. Automobile piston in aluminium alloy	**1.** Pressure die-casting
B. Engine crankshaft in spheroidal graphic iron	**2.** Gravity die-casting
C. Carburettor housing in aluminium alloy	**3.** Sand casting
D. Cast titanium blades	**4.** Precision investment casting
	5. Shell moulding

Codes:

	A	B	C	D
(a)	2	3	1	5
(b)	3	2	1	5
(c)	2	1	3	4
(d)	4	1	2	3

32. Which one of the following pairs is not correctly matched?
(a) Aluminium alloy piston .... pressure die casting
(b) Jewellery............ lost wax process
(c) Large pipes............ centrifugal casting
(d) Large bells............ loam moulding

33. Which of the following pattern-materials are used in precision casting?
1. Plaster of paris
2. Plastics
3. Anodized aluminium alloy
4. Frozen mercury

Select the correct answer using the codes given below:
(a) 1 and 2
(b) 2 and 4
(c) 3 and 4
(d) 1 and 3

34. Assertion(A): Semi-centrifugal casting process is similar to true centrifugal casting except that the central core is used in it from inner surface
Reason(R): In semi-centrifugal casting process the Axis of spin is always vertical
(a) Both A and R are individually true and R is the correct explanation of A
(b) Both A and R are individually true and R is not the correct explanation of A
(c) A is true but R is false
(d) A is false but R is true

35. Match **List-I** with **List-II** and select the correct answer using the codes given below the lists:

List-I	**List-II**
A. Die casting	**1.** Molten metal is forced into the die under pressure
B. Centrifugal casting	**2.** Axis of rotation doesnot coincide with axis of mould
C. Centrifuging	**3.** Metal solidifies when mould is rotating
D. Continuous casting	**4.** Continuously pouring molten metal into mould

Codes:

	A	B	C	D
(a)	1	3	2	4
(b)	4	3	2	1
(c)	1	2	3	4
(d)	4	2	3	1

36. Assertion(A): die casting yields a product of good accuracy and finish
Reason(R): low melting alloys are used in die casting
(a) both A and R individually true and R is the correct explanation of A.
(b) both A and R individually true and R is not the correct explanation of A
(c) A is true but R is false
(d) A is false but R is true

37. Consider the following statements:
Assertion(A): In CO_2 casting process, the mould or core attains maximum strength
Reason(R): The optimum passing time of CO_2 through the mould or core forms Silica Gel which imparts sufficient strength to the mould or core.
(a) Both A and R are individually true and R is the correct explanation of A
(b) Both A and R are individually true and R is not the correct explanation of A
(c) A is true but R is false
(d) A is false but R is true

38. Match the items of **List-I** (Equipment) with the items of **List-II** (Process) the process answer using the given codes

	List-I		List-II
P.	Hot chamber machine	**1.**	Cleaning
Q.	Muller	**2.**	Core making
R.	Dielectric Baker	**3.**	Die casting
S.	Sand Blaster	**4.**	Annealing
		5.	Sand mixing

(a) P-2, Q-1, R-4, S-5
(b) P-4, Q-2, R-3, S-5
(c) P-4, Q-5, R-1, S-2
(d) P-3, Q-5, R-2, S-1

39. Match **List-I** with **List-II** and select the correct answer using the codes given below the lists:

	List-I		List-II
A.	Hollow state	**1.**	Centrifugal casting
B.	Dentures	**2.**	Investment casting
C.	Aluminium alloy pistons	**3.**	Slush casting
D.	Rocker arms	**4.**	Shell moulding
		5.	Gravity die casting

Codes:

	A	B	C	D
(a)	3	2	4	5
(b)	1	3	4	5
(c)	1	2	3	4
(d)	3	2	5	4

40. Which of the following are produced by slush Casting?
(a) Hollow castings with thick walls
(b) Hollow castings with thin walls
(c) Thin castings
(d) Thick castings

41. In shell moulding, how can the shell thickness be Accurately maintained?
(a) by controlling time during which the pattern is in contact with mould
(b) by controlling time during which the pattern is heated
(c) by maintaing the temperature of the pattern in the range of 175°C-380°C
(d) by the type of binder used

42. An expandable pattern is used in
(a) slush casting
(b) squeeze casting
(c) centrifugal casting
(d) investment casting

43. Which one of the following processes produces a casting when pressure forces the molten metal into the mould cavity?
(a) shell moulding
(b) investment casting
(c) die casting
(d) continuous casting

44. Which of the following engineering materials is the most suitable candidate for hot chamber die casting?
(a) low carbon steel
(b) titanium
(c) copper
(d) tin

45. Shell moulding can used for
(a) producing milling cutters
(b) making gold ornaments
(c) producing heavy and thick walled casting
(d) producing thin casting

46. Which one of the following processes consist of central sprue to feed metal into cavities through a number of radial gates?
(a) centrifuging
(b) semi-centrifugal casting
(c) true centrifugal casting
(d) precision casting

47. Which of the following materials are used for making patterns in investment casting methods?
1. Wax
2. Rubber
3. Wood
4. Plastic

Select the correct answer using the codes given below:
(a) only 1 and 3
(b) only 2 and 3
(c) only 1, 2 and 4
(d) only 2, 3 and 4

48. Consider the following statements:
1. Zinc die casting have low strength
2. In the die casting process, very thin sections and complex shapes can be obtained easily

Which of the statements given above is /are correct?
(a) 1 only
(b) 2 only
(c) both 1 and 2
(d) neither 1 nor 2

49. Which one of the following is the correct Statements in a centrifugal method
(a) no core is used
(b) core may be made of any metal
(c) core is made of sand
(d) core is made ferrous metal

50. The mould in shell moulding process is made up of which of the following?
(a) gypsum + setting agents (b) green sand + clay
(c) sodium silicate + dried sand (d) dried silica + phenolic resin

51. Which of the following are employed in shell moulding?
1. Resin binder
2. Metal pattern
3. Heating coils

Select the correct answer using the code given below
(a) 1 and 2 only (b) 1 and 3 only
(c) 2 and 3 only (d) 1 , 2 and 3

52. Consider the following statements in respect of investment casting :
1. The pattern or patterns is/are joined to a stalk or sprue also of wax to form a tree of patterns
2. The prepared moulds are placed in an oven and Heated gently to dry off the invest and melt out the bulk of wax
3. The metal is usually poured by placing the moulds in a vaccum chamber

which of the statements given above are correct
(a) 1 and 2 only (b) 1 and 3 only
(c) 2 and 3only (d) 1, 2 and 3

53. In sand moulding the bottom most part of the flask is called
(a) cope (b) cheek
(c) drag (d) flask bottom

54. Moulding sands can contain following percentage of maximum quantity of moisture
(a) 2.5% (b) 5%
(c) 8% (d) 12%

55. The purpose of chaplets is
(a) just like chills to ensure directional solidification
(b) to provide efficient venting
(c) to support the cores
(d) to join lower and upper parts of the moulding box

56. Fluidity is greatly influences by
(a) carbon content of molten metal (b) melting temperature of molten metal
(c) inoculant addition (d) pouring temperature of molten metal

57. The chief advantage of die casting is
(a) possibility of incorporating thick sections in small castings
(b) casting of inserts is possible
(c) wide tolerances are possible
(d) high production rates are possible

58. The purpose of inoculation is
(a) to clean the casting
(b) to decrease the melting temperature of a cast metal
(c) to alter the chemical composition of a cast metal
(d) to modify the structure and properties of a cast metal

59. The main advantage of shell moulding is that
(a) a metallic pattern is used (b) the moulds are strong
(c) thin sections can be easily obtained (d) highly complex sections can be easily obtained

60. Draft on pattern for casting is
(a) shrinkage allowance
(b) taper to facilitate its removal from mould
(c) increase in size of cavity due to shaking of pattern
(d) for machining allowance

61. Strength and permeability of green sand and are related
(a) grain size (b) clay-content
(c) moisture content (d) type of sand

62. Least shrinkage allowance is provided in the case of following
(a) brass (b) aluminium
(c) cast iron (d) white cast iron

63. Cores are used to
(a) make desired recess in castings (b) strengthen moulding sand
(c) support loose pieces (d) remove pattern easily

64. Shrinkage allowance is made
(a) adding to external and internal dimensions
(b) subtracting from external and internal dimensions
(c) subtracting from external dimensions and adding to internal dimensions
(d) adding to external dimensions and subtracting from internal dimensions

65. The purpose of gate is to
(a) feed the casting at a rate consistent with the rate of solidification
(b) act as reservoir for molten metal
(c) help feed the casting until all solidification takes place
(d) feed molten metal from pouring basin to gate

66. Pick up the correct statement
(a) loose piece patterns are used when the pattern can be drawn from the mould
(b) sweep patterns eliminate the need for three dimensional patterns
(c) match plate patterns are made by fastening each half of a split pattern on the same side of one plate
(d) cope and drag patterns are solid one piece patterns.

67. The impurities in true centrifugal casting
(a) get collected at outer surface
(b) mix up thoroughly throughout
(c) get collected at the inner surfaces
(d) get collected in the middle portion in between inner and outer surface

68. The purpose of pouring basin is to
(a) feed the casting at a rate consistent with the rate of solidification
(b) act as reservoir of molten metal
(c) help feed the casting until all solidification takes place
(d) feed molten metal from pouring basin to gate

69. As the size of castings increase, it is often better to use increasingly
(a) fine grain (b) medium grain
(c) coarse grain (d) any one of the above

70. The purpose of riser is to
(a) feed the casting at a rate consistent with the rate of solidification
(b) act as a reservoir for molten metal
(c) help feed the casting until all solidification takes place
(d) feed molten metal from pouring basin to gate

71. Freezing ratio or relative freezing time according to Caine's equation is
(a) $\frac{A_C / V_C}{A_R / V_R}$ (b) $\frac{A_R / V_R}{A_C / V_C}$
(c) $\frac{V_C / A_C}{V_R / A_R}$ (d) $\frac{V_R / A_R}{V_C / A_C}$

72. Hot tear refers to
(a) casting defect (b) process of fabrication
(c) process of heat treatment (d) weathering of non-ferrous materials

73. True centrifugal casting
(a) is used to ensyre purity and density at extremeties of a casting
(b) is used to cast symmetrical objects
(c) is used to obtain high density and pure castings
(d) uses a heavy cast iron mould to act as chill

74. Investment casting is used for
(a) shapes which are made by difficulty using complex patterns in sand casting.
(b) mass production
(c) shapes which are very complex and intricate and can't be cast by any other method.
(d) there is nothing like investment casting

75. The materials of pattern in the case of investment casting is
(a) thermosetting resin (b) synthetic sand
(c) wax (d) mercury

76. Which of the following process would produce strongest components?
(a) die casting (b) hot rolling
(c) cold rolling (d) forging

77. The hot chamber die casting method is used to cast
(a) brass (b) aluminium
(c) alloys of lead, tin and zinc (d) all of the above

78. In hot chamber method of die casting
(a) only low melting point metals can be used
(b) high melting point metals can be cast
(c) die is kept hot by electrical heating
(d) die is kept cold by circulating water

79. Blind risers
(a) assist in feeding the metal into casting properly
(b) help to trap slag or other lighter particles
(c) supply the hottest metal when pouring is completed
(d) do not exit

80. In slush casting process
- (a) molten metal is fed into the cavity in metallic mould by gravity
- (b) metal is poured into die cavity, and after a predetermined time the mould is inverted to permit a part of metal is till in molten state to flow out of cavity.
- (c) cavity is filled with a perculated quantity of metal and a core or plunger is inserted to force the metal into cavity
- (d) metal is forced into mould under high pressure

81 Ornamental objects, statues, toys etc. are cast by
- (a) die casting
- (b) pressed casting
- (c) centrifugal casting
- (d) slush casting

82. Steel and cast iron pipes are cast by
- (a) die casting
- (b) continuous casting
- (c) true centrifugal casting
- (d) centrifuging

83. Scabs or buckles are the casting defects
- (a) which occur due to some sand shearing from the cope
- (b) which take the form of internal voids or surfaces depression due to excessive gaseous material not able to escape
- (c) which occur due to discontinuity in metal casting resulting from hindered contraction
- (d) caused by two streams of metals that are too cold to fuse property

84. Drossing in foundry practice refers to
- (a) a method of cleaning the castings
- (b) an inspection method for castings
- (c) a method of de-oxidation of molten metal
- (d) the formation of oxides on the molten metal surface

85. Cold shuts are casing defects
- (a) which occur due to some sand shearing from the cope surfaces
- (b) which take the form of internal voids or surface depression due to excessive gaseous material not able to escape
- (c) which occur due to discontinuity in metal casting resulting from hindered contraction
- (d) caused by two streams of metals that are too cold to fuse properly

86. Slag inclusion in casting is a
- (a) surface defect
- (b) internal defect
- (c) crack
- (d) notch

87. Jolt machines produce
- (a) uniform ramming about the pattern
- (b) uniform ramming about the flask
- (c) uniform distribution throughout
- (d) pack sand loosely all around

88. Blow holes are casting defects
- (a) which occur due to some sand shearing from the cope surface
- (b) which take the from of internal voids or surfaces depression due to excessive gaseous material not able to escape
- (c) which occur due to discontinuity in metal casting resulting from hindered contraction
- (d) caused by two streams of metals that are too cold to fuse properly

■■■■

ANSWERS

CASTING

1. (c)	**2.** (d)	**3.** (c)	**4.** (a)	**5.** (b)	**6.** (c)	**7.** (a)
8. (a)	**9.** (d)	**10.** (b)	**11.** (b)	**12.** (a)	**13.** (b)	**14.** (a)
15. (a)	**16.** (c)	**17.** (d)	**18.** (a)	**19.** (a)	**20.** (b)	**21.** (d)
22. (c)	**23.** (b*)	**24.** (c)	**25.** (b)	**26.** (b)	**27.** (a)	**28.** (d)
29. (a)	**30.** (d)	**31.** (a)	**32.** (b)	**33.** (b)	**34.** (b)	**35.** (a)
36. (b)	**37.** (a)	**38.** (d)	**39.** (d)	**40.** (b)	**41.** (a)	**42.** (d)
43. (c)	**44.** (d)	**45.** (d)	**46.** (a)	**47.** (c)	**48.** (a)	**49.** (a)
50. (d)	**51.** (d)	**52.** (d)	**53.** (c)	**54.** (c)	**55.** (c)	**56.** (d)
57. (d)	**58.** (d)	**59.** (c)	**60.** (b)	**61.** (c)	**62.** (c)	**63.** (a)
64. (d)	**65.** (a)	**66.** (b)	**67.** (c)	**68.** (d)	**69.** (c)	**70** (b)
71. (d)	**72.** (a)	**73.** (b)	**74.** (c)	**75** (c)	**76.** (d)	**77.** (c)
78. (a)	**79.** (a)	**80.** (b)	**81.** (d)	**82.** (c)	**83.** (a)	**84.** (d)
85. (d)	**86.** (b)	**87.** (a)	**88.** (b)			

* Chaplets are used to support internal chills and cores.In case of internal chill both the chill and chaplet are supossed to fuse with the liquid metal. In such cases chapltes also leads to directional solidification

CHAPTER 4

Metal Cutting

In machining the total energy required is summation of energy required in plastic deformation to break crystal structure and in overcoming the friction. Cutting tools are classified into two major groups:

1. Single point cutting tools
2. Multipoint cutting tools.

Multipoint cutting tool : They have more than one cutting edge to remove excess material from the work piece. Milling cutters, drills, reamers, broaches and grinding wheels are multi point cutting tools.

Single point cutting tool : The tool terminating in a single point has been termed as single point cutting tool. The tool is made of either high carbon steel, high speed steel or carbide bar. The cutting edge is prepared by grinding. It has one cutting edge which is partially formed by the end cutting edge and largely by the side cutting edge. Its nose is given a small radius and it is never a sharp point. This imparts strength to the cutting edge and reduces the magnitude of stress over the cutting edge. The nature of single point cutting tool are presented in Fig. 4.1 and 4.2.

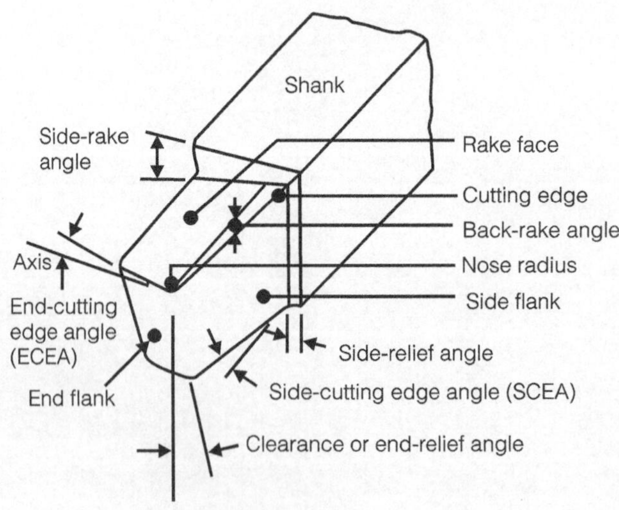

Fig. 4.1 : Single point cutting tool

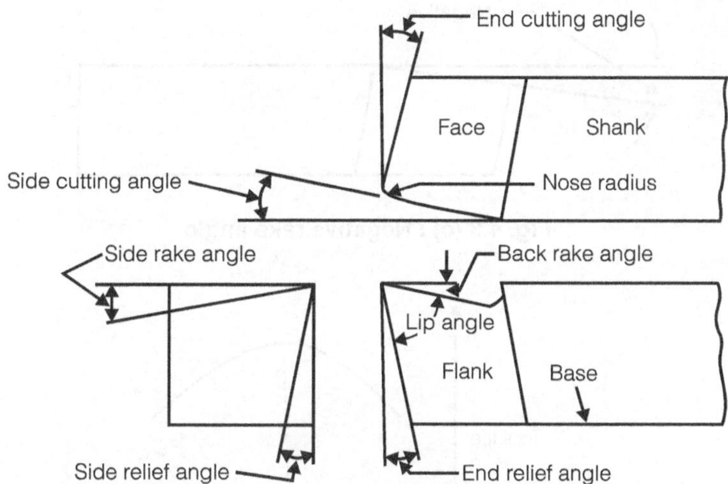

Fig. 4.2 : Representation of various angles in different views of single point cutting tool

4.1 TERMINOLOGY OF SINGLE POINT CUTTING TOOL (AS PER ASA)

4.1.1 Back rake angle

It is the angle between the line parallel to the tool axis passing through the tip and the rake face and angle is measured in a plane perpendicular to the base.

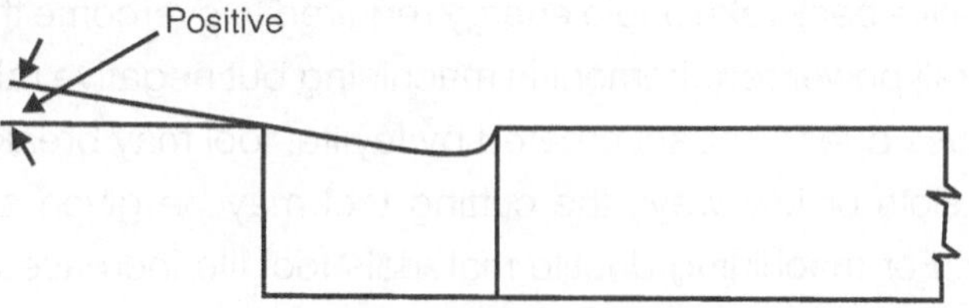

Fig. 4.3 (a) : Positive Rake angle

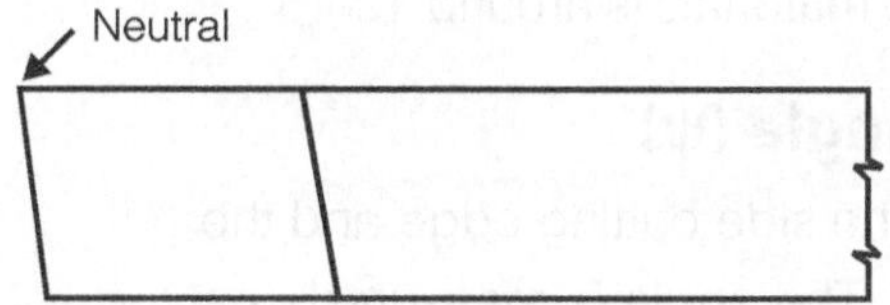

Fig. 4.3 (b) : Neutral Rake angle

After plastic deformation chips flow over the rake face and heavy drag exists between chip and rake face. Due to this, temperature continues to develop and the maximum temperature will appear 2-3 mm away from the cutting edge. At high temperature, carbon starts diffusing from the tool to the chip and as a result of that tool becomes weaker and weaker. Due to the drag between the chip and the tool, a portion of the tool will be carried away by the chip. This phenomenon is called diffusion wear. By increasing the Back Rake angle, chip flow will be easier, that is drag will decrease and hence the tool wear. So initially by increasing back rake angle tool life increases.

Negative

Fig. 4.3 (c) : Negative rake angle

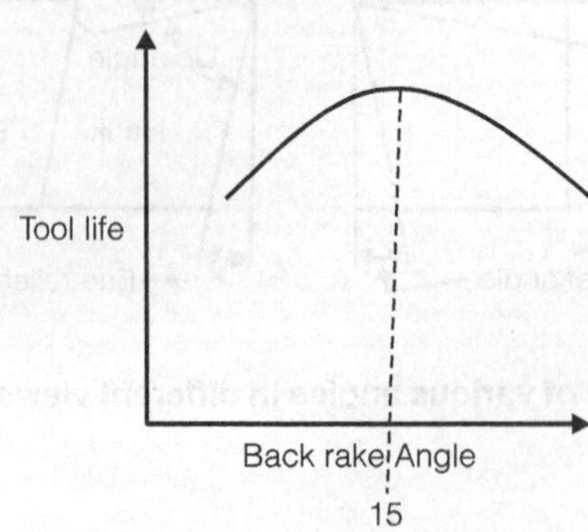

Fig. 4.3 (d) : Variation of Back rake angle with respect to tool life

While machining stronger or brittle materials, smaller rake angles are used. For ex: machining brass, zero degree rake angles are chosen. If the tool material is brittle like ceramics and carbides, negative back rake angles have to be provided because we want to make the tool stronger. By providing the negative back rake angle energy required to overcome the friction increases which increases the overall power requirement in machining but negative rake angles are required for tool stability because due to impact created by faylite, tool may break.

When machining slots or key ways the cutting tool may be given a negative back rake and positive side rake. For machining ductile materials tool life increases by increasing back rake angle because of decrease in contact length between chip and tool over the rake face. But when this angle becomes more than a particular value, lip angle of the tool decreases. This decreases the strength of tool hence tool life decreases. This is represented in Fig. 4.3 (d). The optimum back rake angle of ductile materials is around 10-15°.

4.1.2 Side cutting edge angle (ψ)

It is the angle between the side cutting edge and the line extending the shank. The angle is measured in a plane parallel to base. Feed represents uncut chip thickness and depth of cut represents width of chip.

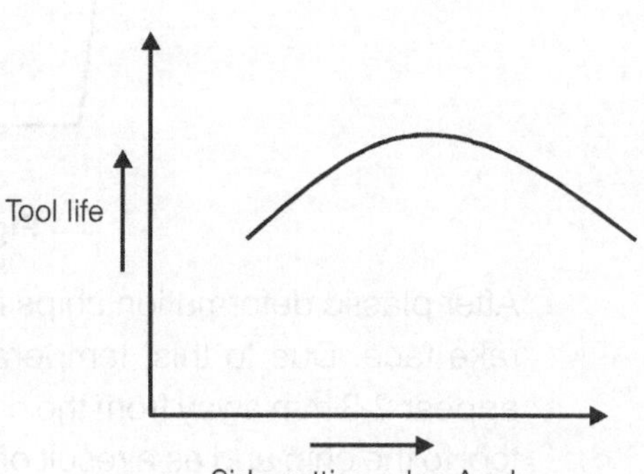

Fig. 4.4 : Variation of tool life with respect to side cutting edge angle

$\theta \propto \sqrt{f_t}$

θ = Maximum temperature on Rake face.

f_t = true feed or feed experience by the cutting edge in perpendicular direction and it is equal to the uncut chip thickness

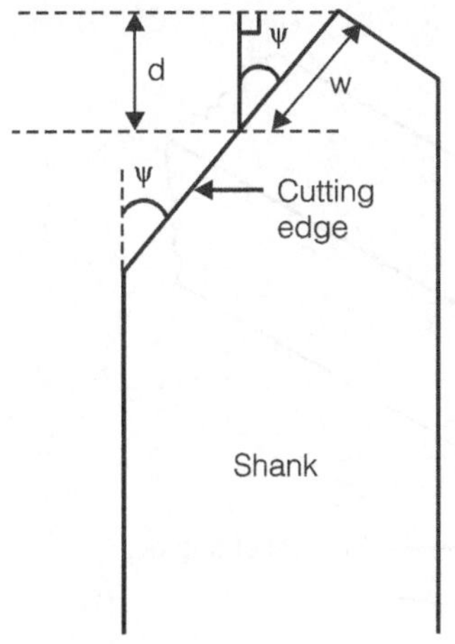

Fig. 4.5

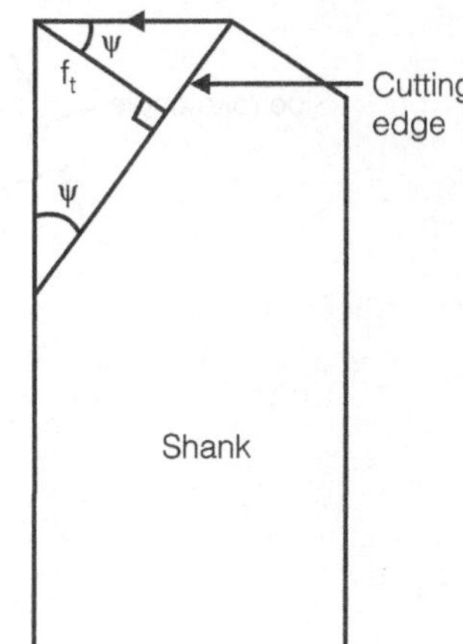

Fig. 4.6

From Fig. 4.5

$$w = \frac{d}{\cos\psi}$$

$$d = \text{depth of cut}$$

$$w = \text{width of cut}$$

From Fig. 4.6

$$\frac{f_t}{f} = \cos\psi$$

$$\text{True feed} = f_t = f\cos\psi$$

$$t_1 = \text{uncut chip thickness.}$$

Width of the chip is the length of the cutting edge covered by the chips and it can be observed in the analysis that as side cutting edge angle increases chips become wider. Uncut chip thickness is feed experienced by the side cutting edge in the perpendicular direction and by increasing the side cutting edge angle, chips become thinner.

Wider chips increases the friction force so by increasing the side cutting edge angle cutting force requirement is slightly increased. But, since the heat is distributed over the larger area, peak temperature over the rake face will decrease. This will decrease, the tool wear and hence increases the tool life. Side cutting edge angles increases possibility of chatter vibrations. For machining shoulders a small negative side cutting edge angle of 2° - 5° is given on tools. Normally the value of Side cutting edge angle varies between 15-30°.

4.1.3 Side Rake Angle

It is the angle between the rake face and the line passing through the tip perpendicular to the tool axis and the angle is measured in a plane perpendicular to the base. Normally this angle varies between 5-15°.

4.1.4 Side Relief Angle

It is the angle between the side flank and the line passing through the tip perpendicular to the base and the angle is measured in a plane perpendicular to the tool axis. This angle varies in the range of 5-15°. The work piece material which is going to be removed in the next revolution will try to hit the side flank due to elastic recovery of work piece material. To avoid this rubbing side relief angle is provided.

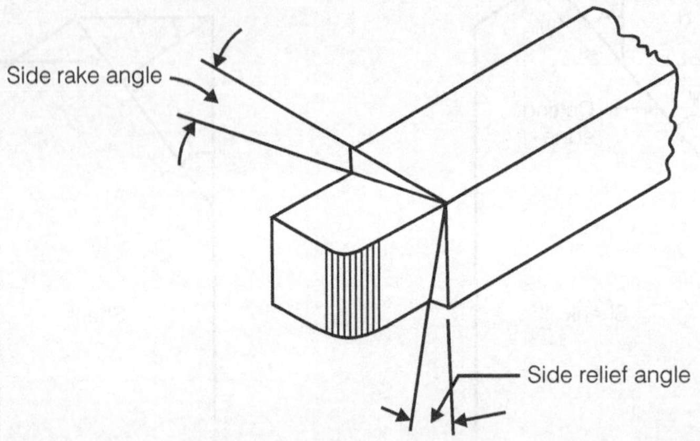

Fig. 4.7 : Representation of side rake and side relief angles

4.1.5 End Cutting Edge Angle

It is the angle between the end cutting edge and the line passing through the tip perpendicular to the tool axis and the angle is measured in a plane parallel to base. At smaller values of the angle larger forces normal to the machine surfaces are produced and the tool may chatter. Side cutting edge angle is made zero for a distance slightly greater than the feed in order to decrease rough marks due to feed marks. The normal value of this angle is in the range of 8-15°.

4.1.6 End Relief Angle (Clearance angle)

It is the angle between the end flank and the line passing through the tip perpendicular to the base and angle is measured in plane parallel to the tool axis. There will be some elastic recovery in the finished work and as a result of that it will try to rub the end flank. While machining ductile materials elastic recovery will be more so larger clearance angle has to be provided. Relief is provided to the side and end flanks in order to minimize physical interference or rubbing contact with the machine surface of work piece. Smaller relief angles do not weaken the cutting edge as much as the larger relief angles.

4.1.7 Nose Radius

Larger nose radius gives better surface finish. Increasing the nose radius also improves the tool life, so that higher cutting speeds can be employed. However the cutting force and possibility of chatter increases when the nose radius is increased excessively.

4.1.8 ASA Tool Signature

Back rake angle - Side rake angle - End relief angle - Side relief angle - End cutting edge angle - Side cutting edge angle - Nose radius.

In this system the geometry of the rake face is expressed in terms of back rake angle and side rake angle.

Surface roughness:

Peak to valley height, $H_{max} = \dfrac{f^2}{8R}$

f = feed, R = nose radius

Center Line Average value, $R_a = \frac{H_{max}}{4}$

$$H_{max} = \frac{f}{\tan\psi + \cot\psi_1}$$

ψ = Side cutting edge angle ψ_1 = End cutting edge angle

4.2 BRITISH SYSTEM

Rake angle in this system defined as the steepest slope over the rake face. It is assumed that taking this value as the rake angle gives accurate results for cutting forces. The main advantage of this system is that the specified rake angle is easy to set on a tool grinding fixture for grinding the rake face. The angle of chip flow is difficult to estimate.

4.3 NORMAL OR ORTHOGONAL RAKE SYSTEM (ORS)

A number of lines drawn perpendicular to the side cutting edge in the horizontal plane and the line which gives the maximum slope is called Normal Rake Angle (α_n). If side cutting edge angle is zero, normal rake angle is equal to the side rake angle. Tool signature in this system is given as: I-α_n- side relief angle - end relief angle - end cutting edge angle - approach angle λ- Nose Radius R. where, I-Angle of inclination

α_n-normal rake angle

λ-approach angle (90° – side cutting edge angle).

Angle of inclination is the angle of tool axis with cross slide motion.

For Orthogonal system I = 0

Conversion from ASA to ORS

$$\tan I = \cos\psi \tan\alpha_{ab} - \sin\psi \tan\alpha_{as}$$

$$\tan\alpha_n = \cos\psi \tan\alpha_{as} + \sin\psi \tan\alpha_{ab}$$

ψ = side cutting edge angle

α_{ab} = back rake angle

α_{as} = side rake angle

4.4 TYPES OF METAL CUTTING PROCESS

The metal cutting processes are of two types

1. **Orthogonal cutting process** (Two dimensional cutting): It occurs when the major cutting edge of the tool is presented to the work piece perpendicular to the direction of feed motion.

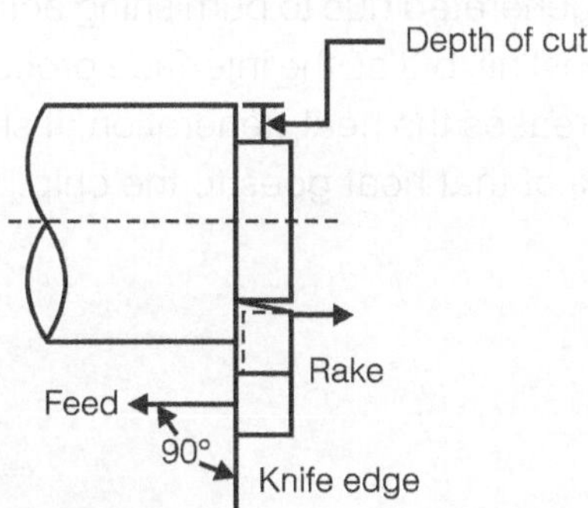

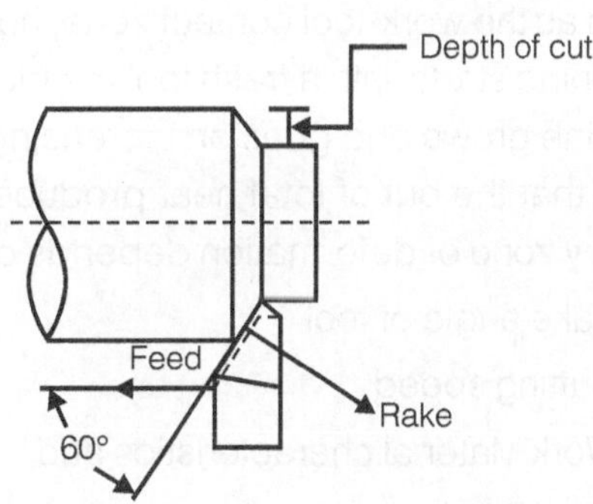

2. **Oblique cutting** (Three dimensional cutting): This form of cutting occurs when the major cutting edge of the tool is presented to the work piece at an angle perpendicular to the direction of feed motion.

4.5 DISTRIBUTION OF HEAT IN METAL CUTTING

1. Primary deformation zone
2. Secondary deformation zone
3. Work-tool contact zone.

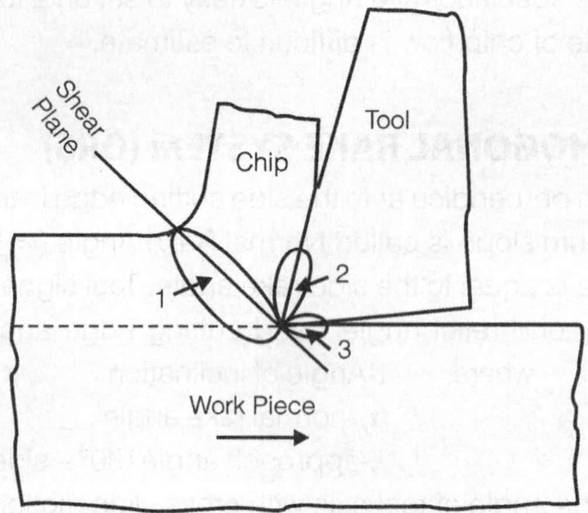

Fig. 4.8 : Representation of shear plane and shear deformation zones

There will be a region in the material in which crystal structures break. This region is called primary deformation zone and around 70-80% of the total energy supplied will be consumed in this region. This is because of the plastic deformation of metal. A portion of this energy will also come out in the form of heat .Around 20-25% energy is consumed to overcome the drag between the chip and rake face. This region is called secondary deformation zone and 100% of energy comes out in the form of heat. Although the primary deformation zone is a region, but by analytical purposes it can be approximated by a line. This plane is called shear plane. By decreasing the feed or uncut chip thickness, the value of shear plane angle will decrease, which will ultimately increase the shear plane area, so larger shear forces will be required to break the crystal structure. This increases the apparent shear strength of material. The third region is known as the work-tool contact zone; here heat is generated due to burnishing action. When the machining starts with a fresh tool the friction is almost nil, but as the interface process continues wear line grows and goes on increasing. This increases the heat generation. It should also be noted that the out of total heat produced, 75-80% of that heat goes to the chip. The extent of primary zone of deformation depends on

1. Rake angle of tool
2. Cutting speed
3. Work Material characteristics and
4. Friction on rake face

4.6 TYPES OF CHIPS

4.6.1 Continuous Chips

Favourable conditions for the continuous chips to form are:

- Ductile materials
- High speed
- Low feed and depth of cut
- High back rake angle

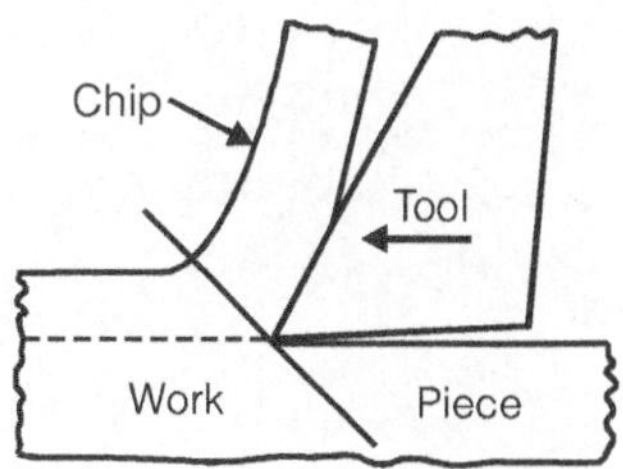

Fig. 4.9 : Continuous Chip

Continuous chips start rotating with the tool and if these chips hit the finished part of the work, it spoils surface finish. So, continuous chips are broken by providing chip breakers. In drilling operation, chips are broken by providing a back feed. High back rake angle provides less resistance to flow of chips, hence favours continuous chips.

4.6.2 Discontinuous Chips

The conditions in which discontinuous chip forms are:

- Brittle materials
- Low speed
- High feed and depth of cut
- Low back rake angle

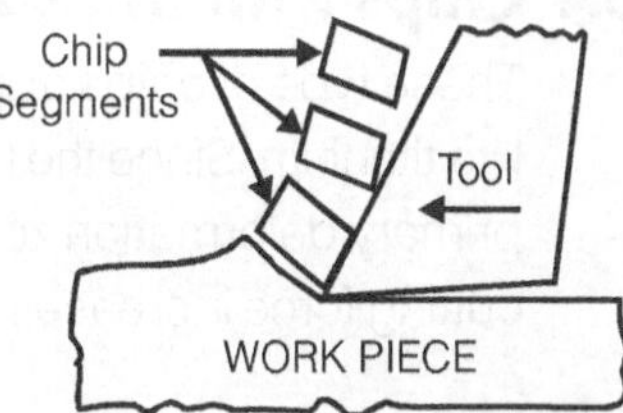

Fig. 4.10 (a) : Discontinuous chip

The slight plastic deformation produced by a small advance of the cutting edge into the job leads to a crack formation in the deforming zone. With further advance, the crack travels and a small lump of materials starts moving up the rake face.

The force and constraints of motion acting on the lump make the crack propagate towards the surface and thus the small segment of the chip gets detached. As most of the heat produced in the chip is carried by the chip, the tool is heated to a longer temperature and hence has longer life.

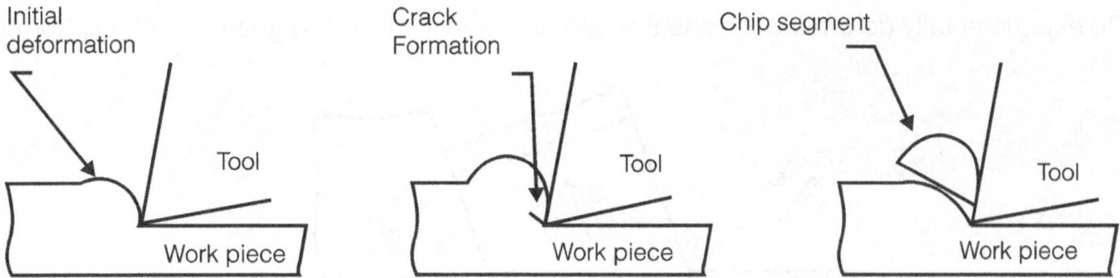

Fig. 4.10 (b) : Formation of discontinuous chips

4.6.3 Chips with build up edge

Favourable conditions for these types of chips are:

- Ductile material
- Low speed
- High feed and depth of cut

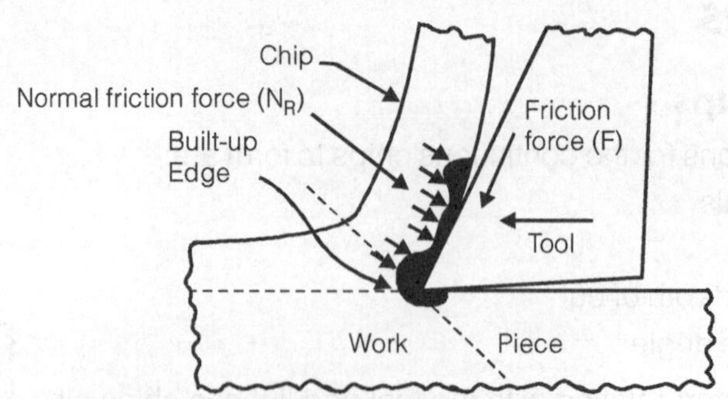

Fig. 4.11 : Chips with build up edge

Due to the low speed, heat is not being dissipated from the machining area. So, a portion of work will stick to the cutting edge called Build up Edge. Although the build up edge gives initial strength to the cutting edge but as the conditions of the build up edge continue, size of build up edge becomes bigger and bigger which changes the geometry of the cutting tool. So, a portion of this build up edge is carried away by the chip and a portion goes away with the work. In this mechanics, a portion of tool material is also lost. This type of tool wear is called Adhesion wear.

4.6.4 Chips with Unusual Strains

These type of chips appear, while machining the work piece materials with low thermal conductivity like titanium. Since the heat is not dissipated from the machining area, chips get accumulated in the primary deformation zone. As a result of that, shear plane angle decreases, and the requirement of cutting force increases.

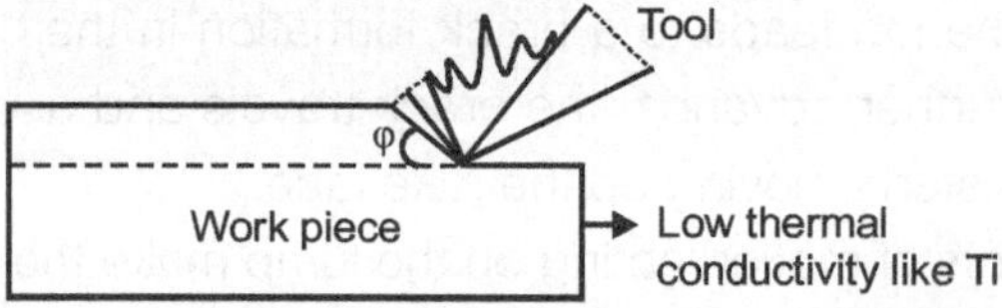

Fig. 4.12 : Chips with unusual strains

4.7 MERCHANT'S ANALYSIS FOR CHIP THICKNESS RATIO

To experimentally determine the shear angle, we have to study the geometry of chip formation..

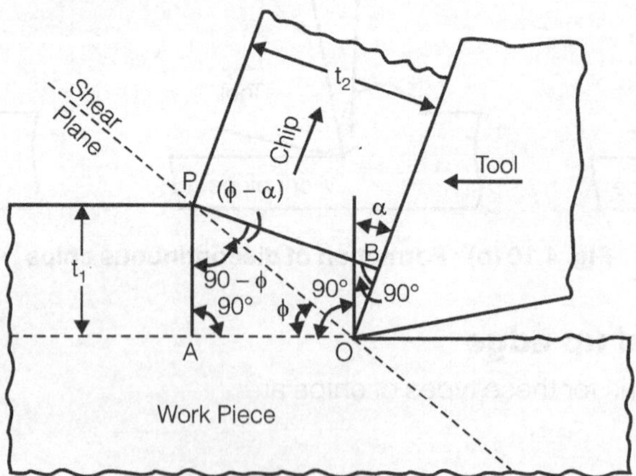

Fig. 4.13 : Orthogonal cutting operation

t_1 = Chip thickness before cutting (uncut chip thickness)
t_2 = Chip thickness after cutting
ϕ = Shear plane angle
α = Back rake angle

$$\Delta OAP : \frac{t_1}{OP} = \sin\phi \qquad ...(1)$$

$$\Delta OPB : \frac{t_2}{OP} = \sin(90 - \phi + \alpha) \qquad ...(2)$$

Divide Eq. (2) by Eq. (1)

$$\Rightarrow \quad \frac{t_2}{t_1} = \frac{\sin(90^\circ - \phi + \alpha)}{\sin\phi}$$

$$\Rightarrow \quad \frac{t_2}{t_1} = \frac{\sin(90^\circ - (\phi - \alpha))}{\sin\phi}$$

$$\Rightarrow \quad \frac{t_2}{t_1} = \frac{\cos(\phi - \alpha)}{\sin\phi}$$

$$\Rightarrow \quad \frac{t_2}{t_1} = \frac{\cos\phi\cos\alpha + \sin\phi\sin\alpha}{\sin\phi}$$

$$\Rightarrow \quad \frac{t_2}{t_1} = \cot\phi\cos\alpha + \sin\alpha$$

$$\Rightarrow \quad \frac{t_2}{t_1} - \sin\alpha = \cot\phi\cos\alpha$$

$$\Rightarrow \quad \frac{t_2}{t_1} - \sin\alpha = \frac{\cos\alpha}{\tan\phi}$$

$$\Rightarrow \quad \tan\phi = \frac{\cos\alpha}{\frac{t_2}{t_1} - \sin\alpha}$$

Always $t_2 > t_1 \rightarrow$ Chip thickness > uncut chip thickness

$$\frac{t_2}{t_1} = \text{chip thickness ratio} > 1$$

$$\frac{t_1}{t_2} = \text{chip reduction coefficient} < 1$$

Constancy of volume

$$l_1 t_1 w_1 = l_2 t_2 w_2$$

$w_1 = w_2$ Width kept constant

$$l_1 t_1 = l_2 t_2$$

D = diameter of work

Uncut chip length $l_1 = \pi D$

Due to Compressive stresses while cutting,

$$l_2 < l_1$$

$$\Rightarrow \quad t_2 > t_1$$

4.8 DETERMINATION OF CUTTING FORCES

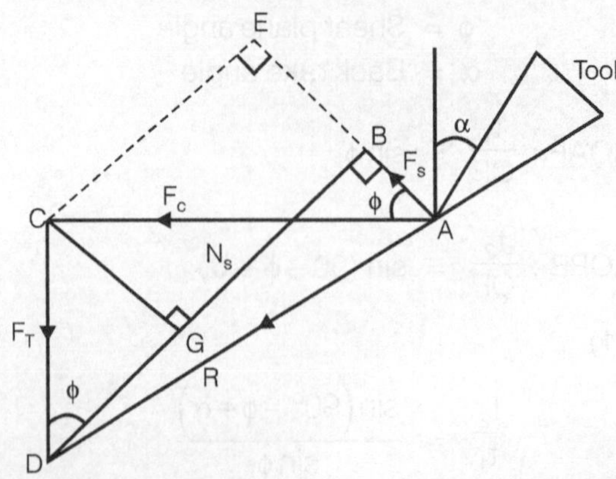

Fig. 4.14 : Representation of cutting forces

F_T = thrust force

F_C = Cutting force

Cutting force always comes in the direction of cutting and in turning operation and it is tangential force acting in the vertically upward direction. Thrust force acts perpendicular to the cutting edge and enables the cutting edge to have physical contact with the work piece material. F_C and F_T are perpendicular to each other. By resolving F_T in feed direction, feed force (f_f) comes and Radial force is resolving F_T is the direction of cross slide movement i.e. toward radial direction as shown in Fig. 4.15.

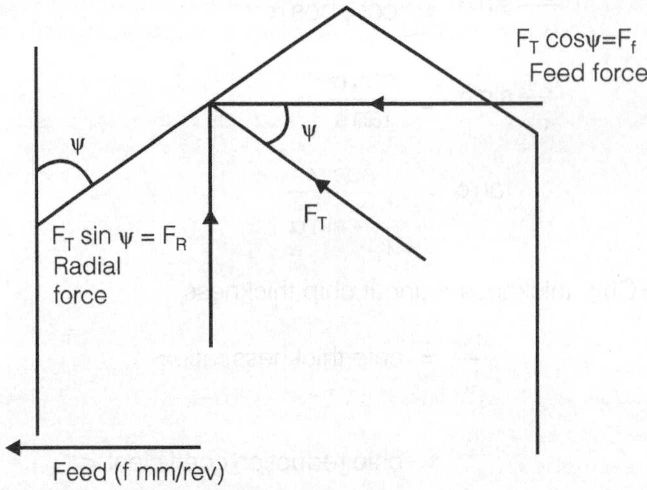

Fig. 4.15 : Feed and Radial Forces

In a true orthogonal cutting, $F_R = 0$

F_C, F_R, and F_f are perpendicular to each other.

From Fig. : 4.14

$$F_S = AE - BE$$
$$= AE - CG$$
$$AE = F_C \cos \phi$$
$$CG = F_T \sin \phi$$

Shear Force

$$F_S = F_c \cos\phi - F_T \sin\phi$$

$$N_S = DG + GB$$

$$= DG + CE$$

$$DG = F_T \cos\phi$$

$$CE = F_C \sin\phi$$

Normal Shear Force

$$N_S = F_T \cos\phi + F_C \sin\phi$$

4.9 DETERMINATION OF AREA OF SHEAR PLANE AND SHEAR STRENGTH

$$\text{Area of shear plane} = w \times \frac{t_1}{\sin\phi}$$

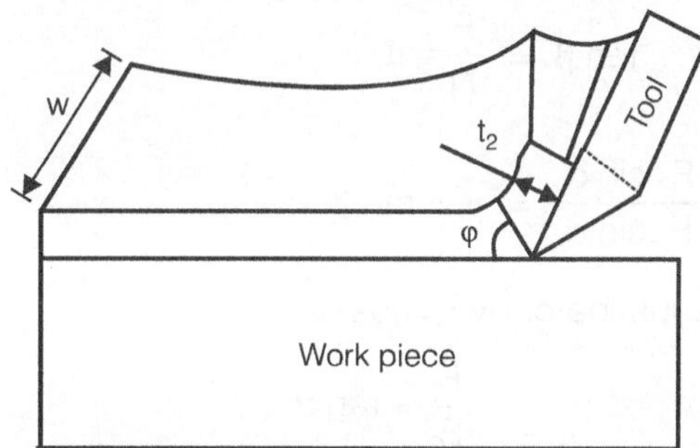

Fig. 4.16

t_2 is perpendicular to rake face and t_1 is perpendicular to cutting

$$\frac{F_S}{\frac{w.t_1}{\sin\phi}} = \tau_s = \text{Shear strength of the material}$$

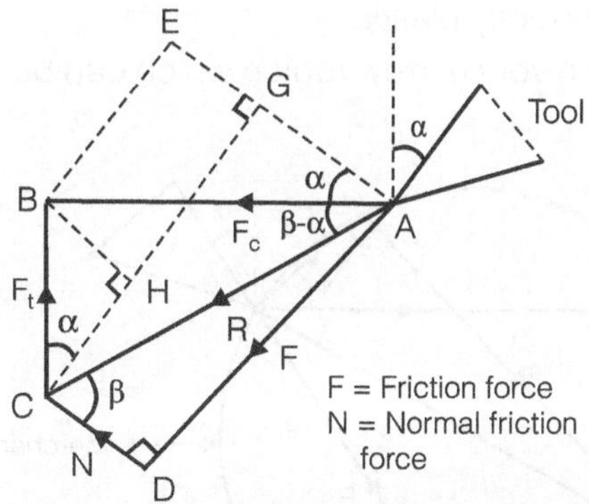

Fig. 4.17 : Cutting forces with friction and normal friction forces

N = Normal friction force

F = friction force, $\tan\beta = \frac{F}{N} = \mu$

β = Friction angle

From Fig. : 4.17

$$CG = CH + HG$$
$$= CH + BE$$
$$CH = F_T \cos\alpha$$
$$BE = F_C \sin\alpha$$
$$\Rightarrow \quad F = F_T \cos\alpha + F_C \sin\alpha$$
$$N = CD = AG$$
$$AG = AE - EG$$
$$= AE - BH$$
$$AE = F_C \cos\alpha$$
$$BH = F_T \sin\alpha$$
$$\Rightarrow \quad N = F_C \cos\alpha - F_T \sin\alpha$$
$$\tan\beta = \frac{F}{N} = \mu$$
$$\Rightarrow \quad \frac{F_T \cos\alpha + F_C \sin\alpha}{F_C \cos\alpha - F_T \sin\alpha} = \mu = \tan\beta$$

Divide Numerator and denominator by $F_C \cos\alpha$

$$\Rightarrow \quad \tan\beta = \frac{\frac{F_T}{F_C} + \tan\alpha}{1 - \frac{F_T}{F_C}\tan\alpha}$$

4.10 ASSUMPTIONS OF MERCHANT'S CIRCLE

1. Cutting edge is straight and sharp.
2. Material is homogeneous.
3. Cutting is orthogonal.
4. Material is rigid and perfectly plastic.
5. Shear zone extends in a very narrow region which can be approximated by a straight line.

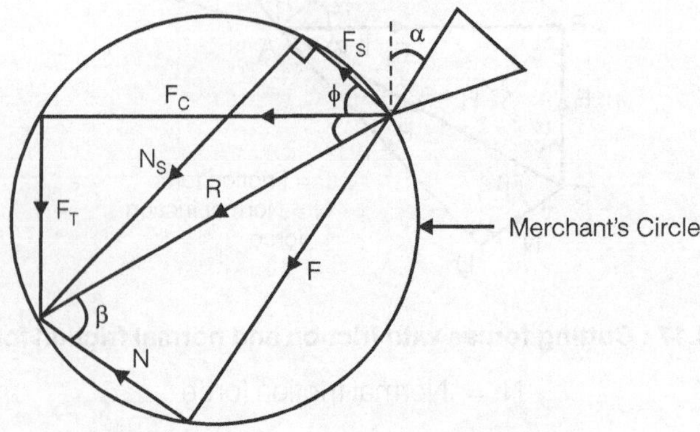

Fig. 4.18 : Merchant's circle

As soon as the cutting edge comes in contact with the work material, cutting edge will try to round off. Due to this area contact, there will be a drag between tool and work piece called PLOUGHING FORCE. These ploughing forces are proportionately insignificant when feeds and depths of cut are high but for smaller feeds; ploughing forces are proportionately very high and cannot be neglected. These theories do not consider the ploughing forces. Every material exhibits certain extent of work hardening which is also not considered in these theories.

4.11 DETERMINATION OF SHEAR ANGLE

$$\gamma = \frac{AD + DB}{CD}$$

$$\gamma = \cot(90 - \phi + \alpha) + \cot\phi$$

So, shear strain is $\boldsymbol{\gamma = \tan(\phi - \alpha) + \cot\phi}$

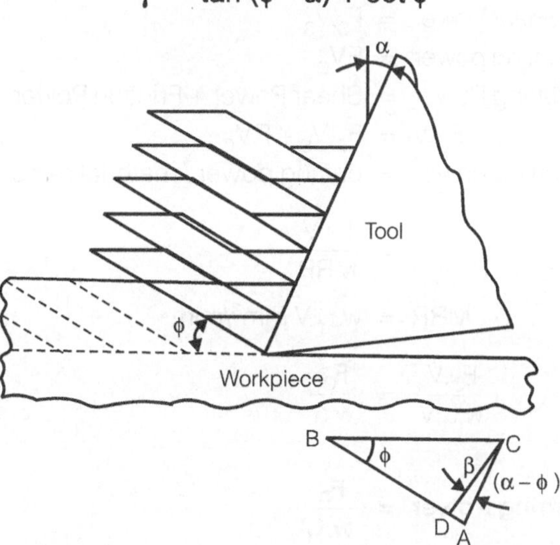

Fig. 4.19

4.12 VELOCITY TRIANGLE

Let us assume that the work material is moving against the cutting tool with a velocity V. The chip velocity V_C is the vector sum of V the velocity of uncut chip and V_s shear velocity.

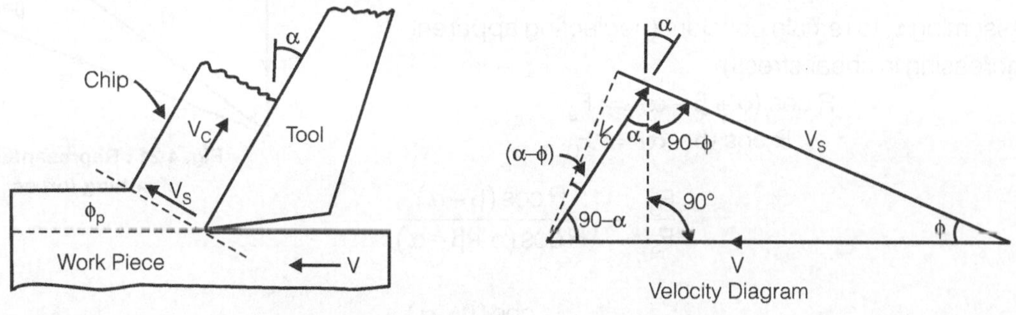

Fig. 4.20 : Representation of different velocities

$$V = \frac{\pi DN}{60}$$

V_S = Shear velocity

V_C = Chip velocity

Applying the sine rule:

$$\frac{V}{\sin(90^\circ - \phi + \alpha)} = \frac{V_C}{\sin\phi} = \frac{V_S}{\sin(90^\circ - \alpha)}$$

$$= \frac{V}{\cos(\phi - \alpha)} = \frac{V_C}{\sin\phi} = \frac{V_S}{\cos\alpha}$$

Cutting Power = $F_C.V$

Shear Power = $F_S.V_S$

Friction power = $F.V_C$

Cutting Power = Shear Power + Friction Power

$\Rightarrow$ $F_C.V = F_S.V_S + F.V_C$

Specific cutting power = cutting power / material removal rate (MRR)

$$= \frac{F_C.V}{MRR}$$

MRR = $w.t_1.V$ mm³/min

$$\Rightarrow \quad \frac{F_C.V}{w.t_1.V} = \frac{F_C}{w.t_1}$$

Specific Cutting Power = $\mathbf{\frac{F_C}{w.t_1}}$

4.13 ERNEST AND MERCHANT THEORY

It is the relationship between ϕ, α and β for minimum power consumption during machining.

Minimum Power is $P_C = F_C.V$

Shear force is $\tau_S \dfrac{wt_1}{\sin\phi} = F_S$

Assuming τ_s to remain constant (neglecting apparent increasing in shear stress)

$$R\cos(\phi + \beta - \alpha) = F_s$$

$$R\cos(\beta - \alpha) = F_C$$

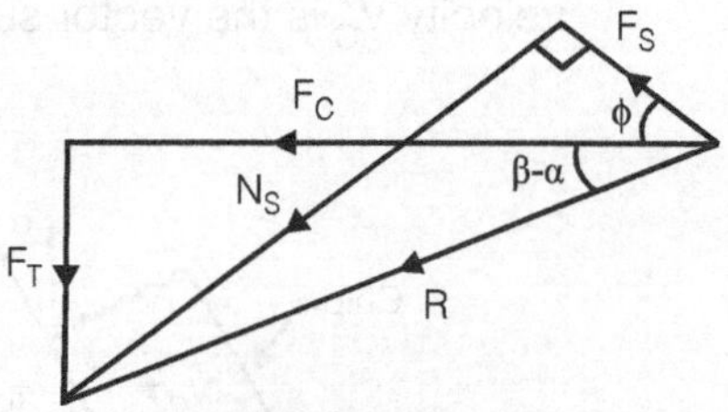

Fig. 4.21 : Representation of cutting forces

$$\frac{F_C}{F_S} = \frac{R\cos(\beta - \alpha)}{R\cos(\phi + \beta - \alpha)}$$

$$F_C = F_S \frac{\cos(\beta - \alpha)}{\cos(\phi + \beta - \alpha)}$$

$$F_C = \tau_s \frac{wt_1}{\sin\phi} \frac{\cos(\beta-\alpha)}{\cos(\phi+\beta-\alpha)}$$

$$P_C = F_C V = \tau_s \frac{wt_1}{\sin\phi} \frac{\cos(\beta-\alpha)}{\cos(\phi+\beta-\alpha)}.V$$

To minimize power consumption denominator is to be maximized.

Let $X = \sin\phi\cos(\phi+\beta-\alpha)$

$$\frac{dX}{d\phi} = \cos\phi\cos(\phi+\beta-\alpha)+\sin\phi[-\sin(\phi+\beta-\alpha)] = 0$$

$$\cos(2\phi+\beta-\alpha) = 0$$

$2\phi+\beta-\alpha = \frac{\pi}{2}$......Ist analysis

IInd Analysis: $2\phi+\beta-\alpha = \cot^{-1}(K)$

K= machining constant

These are the procedures to calculate the shear plane angle and these formulas should be used only when sufficient data is not given in the problem to find out ϕ in a normal way.

4.14 LEE AND SHAFFER (SLIP LINE FIELD THEORY)

The theory of Lee and Shaffer was the result of an attempt to apply the plasticity theory to the orthogonal metal cutting. The assumptions are:

1. The work material ahead of the tool behaves as ideal plastic mass.
2. There exists a shear plane which separates the chip and work piece.
3. No hardening in chip occurs

Based on the following assumptions the relationship obtained is

$$\phi+\beta-\alpha = 45°$$

Slip lines meet the surface at 45°

Example : 4.1

During turning process with 7 – α – 6 – 6 – 8 – 30 – 1 mm (ASA) tool the undeformed chip thickness of 0.2 mm and width of cut of 2.5 mm were used. The side rake angle of the tool was choosen such that the machining operation could be approximated to be orthogonal cutting. The tangential cutting force and thrust force were 1177 N and 560 N respectively. Calculate :

1. **The side rake angle**
2. **Coefficient of friction at the rake face.**
3. **The dynamic shear strength of the work material.**

Solution :

$$t_1 = 0.2 \text{ mm}$$
$$w = 2.5 \text{ mm}$$
$$I = 0$$

$$F_C = 1177\text{ N}$$
$$F_T = 560\text{ N}$$

1. Side rake angle

$$\tan I = \cos\psi \tan\alpha_{ab} - \sin\psi \tan\alpha_{as}$$

Given data

$$\psi = 30$$
$$\alpha_{ab} = 7$$
$$I = 0$$

$$\Rightarrow \quad \tan\alpha_{as} = \frac{\cos\psi \tan\alpha_{ab} - \tan I}{\sin\psi}$$

$$\alpha_{as} = 12° = \text{Side rake angle}$$

2. Coefficient of friction at rake angle

$$\mu = \tan\beta = \frac{\frac{560}{1177} + \tan 7°}{1 - \frac{560}{1177}\tan 7°}$$

$$\mu = 0.6357$$
$$\beta = \tan^{-1}(\mu) = 32.44°$$

3. Dynamic shear strength of work material

$$\frac{F_S}{\left(\frac{w.t_1}{\sin\phi}\right)} = \tau_S$$

Using Ernest and Merchant's theory 1st analysis

$$\phi = \frac{90+\alpha-\beta}{2} = \frac{90+7-32.44}{2} = 32.28°$$

$$\frac{F_C}{F_S} = \frac{\cos(\beta-\alpha)}{\cos(\phi+\beta-\alpha)} = \frac{\cos(32.44-7)}{\cos(32.28+32.44-7)}$$

$$F_s = \frac{F_C}{1.6909}$$

$$\Rightarrow \quad F_s = 696\text{ N}$$

$$\tau_s = \frac{696\sin(32.28°)}{(2.5\times 0.2)} = 743.5\text{ MPa}$$

Example : 4.2

In an orthogonal cutting operation, the following data have been observed:

Uncut chip thickness = 0.127 mm
Width of cut = 6.35 mm
Cutting speed = 2 m/s
Rake angle = 10°
Cutting force = 567 N
Thrust force = 227 N
Chip thickness = 0.228 mm

Determine the shear angle, the friction angle, shear stress along the shear plane, and the power for the cutting operation. Also find the chip velocity, shear strain in chip, and the strain rate.

Solution :

Given data in a cutting operation is

$$t_1 = 0.127 \text{ mm}$$
$$w = 6.35 \text{ mm}$$
$$v = 2 \text{m/s}$$
$$\alpha = 10^\circ$$
$$F_C = 567 \text{ N}$$
$$F_T = 227 \text{ N}$$
$$t_2 = 0.228 \text{ mm}$$

1. Shear plane angle (ϕ)

$$\tan\phi = \frac{\cos\alpha}{\dfrac{t_2}{t_1} - \sin\alpha}$$

$$\tan\phi = \frac{\cos 10}{\dfrac{0.228}{0.127} - \sin 10} = 0.6073$$

$$\phi = 31.27^\circ$$

2. Friction angle (β)

$$\tan\beta = \frac{\dfrac{F_T}{F_C} + \tan\alpha}{1 - \dfrac{F_T}{F_C}\tan\alpha}$$

$$\tan\beta = \frac{\dfrac{227}{567} + \tan 10}{1 - \dfrac{227}{567}\tan 10}$$

$$\tan\beta = 0.62$$
$$\beta = 31.82^\circ$$

3. Shear stress along shear plane

$$\frac{F_S}{\left(\dfrac{w.t_1}{\sin\phi}\right)} = \tau_S$$

$$F_S = F_C\cos\phi - F_T\sin\phi$$
$$F_S = 567\cos 31.27 - 227\sin 31.27$$
$$F_S = 366.8 \text{N}$$

$$\tau_s = \frac{366.8 \times \sin 31.27}{(0.127 \times 6.35)}$$

$$\tau_s = 236.1 \text{ Mpa}$$

4. Power for cutting operations

$$P_C = F_C.V$$
$$P_C = 2 \times 567 = 1134 \text{ Nm/s}$$
$$P_C = 1134 \text{ W}$$

5. Chip velocity (V_C)

$$\frac{V_C}{\sin\phi} = \frac{V}{\sin(90^o - \phi + \alpha)}$$
$$V_C = \frac{2 \times \sin 31.27^o}{\sin(90^o - 31.27 + 10^o)}$$
$$V_C = 1.114 \text{m/s}$$

6. Shear strain in chip

$$\gamma = \tan(\phi - \alpha) + \cot\phi$$
$$\gamma = \tan(31.27 - 10) + \cot 31.27$$
$$\gamma = 2.036$$

7. Strain rate

Strain rate is the shear velocity divided by length of shear plane = $\dfrac{V_S}{\text{length of shear plane}}$

Note: If this length is not given in the problem, we generally take 10% of $\dfrac{t_1}{\sin\phi}$.

$$\frac{V_S}{\sin(90 - \alpha)} = \frac{V}{\sin(90 - \phi + \alpha)}$$
$$V_S = \frac{2\sin(90 - 10)}{\sin(90 - 31.27 + 10)}$$
$$V_S = 2.1136 \text{ m/s}$$
$$\frac{V_S}{0.1 \times \dfrac{t_1}{\sin\phi}} = \frac{2.1136 \times \sin 31.27}{0.1 \times 0.127 \times 10^{-3}} = 86386.2 \text{ per sec}$$

Example : 4.3

The following data from the orthogonal cutting test is available :
Rake angle = 10°, chip thickness ratio = 0.35, uncut chip thickness = 0.51, width of cut = 3 mm, yield shear stress of work material = 285 N/mm², Mean friction coefficient on tool face = 0.65. Determine the (i) Cutting force (ii) Radial force (iii) Normal force on tool and (iv) Shear force on the tool.

Solution :

$$\alpha = 10^o$$
$$\frac{t_1}{t_2} = 0.35$$
$$t_1 = 0.51 \text{ mm}$$
$$w = 3 \text{ mm}$$
$$\tau_s = 285 \text{ MPa}$$

$$\mu = 0.65 \qquad \text{Find } F_c, F_R, N, F_s$$

$$t_2 = \frac{t_1}{0.35} = \frac{0.51}{0.35} = 1.457 \text{ mm}$$

$$F_R = 0$$

$$\tan\phi = \frac{\cos 10^\circ}{\frac{1.457}{0.51} - \sin 10^\circ} = 0.367$$

$$\phi = 20.152^\circ$$

$$\tau_S \frac{wt_1}{\sin\phi} = F_S$$

$$F_S = 1265.675 \text{ N}$$

$$F_C = \frac{F_S \cos(\beta - \alpha)}{\cos(\phi + \beta - \alpha)}$$

$$\tan\beta = \mu = 0.65$$

$$\beta = 33.024^\circ$$

$$F_C = \frac{1265.675 \cos(33.034 - 10)}{\cos(20.152 + 33.024 - 10)}$$

$$F_C = 1597.32 \text{ N}$$

$$N = R\cos\beta$$

$$F_S = R\cos(\phi + \beta - \alpha)$$

$$N = F_S \frac{\cos\beta}{\cos(\phi + \beta - \alpha)}$$

$$N = 1455.177 \text{ N}$$

Example : 4.4

During turning a carbon steel rod of 160 mm diameter by a carbide tool of geometry; 0°, 0, 10°, 8°, 15°, 75°, 0 (mm) at speed of 400 rpm, feed of 0.32 mm/rev and 4.0 mm depth of cut, the following observations were made :

Tangential component of the cutting force, $P_Z = 1200$ N

Axial component of the cutting force, $P_X = 800$ N

Chip thickness (after cut), $t_2 = 0.8$ mm

For the above machining condition determine the value of

1. **Friction force, F and normal force, N acting at the chip-tool interface.**
2. **Yield shear strength of the work material under this machining condition.**
3. **Cutting power consumption in kW.**

Solution :

1. Since the second last value in the tool signature is very high and the side cutting edge angle as per ASA can not be so high. So, ORS tool signatures are given in the above problem. So approach angle = 75°

$$N = 400 \text{ rpm}$$

$$f = 0.32 \text{ mm/s}$$

$$d = 4 \text{ mm}$$
$$F_C = 1200 \text{ N}$$
$$F_f = 800 \text{ N}$$
$$t_2 = 0.8 \text{ mm}$$

Find : (i) F, N (ii) τ_S (iii) P_C

$$\alpha = 0$$
$$V = \frac{\pi DN}{60} = \frac{\pi \times 400 \times 0.16}{60}$$
$$V = 3.35 \text{ m/s}$$
$$F_R = F_T \sin\psi$$
$$F_F = F_T \cos\psi$$
$$\psi = 90° - 75° = 15°$$
$$800 = F_T \cos 15°$$
$$F_T = 828.22 \text{ N}$$
$$F = F_T \cos\alpha + F_C \sin\alpha$$
$$\Rightarrow \quad F = 828.22\text{N}$$
$$N = F_C \cos\alpha - F_T \sin\alpha$$
$$N = 1200 \text{ N}$$

2.
$$\frac{F_S}{\left(\frac{w.t_1}{\sin\phi}\right)} = \tau_s$$
$$\tan\phi = \frac{\cos\alpha}{\frac{t_2}{t_1} - \sin\alpha}$$

As $\alpha = 0$ then $\tan\phi = \frac{t_1}{t_2}$

$$t_1 = f\cos\psi = 0.31 \text{ mm}$$
$$w = \frac{d}{\cos\psi} = 4.14 \text{ mm}$$
$$\tan\phi = \frac{0.31}{0.8}$$
$$\Rightarrow \quad \phi = 21.18°$$
$$F_S = F_C \cos\phi - F_T \sin\phi$$
$$F_S = 1200 \cos 21.18° - 828.22 \sin 21.18$$
$$F_S = 819.7 \text{ N}$$
$$\tau_S = \frac{819.7 \sin 21.18}{0.31 \times 4.14}$$
$$\tau_S = 230.76 \text{ MPa}$$

3.
$$P_C = F_C V$$
$$P_C = 1200 \times 3.35$$
$$P_C = 4020 \text{ W}$$

Example : 4.5

While machining steel with zero rake angle prove the following expression where τ_S is shear strength of material P_C is specific cutting power and r is chip reduction coefficient (t_1/t_2)

$$\frac{\tau_S}{P_C} = \frac{r(1-\mu r)}{1+r^2}$$

Solution :

$$P_C = \text{Specific cutting power}$$

$$P_C = \frac{F_C}{w.t_1}$$

$$r = \frac{t_1}{t_2}$$

If sufficient data is not provided assume $\alpha = 0$

$$\frac{F_S}{\left(\frac{w.t_1}{\sin\phi}\right)} = \tau_S$$

$$\frac{\tau_S}{P_C} = \frac{F_S}{F_C}\sin\phi$$

$$\frac{F_S}{F_C} = \frac{\cos(\phi+\beta-\alpha)}{\cos(\beta-\alpha)}$$

$$\frac{\tau_S}{P_C} = \frac{\cos(\phi+\beta-\alpha)}{\cos(\beta-\alpha)}\sin\phi$$

As $\alpha = 0$

$$\frac{(\cos\phi\cos\beta-\sin\phi\sin\beta)}{\cos\beta}\sin\phi$$

$$= (\cos\phi - \sin\phi\tan\beta)\sin\phi$$

$$= \cos\phi\sin\phi - \mu\sin^2\phi$$

$$r = \tan\phi = \frac{\cos\alpha}{\frac{1}{r}-\sin\alpha}$$

$$\Rightarrow \quad \frac{\sin\phi}{\cos\phi} = r$$

$$\Rightarrow \quad \sin^2\phi + \cos^2\phi = 1$$

$$\Rightarrow \quad \tan^2\phi + 1 = \frac{1}{\cos^2\phi} = 1+r^2$$

$$\cos\phi = \frac{1}{\sqrt{1+r^2}}$$

$$\Rightarrow \quad \sin\phi = \frac{r}{\sqrt{1+r^2}}$$

$$\Rightarrow \quad \frac{\tau_S}{P_C} = \left(\frac{1}{\sqrt{1+r^2}} - \frac{r}{\sqrt{1+r^2}}\mu\right)\frac{r}{\sqrt{1+r^2}}$$

$$\Rightarrow \quad \frac{r}{1+r^2} - \frac{r^2}{1+r^2}\mu$$

$$\Rightarrow \quad \frac{\tau_S}{P_C} = \frac{r(1-\mu r)}{1+r^2}$$

Example : 4.6

During orthogonal machining with an HSS Tool, the rake angle was 5°, the undeformed chip thickness was 0.25 mm and width of cut was 4 mm. Assuming shear strength of work material to be 350 N/mm² and coefficient of friction to be 0.5, estimate cutting and thrust force.

Solution :

$\alpha = 5°$, $t_1 = 0.25$ mm, $w = 4$ mm, $\tau_s = 350$ N/mm², $\mu = 0.5$

Friction angle, $\beta = \tan^{-1}(\mu) = \tan^{-1}(0.5) = 26.565°$

∵ Merchant condition for minimum power consumption

$$2\phi + \beta - \alpha = 90$$

$$2\phi + 26.565 - 5 = 90$$

$$2\phi = 68.435°$$

Shear Angle, $\phi = 34.22°$

$$\text{Shear force, } F_s = \frac{\tau_S . w . t_1}{\sin\phi} = \frac{350 \times 4 \times 0.25}{\sin 34.22} = 622.4 \text{ N}$$

$$\because \quad F_s = R\cos(\beta - \alpha + \phi)$$

$$\therefore \quad R = \frac{F_S}{\cos(\beta - \alpha + \phi)}$$

$$\therefore \quad \text{Cutting force, } F_c = R\cos(\beta - \alpha) = \frac{F_S\cos(\beta-\alpha)}{\cos(\beta-\alpha+\phi)}$$

$$= \frac{622.4\cos(26.565 - 5)}{\cos(26.565 - 5 + 34.22)} = 1029.4 \text{ N}$$

$$\text{Thrust force, } F_T = \frac{F_S\sin(\beta-\alpha)}{\cos(\beta-\alpha+\phi)}$$

$$= \frac{622.4\sin(26.565 - 5)}{\cos(26.565 - 5 + 34.22)} = 406.8 \text{ N}$$

Example : 4.7

Mild steel is being machined at a cutting speed of 200 m/min with a tool rake angle of 10°. The width of cut and uncut chip thickness are 2 mm and 0.2 mm respectively. If the avg. value of coefficient of friction between the tool and chip is 0.5 and shear strength of work material is 400 N/mm^2. Determine

(i) Shear angle.

(ii) Cutting and thrust components of the machining forces.

Solution :

Cutting speed, V_c = 200 m/min, α = 10°, w = 2 mm, t_1 = 0.2 mm, μ = 0.5, τ_s = 400 N/mm^2

Friction angle,
$$\beta = \tan^{-1}(\mu) = \tan^{-1}(0.5)$$
$$\beta = 26.565°$$

We know that for metal cutting Lee-shaffer condition
$$\phi + \beta - \alpha = 45$$
$$\phi + 26.565 - 10 = 45$$
$$\phi = 28.435° \quad \text{... shear angle}$$

Shear force
$$F_s = \frac{\tau_s . w . t_1}{\sin \phi}$$
$$= \frac{400 \times 2 \times 0.2}{\sin(28.435)} = 336.02 \text{ N}$$

$\because$
$$F_s = R.\cos(\phi + \beta - \alpha)$$

$\therefore$
$$R = \frac{336.02}{\cos(28.435 + 26.565 - 10)} = 475.204 \text{ N}$$

where R is resultant force of

(i) Cutting force & thrust force or

(ii) Normal & friction force at tool chip interface.

$\therefore$ Cutting force
$$F_c = R\cos(\beta - \alpha)$$
$$= 475.2 \times \cos(26.565 - 10) = 455.48 \text{ N}$$

Thrust force
$$F_T = R\sin(\beta - \alpha)$$
$$= 475.2 \sin(26.565 - 10) = 135.48 \text{ N}$$

Example : 4.8

Following data from the orthogonal cutting test is available rake angle α = 10°, chip thickness ratio r = 0.35, uncut chip thickness t_1 = 0.51 mm width of cut w = 3 mm, yield shear stress of work material = 285 N/mm^2. Coefficient of friction at tool-chip interface = 0.65. Determine

1. Cutting force.

2. Radial force.

3. Normal force on tool chip in interface.

4. Shear force on tool.

Solution :

1. Cutting force

$\therefore$ Shear angle, ϕ

$$\tan\phi = \frac{r\cos\alpha}{1 - r\sin\alpha}$$

$$= \frac{0.35\cos 10}{1 - 0.35\sin 10} = 0.367$$

$$\phi = 20.15°$$

$$\text{Friction angle, } \beta = \tan^{-1}(\mu) = \tan^{-1}(0.65)$$

$$\beta = 33.024°$$

$$\text{Shear force, } F_s = \frac{\tau_s.w.t_1}{\sin\phi}$$

$$= \frac{285 \times 3 \times 0.51}{\sin(20.15°)} = 1265.8 \text{ N}$$

$\because$ $(\beta + \phi - \alpha) = 33.024 + 20.15 - 10 = 43.174°$

$\because$ Net force on the tool

$$R = \frac{F_s}{\cos(\beta + \phi - \alpha)}$$

$$= \frac{1265.8}{\cos(43.174)} = 1735.7 \text{ N}$$

$\therefore$ Cutting force

$$F_c = R\cos(\beta - \alpha)$$

$$= 1735.7 \times \cos(33.024 - 10) = 1597.4 \text{ N}$$

Radial force i..e thrust force

$$F_T = R\sin(\beta - \alpha)$$

$$= 1735.7 \times \sin(23.024) = 679.3 \text{ N}$$

Normal force on tool

$$N = R\cos\beta$$

$$= 1735.7\cos(33.024) = 1455.3 \text{ N}$$

Shear force on the tool

$$F_s = 1265.8 \text{ N}$$

Example : 4.9

A seamless tube 32 mm outer diameter is turned on the lathe the cutting velocity of tool relative to workpiece is 10 m/min. Rake angle α = 35°, depth of cut t_1 = 0.125 mm, length of chip = 60 mm. Horizontal cutting force of the tool on work piece = 200 N vertical cutting force required to hold the tool against work = 80 N. Calculate

1. **Coefficient of friction**
2. **Chip thickness ratio.**
3. **Shear plane angle.**
4. **Velocity of chip relative to tool.**
5. **Velocity of chip relative to work piece.**

Solution :

$V = 10$ m/min, $\alpha = 35°$, $l_c = 60$ mm

Assuming no expansion along width

$$\therefore \quad t_1.\pi D = l_c.t_2$$

$$\frac{t_1}{t_2} = \text{r–chip thickness ratio} = \frac{l_c}{\pi D}$$

$$\therefore \quad r = \frac{60}{\pi \times 32} = 0.597$$

$F_c = 200$ N, $F_T = 80$ N

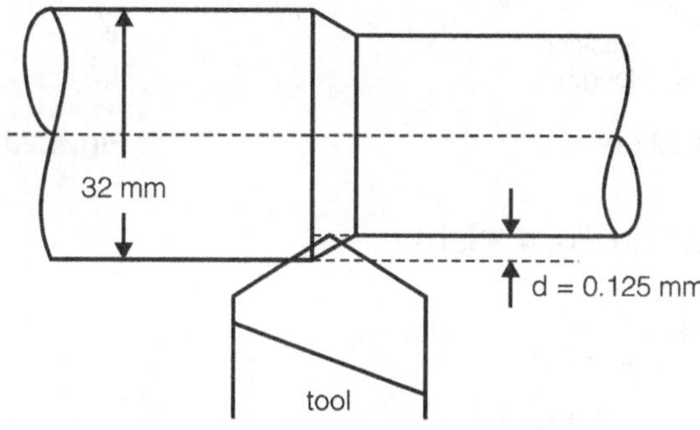

Fig. 4.22

1. Coefficient of friction

$$\mu = \frac{F_T + F_c \tan \alpha}{F_c - F_T \tan \alpha} = \frac{80 + 200 \tan 35}{200 - 80 \tan 35}$$

$$= \frac{220.04}{143.98} = 1.5282$$

2. Chip thickness ratio $r = 0.597$
3. Shear plane angle (ϕ)

$$\tan \phi = \frac{r \cos \alpha}{1 - r \sin \alpha} = \frac{0.597 \cos 35}{1 - 0.597 \sin 35}$$

$$\phi = 36.64°$$

4. Velocity of chip relative to tool

$$V_c = r.V = 0.597 \times 10$$

$$= 5.97 \text{ m/min}$$

5. Velocity of chip relative to workpiece

$$V_s = \frac{V \cos \alpha}{\cos (\phi - \alpha)} = \frac{10 \cos 35}{\cos (36.64 - 35)}$$

$$= 8.195 \text{ m/min}$$

4.15 CONTACT LENGTH

After machining as the chips flow over the rake face, the temperature of material will be very high. At high temperature coefficient of friction increases and hence shear stress. Once the shear stress reaches yield strength of shear, sticking will take place between the materials. So over the rake face initially there will be sticking and there is slipping.

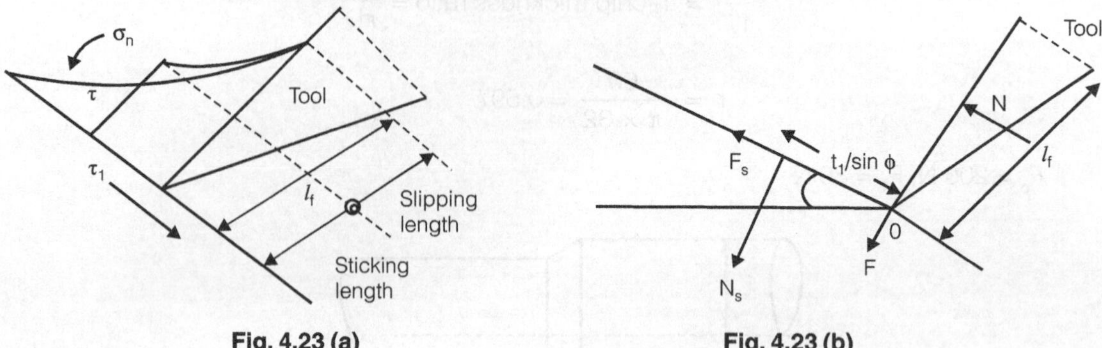

Fig. 4.23 (a) **Fig. 4.23 (b)**

Taking moments about O in Fig. 4.23 (b)

$$\Rightarrow \quad N_S \times \frac{1}{2}\frac{t_1}{\sin\phi} = N \times \frac{1}{2} l_f$$

$$\Rightarrow \quad N_S \times \frac{t_1}{\sin\phi} = N l_f$$

$$N_S = R\sin(\phi + \beta - \alpha)$$

$$N = R\cos\beta$$

$$\Rightarrow \quad R\sin(\phi + \beta - \alpha)\frac{t_1}{\sin\phi} = R l_f \cos\beta$$

So contact length l_f is

$$\Rightarrow \quad l_f = t_1 \frac{\sin(\phi + \beta - \alpha)}{\sin\phi\cos\beta}$$

4.16 MEASUREMENT OF TEMPERATURE

4.16.1 Tool work piece thermocouple

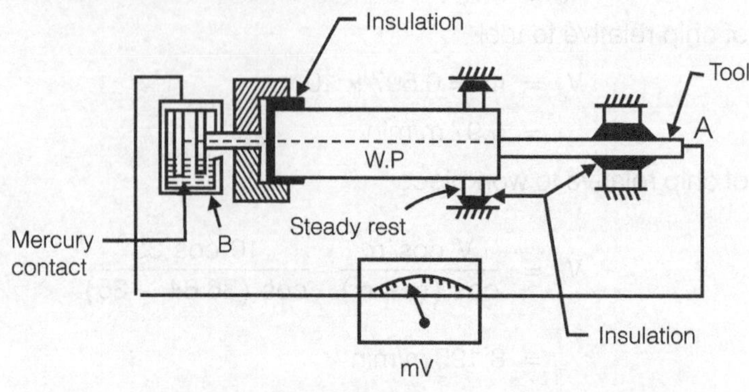

Fig. 4.24 (a)

It is based on the principle of Seeback effect that is when a combination of materials is kept at a temperature gradient there will be a flow of current which will be proportional to the temperature difference. Tool work piece junction acts as a hot junction point and mercury bath serves as an cold junction point. Mercury bath is also provided to have a metal contact. If there is a build up edge formation of oxide layer measured temperatures are not accurate.

4.16.2 Inserted thermocouple

A hole is drilled from the back side of the tool and a thermocouple is inserted into it. Although the temperature measurements are unaffected by build up edge or oxide layer formation over the tool, but it has to be calibrated before use.

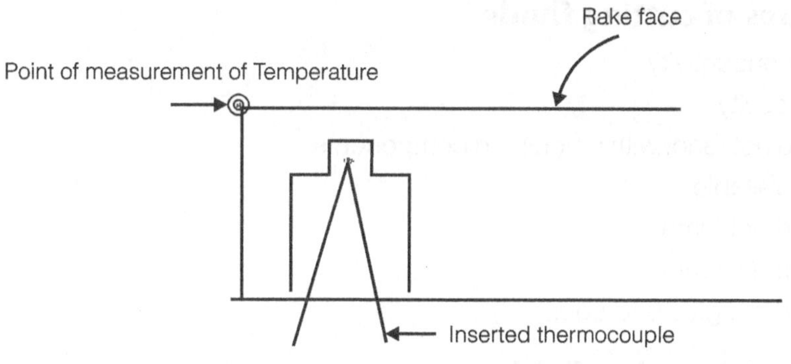

Fig. 4.24 (b)

4.16.3 Infrared measurement

The infrared radiation from the cutting zone may also be monitored with a radiation pyrometer or probes. These probes are very expensive. However the technique indicates only surface temperatures. Depending on the temperatures certain wave length radiations will come out from the machining area and the radiations are captured by probes. The accuracy of the results depends on the emissivity of the surfaces, which is difficult to determine accuracy.

4.16.4 Hardness and microstructure studies

Samples of known steel composition are inserted over the rake face. After machining, hardness and microstructure of the specimens are tested. By analysing the microstructure it is evaluated that what was the temperature over the rake face.

4.17 CUTTING FLUIDS

Cutting fluids are used for decreasing power requirement and increasing heat dissipation. The main factors likely to cause excessive heat during metal cutting are

- Cutting speed is too high.
- Poor surface finish on the cutting face of the tool.
- Worn or incorrectly ground cutting tool.
- Formation of a built up edge on cutting face of the tool.
- Friction between tool and work piece.

There are three zones at which heat is generated they are Shear zone, Friction zone and Tool work piece zone. Cutting fluids generally used are:

1. Neat oils + Extreme pressure additives (EP additives)
2. Water emulsions

EP (Extreme Pressure) additives like Chlorine, Sulphur, and Phosphorus are added into Neat oils. Since Manganese is present in the material in the form of impurity EP additives react with the manganese and produce low shear strength film. This decreases the power consumption in machining operation. When the cutting speeds are higher, there will not be any time for the reaction to take place, so water emulsions are used to decrease the viscosity of the cutting fluid so that it takes away heat from the machining area.

4.17.1 Properties of cutting fluids

1. Higher conductivity
2. Low viscosity
3. It should not react with machining components
4. Easily available
5. It should not fume
6. It should not foam
7. It should not give bad odor.

4.17.2 Selection of cutting fluids

We can apply cutting fluids in two ways.

- Mist application
- Flood application

In mist application cutting fluids come out in the form of spray and penetrates into the machining area which increases the probability of reaction. But at higher speeds flood applications are used.

Cutting speeds :	Low speeds	:	Neat oils + EP additives.
	Medium speeds	:	1 : 10 water emulsion + EP additives
	High speeds	:	1 :100 water emulsions.

4.17.3 Work piece material

1. **Cast iron (CI) :** While machining cast iron cutting fluids are not used because, it produces graphite flakes, this mixes with the cutting fluid and affects the machining area. So, either no cutting fluid is used or compressed air is used.
2. **Steel :** For low cutting speeds cutting fluids used are Neat oil and EP additives. For medium cutting speeds cutting fluids used are water emulsions in the ratio 1:10.For high cutting speeds cutting fluids used are water emulsions in the ratio 1:100.
3. **Aluminium :** Cutting fluids used are Neat oil and EP additives (Kerosene with EP additives) Aluminium is a very soft material so, at a high speed nothing is required, but at low speed, there will be a tendency of build up edge formation, so neat oils with EP additives are used.
4. **Magnesium :** Magnesium reacts with water at high temperature and burns, so only neat oils are used.
5. **Brass and Bronze :** EP additives present in the cutting fluid reacts with the material and produces dull surface. So, only Neat oils like kerosene is used.

4.18 MECHANISM OF TOOL WEAR

1. **Diffusion wear :** The favorable condition for the diffusion is provided by the localized temperature over the actual area between the chip underside and the tool face. In that condition the metal atoms will transfer from the tool material to the chip material at the points of contact. This weakens the surface structure of the cutting tool and may ultimately lead to tool failure. The amount of diffusion depends upon temperature, period of contact between tool face and the chip and the bonding affinity between the materials of the tool and chip.

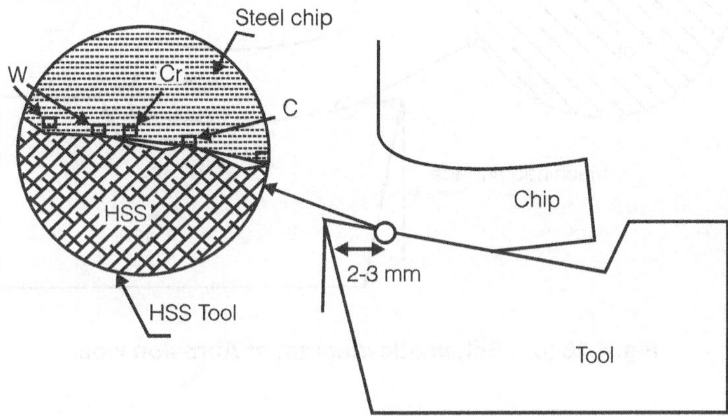

Fig. 4.25 (a) : Schematic diagram of Diffusion wear

2. **Adhesion wear :** Due to the excessively high temperature at the chip-tool interface a metallic bond takes place between the chip material and tool material at the contact points, in the form of spot welds are formed. When the chip slides, these small welds are broken and small amount of welded tool carried away by the sliding chip. Thus small particles will continue to separate through this phenomenon and carried away by the chip by adhesion to its underside.

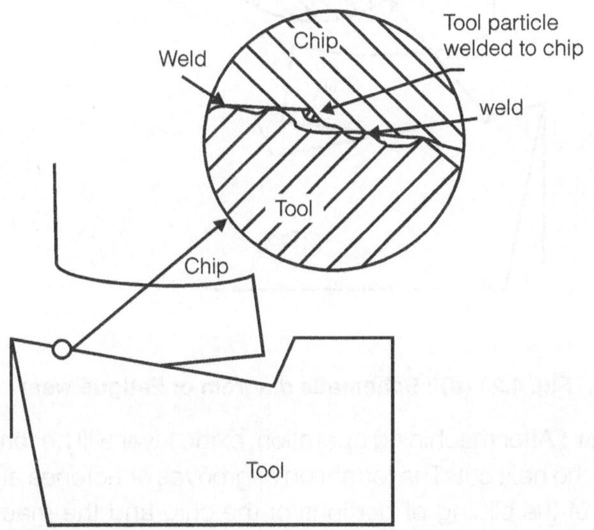

Fig. 4.25 (b) : Schematic diagram of Adhesion wear

3. **Abrasion wear :** Since there will be thousands of faylite pockets inside the work material and as these faylite pockets come in contact with the cutting edge, there will be a shock. As a result of that, a portion of cutting edge will be eroded.

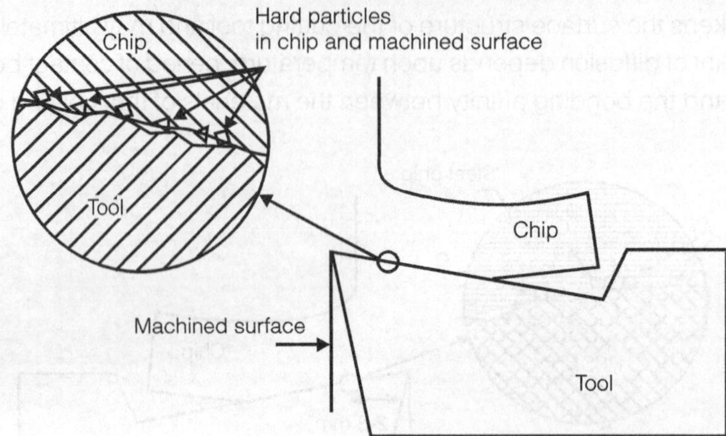

Fig. 4.25 (c) : Schematic diagram of Abrasion wear

4. **Fatigue wear :** On a microscopic level, hills of the tool will be eroded and fresh hills will be formed by the interaction with the work piece. This eroded material is called fatigue wear.

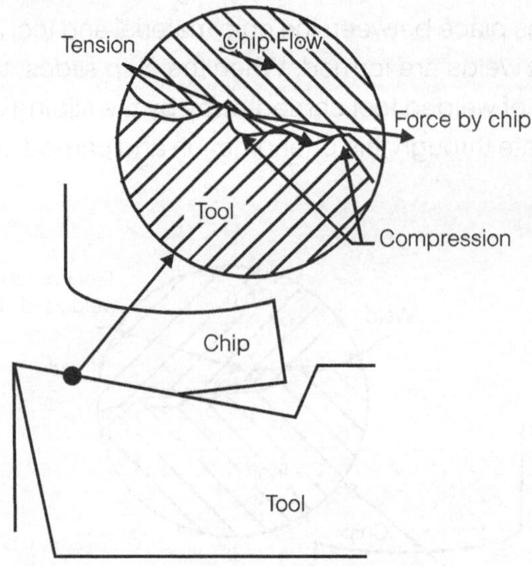

Fig. 4.25 (d) : Schematic diagram of Fatigue wear

5. **Oxidation wear :** After machining operation, oxide layer will be formed over the tool which will be removed in the next cut. The formation of grooves or notches at the rake face and the flank is on account of the sliding of portions of the chip and the machined surface which have reacted with the oxygen in the atmosphere to form abrasive oxides. This causes oxidation wear.

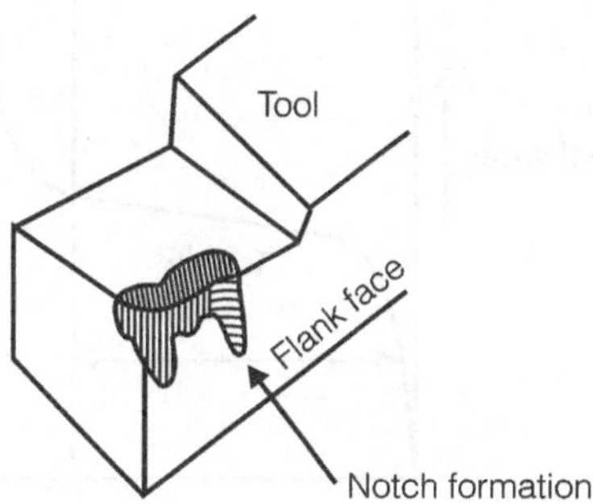

Fig. 4.25 (e) : Schematic diagram of Oxidation wear

4.18.1 Types of Tool Wear

1. **Flank wear :** Flank wear is more at the corner and notch because:
   (a) There is sudden entry and sudden exit at the corner and at the notch.
   (b) There is also a work hardening of the material in the width direction.

   A stage reaches when diffusion becomes the predominant wear mode on the flank. After a critical wear land has formed, further wear takes place at an accelerating rate. It is advisable to change the tool in order to avoid the tool failure. Flank wear is caused mainly through abrasion, where as during the rapid wear phase it is caused by diffusion.
2. **Crater wear :** Crater wear is mainly due to the diffusion and abrasion between the chip and the face of the tool, a short distance from the cutting edge. The crater is formed on the surface of the tool by the action of chip particles flowing over it because of very high temperature. When crater becomes excessive, the cutting edge may break from the tool. This is generally observed while machining ductile materials, which produces continuous chips.

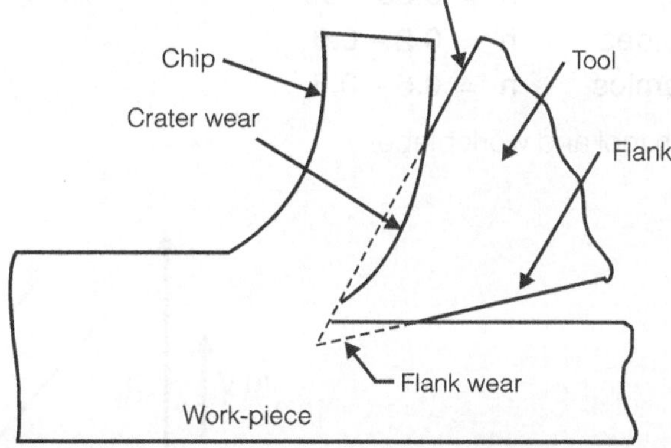

Fig. 4.26 (a) : Schematic diagram of Flank wear and Crater wear

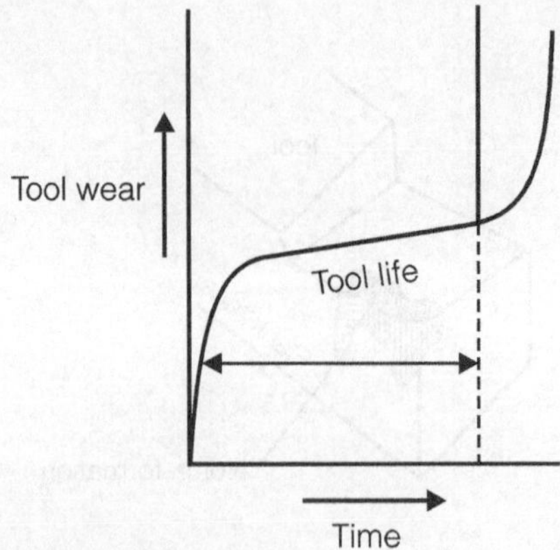

Fig. 4.26 (b) : Variation of tool wear with respect to time

V_B = 0.3, it is the depth of Flank wear.

K_T = 0.3, it is the depth of crater wear.

Initially, the cutting edge is sharp and as soon as it comes in contact with the work piece, sharp edges try to round off. That is why; the tool wear is more in the beginning. After some amount of tool wear takes place, tool loses its geometry and causes the drag between the tool and the work piece increases. This results in tool breakdown. Before this condition comes, tool has to be with drawn from the work shop and regrinded. This time between two regrindings is called **TOOL LIFE.**

4.18.2 TAYLOR'S TOOL LIFE EQUATION

The tool life is mainly affected by cutting speed, means higher the cutting speed the smaller the tool life. Taylor gave the relation between cutting speed and tool life that is

$$VT^n = C$$

V = cutting speed

T = tool life.

C = machining constant.

n = Tool life exponent (depends only on tool material)

For HSS, n = 0.08 – 0.2

For carbides n = 0.2 – 0.6

For ceramics n = 0.5 – 0.8

C depends upon both tool and work piece.

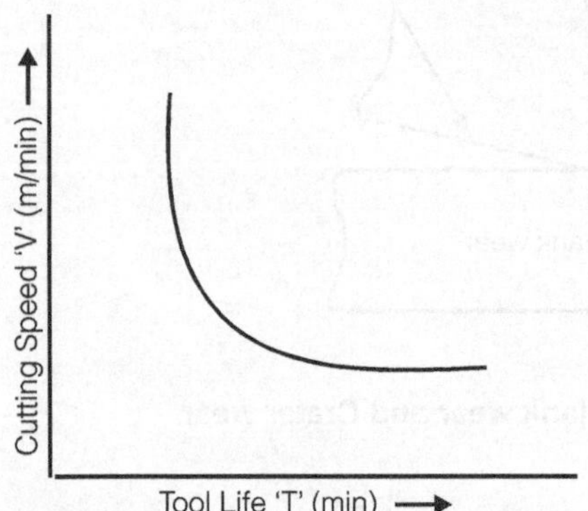

Fig. 4.27 : Variation of velocity with respect to time

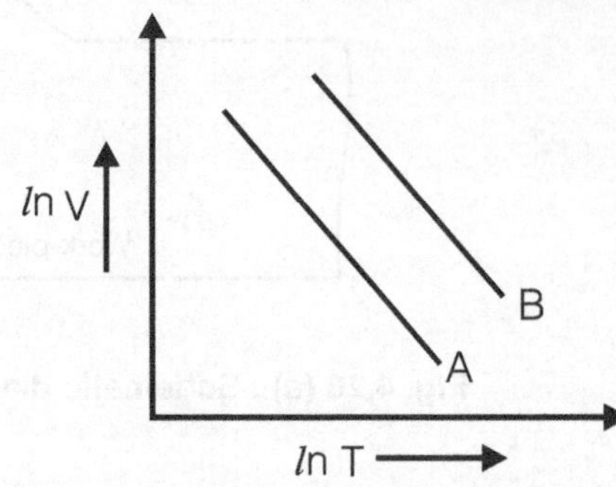

Fig. 4.28 : Two curves having same slope

Hyperbola

$$VT^n = C$$

$$\log V + n \log T = \log C$$

As the two lines are parallel slope (n) is same for both. B makes higher intercept on the axis. 'C' for B is higher than that for A.

Example : 4.10

Following is the data available on cutting speed and tool life.

V = 150 m/min T = 60 min

V = 200 m/min T = 23 min

Determine the Taylors constant and tool life exponent.

Solution :

$$VT^n = C$$

$$150\,(60)^n = C$$

$$200\,(23)^n = C$$

$$\Rightarrow \quad 150\,(60)^n = 200\,(23)^n$$

$$\Rightarrow \quad \left(\frac{60}{23}\right)^n = \frac{200}{150}$$

$$\Rightarrow \quad (2.6087)^n = 1.33$$

$$\Rightarrow \quad n \log (2.6087) = \log 1.33$$

$$\Rightarrow \quad n = 0.3$$

$$C = 512.31$$

Example : 4.11

Spindle speed (N)	Feed (f)	Components
250	0.1	311
250	0.125	249
300	0.125	144

This is the record of number of identical components produced at different speeds and feeds. Estimate the number of components that can be produced at a spindle speed of 310 rpm and feed of 0.15 mm/rev

Solution :

$$Nf^a C_1^{\,b} = C$$

$$250\,(0.1)^a\,(311)^b = C$$

$$\ln 250 + a \ln 0.1 + b \ln 311 = \log C \qquad ...(1)$$

$$\ln 250 + a \ln 0.125 + b \ln 249 = \log C \qquad ...(2)$$

$$\ln 300 + a \ln 0.125 + b \ln 144 = \log C \qquad ...(3)$$

Solving (1) and (2)

$$a\,(\ln 0.1 - \ln 0.125) + b\,(\ln 311 - \ln 249) = 0$$

$$\Rightarrow \quad -0.223\,a + 0.2223\,b = 0$$

$$\Rightarrow \quad 0.223\,a = 0.2223\,b$$

Solving (2) and (3)

$$(\ln 250 - \ln 300) + b\,(\ln 249 - \ln 144) = 0$$

$$b = 0.333$$
$$a = 0.3318$$
$$C = 787.46$$

So $$350\,(0.15)^{0.3318}C_1^{\,0.333} = 787.46$$

$$C_1^{0.333} = 4.222 \Rightarrow C_1 = 4.222^{(1/0.333)}$$

So $$C_1 = 75.59$$

So 75 components will be produced.

Example : 4.12

In drilling of 20 mm diameter hole in a M.S. solid block of 25 mm thickness, the feed rate used is 0.25 mm/rev. The drill must be changed after making 10 holes if it is rotated at 250 rpm. But the drill can produce 50 holes if it is rotated at 200 rpm. How many holes can be produced if its rpm is 150?

Solution :

Time to drill 10 holes at 250 rpm

$$T_1 = \frac{10 \times 25}{0.25 \times 250} = \frac{1000}{250} = 4 \text{ minutes}$$

Time to drill 50 holes at 200 rpm

$$T_2 = \frac{50 \times 25}{0.25 \times 200} = \frac{50 \times 25 \times 100}{25 \times 200}$$

$$T_2 = 25 \text{ minutes}$$

Taylor Tool life equation

$$T_1^n N_1 = T_2^n N_2 = C \qquad \text{... (1)}$$

Taking logarithmic of both sides

$$n\, ln\, T_1 + ln\, N_1 = n\, ln\, T_2 + ln\, N_2$$

$$n = \frac{(ln\, N_1 - ln\, N_2)}{(ln\, T_2 - ln\, T_1)}$$

$$n = \frac{ln\, 250 - ln\, 200}{ln\, 25 - ln\, 4}$$

$$= \frac{0.22314}{1.8326} = 0.1218$$

Tool life at $N_3 = 150$ rpm

$$\therefore \quad T_1^n N_1 = T_3^n N_3$$

$$4^{0.1218} \times 250 = T_3^{0.1218} \times 150$$

$$T_3 = 4 \times \left(\frac{250}{150}\right)^{\frac{1}{0.1218}}$$

$$= 265.142 \text{ minutes}$$

Time to drill a hole

$$T_1 = \frac{25}{0.25} \times \frac{1}{150} = \frac{100}{150} \text{ min}$$

$$\text{No. of holes} = \frac{T_3}{T_1}$$

$$= \frac{265.143}{100} \times 150$$

$$= 397.7145 = 398 \text{ holes}$$

Example : 4.13

A carbide tool with MS workpiece was found to give tool life of 2 hrs while cutting at 0.5 m/min. Compute the tool life if the same tool is used at speed of 25% higher than previous one. Also determine the value of cutting speed if the tool is required to have tool life of 3 hrs. Assume Taylor's exponent 'n' to be 0.27?

Solution :

Taylor's Tool life

$VT^n = C$, where V-m/min, T-minutes

n = 0.27-Taylor's exponent

at V_1 = 0.5 m/min, T = 2 hrs = 120 min

$$\therefore \quad 0.5 \times (120)^{0.27} = c$$

$$\therefore \quad c = 1.82116$$

Now the speed is 25% higher

$$\therefore \quad V_2 = 1.25 \times 0.5 = 0.625 \text{ m/s}$$

$$\therefore \quad V_2 T_2^{0.27} = 1.82116$$

$$0.625.T_2^{0.27} = 1.82116$$

$$T_2 = 52.51 \text{ min}$$

Again the tool life, T_3 = 3 hrs = 180 min

$$\therefore \quad V_3 T_3^{0.27} = 1.82116$$

$$V_3 = \frac{1.82116}{(180)^{0.27}}$$

$$V_3 = 0.44815 \text{ m/min}$$

Example : 4.14

In a drilling operation under a given condition, the tool life found to decrease from 20 min to 5 min, due to increase in drilling speed from 200 rpm to 400 rpm. What will be the tool life of that drill under the same condition if the drill speed is 300 rpm.

Solution

$\because$ T_1 = 20 min at N_1 = 200 rpm

T_2 = 5 min at N_2 = 400 rpm

But Taylor's tool life equation for drill

$$N_1 T_1^n = N_2 T_2^n = \text{constant} \quad \dots (1)$$

$$\therefore \quad 200(20)^n = 400(5)^n$$

$$4^n = 2$$

$$\therefore \quad n = 0.5$$

$$\therefore \quad \text{Tool life at } N_3 = 300 \text{ rpm}$$

$$\therefore \quad N_1 T_1^n = N_3 T_3^n$$

$$200(20)^{0.5} = 300.T_3^{0.5}$$

$$T_3 = 20\left(\frac{200}{300}\right)^2 = 8.89 \text{ min}$$

Example : 4.15

A generalized Taylor's tool life equation was obtained for HSS tool;

$$V = \frac{C}{T^{0.13}\ f^{0.6}\ d^{0.3}}$$

A 60 min tool life was obtained using the following cutting parameters, V = 40 m/min, f = 0.25 mm, d = 2.00 mm. Compute the tool life if cutting speed, feed and depth of cut are increased by 25% and also if they are increased individually by 25% where V-cutting speed, f = feed, d = depth of cut.

Solution :

$$\because \quad V = \frac{C}{T^{0.13}\ f^{0.6}\ d^{0.3}}$$

$$\therefore \quad T = \frac{C^{\frac{1}{0.13}}}{V^{\frac{1}{0.13}}.f^{\frac{0.6}{0.13}}.d^{\frac{0.3}{0.13}}}$$

$$T = \frac{C^{7.7}}{V^{7.7}.f^{4.6}.d^{2.3}} \quad \text{... (1)}$$

This expression shown that tool life depend upon speed most, then feed and then least on depth of cut.

$$\therefore \quad 60 = \frac{C^{7.7}}{40^{7.7}.(0.25)^{4.6}.(2)^{2.3}}$$

$$C = 36.5 \text{ units}$$

Now new set of parameter

$$V' = 1.25 \times 40 = 50 \text{ m/min}$$

$$f' = 1.25 \times 0.25 = 0.3125 \text{ mm}$$

$$d' = 1.25 \times 2 = 2.5 \text{ mm}$$

$$\therefore \quad T' = \frac{(36.5)^{7.7}}{(50)^{7.7}.(0.3125)^{4.6}.(2.5)^{2.3}}$$

$$= 2.27 \text{ minutes}$$

Now speed increased

$$= \left(\frac{40}{50}\right)^{7.7} \times 60 = 10.76 \text{ min}$$

Now feed is increased

$$= \left(\frac{0.25}{0.3125}\right)^{4.6} \times 60 = 21.5 \text{ min}$$

Now depth of cut is increased

$$\therefore \quad T_d' = 60 \times (0.8)^{2.3}$$

$$= 36 \text{ min}$$

Hence decreasing order of effect of various parameters on tool life are

speed → feed → depth of cut.

Example : 4.16

A turning tool with sharp corner and a major cutting edge angle of 60° is to be used at a feed of 0.05 mm. What minor cutting edge angle should be provided to obtain an arithmetical mean surface roughness of 3 μm. under ideal condition?

Solution :

ψ major cutting edge angle = 60

ψ_1 minor cutting edge angle = ?

∴ Maximum height of roughness

$$\therefore \quad H_{max} = \frac{f}{(\tan\psi + \cot\psi_1)} \quad \text{... (1)}$$

$$R_a = \frac{H_{max}}{4}$$

Given that

$$f = 0.05 \text{ mm}$$

$$R_a = 3 \times 10^{-3} \text{ mm}$$

So

$$R_a = \frac{f}{4(\tan\psi + \cot\psi_1)}$$

$$3 \times 10^{-3} = \frac{0.05}{4(\tan\psi + \cot\psi_1)}$$

$$\tan\psi + \cot\psi_1 = \frac{50}{4 \times 3} = 4.1607$$

$$\tan 60 + \cot\psi_1 = 4.1667$$

$$\cot\psi_1 = 4.1667 - 1.732$$

$$= 2.4316$$

$$\tan\psi_1 = 0.41074$$

$$\psi_1 = 22.33°$$

4.19 EFFECT OF PARAMETERS ON TOOL LIFE

1. **Cutting Speed :** Higher the cutting speed more will be the temperature over the rake face and faylite pockets will create more impact over the cutting edge. So tool life will decrease.

2. **Feed and Depth of cut :** By increasing the feed and depth of cut, tool life will decrease because it increases the cutting forces.
   The imperical formulae is given by

$$V = \frac{257}{T^{0.19} f^{0.36} t^{0.8}}$$

$$V = \text{cutting speed in m/min}$$
$$T = \text{tool life in min}$$
$$f = \text{feed rate in mm/min}$$
$$t = \text{depth of cut in mm}$$

   For a given tool life the relationship among other variables is also given by the following imperical formulae

$$V = \frac{C}{f^a t^b}$$

$$V = \text{cutting speed for given tool life}$$
$$C = \text{constant.}$$

   The exponents a and b depends on the mechanical properties of the material.

3. **Size and Structure :** When the work piece has fine grain structure, tool life will decrease and when the tool material has fine grain structure, tool life will increase.
4. By increasing the intermittent cuts tool life decreases.

4.20 MACHINABILITY

The ease with which a given material may be worked with a cutting tool is Machinability. Factors affect the Machinability are :

1. **Tool life :** The longer the tool life it enables at a given cutting speed better is the Machinability
2. **Surface finish :** Two materials are machined under identical cutting conditions and material which produces good finish is considered to be more machinable material. This criterion is used in finished cuts.
3. **Cutting Forces :** Two materials are machined under identical cutting conditions and the material which requires smaller cutting forces is considered to be more machinable. This criterion is used in smaller and old machines.

Machinability index :

V_S = Cutting speed of standard free-cutting steel for 1 min tool life.

V_t = Cutting speed of metal for 1 min tool life.

Standard work piece material is machined at a cutting speed so that tool life 'T' is observed. Keeping other parameters same, test work piece is machined to get the same tool life 'T'. Machinability index indicates that the material is more machinable.

$$\text{Machinability index} = \frac{V_t}{V_S} \times 100$$

Table : Representative Machinability indexes for some material

Material	Machinability Index
Stainless steel	25
Low carbon steel	55-65
Copper	70
Red brass	180
Aluminium alloys	300 – 1500
Magnesium alloys	500 – 2000
C–45 (0.45%C)	100
Mn steel	60
Free cutting steels	200
C–14 (0.14%C)	160

4.21 ECONOMICS OF MACHINING

One of the main objective in metal cutting is optimizing the tool life. In order to minimize the production cost, maximize the production rate etc. optimum cutting speed must be selected. For Example if cutting speed reduced in order to enhance the tool life the metal removal rate is also reduced and, therefore the production cost increases similarly to increase the tool life by reducing the feed rate and depth of cut. If metal removal rate increased, it results in increase of the cutting speed and depth of cut, the tool life shortens and therefore tooling cost increases and hence total production cost increases. So a balance is therefore required to get optimum cutting speed which ensures economical production.

As seen in figure 4.29 it is observed that the tooling cost increases while the machining cost decreases with increase in cutting speed. Also there is a point 'p' on the total cost curve which indicates the minimum cost of production. The cutting speed corresponding to this point gives the optimum cutting speed for economical production.

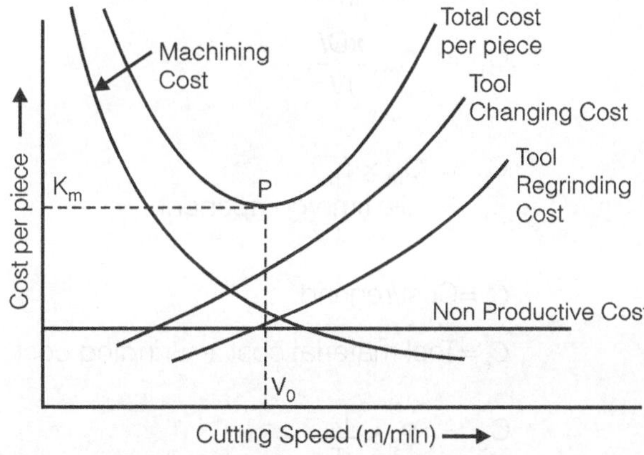

Fig. 4.29 (a) : Effect of variations in cutting speed on various cost factors

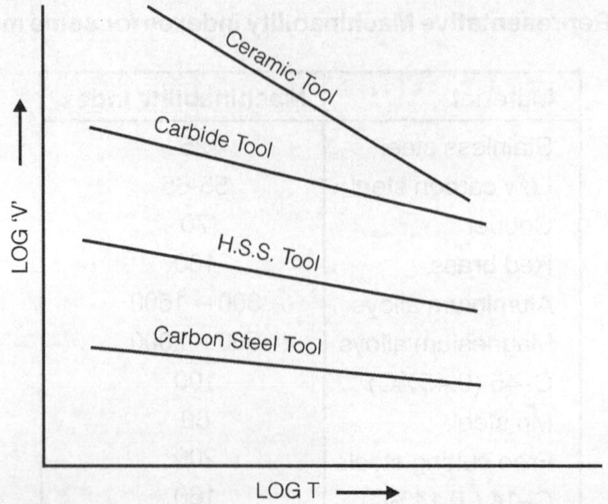

Fig. 4.29 (b) Variation of velocity with respect to time for different materials

$$C_m = \text{Machine cost in Rs/time}$$

$$T_m = \text{Machining time}$$

$$T_m = \frac{l}{fN}$$

$$l = \text{total length of cut}$$

$$f = \text{feed}$$

$$N = \text{rpm}$$

$$V = \pi DN$$

$$N = \frac{V}{\pi D}$$

$$\Rightarrow \quad T_m = \frac{\pi D l}{fV}$$

1. Machining cost

$$C_1 = C_m . T_m$$

$$C_1 = C_m . \frac{\pi D l}{fV} \quad \text{...(1)}$$

2. Idle cost

$$C_2 = C_m \times T_h \quad \text{...(2)}$$

$$T_h = \text{idle time/component}$$

3. Tool cost

C_e =Cost/regrind

C_e =Tool material cost + Grinding cost

$$C_3 = C_e \times \frac{T_m}{T}$$

$$VT^n = C \quad \Rightarrow \frac{1}{T} = \left(\frac{V}{C}\right)^{\frac{1}{n}}$$

$$C_3 = C_e \times \left(\frac{\pi Dl}{fV}\right)\left(\frac{V}{C}\right)^{\frac{1}{n}} \quad ...(3)$$

4. Tool Changing Cost

$$C_4 = C_m \times \left(\frac{T_m}{T}\right) \times T_C$$

$$T_C = \text{tool changing time}$$

$$C_4 = C_m \cdot \left(\frac{\pi Dl}{fV}\right) \times \left(\frac{V}{C}\right)^{\frac{1}{n}} \times T_C \quad ...(4)$$

The total cost can be found by adding the four costs.

$$C = C_1 + C_2 + C_3 + C_4$$

1. For total minimum cost

$$\frac{dC}{dV} = \frac{dC_1}{dV} + \frac{dC_2}{dV} + \frac{dC_3}{dV} + \frac{dC_4}{dV}$$

$$= \frac{d}{dV}\left(C_m \times \frac{\pi Dl}{fV}\right) + \frac{d}{dV}(C_m \times T_h) + \frac{d}{dV}\left(C_e \times \frac{\pi Dl}{fV}\left(\frac{V}{C}\right)^{\frac{1}{n}}\right) + \frac{d}{dV}\left(C_m \times \frac{\pi Dl}{fV} \times \left(\frac{V}{C}\right)^{\frac{1}{n}} \times T_C\right) = 0$$

$$\Rightarrow \frac{-C_m \pi Dl}{fV^2} + \frac{d}{dV}\left(\frac{C_e \pi Dl}{fC^{1/n}} \times (V)^{\frac{1}{n}-1}\right) + \frac{d}{dV}\left(C_m \times \frac{\pi Dl}{fC^{1/n}} T_c\, V^{\frac{1}{n}-1}\right) = 0$$

$$\Rightarrow \frac{\pi Dl}{f}\left(\frac{-C_m}{V^2} + \frac{C_e}{C^{1/n}}\left(\frac{1-n}{n}\right)V^{\frac{1}{n}-2} + \frac{C_m}{C^{1/n}} T_C \left(\frac{1-n}{n}\right)V^{\frac{1}{n}-2}\right) = 0$$

$$\Rightarrow \frac{-C_m}{V^2} + \frac{C_e}{C^{1/n}}\left(\frac{1-n}{n}\right)V^{\frac{1-2n}{n}} + \frac{C_m}{C^{1/n}} T_C \left(\frac{1-n}{n}\right)V^{\frac{1-2n}{n}} = 0$$

$$\Rightarrow \frac{-C_m}{V^2} + \frac{C_e}{C^{1/n}}\left(\frac{1-n}{n}\right) \times \frac{1}{V^{\frac{2n-1}{n}}} + \frac{C_m}{C^{1/n}} T_C \left(\frac{1-n}{n}\right) \times \frac{1}{V^{\frac{2n-1}{n}}} = 0$$

$$\Rightarrow \frac{1}{V^2}\left(-C_m + \frac{C_e}{C^{1/n}}\left(\frac{1-n}{n}\right) \times \frac{1}{V^{\frac{1}{\frac{2n-1-2n}{n}}}} + \frac{C_m}{C^{1/n}} T_C \left(\frac{1-n}{n}\right) \times \frac{1}{V^{\frac{1}{\frac{2n-1-2n}{n}}}}\right) = 0$$

$$\Rightarrow \left(-C_m + \frac{C_e}{C^{1/n}}\left(\frac{1-n}{n}\right) \times \frac{1}{V^{-1/n}} + \frac{C_m}{C^{1/n}} T_C \left(\frac{1-n}{n}\right) \times \frac{1}{V^{\frac{-1}{n}}}\right) = 0$$

$$\Rightarrow C_m = \frac{1}{V^{-1/n}}\left(\frac{1-n}{n}\right)\frac{1}{C^{1/n}}(C_e + T_C C_m)$$

$$\Rightarrow V^{-1/n} = \left(\frac{1-n}{n}\right)\frac{1}{C^{1/n}C_m}(C_e + T_C C_m)$$

$$\Rightarrow V^{\frac{1}{n}} = \frac{n}{1-n}\frac{C^{1/n}C_m}{(C_e + T_C C_m)}$$

$$\Rightarrow V = \left[\frac{n}{1-n}\frac{C^{1/n}C_m}{(C_e + T_C C_m)}\right]^n$$

$$V_{opt} = \frac{C}{\left[\left(\frac{C_e}{C_m} + T_C\right)\left(\frac{1}{n} - 1\right)\right]^n}$$

$$T_{opt} = \left[\left(\frac{C_e}{C_m} + T_C\right)\left(\frac{1}{n} - 1\right)\right]$$

2. For maximum production rate

Time taken for 1 product

$$T = T_m + T_h + \frac{T_m}{T} T_C$$

No. of products per unit time

$$R_p = \frac{1}{T}$$

$$T = \frac{\pi D l}{fV} + T_h + \left(\frac{\pi D l}{fV}\right)\left(\frac{V}{C}\right)^{1/n} T_C$$

$$\frac{dT}{dV} = 0$$

$$= \frac{-\pi D l}{fV^2} + \frac{\pi D l}{fC^{1/n}} V^{\frac{1}{n}-1-1}\left(\frac{1}{n} - 1\right)T_C = 0$$

$$\Rightarrow \frac{\pi D l}{f}\left[\frac{-1}{V^2} + \frac{1}{C^{1/n}} V^{\frac{1}{n}-2}\left(\frac{1}{n} - 1\right)T_C\right] = 0$$

$$\Rightarrow \frac{1}{V^2} = \frac{1}{C^{1/n}} V^{\frac{1}{n}-2}\left(\frac{1}{n} - 1\right)T_C$$

$$\Rightarrow V^{\frac{1}{n}-2+2} = \frac{C^{1/n}}{\left(\frac{1}{n} - 1\right)T_C}$$

$$\Rightarrow V^{\frac{1}{n}} = \frac{C^{1/n}}{\left(\frac{1}{n}-1\right)T_C}$$

$$\Rightarrow V_{opt} = \frac{C}{\left[\left(\frac{1}{n}-1\right)T_C\right]^n}$$

$$\Rightarrow T_{opt} = \left(\frac{1}{n}-1\right)T_C$$

3. For maximum profit rate

$$\text{Profit rate} = \frac{R - C_0}{T_m + T_h + \frac{T_m}{T} \times T_c}$$

$$R = \text{selling price}$$

(V_{opt}) production rate > (V_{opt}) profit rate > (V_{opt}) minimum unit cost

When V is increased beyond max productivity, tool life decreases. Frequent grinding is required that increases and hence productivity decreases.

Example : 4.17

The results of machining steel with two grades of tool are given below

Tool	**Taylor's experiment**	**Cutting speed for 1 min tool life, m/min**
A	**0.20**	**100**
B	**0.25**	**120**

For a 180 min tool life which tool is recommended and why ? If the tool grinding and changing time is 15min which cutting speeds for tool A has to be chosen from two available speeds 45 m/min and 50 m/min?

Solution :

For

$$A \rightarrow VT^{0.2} = 100$$

$$B \rightarrow VT^{0.25} = 120$$

1. If tool life = 180 min

$$V_A = \frac{100}{180^{0.2}} = 35.39 \text{ m/min}$$

$$V_B = \frac{120}{180^{0.25}} = 32.76 \text{ m/min}$$

Tool A : as $V_A > V_B \Rightarrow$ higher speed leads to higher productivity

2. $T_C = 15$ min

Calculation of optimum cutting speed for maximum productivity.

$$V_{opt} = \frac{C}{\left[\left(\frac{1}{n}-1\right)T_C\right]^n}$$

$$\Rightarrow \quad V_{opt} = \frac{100}{\left[\left(\frac{1}{0.2}-1\right)15\right]^{0.2}}$$

$$= 44.093 \text{ m/min}$$

45 m/min has to be chosen

Example : 4.18

In a cutting operation with a cutting speed 50m/min tool life of 45min was observed. When the cutting speed was increased to 100 m/min tool life decreases to 10 min. Estimate the cutting for maximum productivity if tool change is 2min?

Solution :

V	T
50 m/min	45 min
100 m/min	10 min

(V_{opt}) max productivity, $T_C = 2\text{min}$

$$VT^n = C$$

$$\Rightarrow \quad (50)(45)^n = C$$

$$\Rightarrow \quad (100)(10)^n = C$$

$$\Rightarrow \quad (50)(45)^n = (100)\ (10^n)$$

$$\Rightarrow \quad n\, ln\, 45 = ln\, 2 + n\, ln\, 10$$

$$\Rightarrow \quad n\,[ln\, 45 - ln\, 10] = ln\, 2$$

$$\Rightarrow \quad n = 0.46$$

$$C = 50\,(45)^{0.46}$$

$$C = 100\,(10)^{0.46}$$

$$C = 288.036$$

$$V_{opt} = \frac{C}{\left[\left(\frac{1}{n}-1\right)T_C\right]^n}$$

$$\Rightarrow \quad = \frac{288.036}{\left[\left(\frac{1}{0.46}-1\right)2\right]^{0.46}}$$

$$\Rightarrow \quad V_{opt} = 194.51 \text{ m/min}$$

Example : 4.19

A cylindrical bar is to be turned. The max. allowable feed is 0.2 mm per revolution and at this feed rate Taylor's tool life equation is $VT^{0.25} = 75$ where V is cutting speed m/min and T is tool life in minutes. The labour cost and overheads is Rs 8/- per minute and regrinding cost is Rs 120/-. On an average it takes about 2 minutes to change the tool. Estimate cutting speed for minimum cost.

Solution

Total cost = Machining cost + tool changing cost + tool regrinding cost

C = Machining time × labour cost + changing time × labour cost + regrinding time × time

$$= \frac{\pi D}{1000\ v} \times \frac{L}{f}\ C_1 + t_{ct} \times \frac{\pi D}{1000\ v} \times \frac{L}{f} \times \frac{1}{T_{tool\ life}} \times C_1 + C_2\ \frac{\pi D}{1000\ f} \cdot \frac{L}{V} \times \frac{1}{T_{tool\ life}}$$

$$C = \frac{\pi D.L}{1000\ f}\left[\frac{C_1}{V} + \frac{t_{ct}}{V}C_1\left(\frac{V}{75}\right)^{\frac{1}{0.25}} + \frac{C_2}{V}\left(\frac{V}{75}\right)^{\frac{1}{0.25}}\right]$$

$$= \frac{\pi DL}{1000\ f}\left[\frac{C_1}{V} + \frac{C_1\ t_{ct}\ V^3}{75^4} + \frac{C_2\ V^3}{75^4}\right] \quad \text{... (1)}$$

where C_1 is labour cost = Rs. 8.0 per min, C_2 is tool regrinding cost = Rs. 120

For minimum cost

$$\left(\frac{dC}{dV}\right) = 0 = \left[-\frac{C_1}{V^2} + \frac{3\ C_1\ t_{ct}\ V^2}{75^4} + \frac{3C_2V^2}{75^4}\right]$$

$$\frac{C_1 \times 75^4}{(3C_1\ t_{cf} + 3C_2)} = V^4$$

$$V = 75\left(\frac{C_1}{3C_1\ t_{ct} + 3C_2}\right)^{1/4}$$

$$= 75\left(\frac{8}{3 \times 8 \times 2 + 3 \times 120}\right)^{0.25}$$

$$= 75 \times 0.3742$$

$$V = 28.065 \text{ m/s}$$

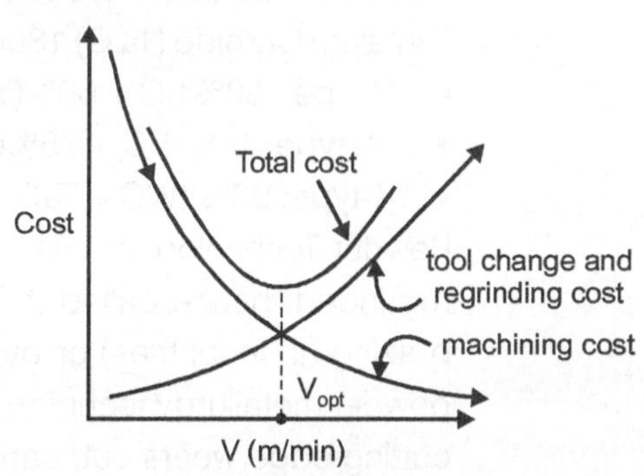

Fig. 4.30 : Graphical Representation of Various Cost

4.22 TOOL MATERIALS

Properties of cutting tool materials

1. **High hot hardness or Red hardness :** It is the property of the material by which it retains hardness at elevated temperatures.
2. **Toughness :** Toughness gives the material, ability to absorb shock.
3. **High thermal conductivity :** To dissipate heat.
4. Low coefficient of friction, at the chip- tool interface, the surface finish is good and wear is minimum.
5. Low coefficient of expansion.

Following are important tool materials:

1. **High carbon steel :** Hardness 750 HV (Vickers Hardness), at cutting speed of 5-6 m/min, and as cutting speed increases its hardness will come down to 150 HV. It is primarily used in wood working.
2. **High Speed Steel :** Normal cutting speed by this tool material is 28 – 30 m/min and its hardness is 850 HV.

   18% W + 4% Cr + 1% V (18 – 4 – 1 HSS) was the initial material developed in this category but at present HSS is broadly classified into T-series (W major alloying element) and M-series (Mo major impurity). Some of the important material in HSS are:

T-series Symbol	Percentage			
	W	Cr	V	CO
T - 1	18	4	1	0
T - 4	18	4	1	5
T - 7	14	4	2	0
T6 (Super HSS)	20	4	2	12

In M-series the most important material is M-2 and its compositions are 0.8% C, 4% Cr, 2% V, 6% W and 5% Mo. Purpose of W is to increase the hot hardness of the material. By addition of Mo decreases the requirement of W and Mo also increases hot hardness.

3. **Cemented Carbides :** Following are the hardness of allowing elements of these tools :
   Tungsten Carbide (WC) 2000 HV,
   Titanium Carbide (Ti C) 3100 HV,
   Tantalum Carbide (Ta C) 1800 HV.
   - P-type : 30%TiC + 60%(WC + TaC) + 10% Co: Hardest
   - M- type: 15% TiC + 75%(WC + TaC)+ 10% Co
   - K-type: 90% (WC + TaC) + 10% Co: Toughest

   Powder Technology is used for preparing Carbide chips. These carbide inserts cannot be regrinded, hence called throw away inserts. Carbide tools are fixed over the tool bit by brazing (in industries) or by fastening (in work shops). Only the carbide chip is made by powder metallurgy technique and it is fixed over the tool bit by brazing operation. Once one cutting edge wears out, carbide chip is removed and again fixed in such a way that other cutting edge comes into action. Once all the cutting edges wear out, carbide chips are discarded. That is why it is called Throw away inserts. Maximum Cutting Speed achieved by these materials is 150 m/min.
4. **UCON :**This is Niobium based alloy developed by union carbide company. UCON consists of 50%columbium 30 % titanium and 20 % tungsten. These are produced by rolling process. Nitrogen is finally diffused into the surface at a temperature of 650°C to produce a hard surface layer. These are used for machining steel at high speeds and feeds. It is not suitable for cast iron and stainless steel.
5. **Cast Hard Alloys (Stellite) :** These are produced by casting process. These alloys retain their hardness up to temperatures as high as 800-900 °C and have low coefficient of friction. This is especially used for making form tools. UCON and cast hard alloys are inferior to carbides and superior to high speed steel. Cobalt (Co) is used as binding agent. The major alloying elements are 53%Co, 33% Cr and 14% W.
6. **Ceramic :** Sintered or powdered metallurgically produced (Aluminium oxide) tools which are superior to carbide attached to tool bit by brazing. They have less tendency to adhere to metals during cutting. So, lower tendencies to form a built up edge. Ceramic tool tips are highly brittle so they are usually attached to the shank by means of epoxy resins. Cutting Speed is 400m/min can be achieved by these tools and its hardness is 2200 HV.
7. **Sialon (Si-Al-O-N) :** It is a recently developed tool material which produces excellent surface finish over the material, but it is costly. This is produced by milling silicon nitride, aluminium nitride, alumina, yitrium oxide. It has higher thermal shock resistance than silicon nitride and is tougher than aluminium. It is recommended for machining cast irons and nickel based

super alloys at intermediate cutting speeds, because of chemical affinity to iron this is not suitable for machining steels. Cutting speed is 300 m/min.

8. **CBN (Cubic Boron Nitride) :** Hardness value is 4700 HV and cutting speed is 600-700 m/min can be achieved by CBN. Boron has Hexagonal structure but when it is heated under contain pressure it converts into cubic, which is a very hard structure. Generally, it is used as an abrasive in grinding wheel. CBN coated tools are used primarily to machine stainless steel and high speed steel. The tool life of CBN is excellent for machining quenched alloy steel for dies.
9. **Diamond :** Its hardness value is 7500 HV. Cutting speed is 1000m/min. Diamond dust is used as an abrasive in grinding wheel. It is harder than any other material and has high thermal conductivity.
10. **Coated Tools :** Coating of HSS cutting tools with thin layers of about 2- 15 μ of a refractory metal carbide or nitride like TiC, TiN, TiCN, TiAIN, and . Coating is done by two methods.
    - Physical Vapor Deposition
    - Chemical Vapor Deposition

4.23 MEASUREMENT OF CUTTING FORCES

Dynamometers : In metal cutting operation the device used for determination of cutting forces is known as tool Dynamometer.

Types of dynamometer:

1. Mechanical Dynamometers.
2. Strain gauge type Dynamometers.
3. Pneumatic and Hydraulic Dynamometers.
4. Electrical Dynamometers.
5. Piezoelectric Dynamometers.

Mechanical Dynamometers are simple type dynamometer, which uses the sensitive dial indicators for directly measuring the tool forces. The dial indicators are calibrated to show readings directly.

Strain Gauge Dynamometer : Strain gauges are mounted on a plate and when the load is applied on the plate, the length of those strain gauges will change. This will in turn change the resistance. Strain gauges are connected to the wheat stone bridge and the change in resistance is recorded by the flow of current. When the strain gauges are bounded with another plate, although such strain gauges will not be affected by the change in environment but the maintenance of such gauges becomes difficult.

4.24 GRINDING

Grinding process is used as a finishing process to get the desired surface finish, correct size and accurate shape of the product.

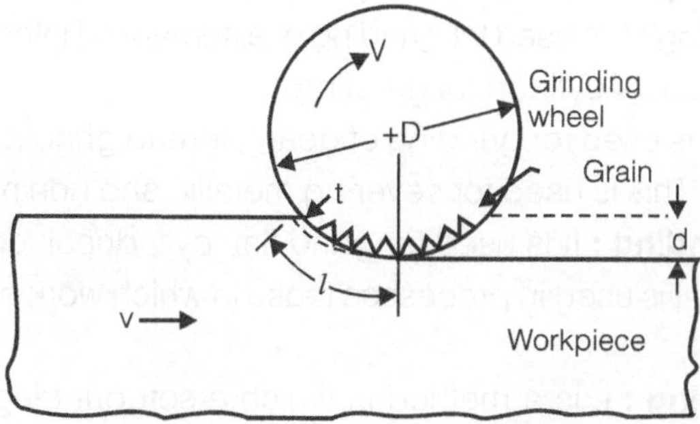

Fig. 4.31 : Grinding Operation

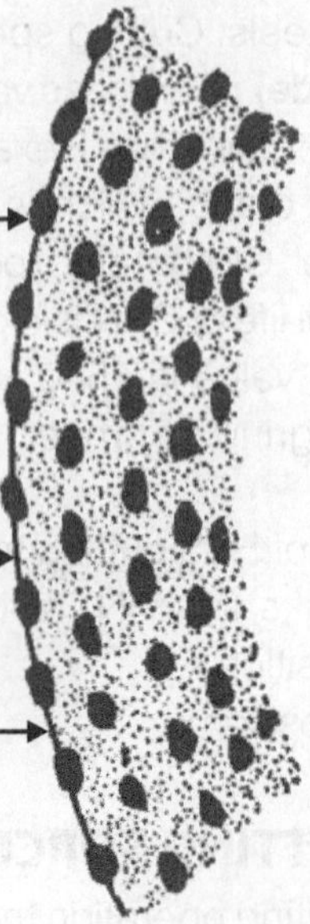

Fig. 4.32 : Cutting Action of Grinding Wheel

Grinding wheel consists of abrasive particles, bonding material and voids. The projecting abrasive particles act like cutting tool tips and remove metal. A properly selected grinding wheel exhibits self sharpening action. As cutting proceeds the abrasive particles, at cutting edge become dulled, and eventually these becomes plane and which resists the cutting action. Thus new cutting edge points are produces which carry out the further cutting action. This process continues till the abrasive grains get worn down till the level of bond. At this position new cutting edges will do further cutting action.

The rate at which grinding wheel enters into the work piece is called In Feed. The in-feed process is used for stepped cylindrical work pieces which cannot be through-feed. And the feed with which, grinding wheel moves along the surface of work is called Through Feed. In High Speed Grinding, infeeds are low and the grinding wheel is not supposed to remove more material. It is meant for super finishing operation. In Creep Feed Grinding, infeeds are high and speeds are low. It is meant for bulk metal removal.

4.24.1 Classification of Grinding Processes

1. **Surface grinding :** Used for grinding flat surfaces.
2. **Cylindrical grinding :** This is used for grinding external and internal cylindrical surfaces.
3. **Center-less grinding :** It is used for grinding of external and internal surfaces. In this, machines are different from that of cylindrical grinding.
4. **Form grinding :** It is used for grinding of gears, thread grinding and grinding of splines etc.
5. **Abrasive cut off :** This is used for severing metallic and non metallic materials.
6. **Abrasive belt grinding :** It is used for grind flat, cylindrical, or curved shapes.
7. **Manual grinding :** it is used in processes those in which work piece or grinding wheel is held by hand.
8. **Creep feed grinding :** it is a method in which a soft grinding wheel is used. In this large amount of metal is removed in a single pass.

Some grinding processes are shown below:

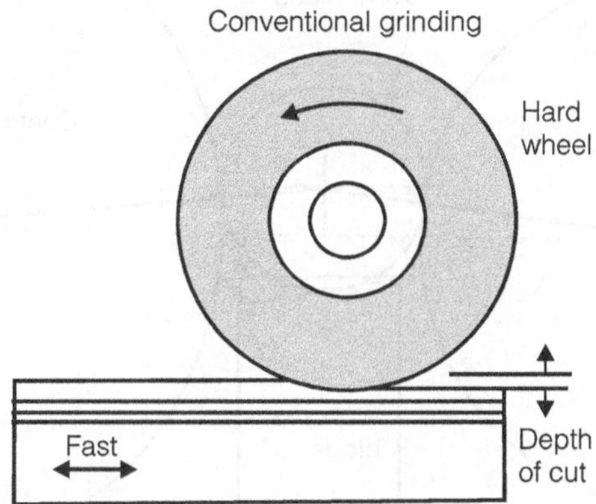

Fig. 4.33 (a) : Surface Grinding

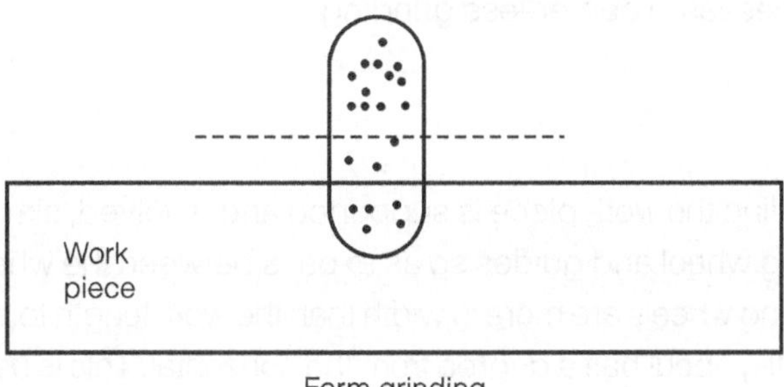

Fig. 4.33 (b) : Form Grinding

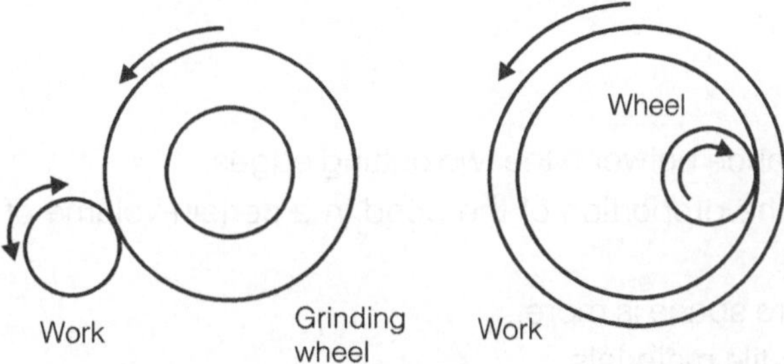

Fig. 4.33 (c) : Internal grinding and external grinding.

Center-less Grinding : Work piece which are smaller in sizes, very fragile or non symmetric can be grinded by this process. Work pieces are kept slightly above the common axis of grinding and the regulating wheel. Regulating wheel is given an inclination so that one component of force provides infeed and the other component provide through feed. As a result of that, work piece will automatically move from the surface of grinding wheel and come out from the other side. Speed of the regulating wheel = 1/20 of the grinding wheel.

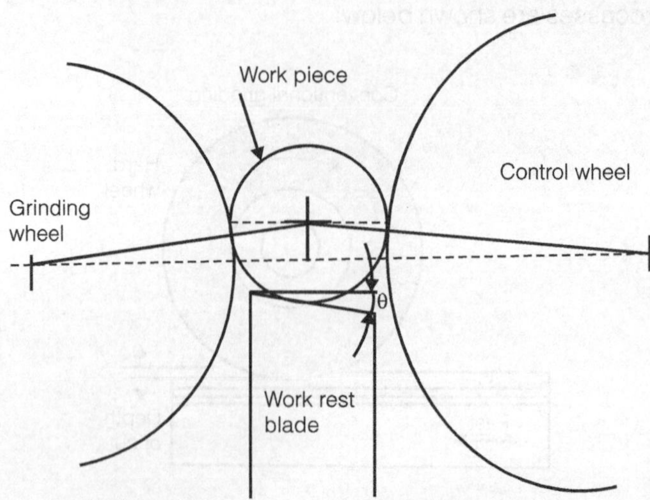

Fig. 4.34 : Center-less through feed grinding.

There are three processes in center-less grinding.

1. Through feed
2. Infeed
3. End feed.

In through feed grinding the work piece is supported and revolved, simultaneously axial movement is given by regulating wheel and guides so as to pass between the wheels. In Infeed grinding both regulating and grinding wheels are more in width than the work length to be ground. Axis of regulating wheel is inclined a little, about half a degree from the horizontal. This is used for grinding, shouldered or formed components. In this the work rest does not carry guides.

In End feed grinding the both wheels, grinding wheel and the regulating wheel are dressed to contain the required shape or form. The work piece is fed longitudinally from the sides of the wheels. As it advances between the revolving wheels its surface is ground till its farther end touches the end stop.

4.24.2 Structure

Structure is the distance between the two cutting edges.

Open structure : The proportion of the bond in a certain volume of the wheel affects the structure.

- In open structure space is more.
- It is used in ductile materials.
- Continuous chips of ductile materials can be accommodated and thrown out of centrifugal action.

Close structure :

- It occupies less space.
- It is used for brittle and hard materials.

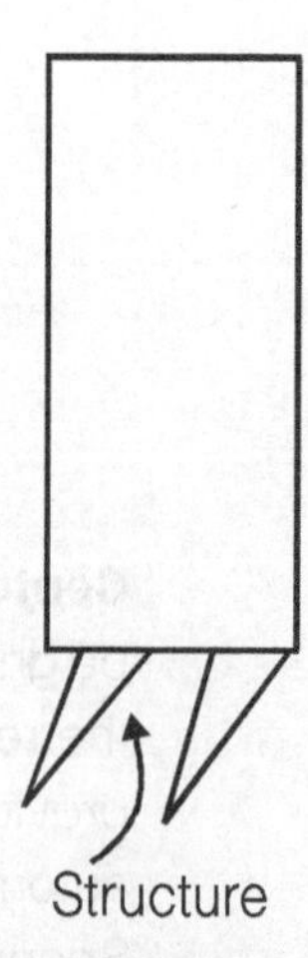

Fig. 4.35

And open structures are used for ductile materials so that long chips goes into the space between the abrasives. When the surface of the wheel which is not in contact with the work pieces, these chips will go away from the space by centrifugal action. When ductile materials are machine with close structure, chips will be forced to enter into the space which is not sufficient. This produces welding between the abrasives. When such process continues, all the voids will be blocked, and wheel will rub over the work piece without any cutting this phenomenon is called loading.

4.24.3 Grade

Grade or hardness indicates the strength, with which the bonding materials holds the abrasive grains in the grinding wheel.

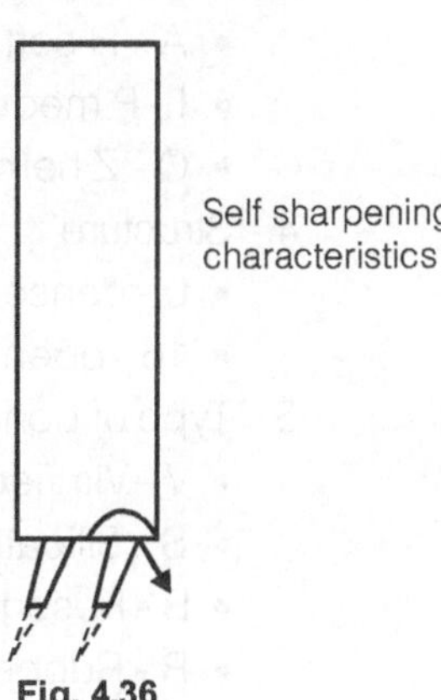

Fig. 4.36

Sharp grains after coming in contact with the work piece become blunt which increases the drag between the abrasive grain and the work piece. If the bonding agent is weak, abrasive grain will be pulled from the wheel and the fresh grain from the back ground will start cutting action. This phenomenon is called self sharpening characteristic. The wheels which have predominating self sharpening characteristics are called Soft Wheels and if the bonding agent is strong such wheels are Hard Wheels. For machining soft materials hard wheels are used and vice versa. If hard materials are machined by hard wheels, gradually all the grains will wear out on the surface of wheel and wheel starts rubbing over the work piece with out cutting. This phenomenon is called glazing.

Depth of cut: It is the thickness of metal removed by the grinding wheel in one traverse stroke.

$$t = \text{Depth of cut}$$

$$\text{depth of cut} = \frac{D_1 - D_2}{2}$$

D_1 = Diameter of work piece before grinding

D_2 = Diameter of work piece after grinding

Feed :

In cylindrical grinding feed is the longitudinal movement of the work piece per revolution.

f = feed in mm per work piece revolution

$= K_1 B$

B = Face width of wheel in mm

= constant

= 0.4 to 0.6 for finish grinding

= 0.6 to 0.9 for rough grinding.

$$\text{Work travel} = \frac{f \times n}{1000} \text{ m/min}$$

4.24.4 ISO Designation of Grinding Wheel

① ② ③ ④ ⑤

51 - C - 30 - M - 5 - V - 20

First and last number : Manufacturer has to give special information about the grinding wheel. According to this system the various elements and characteristics of grinding wheel are represented on all the wheels in a definite sequence as follows.

1. Type of abrasive
   - C - carbide
   - A - Aluminium oxide (Al_2O_3)
   - D - Diamond
2. Grain size
   - 10 - 24 roughening
   - 30 - 60 Medium

- 70 - 180 Finishing
- 220 - 600 Superfinishing operation

Per square inch opening in the mesh. These are number of opening per square inch area.

3. Hardness
   - A - H soft wheels
   - I - P medium wheels
   - Q - Z hard wheels
4. Structure
   - 0 - dense
   - 16 - open
5. Type of bond
   - V - Vitrified bond (glass material which produces strong wheel)
   - S - Silicate
   - B - Resonoid (Thermosetting materials which are flexible than vitrified bond)
   - R - Rubber (Most flexible bond material)
   - M - Metal bond

4.24.5 Wheel Dressing

Wheel wear ratio = Work piece material/Wheel material lost

As soon as new grinding wheel comes in contact with the work, sharp grains become blunt, that is why wheel wear will be more in the beginning. After certain period of time, due to glazing or loading, there is a surface contact between the wheel and work piece, large piece of grinding wheel will start coming out causing break down. Before this situation comes wheel has to be with drawn from the work shop and dressed. Dressing is a procedure to remove a layer of abrasive from the surface of grinding wheel. This time between the two dressings is called wheel life. Due to dressing, wheel may loose its cylindricity and the process of making it cylinder again is called Wheel Truing.

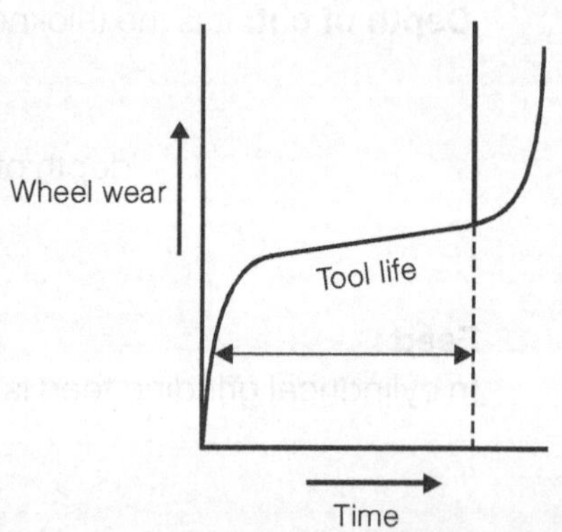

Fig. 4.37 : Variation of wheel wear with respect to time

4.25 UNCONVENTIONAL/NON-TRADITIONAL MACHINING PROCESSES

4.25.1 Electro Chemical Machining

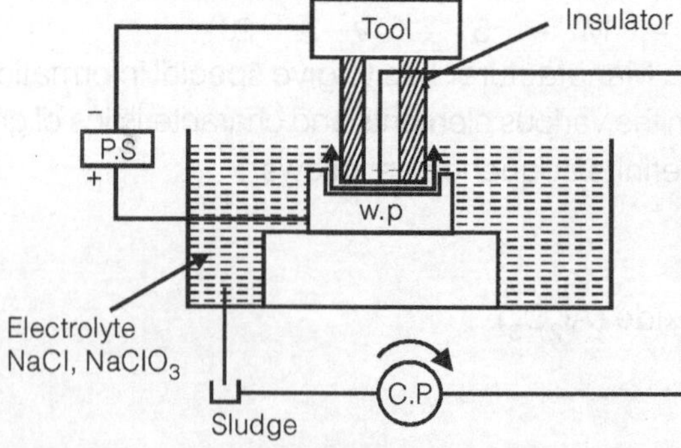

Fig. 4.38 (a) : Machining using Electro chemical machining

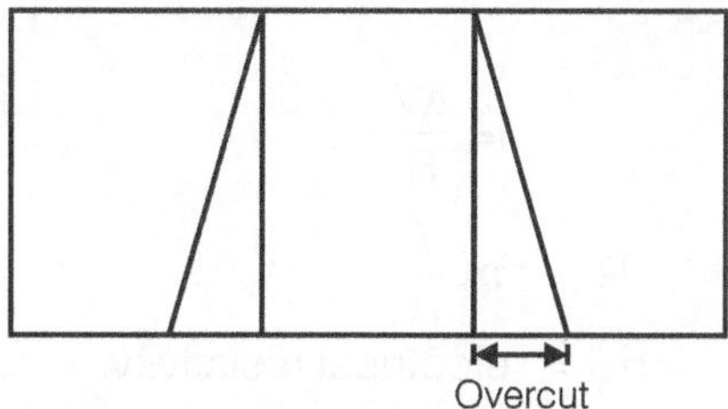

Fig. 4.38 (b) : Showing Over Cut

Tool Material used : Brass, Bronze, Copper, Aluminium.
Gap between electrode and work piece : 0.03 – 0.5 mm
Centre line average value (R_a) 0.2 – 0.8 μ's

$$Fe \rightarrow Fe^{2+} + 2e^-$$
$$NaCl \rightarrow Na^+ + Cl^-$$
$$2e^- + 2H_2O \rightarrow H_2 \uparrow + 2OH^-$$
$$Fe^{2+} + 2OH^- \rightarrow Fe(OH)_2 \downarrow$$

In ECM process, material removal takes place atom by atom, that is, reverse of electrolysis. This is high current, low voltage process. Material removal is proportional to the current settings and the temperature of electrolyte. The electrolyte temperature has to be controlled very precisely because even increase in temperature 5°C produces over cut of 0.2 mm. Electrolyte which is conducting in nature is supplied within the machining area through a hole in the tool. Insulation is provided on the side of the tool so that reaction doesn't take place side ways making the hole tapered. The process produces excellent surface finish and there is no wear of tool and it has main applications in machining steam turbine blades and die sinking. Since, hydrogen gas is produced during the reaction, so there has to be a safe passage for its removal because it can explode under its own weight. Although ECM installations are very expensive but its operational cost is very low and economics is justified by the quality product it produces.
Mass of material removal ∝ charge

$$\text{Mass} \propto q \propto It$$
$$\text{Mass} = ZIt$$
$$Z = \frac{e}{F}$$

Z = Electrochemical Equivalent
F = Faraday's constant = 96500
e = Chemical equivalent (Atomic weight/valency)
I = Current passed(Amp)
t = time(seconds)

Metal Removal Rate (MRR)= $\frac{eI}{F\rho}$ =sI (cm^3/sec)

ρ = density gm/cm^3

s= Specific material removal rate(MRR) = $e/F\rho$

Electrode feed rate = $s \times s_1$

s_1 = current density = I/A

$$I = \frac{\Delta V}{R}$$

$$R = \rho_s \frac{l}{A}$$

ρ_s = electrical resistivity

$$S_1 = \frac{\Delta V}{\rho_s l}$$

l = Gap between tool and workpiece.

Example : 4.20

Calculate the MRR and electrode feed rate when copper is electrochemically machined Under following conditions.

ρ_s = **5Ω cm**

ΔV = 18V

I = 500A

Tool gap = 0.5 mm

Atomic Weight = 56

Valency = 2

ρ = 7.8 g/cm³

Solution :

$$MRR = s\,I$$

$$s = \frac{e}{F\rho}$$

e = Atomic weight/valency

= 56/2 = 28

F = 96500

$$s = \frac{28}{96500 \times 7.8} = 3.72 \times 10^{-5}$$

$$MRR = s\,I = 0.186\ cm^3/s$$

$$\text{Electrode feed rate} = s \times S_1$$

$$S_1 = \text{Current density} = \frac{I}{A} = \frac{\Delta V}{RA} = \frac{\Delta V}{\rho_S l}$$

$$= \frac{18}{5 \times 0.05} = 0.134\ cm/s$$

$$\text{Feed rate} = 3.72 \times 10^{-5} \times 72 cm/s = 2.67 \times 10^{-3} cm/s = 0.0267 mm/s$$

Example : 4.21

A researcher conducts electrochemical machining (ECM) on a binary alloy (density 6000 kg/m³) of iron (atomic weight 56, valency 2) and metal P (atomic weight 24, valency 4). Faraday's constant = 96500 coulomb/mole. Volumetric material removal rate of the alloy is 50 cm³/s at a current of 2000 A. What is the percentage of the metal P in the alloy.

Solution :

The equivalent, chemical equivalent is

$$\frac{1}{e} = \frac{4}{24} \times \left(\frac{x}{100}\right) + \left(\frac{100 - x}{100}\right) \times \frac{2}{56}$$

where the percentage of P in alloy is x.

The MRR in ECM process can be represented by

$$MRR = \frac{eI}{F\rho}$$

but putting the values,

$$50 = \frac{e \times 2000}{96500 \times 6000 \times 10^{-6}}$$

$$e = 14.475$$

putting this value in previous expression we get

$$x = 25.482\%$$

Example : 4.22

Electrochemical machining is performed to remove material from an iron surface of 20 mm × 20 mm under the following conditions:

Inter electrode gap = 0.2 mm

Supply voltage (DC) = 12 V

Specific resistance of electrolyte= 0.2 Ω mm

Atomic weight of Iron = 55. 85

Valency of Iron = 2

Faraday's constant = 96540 Coulombs

What is the material removal rate (in gm/s).

Solution :

$$R = \rho_s . \frac{l}{A}$$

$$V = IR$$

$$MRR = \frac{eI}{F\rho} \text{ m}^3\text{/s}$$

$$R = 2 \times 10^{-1} \times \frac{0.2}{20 \times 20} \Omega$$

$$R = 10^{-4} \Omega$$

$$12 = I \times 10^{-4}$$

$$I = 12 \times 10^{4} A$$

$$e = \frac{55.85}{2} = 27.925$$

$$MRR = \frac{27.925 \times 12 \times 10^4}{96500}$$

$$= 34.725 \text{ gm/s}$$

Example : 4.23

Evaluate MRR of a Nimonic alloy {Co - Ni - cr}

	%	**Atomic Weight**	**Valency**
Co	**18**	**58.93**	**2**
Ni	**62**	**58.71**	**2**
cr	**20**	**51.99**	**6**

I = 500A ρ = 8.28 gm/cm³

Solution :

$$\frac{1}{e} = \frac{\sum \% \times \text{valency}}{\text{atomic weight}}$$

$$= \frac{0.18\times 2}{58.93} + \frac{0.62\times 2}{58.71} + \frac{0.2\times 6}{51.99}$$

$$\Rightarrow \quad \frac{1}{e} = 0.050311$$

$$\text{MRR} = \text{SI}$$

$$\Rightarrow \quad \frac{eI}{F\rho} = \frac{19.876\times 500}{96500\times 8.28}$$

$$\text{MRR} = 0.0124 \text{ cm}^3/\text{sec}$$

4.25.2 Electrode Discharge Machining

Gap between tool and work piece : 0.025 – 0.05

Centre line average value (R_a) : 0.25 μs

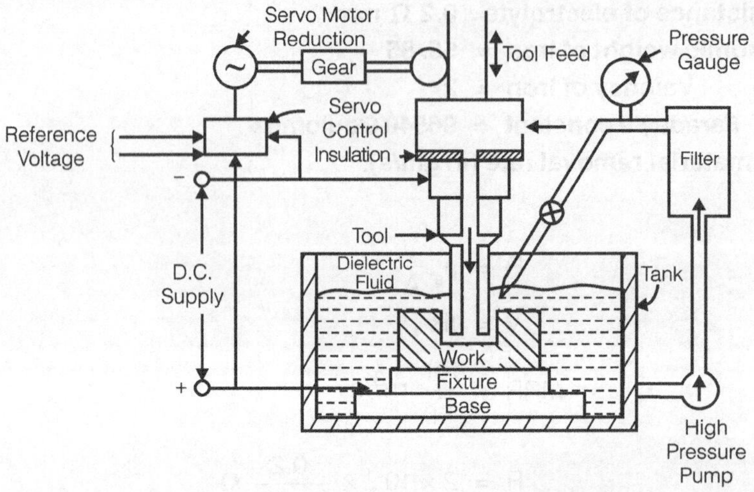

Fig. 4.39 (a) : Electrode discharge machining

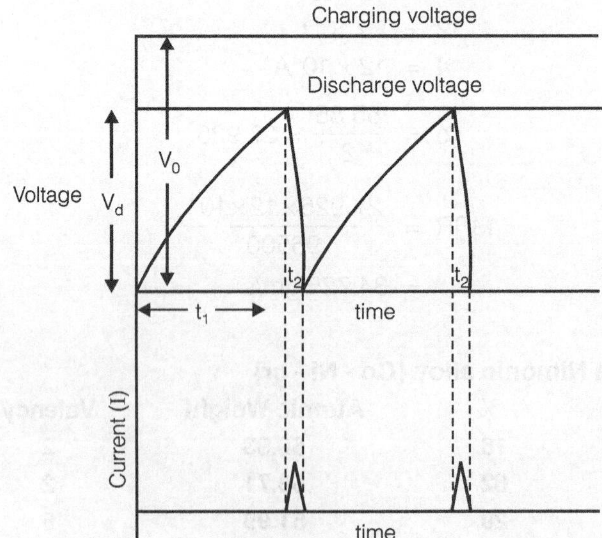

Fig. 4.39 (b) : Variation of current and voltage

EDM process is also called Spark Erosion machining where the material is removed by the erosion of repeated sparks. It is high voltage and low current process. During a major portion of cycle, a capacitor bank charges and once voltage reaches a threshold value, all the capacitors will discharge simultaneously, producing a spark at the tool tip. Dielectric becomes conducting only in that region. When spark hits the work piece, a crater will be produced over it. During the transportation of spark, a portion of kerosene also burns which provides solid lubrication in the machining process. During the machining process, a portion of tool material also melts out. So, materials like graphite are used to make the tools because they not only have higher melting point but also can be machined to a great degree of accuracy. To increase the conductivity copper is added in the graphite. When the thermal conductivity of work piece material increases, MRR decreases.

The removed material starts accumulating in the tool work piece gap, so a servomechanism connected with the tool withdraws it and the flushing mechanism removes the material servo-mechanism again places the tool back to its position. The different flushing systems are servomechanism, solenoid value and steppes motor. Servomechanism maintains a mean spark gap voltage at a constant value.

The major application of the process is in Die Sinking and since the process produces cratoring type surface. These surfaces are used in bearing applications.

If V_d is the discharge voltage, V_0 is charging voltage then

$$V_d = V_0\left(1-e^{-\frac{t}{RC}}\right) \qquad \text{...(1)}$$

The frequency of spark $f = \frac{1}{t}$, so

$$f = \frac{1}{RC\log\left(\frac{V_0}{V_0 - V_d}\right)} \qquad \text{...(2)}$$

It can be seen in Fig. 4.40 (b) that t_1 is the charging time and t_2 is discharge time. So energy released per spark

$$E = \frac{1}{2}CV_d^2$$

Average value of power delivered

$$P_{av} = \frac{E}{t_1 + t_2}$$

t_2 is very less as compared to t_1 so

$$P_{av} = \frac{1}{2}\frac{CR}{Rt_1}V_0^2\left\{1-e^{-\frac{t_1}{RC}}\right\}^2$$

take $\frac{t_1}{RC} = N$

$$P_{av} = \frac{1}{2}\frac{V_0^2}{RN}\left\{1-e^{-N}\right\}^2$$

For maximum power delivery

$$\frac{dP_{av}}{dN} = 0$$

So after solving we get $N = 1.26$

So optimum value of discharge voltage

$$\frac{V_d}{V_0} = 1 - e^{-1.26}$$

$$= 0.72$$

V_d is 72% of charging voltage.

4.25.3 Ultrasonic Machining

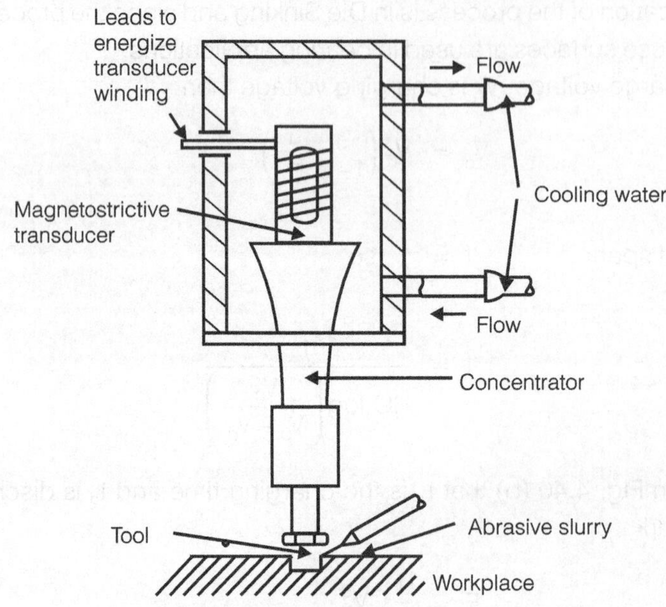

Fig. 4.40 (a) : Machining using ultra sonic machining

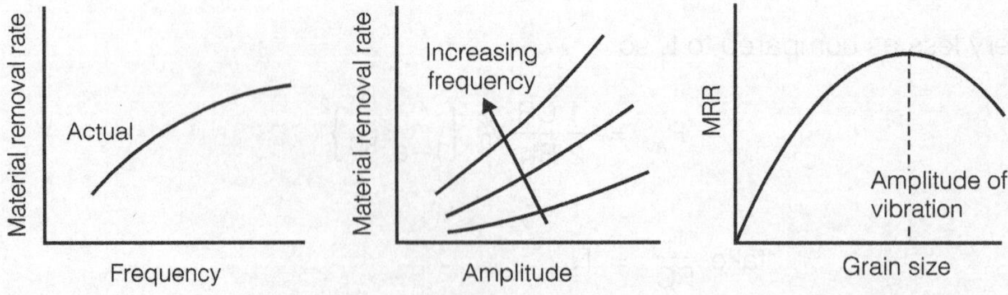

Fig. 4.40 (b) : Variation of MRR with respect to frequency and amplitude

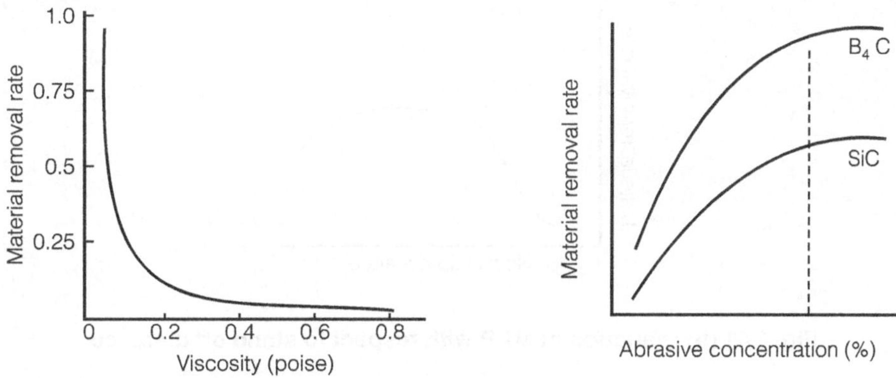

Fig. 4.40 (c) : MRR with respect to Viscosity **Fig. 4.40 (d) : MRR with respect to % of abrasive**

Ra : 0.25-0.75 μs
Amplitude : 0.02-0.1 mm
Frequency : 15-30 kHz
Gap between tool and work piece : 0.25 mm

In Ultrasonic machining, tool is made up of ductile material. Between tool and workpiece, there is slurry of abrasives. Non–conducting brittle materials can be machined by this process like glasses and ceramics. Abrasive gets embedded into the tool and during the downward journey of the tool abrasives hammer the work piece, removing a material. This material will be flushed away from the machining area by the flow of slurry. Tool is made slightly tapered to produce straight holes. Upon increasing the viscosity of the carrier fluid, material removal rate decreases due to difficulty in flushing. By increasing the frequency, MRR will increase because number of impacts per unit time will increase. By increasing the amplitude, MRR will increase due to the increase of momentum of abrasives. By increasing the concentration of abrasives, impact will be there at more number of places which increases MRR. But when the concentration increases beyond a certain value, due to Collision between the abrasives momentum is lost, decreasing the MRR. By increasing the size of abrasive, an impact will appear on the larger area. But when the size increases beyond a certain value, momentum of abrasives will decrease. It should be noted that :

MRR : ECM > EDM > USM
Tool wear : USM > EDM > ECM

4.25.4 Abrasive Jet Machining

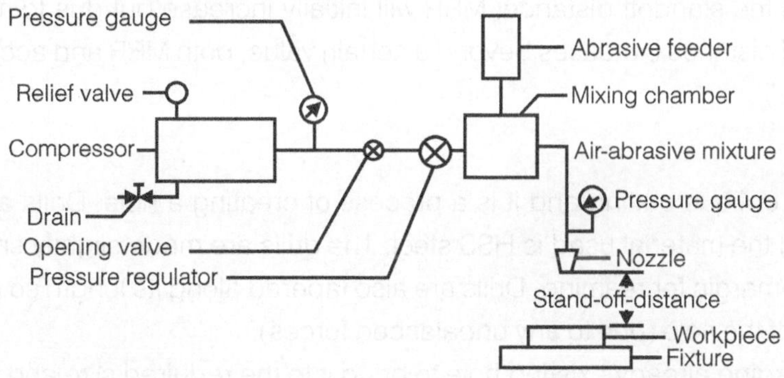

Fig. 4.41 (a) : Machining using abrasive machining

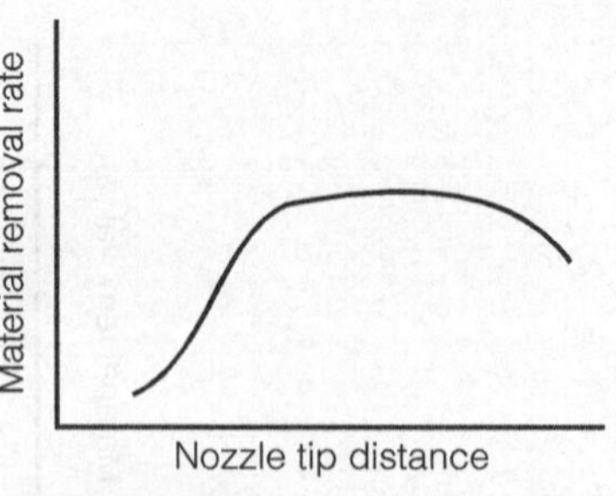

Fig. 4.41 (b) : Variation of MRR with respect to stand off distance

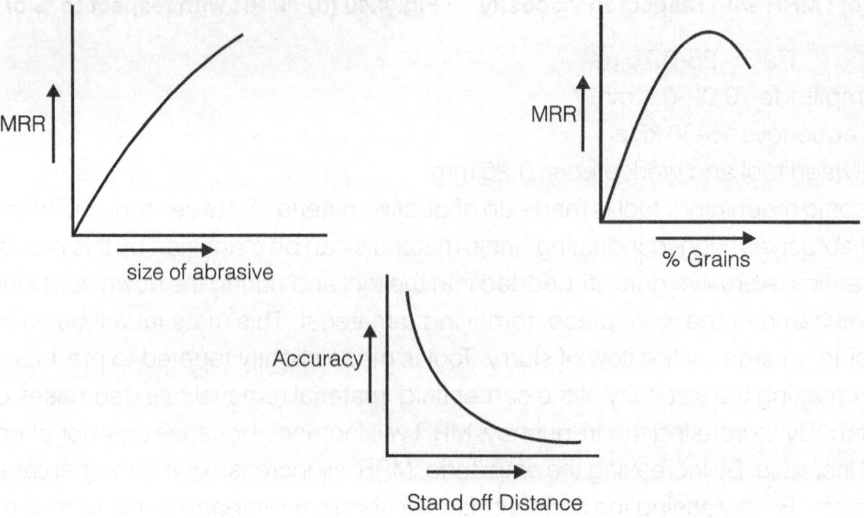

Fig. 4.41 (c) : Variation of MRR with respect to grain size, % of grains and stand off distance

For machining very fine holes of about 0.002 mm size, abrasive jet machining is used. Abrasives and air are mixed and thrown over the surface at a very high velocity. Due to the impact by the abrasive, a crack will be produced over the surface which will be blown by the air. The main applications of the process is, in machining inaccessible areas like inside surface of the bottle, cleaning metallic moulds, producing very fine holes in the tool for the purpose of lubrication, etc. By increasing the standoff distance, MRR will initially increase but due to mushrooming effect, when standoff distance increases beyond a certain value, both MRR and accuracy will decrease.

4.26 DRILLING

Drilling is an oblique cutting and it is a process of creating a hole. Drills are made by forging operation and the material used is HSS steel. The drills are made slightly smaller in size so that there is some margin for reaming. Drills are also tapered along its length so that it doesn't hit the finished part of the hole (due to any unbalanced forces).

Boring : Enlarging already existing hole to bring it to the required size and have a better finish. The accuracy expected is 0.125 mm.

Reaming : It is a process of exacting the hole and finishing it and has a fine surface finish of ±.005 mm.

Broaching : Process for super finishing.

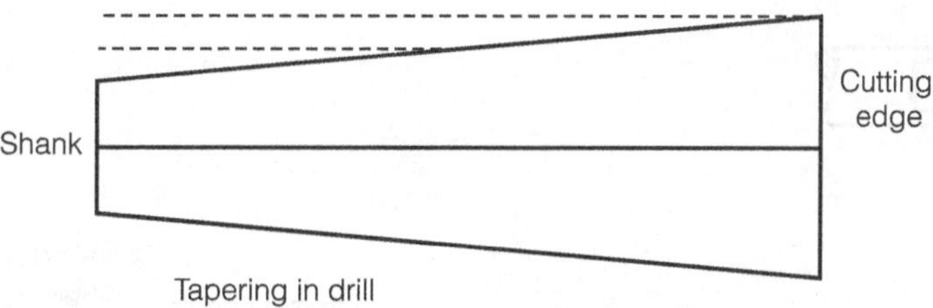

Fig. 4.42 : Showing tapering on a drill

Pack Drilling : Keeping a number of pieces together and drilling simultaneously.

Core Drilling : In casting operation holes are being created by using cores. This process is known as core drilling.

Trepanning : Drill is in the form of a tube at the periphery of which there are cutting edges. Complete material is not removed and initially this process was used in Gun Barrel manufacturing process. [Fig. 4.43 (a)].

Counter boring : Counter boring is making the hole little larger. Counter boring is done by end milling. It is seating place for bolt heads. The enlarged hole forms a square shoulder with the original hole. This is necessary in some cases to accommodate the heads of bolts, studs and pins. The cutting edges have straight or spiral teeth.

Counter sinking : It is the process of making the holes slightly tapered in the beginning. In counter sinking tools are used as shown in Fig. 4.43(c). It is a seating place for screws.

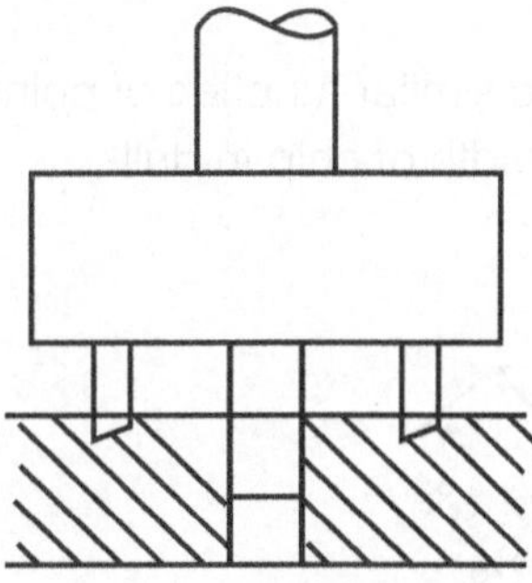

Fig. 4.43 (a) : Trepanning

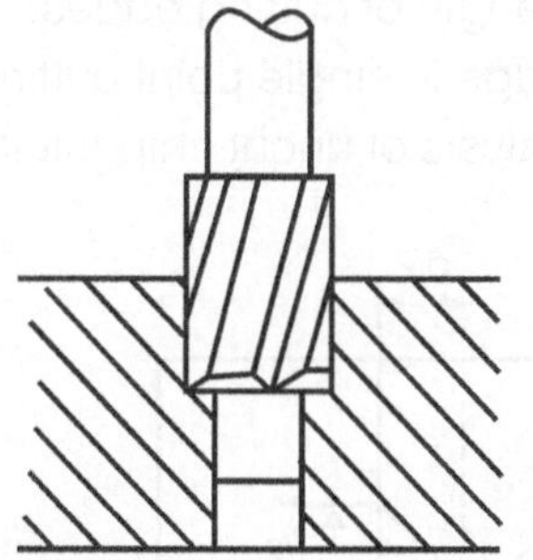

Fig. 4.43 (b) : Counter boring

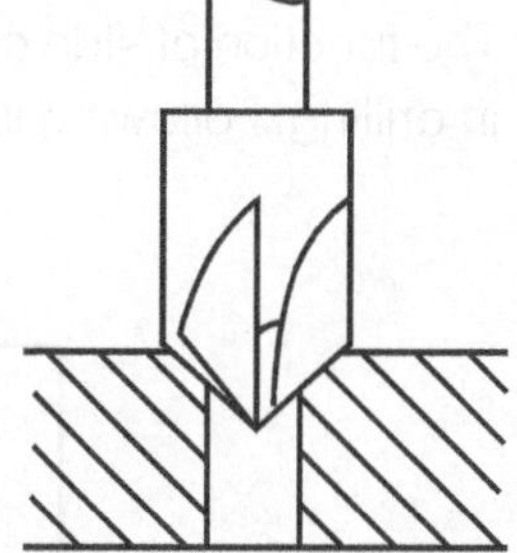

Fig. 4.43 (c) : Counter sinking

4.26.1 Drill Geometry

Following are major actions taking place at the point of a drill.

1. A small hole is pierced by the rotating web.
2. Chips are formed by rotating cutting edges.
3. Chips are conveyed out of the hole by helix.
4. The drill guided in the hole already produced.

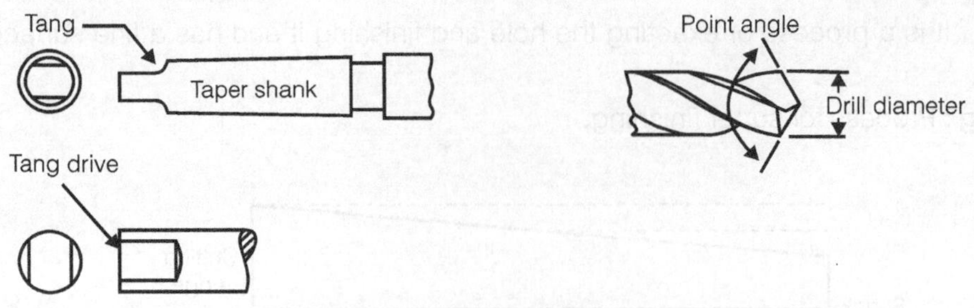

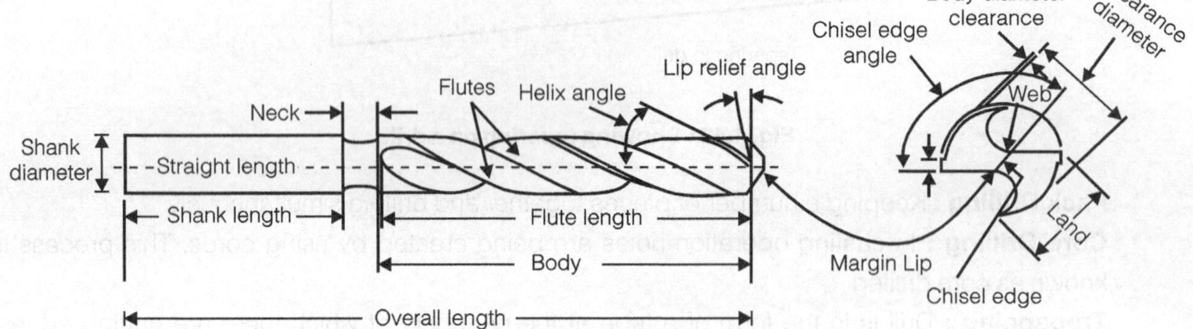

Lip angle + Lip relief angle + Helix angle = 90°

Fig. 4.44 : Drill geometry

Note on drill geometry :

1. A small web to reduce thrust on a drill but a large web for greater resistance to chipping and greater torsional rigidity.
2. Large flutes to provide a larger space for chip transfer but small flutes in the interest of torsional rigidity.
3. An increase in the helix angle to more quickly remove chips but a decrease in helix angle in the interest of greater strength of cutting edges.

The function of side cutting edge in single point cutting tool is the similar function of point angle in drilling. Following is the analysis of uncut chip thickness and width of chip in drill.

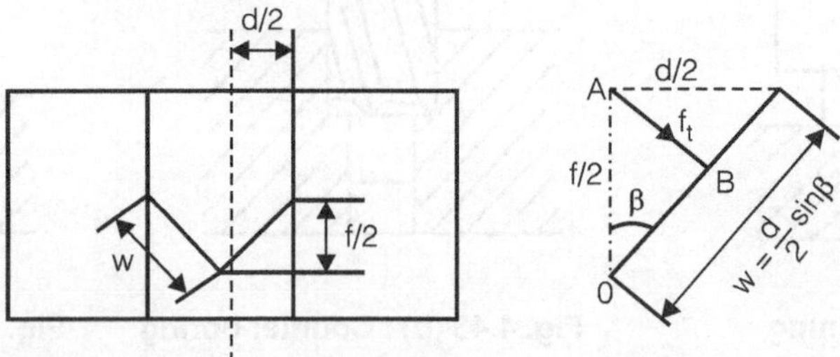

Fig. 4.45

$$W = \frac{d/2}{\sin\beta}$$

$$\Delta OAB : \frac{AB}{OA} = \sin\beta$$

$$AB = \frac{f}{2}\sin\beta = t_1$$

$$2\beta = \text{point angle,}$$

$$\text{feed} = \text{mm/rev}$$

How much angle the helix makes with the horizontal is called helix angle.

4.26.2 Important angles of a drill

Rake angle : It is the angle formed between a plane containing the drill axis and the leading edge of the land. It can have the positive, negative or zero value. For a right hand flute its value is positive, for a left hand flute its value is negative and for parallel flutes it is zero. Higher values are suitable for softer materials and lower values for harder materials. Larger the value of this angle lesser will be the power required in drilling. A line is drawn parallel to the drill axis passing through the cutting edge. The angle this line makes with a rake face is called back rake angle. At the centre, cutting edge merges into the chisel edge which is a flat area, so the back rake angle is zero. Towards the surface back rake angle increases and at the outer surface, since the cutting edge merges into the helix, back rake angle is equal to the helix angle.

Point angle : It is known as cutting angle. It is the angle included between the two opposite lips of a drill measured in a plane containing the axis of the drill and both lips. As it can be seen in the analysis (Fig. 4.44) that smaller is the point angle, chips will be thinner and wider. In case of ductile materials, if chips are thick, it will be jammed in the helix. So chips should be thinner. **Hence smaller point angles are used to drill ductile materials and larger point angles for drilling brittle materials.**

Feed angle : Material on the back ground of cutting edge will form an angle which will try to hit the cutting edge and break it. This angle is called feed angle.

$$\tan\alpha = \frac{f}{2\pi r}$$

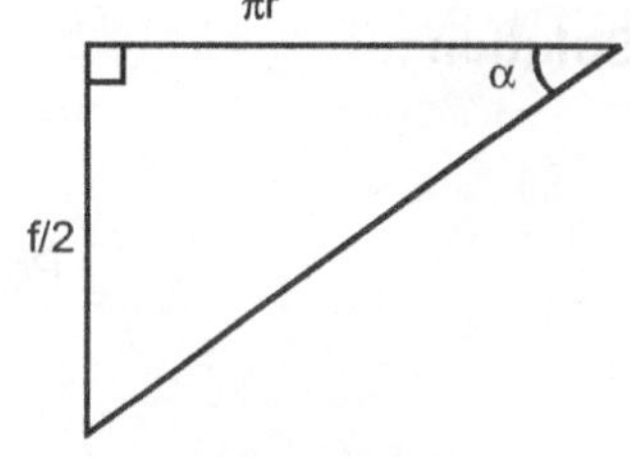

Fig. 4.46

Clearance angle : The angle formed between the flank and a plane normal to the drill axis, measured at the periphery of the drill. This angle is formed as a result of grinding the relief adjacent to the cutting edges to enable easy entry of the drill. Clearance angle is measure on the back side of the cutting edge, the angle which the background of cutting edge makes with the horizontal plane is called clearance angle. In single point tool clearance is provided to take care of elastic recovery. Since soft material exhibit more elastic recovery so larger clearance is provided. Ductile materials should be drilled at lower feeds and with larger clearance angles because such materials exhibit larger elastic recovery and clearance in drill has to compensate the feed angle and elastic recovery.

	Ductile	Brittle
Clearance angle	8-12°	6-9°
Point angle	118°	135°

Chisel edge angle : When the drill is viewed from its end, there appears to be an obtuse angle formed between the lip and the chisel edge, this is known as Chisel edge angle. The greater the angle the larger the clearance.

4.26.3 Machining time

$$T = \frac{L}{N \times f}$$

N = rpm of the drill

L = Length of axial travel of drill in mm

F = feed per rev in mm

T = machining time in min

L = l + a

l = depth or thickness of work piece

a = approach of drill = 0.3d

d = diameter of drill

Cutting speed : $S = \frac{\pi dN}{1000}$ rpm

$$= \frac{\text{Cutting speed in rpm}}{\pi \times d}$$

Example : 4.24

Through holes of 10 mm diameter are to drilled in a steel plate of 20 mm thickness drill spindle speed is 300 rpm feed is 0.2 mm/rev and drill point angle is 120. Assuming drill over travel of 2 mm determine the time to produce a hole.

Solution :

N = 300 rpm

f = 0.2 mm/rev

Point angle 2β = 120°

over travel = 2 mm

Total travel = x + 20 + 2

$$\frac{d/2}{x} = \tan 60°$$

$$x = \frac{10}{2\ \tan 60°}$$

x = 2.88 mm

Total travel = 24.88 mm

Number of revolutions required to make the hole

$$= \frac{24.88}{0.2}$$

= 124.4rev

$$\text{Time} = \frac{124.4}{300}$$

= 0.414 min

= 24.88 s

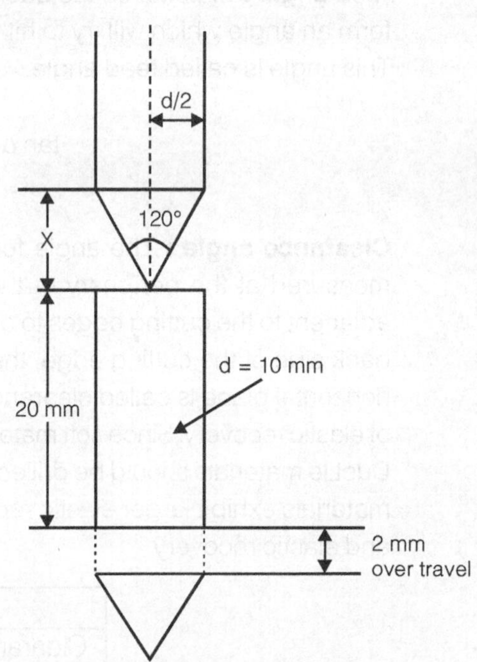

Fig. 4.47

4.27 MILLING

Milling is one of the machining processes of flat surfaces as well as formed surfaces like those of the gears, splines, slots and key ways etc.

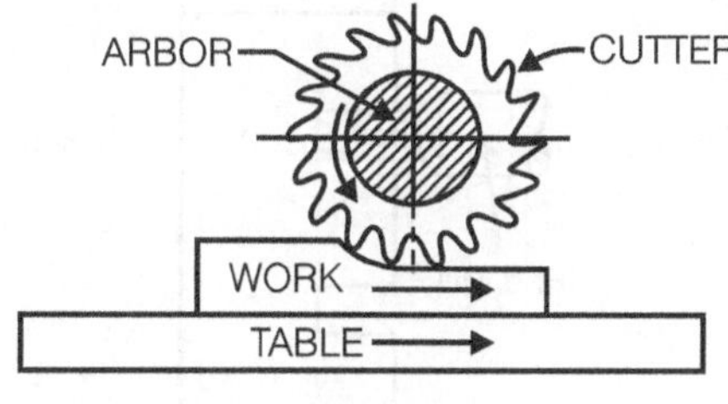

Fig. 4.48

There are two types of milling processes

- Peripheral milling (or slab milling)
- Face milling

4.27.1 Peripheral or slab milling operation

In the peripheral milling cutting teeth are on the periphery of the cutter. Movement of the tool or cutter is generatrix and movement of the work piece is directrix. In the slab milling operation, the axis of rotation of milling cutter is parallel to the work.

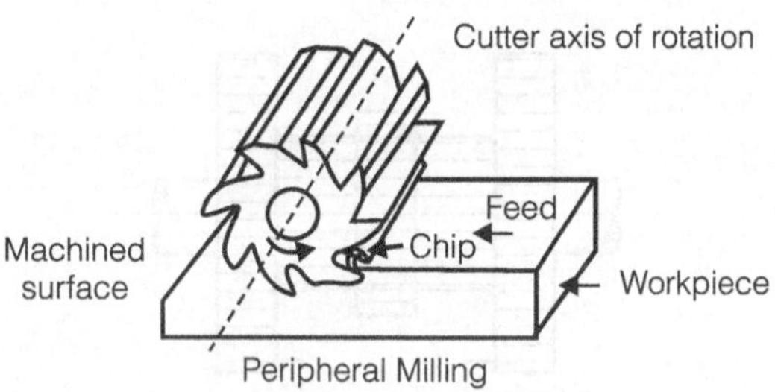

Fig. 4.49 : Schematic of peripheral milling

4.27.2 Face Milling

In the face milling operation, axis of rotation of cutter is perpendicular to the work. Because of the relative motion between the cutting teeth and work piece a face milling cutter leaves feeds marks on machine surface.

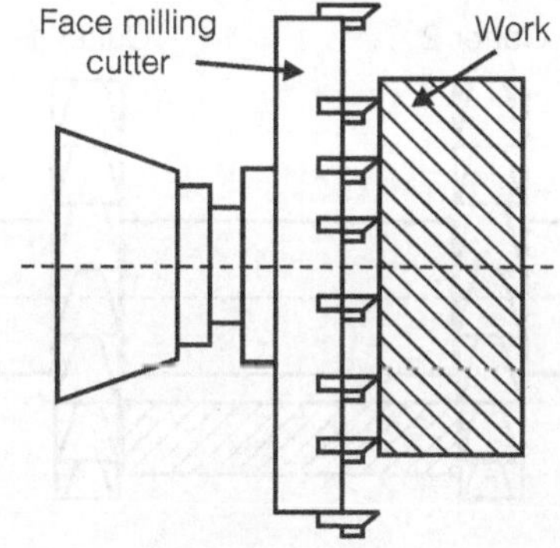

Fig. 4.50 : Schematic of face milling

4.27.3 End Milling

End milling cutter are combination of both slab milling and face milling operation. An end mill cutter is used to machine and produce a flat surface or a pair of parallel flat surfaces.

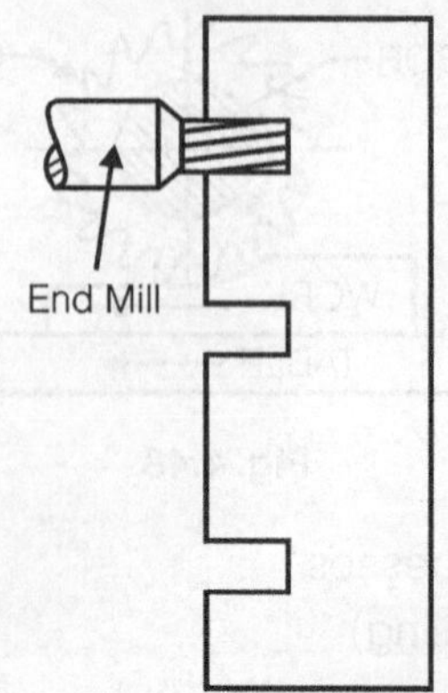

Fig. 4.51 : Schematic of end milling

4.27.4 Gang Milling

When a number of cutters are mounted on the arbor and a number of steps have been cut in the work piece, it is called Gang Milling.

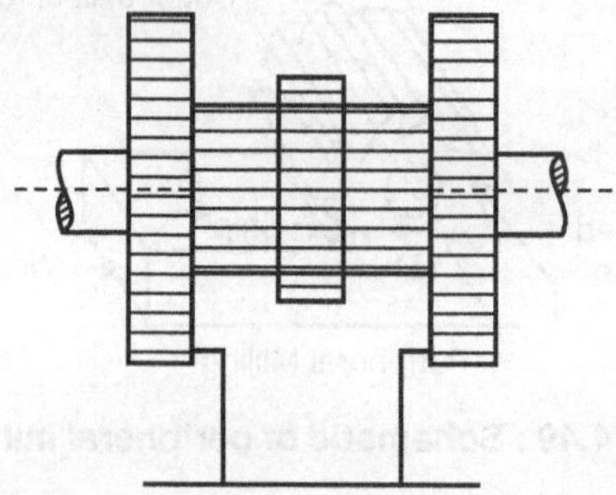

Fig. 4.52 : Schematic of Gang milling

4.27.5 Straddle Milling

When milling is performed on two surfaces simultaneously, it is called STRADDLE MILLING. T- Slot milling is a unique example of straddle milling by a single milling cutter.

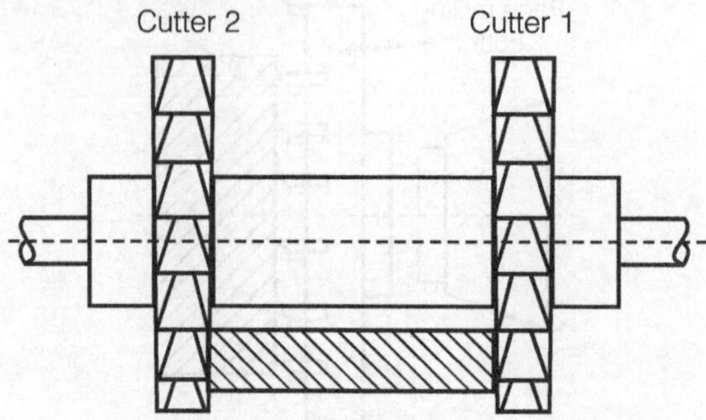

Fig. 4.53 : Schematic of Straddle milling.

4.27.6 Up Milling and Down Milling

In the up milling operation, cutter and the work piece movements are in opposite direction but in the down milling operation, both movements are in the same direction.

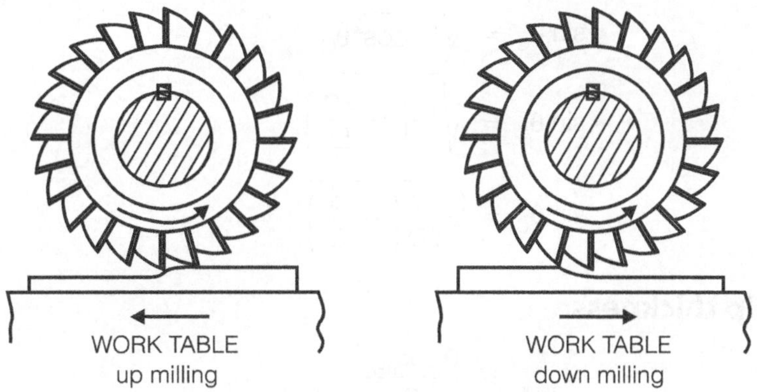

Fig. 4.54 : Up and Down Milling

In the up milling operation, chip thickness varies from minimum to maximum and before machining, cutting edge will rub over the finished part so the surface finish produced is very poor. In up milling operation, hot chips remains in the tooth space and be there for considerable period of time, so due to diffusion, tool life is lower in up milling. If back lash error is there in the fixture it has no significance in up milling because the fixtures are under tension opposite will be the case for down milling operation.

4.27.7 Mechanics of Milling

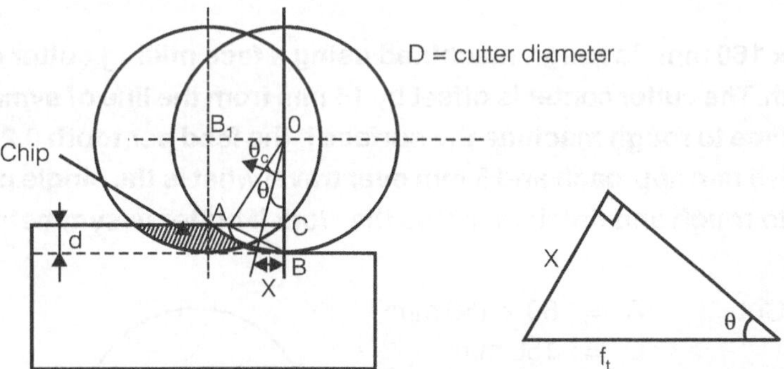

Fig. 4.55 : Cutting Action in Milling

$$f = \text{feed/ rev}$$

$$f_t = \text{feed per tooth} = \frac{f}{n_t}$$

$$n_t = \text{no. of teeth}$$

$$x = f_t \sin\theta$$

At $\theta = \theta_C \quad x = x_{max}$

$$\theta = 0 \quad x = 0$$

$$x_{max} = f_t \sin\theta_C$$

From ΔOAC $\cos\theta_C = \frac{OC}{OA} = \frac{\frac{D}{2}-d}{\frac{D}{2}} \Rightarrow 1-\frac{2d}{D}$

$$\sin\theta_C = \sqrt{1-\cos^2\theta_C}$$

$$\Rightarrow \quad \sin\theta_C = \sqrt{1-\left(1-\frac{2d}{D}\right)^2}$$

$$\Rightarrow \quad \sin\theta_C = 2\sqrt{\frac{d}{D}\left(1-\frac{d}{D}\right)}$$

4.27.8 Mean chip thickness

$$x_{mean} = \frac{0+x_{max}}{2}$$

$$\frac{f_t \sin\theta_C}{2} = f_t\sqrt{\frac{d}{D}\left(1-\frac{d}{D}\right)}$$

Mean chip area $A_{mean} = w \times x_{mean}$

Volume of metal removed

$$wdf = A_m V$$

V = Velocity of cutting

Uncut chip thickness is related to feed per tooth and by increasing the feed since there will be increase in the uncut chip thickness, cutting force requirement will increase. But while milling at the constant speed, cutting forces will be as it is but cutting power requirement will increase.

Example : 4.25

A surface 80 × 160 mm² is rough machined using a face milling cutter of 150 mm diameter having 10 teeth. The cutter center is offset by 15 mm from the line of symmetry of work piece. Estimate the time to rough machine the surface if the feed per tooth 0.25 and cutting speed 20 m/min. with 5 mm approach and 5 mm over travel what is the single pass feed time? Also find the time to rough and finish machine the work if cutter is symmetric.

Solution :

For Rough Cut $A = 80 \times 160\ \text{mm}^2$

$D = 150$ mm

$n_t = 10$

offset = 15 mm

$f_t = 0.25$ mm/tooth

$V = 20$ m/min

Movement = 160 + x

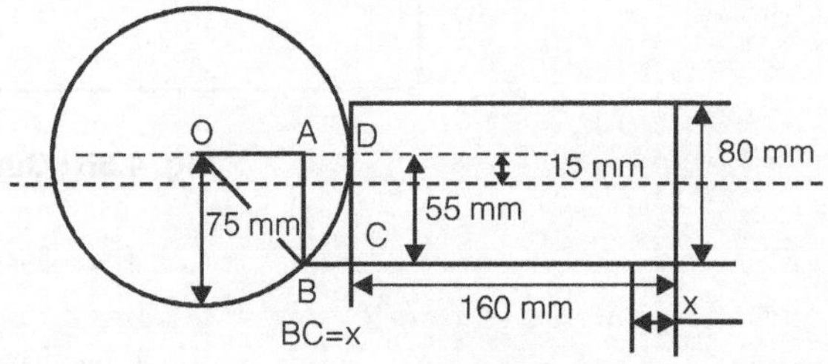

Fig. 4.56

From ΔOAB we get

$$x = 75 - \sqrt{75^2 - 55^2} = 24.0098$$

Total Movement of cutter = 184.0098 mm

Feed/ rev = 0.25 × 10 = 2.5 mm/rev

Number of revolution required for machining

$$= \frac{184}{2.5} = 73.6 \text{ rev}$$

$$V = \pi DN$$
$$20 = \pi \times 0.15 \times N$$
$$\text{rpm (N)} = 42.44 \text{ rev}$$
$$\text{machining time } (T_{min}) = \frac{73.6}{42.44} = 1.7342 \text{ min}$$

For finish cut

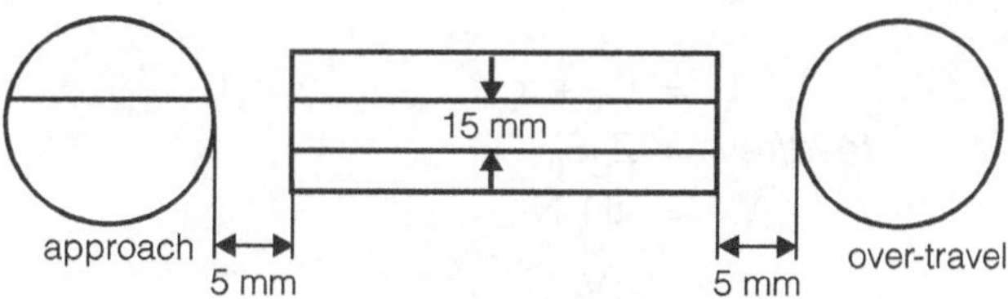

Fig. 4.57

$$\text{Total Movement of cutter} = 5 + 160 + 5 + 150 = 320 \text{ mm}$$
$$\text{Feed/rev} = 2.5 \text{ mm/rev}$$
$$\text{No. of revolutions} = \frac{320}{2.5} = 128 \text{ rev}$$

Time required to finish cut

$$T = \frac{128}{42.44} = 3.016 \text{ min}$$

If cutter axis is symmetrical

$$x = 75 - \sqrt{75^2 - 40^2} = 11.55$$
$$\text{Rough cut Movement} = 160 + x = 171.55 \text{ mm}$$
$$\text{No. of revolutions} = \frac{171.55}{2.5} = 68.62 \text{ rev}$$
$$T = \frac{68.62}{42.44} = 1.617 \text{ min}$$

There will not be any change in finish time.

Example : 4.26

The thickness of a rectangular brass plate of length L_w and width B_w has to be reduced by t mm in one pass by a helical fluted plain or slab milling cutter of length l_c(> B_w), diameter D_c and number of teeth, Z_c at cutting velocity, V_c m/min and feed, S_0 mm/tooth. How will you determine the time that will be required to accomplish the aforesaid work? All other dimensions are in mm.

Solution :

l_c = cutting length
$l_c > B_w$
B_w = width
V_C = m/min
S_0 = feed/tooth

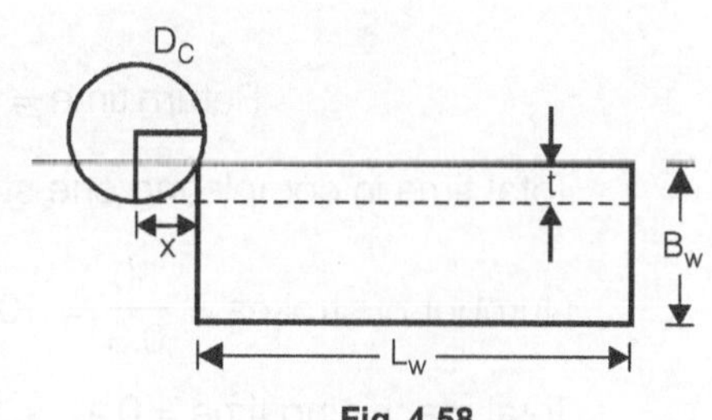

Fig. 4.58

Length of cutter movement

$$x = \sqrt{\left(\frac{D_C}{2}\right)^2 - \left(\frac{D_C}{2} - t\right)^2}$$

$$\sqrt{D_C t - t^2} = \sqrt{t(D_C - t)}$$

Total length

$$L = L_w + x$$

$$\text{feed/rev} = Z_C S_o$$

$$V_C = \pi D_C N$$

$$N = \frac{V_C}{\pi D_C}$$

$$\text{No. of revolutions} = \frac{L_w + \sqrt{t(D_C - t)}}{Z_C S_0}$$

$$\text{Time of machining} = \frac{\left[L_w + \sqrt{t(D_C - t)}\right]\pi D_C}{Z_C S_0 V_C}$$

Example : 4.27

A 600 mm × 30 mm flat surface of a plate is to be finish machined on a shaper. The plate has been fixed with the 600 mm side along the tool travel direction. If the tool over-travel at each end of the plate is 20 mm, average cutting speed is 8 m/min, feed rate is 0.3 mm/stroke and the ratio of return time to cutting time of the tool is 1 : 2 what is the time required for machining.

Solution :

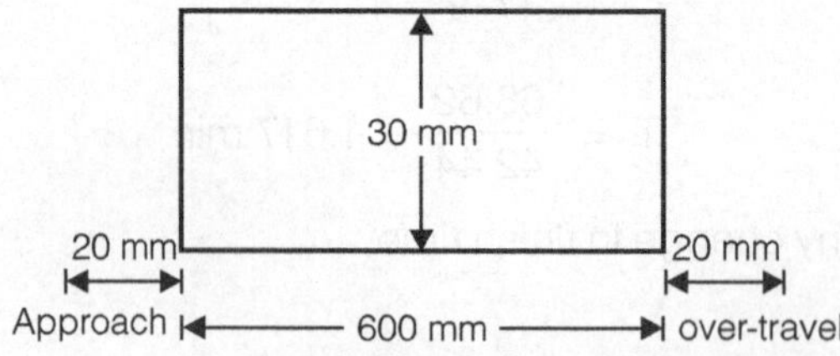

Fig. 4.59

Total forward stroke length = 20 + 600 + 20 = 640 mm

$$\text{Forward time} = \frac{0.640}{8} \text{ min} = 0.08 \text{ min}$$

$$\text{Return time} = \frac{0.08}{2} = 0.04 \text{ min}$$

Total time in completing one stroke = 0.08 + .04 = 0.12 min

$$\text{Number of strokes} = \frac{30}{0.3} = 100 \text{ strokes}$$

Total machining time = 0.12 × 100 = 12 min

Example : 4.28

Estimate the power required during up milling of mild steel block of 20 mm width with using a straight slab cutter with 10 teeth, 75 mm diameter and 10° radial rake. The feed velocity of table is 100 mm/min. The coefficient of friction at rake face and shear strength of material may be assumed to be 0.5 and 400 N/mm² respectively. The depth of cut is 5 mm and cutter rotates at 60 rpm.

Solution

$D = 75$ mm, $\alpha = 10°$, $\mu = 0.5$, $\tau_s = 400$ N/mm²

$N = 60$ rpm, $d = 5$ mm, $Z = 10$, $w = 20$ mm

Feed per teeth per revolution

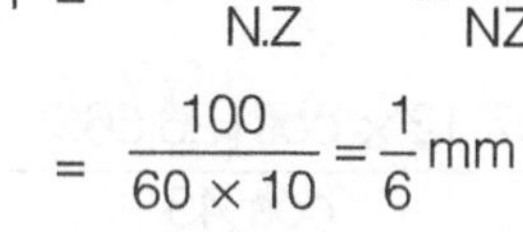

$$f = \frac{\text{Table feed}}{N.Z} = \frac{F}{NZ}$$

$$= \frac{100}{60 \times 10} = \frac{1}{6}\text{ mm}$$

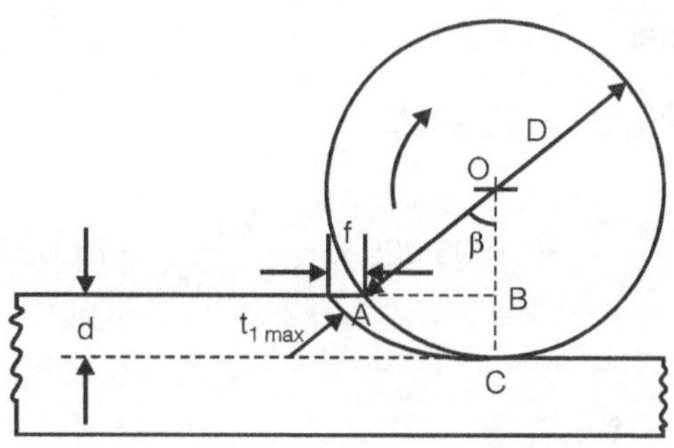

Fig. 4.60

Max. thickness of uncut chip

$$t_{1\,max} = f \sin\beta = \frac{1}{6}\sin(29.93) = 0.083156$$

$$\because \quad (D - d).d = AB^2$$

$$AB = \sqrt{(D-d).d} = \sqrt{(75-5)\times 5} = 18.7 \text{ mm}$$

$$\sin\beta = \frac{AB}{AO} = \frac{18.7}{75/2} = \frac{18.7 \times 2}{75}$$

$$\beta = 29.93°$$

Angle between two teeth, $\theta = \frac{2\pi}{Z} = \frac{360}{10} = 36°$

$\because$ $\theta > \beta$ so the cutting will be intermittent.

Max. shear force, $$F_{s_{max}} = \frac{\tau_s.w.t_{1\,max}}{\sin\phi} \quad \text{... (1)}$$

Friction angle, $\lambda = \tan^{-1}\mu$

$= \tan^{-1} 0.5 = 26.565°$

From Lee-& shaffer condition of optimum power

$$\phi + \lambda - \alpha = 45$$

$$\phi + 26.565 + 10 = 45$$

$$\phi = 28.435°$$

$$\therefore \quad F_{s_{max}} = \frac{400 \times 20 \times 0.083156}{\sin (28.435)}$$

$$= 1397.12 \text{ N}$$

$$\text{Max. cutting force, } F_c = \frac{F_s . \cos (\lambda - \alpha)}{\cos (\phi + \lambda - \alpha)}$$

$$= \frac{1397.12 \times \cos (26.565 - 10)}{\cos 45} = 1893.825 \text{ N}$$

Max. torque on the cutter

$$M_{t_{max}} = F_c \times D/2$$

$$= 1893.825 \times \frac{75}{2 \times 1000} = 71.02 \text{ N-m.}$$

The torque variation is

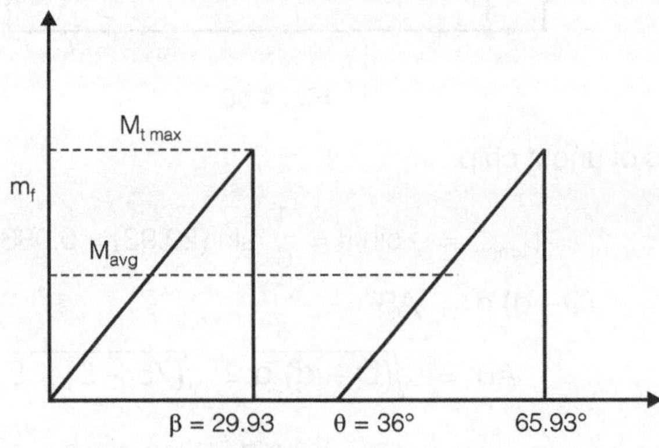

Fig. 4.61

Average torque

$$M_{avg} \times \theta = \frac{1}{2} . M_{max} . \beta$$

$$M_{avg} = \frac{M_{max} \times \beta}{2\theta} = \frac{71.02 \times 29.93}{72}$$

$$= 29.523 \text{ N-m}$$

$$\text{Power required} = M_{avg} \times \omega = 29.523 \times \frac{2\pi \times 60}{60} = 185.5 \text{ W}$$

Example : 4.29

In a slab milling operation with straight teeth cutter, the cutter has 15 teeth with 10° rake angle and rotates at 200 rpm. The diameter of cutter is 80 mm and table feed is 75 mm/min, the depth of cut is 5 mm. The width of ms slab is 50 mm and ultimate shear stress of work material is 420 N/mm². Assuming the coefficient of friction between chip and cutter to be 0.7 and using Lee and shaffer relation, plot variation of resultant torque with cutter rotation and estimate avg. power consumption.

Solution :

$Z = 15$, $\alpha = 10°$, $N = 200$ rpm, $D = 80$ mm

table feed, $F = 75$ mm/min, $d = 5$ mm

$\tau_s = 420$ N/mm², $\mu = 0.7$, $w = 50$ mm

Lee and shaffer relation

$$\phi + \lambda - \alpha = 45 \quad \text{... (1)}$$

$$\text{Friction angle, } \lambda = \tan^{-1}(\mu) = \tan^{-1}(0.7) = 35°$$

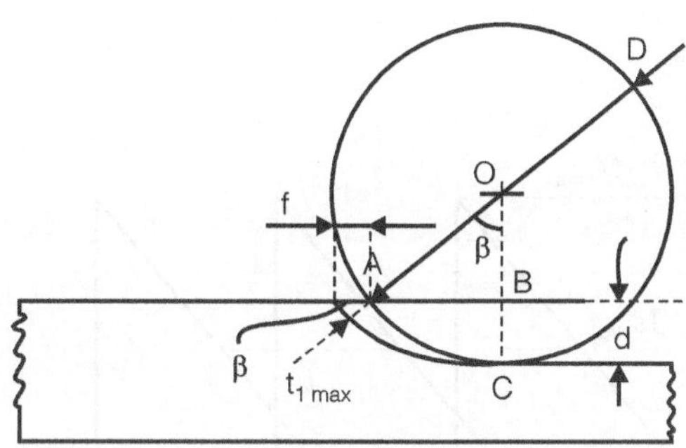

Fig. 4.62

∴ Shear angle from (1)

$$\phi + 35 - 10 = 45$$

$$\phi = 20°$$

∵ $$(D - d).d = AB^2$$

$$AB = \sqrt{75 \times 5} = 19.365 \text{ mm}$$

$$\sin\beta = \frac{AB}{OA} = \frac{19.365}{40}$$

$$\beta = 28.96°$$

Feed per teeth per revolution

$$f = \frac{F}{N.Z} = \frac{75}{200 \times 15} = \frac{1}{40} \text{mm}$$

Maximum thicnkness of uncut chip

$$t_{1\ max} = f \sin\beta$$

$$= \frac{1}{40}\sin(28.96) = 0.0121 \text{ mm}$$

Maximum shear force

$$F_{s\,max} = \frac{\tau_s . w . t_{1\,max}}{\sin\phi}$$

$$= \frac{420 \times 50 \times 0.0121}{\sin 20} = 742.938 \text{ N}$$

Maximum cutting force,

$$F_{c\,max} = \frac{F_s \cos(\lambda - \alpha)}{\cos(\phi + \lambda - \alpha)}$$

$$F_{c\,max} = \frac{742.938 \times \cos(35 - 10)}{\cos 45} = 952.233 \text{ N}$$

Angle between torque teeth

$$\theta = \frac{360}{15} = 24°$$

$\because$ $\theta < \beta$ so there will be continuous cutting i.e. there will be one teeth in contact always. The variation is as

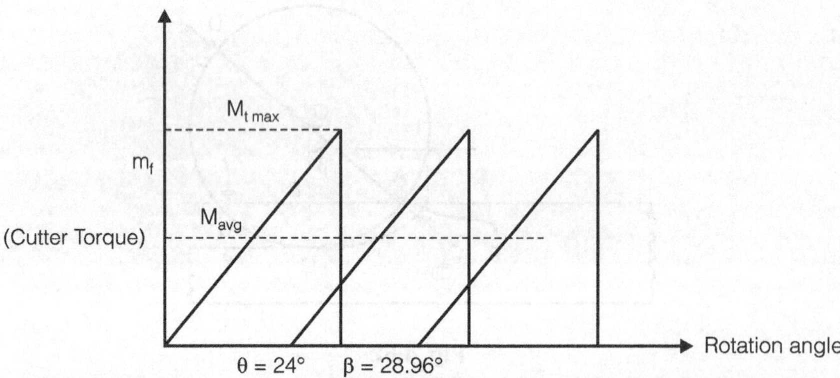

Fig. 4.63

torque due to one tooth

$$M_{t\,max} = F_{c\,max} . D/2$$

$$= 952.33 \times \frac{40}{1000} = 38.0932 \text{ N-m}$$

Average Torque $\quad M_{avg}\,\theta = \frac{1}{2} M_{max} . \beta$

$$M_{avg} = \frac{1}{2} M_{max} . \frac{\beta}{\theta}$$

$$= \frac{38.0932}{2} \times \frac{28.96}{24} = 22.982 \text{ N-m}$$

(i) Power consumption $= M_{avg} . \frac{2\pi N}{60}$

$$= 22.982 \times 2\pi \times \frac{200}{60} = 481.426 \text{ W}$$

Example : 4.30

A slot 25 mm deep is to be cut through a workpiece 200 mm long with the help of HSS side and face cutter whose diameter is 150 mm and has 10 teeth. The cutting speed is 50 m/min and feed is 0.25 mm per teeth find

(i) Table feed in mm/min.

(ii) Total cutter travel.

(iii) Time required to machine the slot.

Solution :

BC = 25 mm, D = 150 mm

l = 200 mm, Z = 10

Cutting speed, V = 50 m/min

Cutter over travel - AB

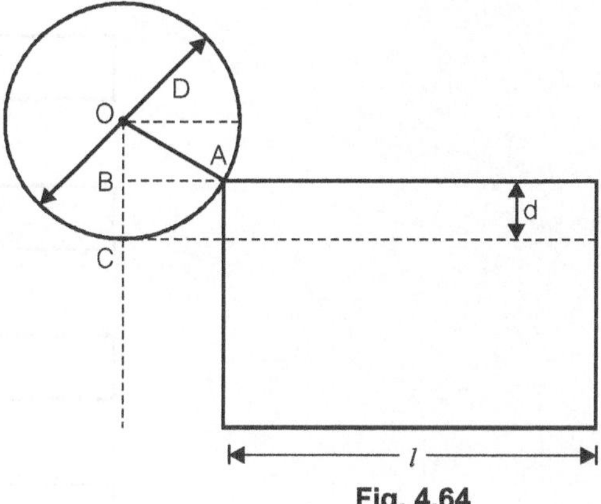

Fig. 4.64

$$\therefore \quad AB = \sqrt{\left(\frac{D}{2}\right)^2 - \left(\frac{D}{2} - d\right)^2}$$

$$= \sqrt{\left(\frac{D^2}{4} - \frac{D^2}{4} - d^2 + D.d\right)}$$

$$= \sqrt{\left(D.d - d^2\right)}$$

$$= \sqrt{\left(150 \times 25 - 25^2\right)} = 55.9 \text{ mm}$$

(i) Table feed = Feed per tooth × no. of tooth × rpm

$$\because \quad \text{Cutting speed, } V = \frac{\pi D N}{1000} \text{ m/min}$$

$$50 = \frac{\pi \times 150 \times N}{1000}$$

$$N = 106.1 \text{ rpm}$$

$$\therefore \quad \text{Table feed} = 0.25 \times 10 \times 106.1$$

$$= 265.26 \text{ mm/min}$$

(ii) Total cutter travel = length of W.P + Over travel

$= l + AB$

$= 200 + 55.9$

$= 255.9$ mm

(iii) Slot machining time, $T = \dfrac{\text{Total cutter travel}}{\text{Table feed}} = \dfrac{255.9}{265.26}$

$= 0.9645$ min $= 57.88$ s

4.28 POWDER METALLURGY

Powder metallurgy technique is used to mix the material which does not form an alloy. Some of the major applications of the process are:

1. Filament of a bulb, tube.
2. Cutting tool and grinding wheel.
3. Nozzles for abrasive jet machining.
4. Porus bearing or self lubricating bearing.
5. Filters used in casting process.
6. Friction material in anti-lock braking system (ABS).

Sequence of Processes in powder metallurgy:

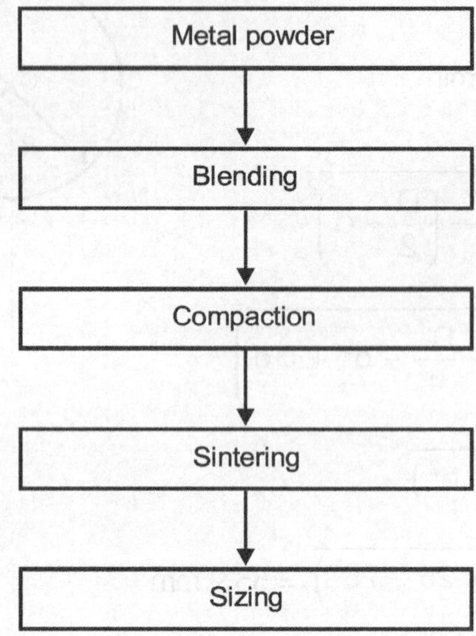

1. **Methods of Making Metal Powders :**
   - **Metal Crushing and pulverizing :** Brittle materials can be converted into powder by crushing. If the material is not sufficiently brittle, its temperature is decreased.
   - **Atomization :** Ductile materials or the materials having low melting point, can be converted into powder by this method. Liquified material is poured over a high speed rotating disk. Due to the centrifugal action, liquid metal comes down in the form of very fine droplets. The material can also be sprayed by using a plasma torch.
   - **Corrosion :** When stainless steel is kept in the environment of sulphuric acid and copper sulphate, it dissolves and settles down at the bottom of tank. But this method takes time.
2. **Blending :** By mixing the lubricant with powders, a layer of lubricant will be deposited over the particles. This will increase the interaction between them and as a result of that powder can be given some shape called green compact. This is called as green because it is freshly prepared.
3. **Compaction :** It is also called as Iso-static pressing. To give initial strength to the green compact, powder is pressed on a press. Variation in the properties of compact will be more when the compaction is done on a single or a double action press. But on a double action press, properties will be more uniform. Smaller is the size of particle, better will be under diffusion and the compact will be stronger. If the particle size is smaller, the strength will be better because more area will be under diffusion.

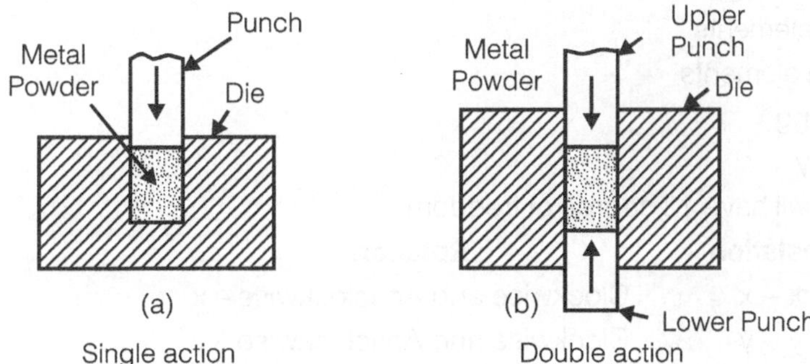

Fig. 4.65 : Compacting the metal powder using single and two punches

4. **(a) Pre-sintering :** When heating is done along with compaction it is called Hot Iso-static pressing (HIP). Compact is heated and due to that lubricant on the surface of particle will burn. Characteristic of lubricant should be such that after burning, it should not leave any residue. This provides the localized heating which increases the extent of diffusion between the particles. When Pre-sintering is clubbed compaction, it is called Hot Iso-static pressing. Rough particles give better strength as interlocking among the particles is better.

   (b) Sintering : In the sintering stage, compact heated 60-70% of the melting point of the base metal. This increases the extent of diffusion between the particles. Those powder which could not form interlocking and whose melting point are below this temperature, they will be liquefied and will fill up the voids. When the compact is a mixture of large number of powders having large difference in their melting point two or three stage melting process is performed. After sintering, product appears to be very hard and brittle, so normally no machining is advisable but to give some simple shapes to the part sizing is performed. Powder metallurgy can only be applied for mass production because of expensive tooling arrangement.

4.29 JIGS AND FIXTURES

Jigs and fixtures are the auxiliary devices used in mass production. Fixture locate the work piece and Jig along with location, guides the tool. Jigs are used in drilling, boring, reaming etc. and fixtures are used in milling, shaping etc.

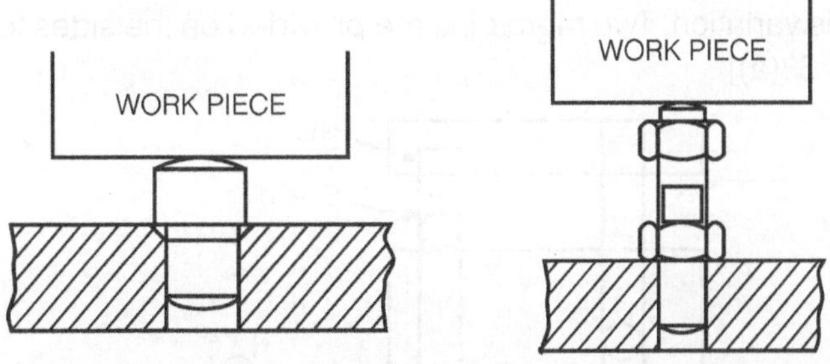

Fig. 4.66 : Fixed type support pin and adjustable type support pin

If deep holes are to be made in the work piece, the drill acts like a cantilever with point load (cutting forces). The hole becomes oversize in the beginning. Providing carbon bush makes it a simply supported beam and hence the drill does not get deflected.

Components of Jig and Fixture :

1. Locating elements
2. Clamping elements
3. Tool guiding
4. Main body

A work piece will have 12 degree of freedom

Translation	Rotation
+ x – x	Clockwise and Anticlockwise – x
+ y – y	Clockwise and Anticlockwise – y
+ z – z	Clockwise and Anticlockwise – z

Cylinder Sphere

Fig. 4.67 : V block

The number of degree of freedom restricted when cylinder placed in a V block and are 7 when sphered placed, only 3 degrees of restricted

4.29.1 Principle of Location

3-2-1 Principle : There are 12 degree of freedom of a work piece i.e., movement about negative and positive axis and clock-wise and anti-clockwise rotation about the three axis. By providing three pins in the base, five degree of freedom will be arrested i.e. movement about negative Z-axis and the clockwise and anti clockwise rotation about X and Y axis. These three pins also ensure that machining always takes place on single plane perpendicular to the base, along the length direction. This arrests movement of work piece along negative Y-direction in clock-wise and anti -clock wise rotation about Z-axis. A sixth pin is provided on a plane perpendicular to the previous two planes, which also arrests movement about negative X-axis. By providing six pins, nine degree of freedom is arrested. Two more Degree of freedom will be arrested by providing clamping elements and from 12th direction, tool comes for machining. A smaller size of pin is called button and are used in walls.

Radial location : This principle of location is used when there is already one drilled hole in the work piece. Since, there will be variation in the size of hole, conical pins are inserted in the hole to accommodate this variation. Two more pins are provided on the sides to arrest two more degree of freedom [Fig. 4.68(a)].

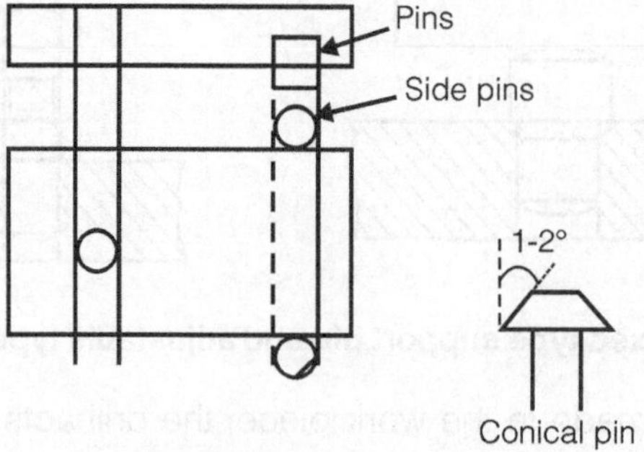

Fig. 4.68 (a) : Method of radial location

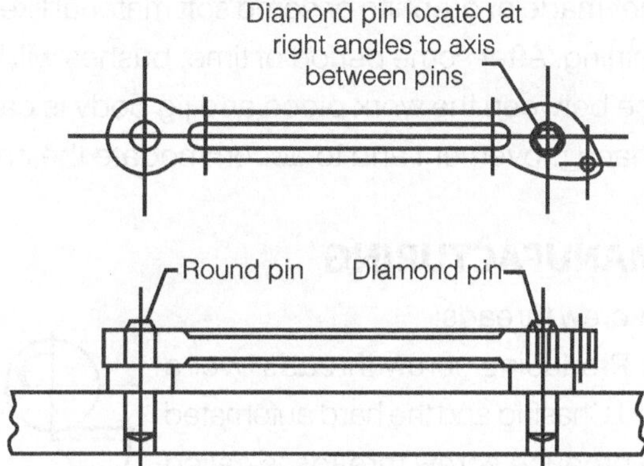

Fig. 4.68 (b) : Use of one round and one diamond locating pin

If there are two already drilled holes, in one hole a conical pin is provided and in the other hole, a diamond pin is provided to accommodate the variation in the centre distance Fig. 4.68(b). Two sides of the diamond pin are relieved and the diamond pins are always kept in the larger hole. Purpose of diamond pin is to arrest the variation in centre to centre distance between holes.

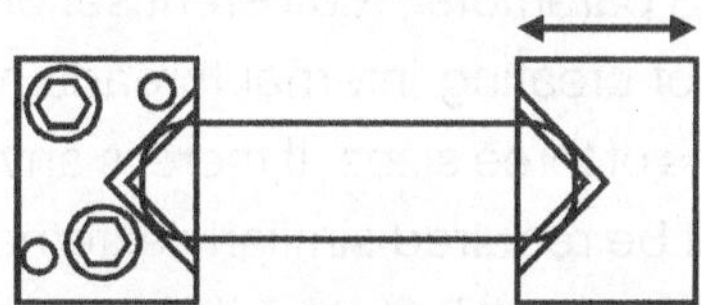

Fig. 4.69 : Method of V location

V-location : Radial locations are used when work piece is either cylindrical or spherical and when it is held between the 2-V jaws all the degrees of freedom will be arrested. One V is movable and the other one is fixed.

4.29.2 Steps in designing the jigs

- Draw the plan, elevation and end view with red pencil.
- Identify the areas where the machining has to be performed.
- Identify the type of location and where the locating element has to be given.
- Decide about the clamping element.

Outline diagram if Jig body:

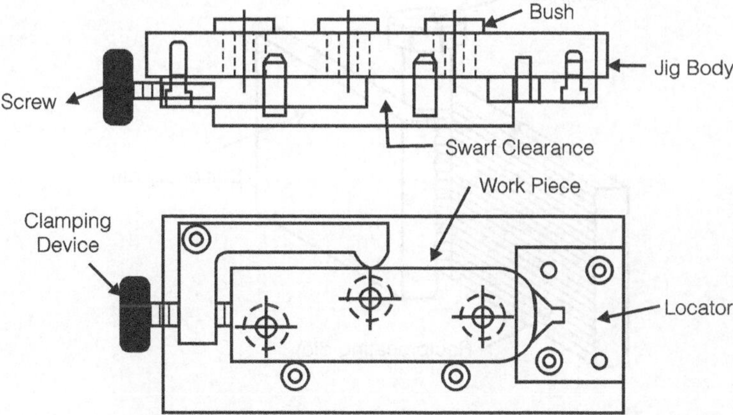

Fig. 4.70 : Showing the body of plate jig

Carbon bushes are either made of graphite or some soft material like aluminium so that it provides lubrication during machining. After some period of time, bushes will be consumed and have to be replaced. The clearance between the work piece and jig body is called swarf clearance and it is given to allow the free hand movement and to accommodate the chips.

4.30 SCREW THREAD MANUFACTURING

Methods of producing screw threads:

1. **Thread Chasing :** Producing screw threads over a lathe is called Thread Chasing and the hard automated lathe dedicated to produce screw threads is called Swiss Automat. A wire is fed through the spindle. It is turned, threads are being produced and by using a parting tool. The screw is cut off from the wire.

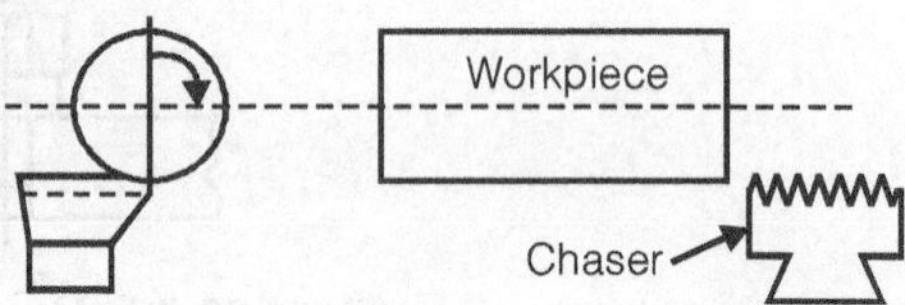

Fig. 4.71 : Thread chasing operation

2. **Die Threading :** External screw threads over pipes are cut by Die Threading. If the size of pipe increases beyond a certain value, the process cannot be used to create threads. If there is any change in the thread parameter, a different set of dies will be required.
3. **Tapping :** It is a process of creating internal threads manually. Different sizes of taps are available and a set consists of three sizes. If there is any change in the pitch or thread angle a separate sets of dies will be required similarly external threads can be cut by using dies. Relatively accurate threads are produced by dies.
4. **Thread milling :** Thread milling is employed when the screw threads finish very near to the shoulder of the work or the diameter is too large. In order to start the cutting action, the cutter axis is first swilled and set at an angle equal to the helix angle of the screw. The work and cutter rotated in opposite directions. The work is also made to advance axially.
5. **Thread Rolling :** A work blank is pressed between either two flat dies or three circular die process and threads are produced plastic deformation process. Because of work hardening threads are stronger and accurate but the process is used only for mass production and threads cannot be cut up to the shoulders. Threads produced by soft and ductile materials will be economical. If there is any change in the thread parameter a separate sets of dies will be required.

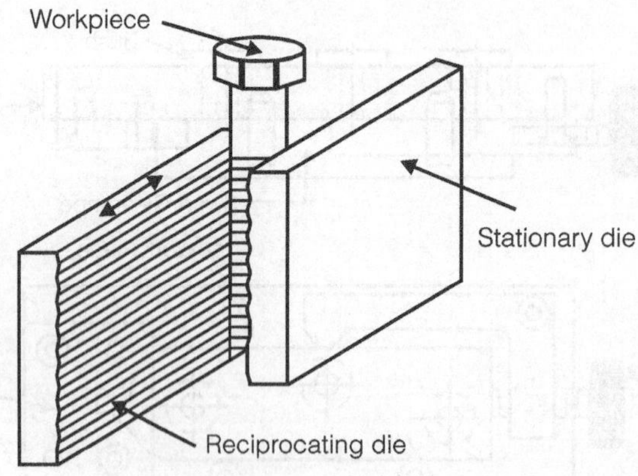

Fig. 4.72 (a) : Preparation of threads using flat rolling

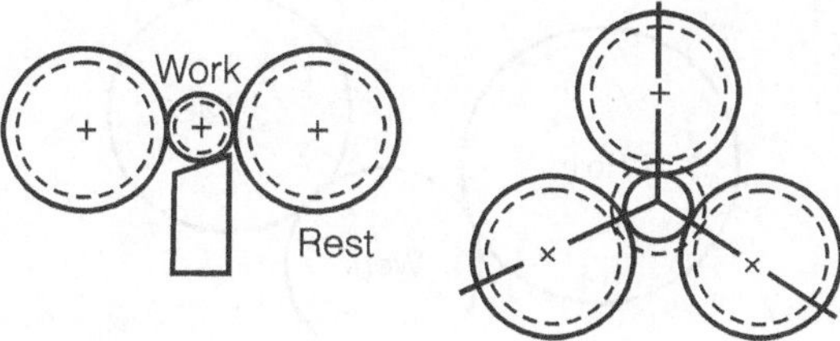

Fig. 4.72 (b) : Preparation of threads using circular rolling

6. **Thread grinding :** It is a type of form grinding with minimum in feed otherwise wheel wear will increase. This process primarily used to produce highly precise screw threads used in gauges.

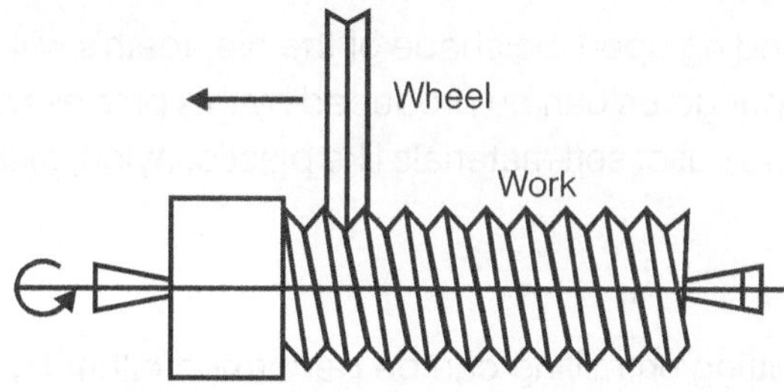

Fig. 4.73 : Preparation of threads using grinding

4.31 GEAR MANUFACTURING

4.31.1 Materials for Gears

1. **Cast Iron :** Large size gears are produced by casting process by this material. Care should be taken that on such gears, fatigue load should not appear.
2. **Steel :** Since the material has good load carrying capacity, such gears are used in power transmission.
3. **Bronze :** In worm gear drives, since heavy sliding loads appear, and bronze is very resistant to sliding loads, it is used to make Worm Gears.
4. **Plastics :** Used in noiseless operations like toys.
5. **Aluminium :** Aluminium is used where lightness of the assembly is the criteria but much power cannot be transmitted by these gears.
6. **Nylon :** Used in Xerox machines. These gears produces noiseless operation but also can carry small amount of loads.

4.31.2 Forming

1. **Rolling:** A number of work blanks are mounted on a common shaft and is presses by a pair of rolls. Teeth's are produced by plastic deformation process so it will be strong. But a separate sets of rolls will be required if there is any change in any parameter of gear. For hard and brittle materials the process becomes very uneconomical.

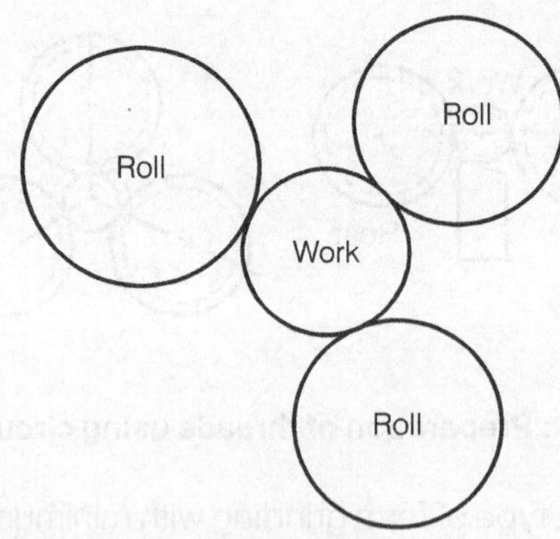

Fig. 4.74 : Preparation of gears using Rolling

2. **Extrusion :** Depending upon the shape of the die, teeth's will be produced over the gears. Both helical and spur gears can be produced by this process and the gear can be cut to the required size. Gears out of soft materials like plastic, nylon, aluminium can be manufactured easily.

4.31.3 Machining

Gear Milling : Gear cutting on milling can be performed either by using a slab milling cutter or end milling cutter. When the required tooth depth is achieved the cutter is withdrawn and the work piece is indexed and moved by'1' pitch, so, that a new position on the work blank comes in contact with its cutter. All types of gears including bevel can be produced by milling operation and since the indexing mechanism is governed by a gear arrangement, gears produced by milling operation are not very accurate.

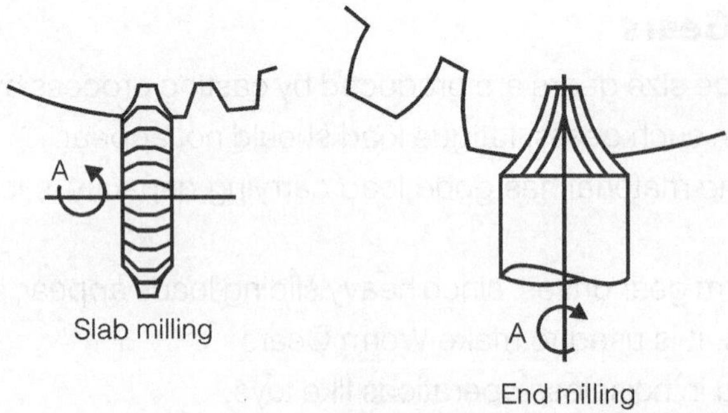

Fig. 4.75 : Preparation of gears using milling

Indexing Mechanism : The milling machine is having an index plate on which there are number of circles. On each circle there are finite number of holes. There is a pin over index plate which is in contact with a worm gear which has exactly 40 number of teeth. This worm gear is in mesh with worm wheel on which work blank is mounted. By giving one complete rotation to the index plate worm gear advances by '1' pitch. So, if 40 number of teeth's has to be produced over the work blank, after cutting one teeth the index plate should be given one rotation.

For example the index plate consists of 12,14,16,18 no of holes. To find the no of revolutions ratio $\frac{40}{N}$ is calculated. Suppose 28 teeths has to be cut on the work blank then:

$$\frac{40}{N} = \frac{40}{28} = 1 + \frac{12}{18} = 1 + \frac{3}{7} = 1 + \frac{6}{14}$$

After cutting each teeth, one complete revolution is given to the index plate and pin is inserted in the 8th hole of 14 hole circle.

If 70 teeth has to be cut: $\frac{40}{70} = \frac{4}{7} = \frac{4 \times 2}{7 \times 2} = \frac{8}{14}$

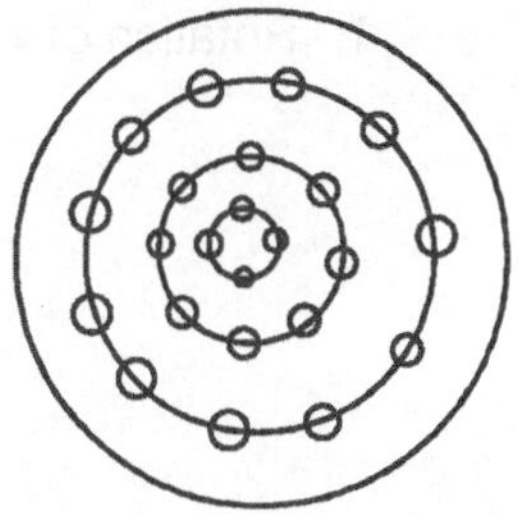

Fig. 4.76 : Indexing plate

The pin is provided in 8th hole in 14th hole circle after cutting each teeth.

Gear Shaping : In the rack type cutter, once all the cutting edges are covered by the work blank, rack is with drawn and the work is indexed. So, the gears produced are not very accurate. But with pinion type cutter, once the required tooth depth is achieved, rotary motions are given to both pinion and work and as a result of that, it is a continuous process. So, the gears produced are very accurate. Internal gears are also cut by this method. Both helical and spur gears can be cut by this procedure but to cut helical gears, work has to be mounted on a helical drive.

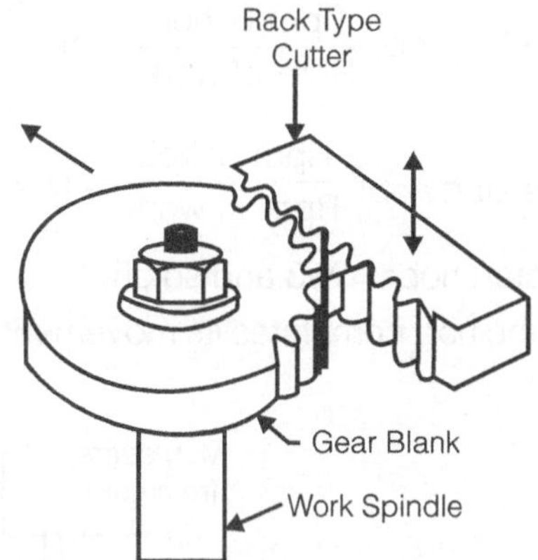

Fig. 4.77 : (a) Gears shaping using pinion cutting (b) Preparation of gears using shaper

Broaching : Splines broaches are used to produce gears. Initially, a pilot hole has to be drilled and a broach is pulled through it. All the teeth will be produced simultaneously for producing external gears, pot broaching is used. Both helical and spur gears can be produced.

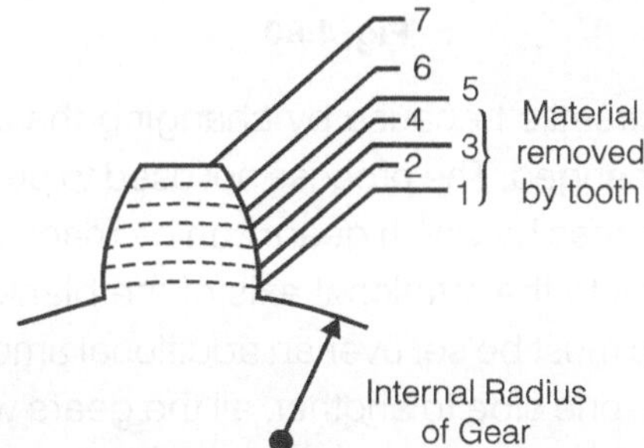

Fig. 4.78 : Preparation of gears using broaching

Gear Hobbing : Hob is in the shape of splined screw thread with a leading edge. A number of work blanks are mounted on a common shaft and the hob is lowered till the required tooth depth is achieved. After that three motions started simultaneously.

1. Rotation of work blank, 2.Rotation of hob, 3. Axial moment of work along the axis of hob.

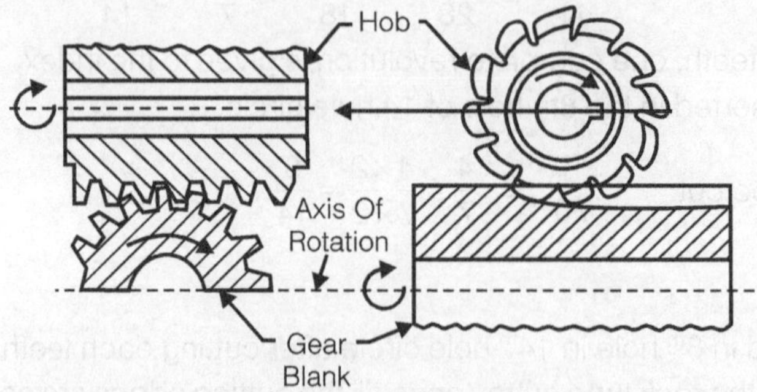

Fig. 4.79 (a) : Preparation of gears using hobbing

Initially the hob is in contact with the work with no rotation is given to the work. When the required tooth depth is achieved, all the three motions are started simultaneously. If it is a single start hob, in one rotation of the hob work piece should advance by '1'pitch. For double start hob, in half rotation of hob, the work should advance by '1' pitch.

Single start hob; $\dfrac{\text{Rpm of hob}}{\text{Rpm of work}} = N$

Double start hob; $\dfrac{\text{Rpm of hob}}{\text{Rpm of work}} = N/2$

Triple start hob; = N/3 and so on.

When the hob completes its movement along width of work blank all the teeth's are modelled.

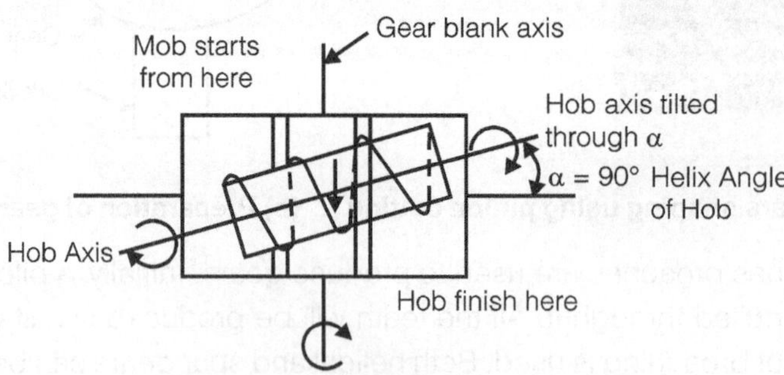

Fig. 4.80

Hobbing process is very versatile because by changing the angle between the work and hob helix angle of work can be changed. The process not used to produce internal gears and gears up to shoulders. It is fastest process by which gears can be machined. To cut spur gears, the axis of hob is setoff from the normal to the rotational axis of the blank by the helix angle of the hob. In cutting helical gears the hob must be set over an additional amount equal to the helix angle of the gear. Once hob moves from one side to another, all the gears will be produced.

For spur gear; $\alpha = 90° - \psi$

For helical gear; $\alpha = (90° - \psi)$ + helix angle of helical gear to be cut.

■■■■

PRACTICE QUESTIONS

1. The angle between the face and the flank of the single point cutting tool is known as:

(a) rake angle (b) clearance angle
(c) lip angle (d) point angle

2. Single point thread cutting tool should ideally have

(a) zero rake angle (b) positive rake angle
(c) negative rake angle (d) normal rake angle

3. Consider the following statements: Chipping of a cutting tool is due to

1. Tool material being too brittle
2. Hot hardness of the tool material is low
3. High positive rake angle of the tool

Which of these statements are correct?

(a) 1, 2 and 3 (b) 1 and 3
(c) 2 and 3 (d) 1 and 3

4. Consider the following statements about nose radius:

1. It improves tool life
2. It reduced the cutting force
3. It improves the surface finish

Which of these statements are correct?

(a) 1 & 2 are correct. (b) 2 & 3 are correct
(c) 1 & 3 are correct. (d) 1, 2 & 3 are correct

5. Consider the following characteristics

1. The cutting edges is normal to the cutting velocity.
2. The cutting forces occur in two directions only.
3. The cutting edge is wider than the depth of cut

The characteristics applicable to orthogonal cutting would include

(a) 1 and 2 (b) 1 and 3
(c) 2 and 3 (d) 1, 2 and 3

6. The primary tool force used to calculate the total power consumption in machining in the

(a) radial force (b) tangential force
(c) axial force (d) frictional force

7. Chip equivalent is increased by:

(a) an increase in side-cutting edge angle of tool
(b) an increase in nose radius and Side cutting edge angle of tool
(c) increase the area of cut
(d) increasing the depth of cut

8. In orthogonal turning of a low carbon steel bar of diameter 150mm with uncoated carbide tool, the cutting velocity is 90m/min. The feed is 0.24mm/rev and the depth of the cut is 2mm. The chip thickness obtained is 0.48 mm. If the orthogonal rake angle is zero and the principal cutting edge angle is 0^0, the shear angle in degree is

(a) 20.56 (b) 26.56
(c) 30.56 (d) 36.56

9. In orthogonal cutting, the feed is 0.5 mm at a cutting speed of 2m/s. If the chip thickness is 0.75 mm, the chip velocity is: (consider the rake angle as zero)
(a) 1.33 m/s (b) 2 m/s
(c) 2.5 m/s (d) 3 m/s

10. A batch of 10 cutting tools could produce 500 components while working at 50 rpm with a tool feed of 0.25mm/rev and depth of cut of 1mm. A similar batch of 10 tools of the same specification could produce 122 components while working at 80 rpm with a feed of 0.25mm/rev and 1mm depth of cut. How many components can produced with one cutting tool at 60 rpm?
(a) 29 (b) 31
(c) 37 (d) 42

11. Consider the following elements:
1. Nose radius 2. Cutting speed
3. Depth of cut 4. Feed
The correct sequence of these elements in DECREASING order of their influence on tool life is:
(a) 2, 4, 3, 1 (b) 4, 2, 3, 1
(c) 2, 4, 1, 3 (d) 4, 2, 1, 3

12. Consider the following statements with respect to relief angle of cutting tool.
1. This affects the direction of chip flow.
2. This reduces excessive friction between the tool & work piece
3. This affects tool life
4. This allow better access of coolant the tool-workpiece.
Which of the following statements given are correct?
(a) 1 & 2 (b) 2 & 3
(c) 2 & 4 (d) 3 & 4

13. In a single point turning tool, the side rake angle and orthogonal rake angel are equal. ϕ is the principal cutting edge angle and its range is $0° \leq \phi \leq 90°$. The chip flows in the orthogonal plane. The value of ϕ is closest to
(a) 0° (b) 45°
(c) 60° (d) 90°

14. In turning operation, the feed could be doubled to increase the metal removal rate. To keep the same level of surface finish, the nose radius of the tool should be:
(a) halved (b) kept unchanged
(c) doubled (d) made 4 times

15. The radical force in single-point tool during turning operation varies between:
(a) 0.2 to 0.4 times the main cutting force
(b) 0.4 to 0.6 times the main cutting force
(c) 0.6 to 0.8 times the main cutting force
(d) 0.5 to 0.6 times the main cutting force

16. **Assertion(A):** In metal cutting, the normal laws of sliding friction are not applicable.
Reason(R): Very high temperature is produced at the tool-chip interface.
(a) Both A and R are individually true and R is the correct explanation of A.
(b) Both A and R are individually true but R is not the correct explanation of A.
(c) A is true but R is false
(d) A is false but R is true

17. In orthogonal turning of medium carbon stel, the specific machining energy is 2.0 J/mm^3. the cutting velocity, feed and depth of cut are 120 m/min, 0.2 mm/rev and 2 mm respectively. The main cutting force in N is
(a) 40 (b) 80
(c) 400 (d) 800

18. In an orthogonal cutting text, the cutting force and thrust force were observed to the 1000 N and 500 N respectively. If the rake angle of tool is zero, the coefficient of friction in chip-tool interface will be:

(a) $\frac{1}{2}$ (b) 2

(c) $\frac{1}{\sqrt{2}}$ (d) $\sqrt{2}$

19. Power consumption in metal cutting is mainly due to:

(a) tangential component of the force (b) longitudinal component of the force

(c) normal component of the force (d) friction at the metal-tool interface

20. Two tool P and Q have signature 8-5°-5°-6°-8°-30°-0° and 5°-5°-7°-7°-8°- 15°-0°(both ASA) respectively. They are used to turn components under the same machining conditions. If h_P and h_Q denote the peak-to-valley heights of surfaces produced by the tool P and Q, the ratio h_P/h_Q will be

(a) $\frac{\tan 8^0 + \cot 15^0}{\tan 8^0 + \cot 30^0}$ (b) $\frac{\tan 15^0 + \cot 8^0}{\tan 30^0 + \cot 8^0}$

(c) $\frac{\tan 15^0 + \cot 7^0}{\tan 30^0 + \cot 7^0}$ (d) $\frac{\tan 7^0 + \cot 15^0}{\tan 7^0 + \cot 30^0}$

21. Consider the following statements: The strength of a single point cutting tool depends upon:

1. Rake angle
2. Clearance angle
3. Lip angle

Which of these statements are correct?

(a) 1 and 3 (b) 2 and 3

(c) 1 and 2 (d) 1, 2 and 3

22. The angle of inclination of the rake face with respect to the tool base measured in a plane perpendicular to the base and parallel to the width of the tool is called:

(a) back rake angle (b) side rake angle

(c) side cutting edge angle (d) end cutting edge angle

23. In orthogonal cutting test, the cutting force=900N, the thrust force=600N and chip shear angle is 30°. Then the chip shear force is

(a) 1079.4 N (b) 969.9 N

(c) 479.4 N (d) 69.6 N

24. The time taken to face a work piece of 72mm length, if the spindle speed is 80rpm and cross-feed is 0.3 mm/rev, is

(a) 1.5 minutes (b) 3.0 minutes

(c) 5.4 minutes (d) 8.5 minutes

25. In a machining operation chip thickness ratio is 0.3 and back rake angle of the tool is 10°. What is the value of shear strain?

(a) 2.24 (b) 0.86

(c) 3.1 (d) 3.34

26. The percentage of total energy dissipated due to friction at the tool-chip interface is

(a) 30% (b) 42%

(c) 58% (d) 70%

27. A medium carbon steel work piece is turned on a lathe at 50m/min cutting speed, 0.8 mm/rev feed and 1.5 mm depth of cut. What is the rate of metal removal?
(a) 1000 mm^3/min
(b) 60000 mm^3/min
(c) 20000 mm^3/min
(d) can not be calculated with the given data.

28. In the selection of optimal cutting conditions, the requirement of surface finish would put a limit on which of the following?
(a) the maximum feed (b) the maximum depth of cut
(c) the maximum speed (d) the maximum number of passes

29. For increasing the material removal rate in turning, without any constraints, what is the right sequence to adjust the cutting parameters?
1. Speed 2. Feed
3. Depth of cut
Select the correct answer using the code given below:
(a) 1-2-3 (b) 2-3-1
(c) 3-2-1 (d) 1-3-2

30. During machining, excess metal is removed in the form of chip as in the case of turning on a lathe. Which of the following are correct? Continuous ribbon like chip is formed when turning
1. at a higher cutting speed 2. at a lower cutting speed
3. a brittle material 4. a ductile material
Select the correct answer using the codes given below?
(a) 1 and 3 (b) 1 and 4
(c) 2 and 3 (d) 2 and 4

31. Consider the following statements
In an orthogonal cutting the cutting ratio is found to be 0.75. the cutting speed is 60m/min and depth of cut 2.4mm. Considering rake angle to be zero.
1. chip velocity will be 45m/min 2. chip velocity will be 80m/min
3. chip thickness will be 1.8mm 4. chip thickness will be 3.2mm
Which of the following are correct?
(a) 1 and 3 (b) 1 and 4
(c) 2 and 3 (d) 2 and 4

32. A cylinder is turned on lathe with orthogonal machining principle. Spindle rotates at 200 rpm. The axial feed rate is 0.25mm per revolution, depth of cut is 0.4 mm. The rake angle is 10°. In the analysis it is found that the shear angle is 27.75°. The thickness of the produced chip is
(a) 0.511 (b) 0.528
(c) 0.818 (d) 0.846

33. In the above problem, the coefficient of friction at the chip total interface obtained using Earnest and merchant theory is
(a) 0.18 (b) 0.36
(c) 0.71 (d) 0.98

34. Consider the following statements
Assertion (A): In a Swiss –type automatic lathe, the turret is given longitudinal feed for each tool in a specific order with suitable indexing.
Reason (R): A turret is a multiple tool holder to facilitate machining with each tool by indexing without the need to change the tools.
(a) Both A and R are individually true and R is the correct explanation of A.
(b) Both A and R are individually true. but R is not the correct explanation of A.

(c) A is true but R is false.
(d) A is false but R is true.

35. The cutting tool material normally used for tunning steel of very high hardness is
(a) HSS (b) Tungsten carbide
(c) CBN (d) Diamond

36. A 400 mm long shaft has a 100 mm tapered step at the middle ware with 4° included angle. The tailstock offset required to produce this taper on a lathe would be:
(a) 400 sin 4° (b) 400 sin 2°
(c) 100 sin 4° (d) 100 sin 2°

37. Apart from hexagonal turret, the element(s) in a turret lathe include(s)
(a) cross-slide tool post (b) cross-slide tool post rear tool post
(c) cross-slide tool post and tail stock (d) Rear tool post and tail stock

38. Which one of the following sets of forces are encountered by a lathe parting tool while groove cutting?
(a) Tangential, radial and axial (b) Tangential and radial
(c) Tangential and axial (d) Radial and axial

39. The spindle speed range in a general-purpose lathe is divided into steps which approximately follow:
(a) Arithmetic progression (b) Geometric progression
(c) Harmonic progression (d) Logarithmic progression

40. Consider the following statements associated with the lathe accessories:
1. Steady rest is used for supporting a long job in between head stock tail stock
2. Mandrel is used for turning small cylindrical Job.
3. Collets are used for turning Disc-shaped job of these statements .
(a) 1 and 2 are correct (b) 2 and 3are correct
(c) 3 alone is correct (d) 1 alone is correct

41. The indexing of the turret in a single-spindle lathe is done using:
(a) Geneva mechanism (b) Ratchet and pawl mechanism
(c) Rack and pinion mechanism (d) Whit worth mechanism

42. The percentage of carbon in gray cast iron is in the range of
(a) 0.25 to 0.75 percent (b) 1.25 to 1.75 percent
(c) 3 to 4 percent (d) 8 to 10 percent

43. In a machine tool gear box, the smallest and largest spindles are 100rpm and 1120 rpm respectively. If there are 8 speeds in all, the forth speed will be:
(a) 400 rpm (b) 280 rpm
(c) 800 rpm (d) 535 rpm

44. Machine tool manufacturers prefer grey cast-iron grade 40 for producing machine columns and tables because grey cast-iron is:
1. Heavy
2. Easily castable
3. Easily weldable
4. Having good damping capacity
Select the correct answer using the code below:
(a) 1 and 2 (b) 2 and 4
(c) 1 and 3 (d) 3 and 4

45. The figure shows an incomplete schematic of a conventional lathe to be used for cutting threads with different pitches. The speed gear box U_v is shown and the feed gear box U_s is to be placed. P,Q,R and S denotes locations and have no other significance. Changes in U_v should NOT affect the pitch of the thread being cut and changes in U_s should not affect the cutting speed.

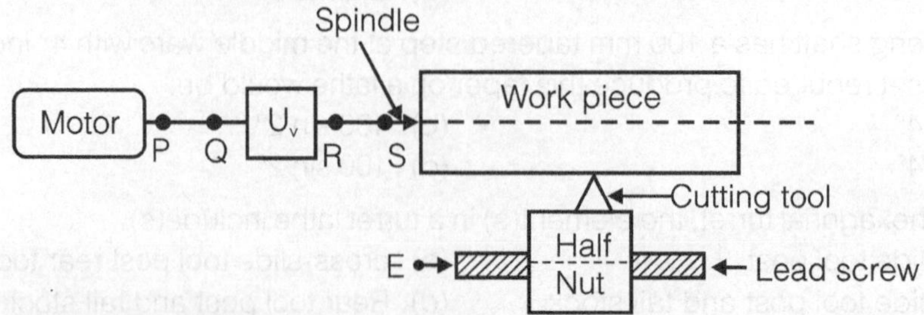

The correct connections and the correct placement of U_s are given by

(a) Q and E are connected U_s is placed between P and Q

(b) S and E are connected, U_s is placed between R and S

(c) Q and E are connected, U_s is placed between S and E

(d) S and E are connected, $U\text{-}_s$ is placed between S and E

46. For taper turning on centre lathes, the method of swiveling the compound rest is preferred for:

(a) Long jobs with small taper angles
(b) Long jobs with steep taper angles
(c) Short jobs with small taper angles
(d) Short jobs with steep taper angles

47. It is required to cut screws threads with double start and 2mm pitch on a lathe having lead screw pitch of 6mm. what is the speed ratio between lathe and screw?

(a) 1:3
(b) 3:1
(c) 2:3
(d) 3:2

48. Assertion (A): in a multi-spindle automatic lathe, the turret tool holder is indexed to engage the cutting tools by one by one for successive machining operations

Reason (R): Turret is a multiple tool holder so that for successive machining operation, the tools need not be changed.

(a) Both A and R are individually true and R is the correct explanation of A.

(b) Both A and R are individually true and R not is the correct explanation of A.

(c) A is true but R is false.

(d) A is false R is true.

49. Assertion (A): A chip breaker is employed to improve the machinability of brass

Reason (R): A chip breaker facilitates the curling and breakage of long continuous chips.

(a) Both A and R are individually true and R is the correct explanation of A.

(b) Both A and R are individually true but R is not the correct explanation of A.

(c) A is true but R is false.

(d) A is false but R is true.

50. Consider the following work materials:

1. titanium
2. mild steel
3. stainless steel
4. Grey cast iron

The correct sequence of these materials in terms of increasing order of difficulty in machining is:

(a) 4,2,3,1
(b) 4,2,1,3
(c) 2,4,3,1
(d) 2,4,1,3

51. Consider the following statement: During third stage of tool-wear, rapid deterioration of tool edge takes place because

1. Flank wear is only marginal.
2. Flank wear is large.
3. Temperature of tool increase gradually.
4. Temperature of tool increase drastically.

(a) 1 and 3 (b) 2 and 4
(c) 1 and 4 (d) 2 and 3

52. The type of wear that occurs due to cutting action of the particles in the cutting fluid is referred to as

(a) adhesion wear (b) diffusion wear
(c) erosive wear (d) corrosive wear

53. Flank ware occurs mainly on which of the following

(a) Nose part and top face
(b) Cutting edge only
(c) Nose part, front relief face, and side relief face of cutting tool
(d) Face of currying tool at a short distance from the cutting edge

54. Assertion (A): Non-ferrous materials are best machined with diamond tools.
Reason (R): Diamond tools are suitable for high speed machining.

(a) Both A and R are individually true and R is the correct explanation of A.
(b) Both A and R are individually true but R is not the correct explanation of A.
(c) A is true but R is false.
(d) A is false but R is true.

55. Assertion (A): The characteristics feature of high speed steel is its red hardness.
Reason (R): Chromium and cobalt in High Speed Steel promote martensite formation when the tools is cold worked

(a) Both A and R are individually true and R is the correct explanation of A.
(b) Both A and R are individually true but R is not the correct explanation of A.
(c) A is true but R is false.
(d) A is false but R is true.

56. The straight grades of cemented carbide cutting tool materials contain:

(a) tungsten carbide only (b) tungsten carbide and titanium carbide
(c) tungsten carbide and cobalt (d) tungsten carbide and coal carbide

57. Assertion (A): During cutting, the diamond is kept flooded with coolant.
Reason (R): The oxidation of diamond starts at about 450°C

(a) Both A and R are individually true and R is the correct explanation of A.
(b) Both A and R are individually true but R is not the correct explanation of A.
(c) A is true but R is false.
(d) A is false but R is true.

58. Crater Wear is predominant in:

(a) carbon steel tools (b) tungsten carbide tools
(c) high speed steel tools (d) ceramic tools

59. A machinist desires to turn a round steel stock of outside diameter 100 mm at 1000 rpm. The material has tensile strength of 75 kg/mm^2. The depth of the cut chosen is 3mm at a feed rate of 0.3 mm/rev. Which one of the following tool materials will be suitable for machining the component under the specified cutting conditions?

(a) Sintered carbides (b) Ceramic
(c) HSS (d) diamond

60. Assertion (A): Cutting tools made of high carbon steel have shorter tool life.
Reason (R): During machining, the top of the cutting tool is heated to 600/700°C which causes the tool tip to lose its hardness.
(a) Both A and R are individually true and R is the correct explanation of A.
(b) Both A and R are individually true but R is not the correct explanation of A.
(c) A is true but R is false.
(d) A is false but R is true.

61. Consider the following statements:
Assertion (A): Ceramic tools, are used for light, smooth and continuous cuts at high speeds.
Reason (R): Ceramics have high wear resistance and high temperature resistance.
(a) Both A and R are individually t true and R is the correct explanation of A.
(b) Both A and R are individually true but R is not the correct explanation of A.
(c) A is true but R is false.
(d) A is false but R is true.

62. Match **List-I** (ISO classification of carbide tools) with **List-II** (Applications) and select the correct answer using the codes given below the lists:

List-I	**List-II**
A. P-10	**1.** Non- ferrous, roughing cut
B. P-50	**2.** Non-ferrous, finishing cut
C. K-10	**3.** Ferrous material, roughing cut
D. K-50	**4.** Ferrous material, finishing cut

Codes:

	A	B	C	D
(a)	4	3	2	1
(b)	3	4	2	1
(c)	4	3	2	1
(d)	3	4	1	2

63. Assertion (A): Cemented carbide Tool tips produced by powder metallurgy.
Reason (R): Carbides cannot be melted and cast.
(a) Both A and R are individually true and R is the correct explanation of A.
(b) Both A and R are individually true but R is not the correct explanation of A.
(c) A is true but R is false.
(d) A is false but R is true.

64. Consider the following statements: for precision machining of non-ferrous alloys, diamond is preferred because it has:
1. low coefficient of thermal expansion
2. high wear resistance
3. high compression strength
4. low fracture toughness

Which of these statements are correct

(a) 1 and 2 (b) 1 and 4
(c) 2 and 3 (d) 3 and 4

65. Consider the following tool materials:

1. HSS
2. Cemented carbide
2. Ceramics
4. Diamond

The correct sequence of these materials in the descending order of their cutting speeds is:

(a) 4, 3, 1, 2 (b) 4, 3, 2, 1
(c) 3, 4, 2, 1 (d) 3, 4, 1, 2

66. Consider the following statements:

Assertion (A): Negative rake is usually provided on carbide tipped tools.

Reason (R): Carbide tools are weaker in compression.

(a) Both A and R are individually true and R is the correct explanation of A.
(b) Both A and R are individually true but R is not the correct explanation of A.
(c) A is true but R is false.
(d) A is false but R is true.

67. Which one of the following is the hardest cutting tool material next only to Diamond?

(a) cemented carbides (b) ceramics
(c) silicon (d) cubic boron nitride

68. The correct sequence of cutting tools in ascending order of their wear resistance is:

(a) HSS-Cast nonferrous alloy (Stellite)-Carbide-Nitride
(b) Cast non-ferrous alloy (Stellite)-HSS-Carbide-Nitride
(c) HSS-Cast non-ferrous alloy (Stellite)-Nitride-Carbide
(d) Cast nonferrous alloy (Stellite)-Carbide-Nitride-HSS

69. Consider the following statements:

An increase in the cobalt content in the straight carbide grades of carbide tools:

1. Increases the hardness
2. Decreases the hardness
3. Increases the transverse rupture strength
4. Lowers the transverse rupture strength

Which of the statements given above are correct?

(a) 1 and 3 (b) 2 and 4
(c) 1 and 4 (d) 2 and 3

70. Consider the following statements:

Assertion (A): carbide tips are generally given negative rake angle.

Reason (R):carbide tips are made from very hard materials.

(a) Both A and R are individually true and R is the correct explanation of A.
(b) Both A and R are individually true but R is not the correct explanation of A.
(c) A is true but R is false.
(d) A is false but R is true.

71. Which of the following indicate better machinability?

1. smaller shear angle
2. higher cutting forces
3. longer tool life
4. better surface finish

(a) 1 and 3 (b) 2 and 4
(c) 1 and 2 (d) 3 and 4

72. Assertion (A): the machinability of a material can be measured as an absolute quantity.
Reason (R): machinability index indicates the ease with which a material can be machined.
(a) Both A and R are individually true and R is the correct explanation of A.
(b) Both A and R are individually true but R is not the correct explanation of A.
(c) A is true but R is false.
(d) A is false but R is true.

73. Poor machinability of centrifugally cast iron pipe is due to:
(a) Chilling (b) Segregation
(c) Dense structure (d) High mould rotation speed

74. The tool life curves for two tools A and B are shown in the figure and they follow the tool life equation $VT^n = C$. Consider the following original statements:
1. Value of n for both the tools is same
2. value of C for both the tools is same
3. value of C for tool A will be greater than that for the tool B.
4. value of C for tool B will be greater than that for the tool

A. which of these statements are correct

(a) 1 and 3 (b) 1 and 4

(c) 2 only (d) 4 only

75. What is the variation of cutting speed with tool life on Log-Log scale?
(a) Parabolic variation (b) Straight line variation
(c) Hyperbolic variation (d) Elliptical variation

76. Consider the following criteria in evaluating machinability:
1. Surface finish
2. Type of chips
3. Tool life
4. Power consumption

In modern high speed CNC machining with coated carbide tools. The correct sequence of these characteristics in DECREASING order of their importance is:
(a) 1, 2, 4, 3 (b) 2, 1, 4, 3
(c) 1, 2, 3, 4 (d) 2, 1, 3, 4

77. In a machining operation, doubling the cutting speed reduces the tool life to 1/8th of the original value. The exponent n in Taylor's tool life equation $VT^n = C$, is
(a) $\frac{1}{8}$ (b) $\frac{1}{4}$
(c) $\frac{1}{3}$ (d) $\frac{1}{2}$

78. In a single-point turning operation to steel with a cemented carbide tool, tailor's tool life exponent is 0.25. If the cutting speed is halved, the tool life will increase by
(a) Two times (b) Four times
(c) Eight times (d) Sixteen times

79. In a tool life test, doubling the cutting speed reduces the tool life to one fourth of the original. the Taylor's tool life index is:

(a) $\frac{1}{2}$ (b) $\frac{1}{3}$

(c) $\frac{1}{4}$ (d) $\frac{1}{4}$

80. Which of the following values of index n is associated with carbide tools when Taylor's tool life equation $V.T^n$ = constant is applied?

(a) 0.1 to 0.15 (b) 0.2 to 0.4

(c) 0.45 to 0.6 (d) 0.65 to 0.9

Statement for Linked Answer Question (81 and 82) :

A low carbon steel bar of 147mm diameter with uncoated carbide insert. The observed tool lives are 24min and 12min for cutting velocities of 90m/min and 120m/min respectively. The feed and depth of cut are 0.2mm/rev and 2mm respectively. Use the unmachined diameter to calculate the cutting velocity.

81. What tool life is 20 min, the cutting velocity in m/min is

(a) 87 (b) 97

(c) 107 (d) 114

82. Neglect over-travel are approach of the tool. when tool life is 20min, the machining time in min for a single pass is

(a) 5 (b) 10

(c) 15 (d) 20

83. Which of the following are machinability criteria?

1. Tool life
2. Cutting forces
3. Surface finish

Select the correct answer using the code given below:

(a) 1, 2 and 3 (b) 1 and 2 only

(c) 1 and 3 only (d) 2 and 3 only

84. What are the reasons for reduction of tool life in a machining operation?

1. Temperature raise of cutting edge
2. Chipping of tool edge due to mechanical impact
3. Gradual wear at tool point
4. Increase in feed of cut at constant cutting force

Select the correct answer using the code given below

(a) 1, 2 and 3 (b) 2, 3 and 4

(c) 1, 3 and 4 (d) 1, 2 and 4

85. Dry and compressed air is used as cutting fluid for machining:

(a) steel (b) aluminum

(c) cast iron (d) brass

86. Consider the following statements:

Assertion (A): in a slow speed high pressure metal cutting process such as gear cutting, chlorinated or sulphonated oils are used .

Reason (R): reduction of coefficient of fraction is achieved by the formation of a solid firm at the rubbing interface of tool and chip.
(a) Both A and R are individually true and R is the correct explanation of A.
(b) Both A and R are individually true and R not is the correct explanation of A.
(c) A is true but R is false.
(d) A is false R is true

87. Assertion (A): For a high-speed turning turning of magnesium alloys, the coolant or cutting fluid preferred is water-miscible mineral fatty oil.
Reason(R): As a rule, water based oils are recommended for high-speed operations in which high temperatures are generated due to high frictional heat. Water being good coolant, the heat dissipation is efficient.
(a) Both A and R are individually true and R is the correct explanation of A.
(b) Both A and R are individually true and R is not the correct explanation of A.
(c) A is true but R is false.
(d) A is false R is true

88. Assertion (A): Oil as a cutting fluid results in a lower coefficient of friction.
Reason (R): Oil forms a thin liquid film between the tool face and chip, and it provides 'hydrodynamic lubrication',
(a) Both A and R are individually true and R is the correct explanation of A.
(b) Both A and R are individually true and R is not the correct ex planation of A.
(c) A is true but R is false.
(d) A is false R is true

89. Which of the following statements are salient features of Hydrostatic extrusion?
1. It is suitable for soft and ductile Material.
2. It is suitable for high-strength Super-alloys.
3. The billet is inserted into the extrusion chamber and the pressure is applied by a ram to extrude the billet through the die.
4. The billet is inserted into the extrusion chamber where it is surrounded by a suitable liquid. The billet is extruded through the die by applying pressure to the liquid. Select the correct answer using the code given below

(a) 1 and 3 (b) 1 and 4
(c) 2 and 3 (d) 2 and 4

90. In turning, the ratio of the optimum cutting speed for minimum cost and optimum cutting speed for maximum rate of producton is always
(a) equal to 1 (b) in the range of 0.6 to 1
(c) in the ranges of 0.1 to 0.6 (d) greater than 1

91. Consider the following approaches normally applied for the economic analysis of machining:
1. Maximum productions rate
2. Maximum profit criterion
3. Minimum cost criterion

The correct sequence in ascending order of optimum cutting speed obtained by these approaches is:
(a) 1, 2, 3 (b) 1, 3, 2
(c) 3, 2, 1 (d) 3, 1, 2

92. Which one of the following does not affect the selection of optimum cutting speed for the minimum cost?
(a) Labour cost (b) Machine maintenance cost
(c) Tool cost (d) Job handling cost

93. The magnitude of the cutting speed for maximum profit rate must be
(a) in between the speeds for minimum cost and maximum production rate
(b) higher than the speed for maximum production rate
(c) below the speed for minimum cost
(d) equal to the speed for minimum cost.

94. In economics of machining, which one of the following costs remain constant?
(a) Machining cost per piece (b) Tool changing cost per piece
(c) Tool handling cost per piece (d) Tool cost per piece

95. A diagram related to machining economics with various cost components is given above. match **List-I** (Cost element) with **List-II** (Appropriate curve) and select the correct answer using the codes given below the lists:

List-I	List-II
A. Machining cost	**1.** Curve-1
B. Tool cost	**2.** Curve-2
C. Tool grinding costs	**3.** Curve-3
D. Non-productive cost	**4.** Curve-4
	5. Curve-5

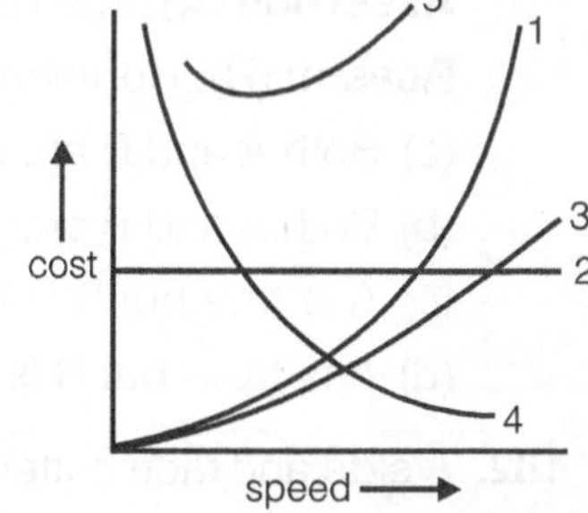

Codes:

	A	B	C	D
(a)	3	2	4	5
(b)	4	1	3	2
(c)	3	1	4	2
(d)	4	2	3	5

96. Consider the following statements:
1. As the cutting speed increases, the cost of production initially reduces and then after an optimum cutting speed, it increases.
2. As the cutting speed increases the cost of production also increases and after a critical value it reduces.
3. Higher feed rate for the same cutting speed increases the cost of production. Which of the following statements are correct?

(a) 1 and 3 (b) 2 and 3
(c) 1 and 4 (d) 3 only

97. In a mechanical shaper, the length of stroke is increased by:
(a) Increasing the centre distance of bull gear and crank pin.
(b) Decreasing the centre distance of bull gear and crank pin.
(c) Increasing the length of the ram.
(d) decreasing the length of the slot in the slotted lever.

98. In a shaping operation, the average cutting is(stroke lengths S, Number of strokes per minute N, Quick return ratio R)
(a) NSR (b) NSR/2
(c) NS(1+R) (d) NS(1+R)/2

99. Which of the following are the advantages of hydraulic shaper over a mechanically driven shaper?
1. More strokes per minute can be obtained at a given cutting speed.
2. The cutting stroke has a definite stopping point.
3. It is simpler in construction.
4. Cutting speed is constant through out most of cutting stroke.

(a) 1 and 2 (b) 1 and 4
(c) 2 and 4 (d) 1, 3 and 4

100. A 600 mm × 30 mm flat surface of plate is to be finish machined on a shaper. The plate has been fixed with the 600 mm side along the tool travel direction. If the tool over-travel at each end of the plate is 20 mm, average cutting speed is 8 m/min, feed rate is 0.3 mm/stroke and the ratio of return time to cutting time of the tool is 1:2, the time required for machining will be
(a) 8 min (b) 12 min
(c) 16 min (d) 20 min

101. Consider the following statements:
Assertion (A): internal gears are cut on a gear shaper.
Reason (R): Hobbing is not suitable for cutting internal gear.
(a) Both A and R are Individually true and R is the correct explanation of A.
(b) Both A and R are individually true but R is not the correct explanation of A.
(c) A is true but R is false.
(d) A is false but R is true.

102. A side and face cutter 125 mm diameter has 10 teeth. If operates at a cutting speed of 14 m/min with a table traverse 100 mm/min.the feed per tooth of the cutter is:
(a) 10 mm (b) 2.86 mm
(c) 0.286 mm (d) 0.8 mm

103. A mild steel plate 400 × 800 × 30 mm is to be shaped along its wider face. The ratio of return time to cutting time is 2:3 and the feed per cycle is 2 mm.tool approach and over–travel respectively are 50 mm latch and cutting speed of tool is 24 m/min.calculate the machining time required for machining the given plate with H.S.S tool:
(a) 5 min (b) 20 min
(c) 13 min (d) 17 min

104. The rake angle in a twist drill:
(a) varies from minimum near the dead centre to a maximum value at the periphery.
(b) is maximum at the dead centre and zero at the periphery.
(c) is constant at every point of the cutting edge.
(d) is a function of the size of the chisel edge.

105. A 31.8 mm H.S.S drill is used to drill a hole in a cast iron block 100 mm thick at a cutting speed 20 m/min and feed 0.3 mm/rev. if the over travel of drill is 4 mm and approach 9mm, the time required to drill the hole is:
(a) 1 min 40 s (b) 1 min 44 s
(c) 1 min 49 s (d) 1 min 53 s

106. Match **List-I** (Mechanism) with **List-II** (associated function) and select the correct answer using code given below the lists:

	List-I		**List-II**
A.	Geneva	**1.**	Feed motion in shape gearing
B.	Rachet and Pawl	**2.**	Feed motion in drilling
C.	Whitworth	**3.**	Indexing of turret
D.	Rack and shaper	**4.**	Quick return motion in Pinion

Codes:

	A	B	C	D
(a)	3	1	2	4
(b)	1	3	2	4
(c)	2	4	1	3
(d)	3	1	4	2

107. The arm of a radial drilling machine is being raised at a speed of 3.9 m/min by a single start square threads of 6 mm pitch and 30 mm diameter. The speed of a screw:
(a) is 650 rpm
(b) is 180 rpm
(c) is 130 rpm
(d) data insufficient

108. Through holes of 10mm diameter are to be drilled in a steel plate of 20 mm thickness. Drill spindle speed is 300 rpm, feed 0.2 mm/rev and drill point angle is 120°. Assuming drill over travel of 2 mm, the time for producing a hole will be
(a) 4 seconds
(b) 25 seconds
(c) 100 seconds
(d) 110 seconds

109. The purpose of helical grooves in a twist drill is to:
1. improve the stiffness
2. save a tool material
3. provide a space for chip removal
4. provide rake angle for the cutting edge

Select the correct answer using the codes given below
(a) 1 and 2
(b) 2 and 3
(c) 3 and 4
(d) 1 and 4

110. Calculate the machining time required for drilling 4 holes of 16 mm dia latch on a flange from the following data. Flange thickness = 30 mm, cutting speed = 22 m/min and feed = 0.2 mm/rev
(a) 1.9 min
(b) 1.47 min
(b) 2 min
(d) 2.3 min

111. Assertion (A): For drilling cast iron, the tool is provided with a point angle smaller than that required for a ductile material.
Reason (R): Smaller point angle results in lower rake angle
(a) Both A and R are individually true and R is the correct explanation of A.
(b) Both A and R are individually true but R is not the correct explanation of A.
(c) A is true but R is false.
(d) A is false but R is true.

112. Consider the following statements:
Assertion (A): Up milling or climb milling is commonly used for machining castings and forgings
Reason (R): Up milling can be done on universal milling machines.
(a) Both A and R individually true and R is the correct explanation for A.
(b) Both A and R individually true and R is not the correct explanation for A.

(c) A is true but R is false.
(d) A is false but R is true

113. Which of the following statements are correct in respect of up-milling and down-milling?

1. In up-milling the cutter rotates in a direction opposite to that of work piece travel whereas in down-milling the cutter rotates in a direction similar to that of work piece travel.
2. In down-milling chip will be thin at the beginning and increase to a maximum at the end of the cut and reverse will be the case of a chip formed by up-milling .
3. Down-milling is desirable with milling cutters having a high radial rake angle when compared to up-milling.
4. Down-milling forces the work-piece against the milling table to exert more pressure while up milling tends to lift the work piece from the table.

(a) 1, 3, 4 (b) 1, 2, 3
(c) 2, 3, 4 (d) 1, 2, 4

114. In a milling operation two side milling cutters are mounted with a desired distance between them so that both sides of work piece can be milled simultaneously. This setup is called
(a) gang milling (b) straddle milling
(c) string milling (d) side milling

115. A milling cutter of 70 mm diameter with 12 teeth is operating at a cutting speed of 22 m/min and a feed of 0.05 mm/tooth. The feed per minute is
(a) 110 mm/min (b) 35 mm/min
(c) 6 mm/min (d) 60 mm/min

116. Consider the following processes of gear manufacture:

1. Milling with form cutter
2. Rack type gear shaper (gear planer)
3. Rotary gear shaper (gear shaper)
4. Gear hobbing

The correct sequence of these processes in increasing order of accuracy of envolute profile of the gear is:
(a) 3, 2, 4, 1 (b) 2, 3, 4, 1
(c) 3, 2, 1, 4 (d) 2, 3, 1, 4

117. Quality screw threads are produced by
(a) thread milling
(b) thread chasing
(c) thread cutting with single point tool
(d) thread casting

118. Gear cutting on a milling machine using an involute profile cutter is a:
(a) gear forming process (b) gear generating process
(c) gear shaping process (d) highly accurate gear producing Process

119. Which of the following statements are true for face milling?

1. Face milling cutter is held on an arbor.
2. It has two rake angles-axial rake and radial rake.
3. The maximum chip thickness equals the feed per tooth.
4. The chip thickness varies from a minimum at the start of cut to a maximum at the end of cut.

Select the correct answer using the codes given below:

(a) 1 and 2 (b) 2 and 3

(c) 2 and 4 (4) 3 and 4

120. Which of the following is a characteristics of the climb milling(down milling) operation?

(a) The work piece is fed in the opposite direction.

(b) Forces are less.

(c) High rigidity of the machine tool is required.

(d) Chip thickness is maximum at the end of the cut.

121. Consider the following statements:

In up-milling process,

1. The cutter starts the cut from the machined surface and proceeds upwards.
2. the cutter starts the cut from the top surface and proceeds downwards
3. The job is fed in a direction opposite to that of cutter rotation.
4. The job is fed in the same direction as that of cutter rotation.

Of these statements:

(a) 1 & 3 are correct (b) 1 & 4 are correct

(c) 2 & 3 are correct (d) 2 & 4 are correct

122. Internal gears can be made by:

(a) hobbing

(b) gear shaping with rack cutter

(c) gear shaping with pinion cutter

(d) gang milling

123. Which of the following motions are needed for spur gear cutting with a hob?

1. Rotary motion of hob.
2. Linear axial reciprocatory motion of hob.
3. Rotary motion of gear blank.
4. Radial advancement of hob.

Select the correct answer using the codes given below:

(a) 1,2 and 3 (b) 1,3 and 4

(c) 1,2 and 4 (d) 2,3 and 4

124. A straight teeth slab milling cutter of 100mm diameter and 10 teeth rotating at 200 r.p.m is used to remove a layer of 3mm thickness from a steel bar. If the table feed is 400 mm/min, the feed per tooth in this operation will be:

(a) 0.2 mm (b) 0.4 mm

(c) 0.5 mm (d) 0.6 mm

125. A 60-teeth gear hobbed on a differential hobber with a two-start hob, the index change gear ratio is governed by which one of the following kinematic balance equations?

(a) 1 revolution of gear blank = 1/60 of hob revolutions

(b) 1 revol;ution of gear blank =1/60 of hob revolutions

(c) 1 revolution of hob = 2/60 of blank revolutions

(d) 1 revolution of hob = 1/60 of blank revolutions

126. Consider the following motions and setting in a Hobbing machine:

1. Hob rotation
2. Job rotation
3. Axial reciprocating job rotation.
4. Tilting of hob to its helix angle.

Which of these motions and setting in a hobbing machine are required to machine a spur gear?
(a) 1, 2 and 3 (b) 2, 3 and 4
(c) 1, 2 and 4 (d) 1, 3 and 4

127. One of the index plates of a milling machine dividing head has the following hole circles:
15 ; 16 ;17; 18; 19; 20
A gear wheel of 34 teeth has to be milled by simple indexing method. To machine each tooth, the index crank has to be rotated through
(a) 17 holes in the 20 hole circle
(b) 18 holes in the 20 hole circle
(c) 1 revolution and 3 holes in 17-hole circle
(d) 1 revolution and 2 holes in 18-hole circle

128. Which one of the following processes of gear manufacture results in best accuracy of the involute gear tooth profile?
(a) Milling (b) Hobbing
(c) Rotary gear shaper (d) Rack type gear shaper

129. Consider the following statements:
Assertion (A): Virtually all modern milling machines are capable of down-milling.
Reason(R): In down-milling the cutter tends to push the work along and lift it upward from the table. This action tends to eliminate any effect in looseness in the feed screw and nut of the milling machine table and results in smooth cut
(a) Both A and R individually true and R is the correct explanation for A.
(b) Both A and R individually true and R is not the correct explanation for A.
(c) A is true but R is false.
(d) A is false but R is true

130. Match **List-I** (Machine tools) with **List-II** (Machine tool parts) and select the correct answer using the correct answer using the codes given below the lists :

List-I	**List-II**
A. Lathe	**1.** Lead screw
B. Milling machine	**2.** Rocker arm
C. Shaper	**3.** Universal indexing
D. Drilling machine	**4.** Flute

Codes:

	A	B	C	D
(a)	4	2	3	1
(b)	1	3	2	4
(c)	4	3	2	1
(d)	1	2	3	4

131. A spur gear of 40 teeth is machined in a gear hobbing machine using double start hob cutter. The speed ratio between the hob and the blank is:
(a) 1 : 20 (b) 1 : 40
(c) 40 : 1 (d) 20 : 1

132. The operation of producing grooves around the periphery of a cylindrical or conical work piece is called
(a) Profile milling (b) Gang milling
(c) Saw milling (d) Helical milling

133. In a plain milling machine, the table can be moved:
(a) Longitudinally (b) Crosswise
(c) Vertically (d) All of these

134. Gear shaping is a process of manufacturing gears. Which one of the following principles is employed by it?
(a) From cutting with cutter
(b) Generating tooth from with a reciprocating cutter
(c) Generating tooth from by a rotating cutter
(d) Generating from with a reciprocating and revolving cutter.

135. Which one of the following Statements is correct?
In up-milling operation, the underformed chip thickness,
(a) is zero at the start of the cut and increases to a maximum value just before the tooth disengages the work place.
(b) increases to the maximum value at the center of the travel and decreases towards the end of the tooth engagement
(c) has a maximum value just after the cut is started and drops to zero at the end of the cut.
(d) Remains unchanged.

136. In helical milling, the ratio of the circumference ot the gear blank to the lead of the helix determines the
(a) proper speed to use
(b) proper feed and depth of cut required
(c) angle setting of the machine table
(d) gear ratio for table screw and dividing head

137. Gang milling is a:
(a) milling process for generating hexagonal surfaces
(b) process of cutting gears
(c) process in which two or more cutters are used simultaneously
(d) milling operation combined with turning

138. In the grinding wheel of A 60 G 7 B 23, B stands for
(a) resinoid bond (b) rubber bond
(c) shellac bond (d) silicate bond

139. Consider the following statements:
The set-up for internal centreless grinding consists of a regulating wheel, a pressure roll and support roll, between which the tubular workpiece is supported with the grinding wheel within the tube, wherein
1. the grinding wheel, workpiece and regulating wheel centres must line on one line.
2. the directions of rotation of workpiece and grinding wheel are same.
3. the directions of rotation or workpiece roll, support roll and regulating wheel are same.
4. the directions of rotation of grinding wheel and regulating wheel are same.
Which of these statements are correct?
(a) 1, 2 and 3 (b) 1, 3 and 4
(c) 2 and 3 (d) 3 and 4

140. Soft material cannot be economically grounded due to:
(a) the high temperature involved (b) frequent wheel clogging
(c) rapid wheel wear (d) low work piece stiffness.

141. Specific energy requirements in a grinding process are more than those in turning for the same metal removal rate because of the:
(a) specific pressures between wheel and work being high.
(b) size effect of the larger contact areas between wheel and work.
(c) high cutting velocities.
(d) high heat produced during grinding.

142. If each abrasive grain is viewed as cutting tool, then which if the following represents the cutting parameters in common griding operations?
(a) Large negative rake angle,low shear angle and high cutting speed
(b) Large positive rake angle,low shear angle and high cutting speed
(c) Large negative rake angle,high shear angle and low cutting speed
(d) Zero rake angle,high shear angle and high cutting speed.

143. A Grinding wheel of 150 mm diameter is rotating at 3000 r.p.m. The grinding speed is:
(a) 7.5π m/s (b) 15π m/s
(c) 45π m/s (d) 450π m/s

144. Which of the following pairs are correctly matched?
1. Drill press : Trepanning
2. Centreless grinding : through feeding
3. Capstan lathe : Ram type turrent
Select the correct answer using the codes given bellow:
(a) 1 and 2 (b) 2 and 3
(c) 1 and 3 (d) 1, 2 and 3

145. Grinding wheel is said to be loaded when the:
(a) Metal particles get embedded in the wheel surface blocking the interspaces between cutting grains.
(b) Bonding material comes on the surface and the wheel becomes blunt.
(c) Work piece being ground comes to a stop in cylindrical grinding.
(d) Grinding wheel stops because of very large depth of cut.

146. A 4 mm thick sheet is rolled with 300 mm diameter rolls to reduce thickness without any change in its width the friction coefficient at the work-roll interface is 0.1. The minimum possible thickness of the sheet that can be produced in a single pass is
(a) 1.0 mm (b) 1.5 mm
(c) 2.5 mm (d) 3.7 mm

147. Which one of the following pairs of parameters and effects is not correctly matched?
(a) large wheel diameter... reduced wheel wear
(b) large depth of cut... increased wheel wear
(c) large work diameter... increased wheel wear
(d) large wheel speed... reduced wheel wear

148. Consider the following reasons:
1. Grinding wheel is soft
2. RPM of grinding wheel is too low
3. Cut is very fine
4. An improper cutting fluid is used

A grinding wheel may become loaded due to reasons stated at

(a) 1 and 4 (b) 1 and 3
(c) 2 and 4 (d) 2 and 3

149. Assertion(A): the ratio of cutting force to thrust force is very high in grinding process as compared to other machinging processes.

Reason(R):Random orientation and effective negative rake angles of abrasive grains increase the cutting force and adversely affect the cutting action and promote rubbing action.

(a) Both A and R individually true and R is the correct explanation for A.
(b) Both A and R individually true and R is not the correct explanation for A.
(c) A is true but R is false.
(d) A is false but R is true

150. Match **List-I** with **List-II** and select the correct answer using the correct answer using the codes given below the lists :

List-I	List-II
A. Quick return	1. Lathe mechanism
B. Apron	2. Milling Mechanism machine
C. Indexing	3. Shaper mechanism
D. Regulating	4. Centerless wheel griding

Codes:

	A	B	C	D
(a)	3	2	1	4
(b)	2	3	4	1
(c)	4	2	3	1
(d)	3	1	2	4

151. Consider the following statements in respect of a grinding wheel of specifications, 51-A-36-L-7-R-23,using the standard alphanumeric codification:

1. Abrasive used in the wheel is aluminium oxide
2. The grain size of abrasive is medium
3. It has an open structure
4. It has resiniod as bonding agent

Which of these statements are correct?

(a) 1, 2 and 3 (b) 1, 3 and 4
(c) 2, 3 and 5 (d) 1, 2 and 4

152. Which is not related term of grinding:

(a) Truing (b) Dressing
(c) Buffing (d) Glazing

153. Which of the following materials are used in Grinding wheel?

1. Alumunium oxide
2. Cubic boron niride
3. Silicon carbide

Select the correct answer using the codes given below:

(a) 1, 2 and 3 (b) 1 and 2
(c) 2 and 3 (d) 1 and 3

154. Match **List-I** (Machine tool/cutting tool) with **List-II** (Characteristics) and select the correct answer using the codes given below the lists:

	List-I		List-II
A.	Screw cutting lathe	1.	Self locking taper
B.	Drill	2.	Chasing dial
C.	End mill	3.	Wiper insert
D.	Grinding wheel	4.	Self relief taper
		5.	Balance weights

Codes:

	A	B	C	D
(a)	4	5	3	2
(b)	2	1	4	5
(c)	4	1	3	5
(d)	2	5	4	1

155. Which one of the following is not a synthetic abrasive material?
(a) Silicon Carbide (b) Aluminium Oxide
(c) Titanium Nitirde (d) Cubic Boron Nitiride.

156. Which one of the following operation is carried out at a minimum cutting velocity if the machines are equally rigid and the tool work material are the same:
(a) Turning (b) Grinding
(c) Milling (d) Boring

157. Consider the following statements in respect of a grinding?
1. The pitch of the grit cutting edges is larger than the pitch of the milling cutter.
2. The cutting angles of the grits have a random geometry
3. The size of the chip cuts is very small for griding

Which of these statements are correct?
(a) 1 and 2 (b) 2 and 3
(c) 1 and 3 (d) 1, 2 and 3

158. Which one of the following is a single point cutting tool?
(a) Hacksaw blade (b) Milling cutter
(c) Grinding wheel (d) Microcomputer

159. In designing the flutes of multiple-edged tools such as broaches and milling cutters, which of the factors are taken into consideration?
1. The work material
2. The size of the tool
3. The degree to which the chip curls
4. The rake angle on the tool

Select the correct answer using the codes given below:
(a) 1, 2 and 3 (b) 1, 3 and 4
(c) 1, 2 and 4 (d) 2, 3 and 4

160. Match **List-I** with **List-II** and select the corect answer using the codes given below the lists:

List - I

A. Reaming
B. Counter-boring
C. Counter-sinking
D. Spot facing

List-II

1. Smoothing and squaring surface around the hole for proper seating
2. Sizing and fishing the hole
3. Enlarging the end of the hole
4. Making a conical enlargement at the end of the hole

Codes:

	A	B	C	D
(a)	3	2	4	1
(b)	2	3	1	4
(c)	3	2	1	4
(d)	2	3	4	1

161. For reaming operation of a blind hole, the type of reamer required is
(a) straight flute reamer (b) right hand spiral fluted reamer
(c) left hand spiral fluted reamer (d) none of the above.

162. Match **List-I** (Drill bits) with **List-II** (Applications) and select the correct answer using the codes given below the

List-I

A. Core drill head
B. Reamer
C. Counter bore drill
D. Tap drill

List-II

1. To enlarge a hole to a certain depth so as to accommodate the bolt head screw
2. To drill & enlarge an already existing hole in a casting
3. To drill hole before making internal thread
4. To improve the surface finish and dimensions accuracy of the already drilled hole.

Codes:

	A	B	C	D
(a)	1	3	2	4
(b)	2	3	1	4
(c)	2	4	1	3
(d)	3	2	4	1

163. Consider the following statements regarding reaming process:
1. Reaming generally produces a hole larger than its own diameter
2. Generally rake angels are not provided on reamers
3. Even number of teeth are preffered in reamer desing.

Which of these statements are correct?
(a) 1 and 2 (b) 2 and 3
(c) 1 and 3 (d) 1, 2 and 3

164. Which one of the following sets of tool or tools and processes are normally for making large diameter holes?
(a) Boring tool
(b) BTA tool(Boring and Trepanning Association) and gun drill
(c) Gun drill and boring drill
(d) Boring tool and trepanning

165. Consider the following statements:

Assertion (A): No separate feed motion is required during broaching.

Reason (R): the broaching machines are Generally hydraulically operated.

(a) Both A and R individually true and R is the correct explanation for A.
(b) Both A and R individually true and R is not the correct explanation for A.
(c) A is true but R is false.
(d) A is false but R is true.

166. Match **List-I** (Machine tool) with **List-II** (Features) and select the corect answer using the codes given below the lists:

List-I	**List-II**
A. Lathe	**1.** Push or pull tool
B. Drilling machine	**2.** Ratchet & pawl mechanism
C. Shaper	**3.** Dividing head
D. Braoching machine	**4.** Hollow tapered spindle
	5. Face plate

Codes:

	A	B	C	D
(a)	2	4	5	1
(b)	5	3	2	4
(c)	2	3	5	4
(d)	5	4	2	1

167. Spline-holes are machined by

(a) Milling
(b) Boring
(c) Drilling
(d) Broaching

168. Among the following machining processes, which can be used for machinging flat surfaces?

1. Shaping
2. Milling
3. Broaching

Select the correct answer using the codes given below:

(a) 1 and 2
(b) 1 and 3
(c) 2 and 3
(d) 1, 2 and 3

169. Match **List-I** (Cutting tools) with **List-II** (Applications) and select the corect answer using the codes given below the lists:

List-I	**List-II**
A. Trepanning tool	**1.** For surface finishing by honing
B. Side milling cutter	**2.** For machining gears
C. Hob cutter	**3.** For cutting keyways in shafts
D. Abrasive sticks	**4.** For drilling large diameter holes

Codes:

	A	B	C	D
(a)	1	3	2	4
(b)	4	3	2	1
(c)	1	2	3	4
(d)	4	2	3	1

170. Which one of the following processing sequences will give the best accuracy as well as surface finish?
(a) Drilling, Reaming and Grinding
(b) Drilling, Boring and Grinding
(c) Drilling, Reaming and Lapping
(d) Drilling, Reaming and Electro-plating

171. Honing process gives surface finish of what order in CLA
(a) 10 μm (b) 1.0 μm
(c) 0.1 μm (d) 0.01μm

172. Given that
S = feed in mm/rev.and
R = nose radius in mm,
The maximum height of surface roughness $H\text{-}_{max}$ produced by a single-point turning tool is given by
(a) $S^2/2R$ (b) $S^2/4R$
(c) $S^2/6R$ (d) $S^2/8R$

173. Consider the following statements in relation to the unconventional machining processes:
1. Different forms of energy directly applied to work piece to have shape transformation of material removal from work surface
2. Relative motion between the work anmt the total is essential
3. Cutting too is not in physical contact with work piece.

Select the correct answer using the codes given below:
(a) 1 and 2 (b) 1,2 and 3
(c) 2 and 3 (d) 1 and 3

174. Consider the following statements:
Assertion (A): Water jet machining uses high pressure and high velocity water stream which acts like a saw and cuts a narrow groove in the material
Reason (R): The force required for cutting is generated in the momentum of the water stream.
Codes:
(a) Both A and R individually true and R is the correct explanation for A.
(b) Both A and R individually true and R is not the correct explanation for A.
(c) A is true but R is false.
(d) A is false but R is true.

175. During ultrasonic machining. The metal removal is achieved by:
(a) high frequency eddy currents
(b) high frequency sound waves
(c) hammering action of abrasive paricles
(d) rubbing action between tool and workpiece

176. Consider the following machining disadvantages:
1. Low metal finish 2. Poor surface cost
3. High tooling cost

Which of these disadvantages are associated with ultrasonic machine(USM)?
(a) 1 and 3 (b) 1 and 2
(c) 2 and 3 (d) 1, 2 and 3

177. Reaming is primarily used for achieving
(a) Higher MRR
(b) Improved dimensional tolerance
(c) Fine surface finish
(d) Improved positional tolerance

178. Consider the following statements:
In electrochemical grinding,
1. A rubber bounded alumina grinding wheel acts as the cathode and the workpiece as the anode.
2. A copper bounded alumina grinding wheel acts as the cathode and the worksiece as the anode.
3. Metal removal takes place due to the pressure applied by the grinding wheel.
4. Metal removal tales place due to electrolysis.

Select the correct answer using the codes given below:
(a) 1 and 3
(b) 2 and 4
(c) 2 and 3
(d) 1 and 4

179. Consider the following statements:
Assertion (A): In ECM, the shape of the cavity is the mirror image of the tool, but unlike EDM, the tool wear in ECM is less.
Reason (R): The tool in ECM is a cathode
Codes:
(a) Both A and R individually true and R is the correct explanation for A.
(b) Both A and R individually true and R is not the correct explanation for A.
(c) A is true but R is false.
(d) A is false but R is true.

180. A researcher conducts electro chemical machining(ECM) on a binary alloy(density 6000 kg/m^3) of iron (atomic weight 56, valency 2) and metal P (atomic weight 24, valency 4). Faraday's constant = 96500 coulomb/mole. Volumetric material removal rate of the alloy is 50 mm^3/s at the current of 2000 A. The percentage of the metal P in the alloy is closest to
(a) 40
(b) 25
(c) 15
(d) 79

181. Which one of the following processes does not cause tool wear?
(a) ultrasonic machining
(b) electrochemical machining
(c) electric discharge machining
(d) anode mechanical machining

182. In electrodischarge machining(EDM), if the thermal conductivity of tools is high and the specific heat of work piece is low, then the tool wear rate and material removal rate are expected to be respectively
(a) high and high
(b) low and low
(c) high and low
(d) low and high

183. Which statements is true for E.D.M
1. M.R.R increases with decreasing resistance
2. M.R.R increases with decreasing capacitance
3. M.R.R increases with increasing capacitance
4. M.R.R increases with optimum spark gap then decreased with increasing spark gap.

(a) 1, 2, 3 and 4
(b) 1, 2 and 3
(c) 1, 3 and 4
(d) 1, 2 and 4

184. As tool and work are not in contact in EDM process
(a) no relative motion occurs between them
(b) no water of tool occurs
(c) no power is consumed during metal cutting
(d) no force between and work occurs.

185. Match **List-I** (Unconvential machining process) with **List-II** (Basic process) and select the corect answer using the codes given below the lists:

	List-I		**List-II**
A.	Electro polishing	**1.**	Thermal
B.	Electrochemical machining	**2.**	Mechanical
C.	Abrasive jet machining	**3.**	Electro-mechanical
D.	Electrical discharge machining	**4.**	Chemical

Codes:

	A	B	C	D
(a)	4	3	2	1
(b)	2	1	4	3
(c)	4	1	2	3
(d)	2	3	4	1

186. Match **List-I** (Machining process) with **List-II** (Associated medium) and select the corect answer using the codes given below the lists:

	List-I		**List-II**
A.	Ultrasonic machining	**1.**	Kerosene
B.	EDM	**2.**	Abrasive slurry
C.	ECM	**3.**	Vacuum
D.	EBM	**4.**	salt solution

Codes:

	A	B	C	D
(a)	2	3	4	1
(b)	2	1	4	3
(c)	4	1	2	3
(d)	2	3	4	1

187. Match **List-I** with **List-II** and select the corect answer using the codes given below the lists:

	List-I		**List-II**
A.	Die sinking	**1.**	Abrasive jet machining
B.	Deburring	**2.**	Laser beam machining
C.	Fine hole drilling	**3.**	EDM
D.	Cutting/sharpening hard materials	**4.**	Ultrasonic machining
		5.	Electrochemical grinding

Codes:

	A	B	C	D
(a)	3	5	4	1
(b)	2	4	1	3
(c)	3	1	2	5
(d)	4	5	1	3

188. Match **List-I** (Machining processes) with **List-II** (Application) and select the corect answer using the codes given below the lists:

List-I	**List-II**
A. EDM	**1.** Holes and cavities in hard and brittle materials.
B. LBM	**2.** Micro-drilling and micro-welding of materials
C. USM	**3.** Shaping of hard metals or reshaping of cemented carbide tools
D. ECM	**4.** Shaping of cemented carbide dies and punches

Codes:

	A	B	C	D
(a)	4	1	2	3
(b)	3	2	1	4
(c)	4	2	1	3
(d)	3	1	2	4

189. Match **List-I** (Materials) with **List-II** (Machining) and select the corect answer using the codes given below the lists:

List-I	**List-II**
A. Machining of conducting materials	**1.** ECM
B. Ruby rod	**2.** EDM
C. Electrolyte	**3.** USM
D. Abrasive slurry	**4.** LBM

Codes:

	A	B	C	D
(a)	4	2	1	3
(b)	4	2	3	1
(c)	2	4	3	1
(d)	2	4	1	3

190. Typical machining operations are to be performed on hard-to-machine materials by using the process combinations

Operations	**Processes**
P. Deburring (internal suface)	**1.** Plasma Arc machining
Q. Die siniking	**2.** Abrasive flow machining
R. Fine hole drilling in thin sheets	**3.** Electric discharge machining
S. Tool sharpening	**4.** Ultrasonic machining
	5. Laser beam machining
	6. Electrochemical machining

(a) P-1, Q-5, R-3, S-4
(b) P-1, Q-4, R-1, S-2
(c) P-5, Q-1, R-2, S-6
(d) P-2, Q-3, R-5, S-6

191. Which of the following components can be manufactured by power metallurgy methods?

1. Carbide tool tips
2. Bearings
3. Filters
4. Brake linings

Select the correct answer using the codes given below:

(a) 1, 3 and 4
(b) 2 and 3
(c) 1, 2 and 4
(d) 1, 2, 3 and 4

192. Which of the following parts can be made economically by powder metallurgy?

1. Gears
2. Automobile cylinders

3. Filters
4. Gun barrels

Select the correct answer using the codes given below:

(a) 1 and 2 (b) 1 and 3
(c) 2 and 3 (d) 2 and 4

193. In powder metallurgy,the operation carried out to improve the bearing proprety of a bush is called :

(a) Infiltration (b) Impregnation
(c) Plating (d) Heat treatment

194. Asserion(A): close dimensional tolerances are NOT possible with isostatic pressing of metal powder in powder metallurgy technique.

Reason(R): In the process os isostaic pressing, the pressure is equal in all directions which permits uniform density of the metal powder.

(a) Both A and R individually true and R is the correct explanation for A.
(b) Both A and R individually true and R is not the correct explanation for A.
(c) A is true but R is false.
(d) A is false but R is true.

195. Consider the following processes for the manufacture of gears:

1. Casting
2. Powder metallurgy
3. Machining from bar stock
4. Closed die forging

The correct sequence in increasing order of bending strength of gear teeth is:

(a) 1, 2, 3, 4 (b) 1, 2, 4, 3
(c) 2, 1, 4, 3 (d) 2, 1, 3, 4

196. Assertion(A): Mechanical Disintegration of a molten metal stream into fine particles by means of a jet of compressed air, is known as atomization.

Reason(R): In atomization process inert-gas or water cannot be used as a substitute for compressed air.

(a) Both A and R individually true and R is the correct explanation for A.
(b) Both A and R individually true and R is not the correct explanation for A.
(c) A is true but R is false.
(d) A is false but R is true.

197. Consider the following statements

Assertion(A): In automization process of manufacture of metal powder the molten metal is forced through a small orifice and broken up by a stream of compressed air.

Reason(R): The metallic power obtained by automization process is quite resistant to oxidation

(a) Both A and R individually true and R is the correct explanation for A.
(b) Both A and R individually true and R is not the correct explanation for A.
(c) A is true but R is false.
(d) A is false but R is true.

198. Consider the following factors:

1. Size and shape that can be produced economically
2. Porosity of parts produced.
3. Available press capacity
4. High density

Which of the following limitations of powder metallurgy

(a) 1, 3 and 4 (b) 2 and 3

(c) 1, 2 and 3 (d) 1 and 2

199. Which of the following are the limitations of metallurgy?

1. High tooling and equipment costs
2. Wastage of material
3. It cannot be automated
4. Expensive metallic powders

Select the correct answer

(a) 1 and 2 (b) 3 and 4

(c) 1 and 4 (d) 1, 2 and 4

200. What are the advantages of powder metallurgy?

1. Extreme purity product
2. Low labour cost
3. Low equipment cost.

Select the correct answer

(a) 1, 2 and 3 (b) 1 and 2

(c) 2 and 3 (d) 1 and 3

201. Match **List-I** (Task) with **List-II** (Recommendation) and select the corect answer using the codes given below the lists:

List-I

A. Three components in a straight line should worked in one loading

B. Unloading of clamp element from jig is essential

C. Clamping of rough surface

D. Need for heavy clamping force

List-II

1. Clamp with floating pad

2. Quick action nut

3. Cam clamp

4. Equalizing clamp

5. Strap clamp

Codes:

	A	B	C	D
(a)	5	2	3	4
(b)	4	2	1	5
(c)	1	4	2	3
(d)	4	1	5	3

202. Match **List-I** (Part) with **List-II** (Method of holding) and select the corect answer using the codes given below the lists:

List-I	List-II
A. Cylindrical parts	**1.** Circular face plate
B. Non-cylindrical parts	**2.** Collet
C. Very complicated shapes	**3.** Three jaw chuck
D. Material in bar form on small lathes	**4.** Four jaw chuck

Codes:

	A	B	C	D
(a)	3	1	4	2
(b)	2	4	1	3
(c)	3	4	1	2
(d)	2	1	4	3

203. If the diameter of the hole is subject to considerable variation, then for locating in jigs and fixtures, the type of locater used is:
(a) conical locater (b) cyclindrical locater
(c) diamond pin locater (d) vee locater

204. Compound die performs:
(a) two or more operations at one station in one stroke
(b) two or more operations at different stations in one stroke
(c) high frequency sound waves
(d) high frequency eddy currents

205. Consider the following statements:
The cutter setting block in a milling fixture:
1. Limits the total travel required by the cutter during machining
2. Limits the total travel required by the cutter during machining
3. Takes location from the location scheme of component.
(a) 1,2 and 3 are correct (b) 1 and 2 are correct
(c) 2 and 3 are correct (d) 1 and 3 are correct

206. Consider the following statements:
Assertion(A): A workpiece with rough unmachined surface can be located in a jig or fixture on three supporting points.
Reason(R): Indexing is made accurate by supporting on three points
(a) Both A and R individually true and R is the correct explanation for A.
(b) Both A and R individually true and R is not the correct explanation for A.
(c) A is true but R is false.
(d) A is false but R is true.

207. Diamond pin location is used in a fixture because:
(a) It does not wear out
(b) it takes care of any variation in centre distance between two holes
(c) it is easy to clamp the part on diamond pins
(d) it is easy to manufacture.

208. consider the following statements:
Assertion(A): spherical washers are used to locate the job in the fixtures.
Reason(R): 3-2-1 principle should be adopted to locate the job.
(a) Both A and R individually true and R is the correct explanation for A.
(b) Both A and R individually true and R is not the correct explanation for A.
(c) A is true but R is false.
(d) A is false but R is true.

209. Match **List-I** (Components used in jigs and fixtures) with **List-II** (Their functions) and select the correct answer using the codes given below the lists:

List-I	**List-II**
A. Jack pin	**1.** To guide the drill during machining
B. V-locater	**2.** For easy removal of the work piece from the jig or fixture after the machining operation is over
C. Bushes	**3.** To locate the ciurcular or semi-circular objects in a jig or fixture
D. Ejectors	**4.** To locate work pieces whose dimensions are subject to variations

Codes:

	A	B	C	D
(a)	3	1	4	2
(b)	3	4	2	1
(c)	4	3	1	2
(d)	4	3	2	1

210. Which of the following sets of elements are quick-acting clamping elements for fixtures?
(a) wedge and cam (b) can and toggle
(c) toggle and wedge (d) wedge, cam and toggle

211. Match **List-I** (An element of jigs and fixtures) with **List-II** (Associating system)and select the correct answer using the codes below the lists:

List-I	**List-II**
A. Bush	**1.** Milling fixture
B. Setting block	**2.** Turning fixture
C. Diamond pin	**3.** Radial location
D. V-bolck	**4.** Cyclindrical location
	5. Drill jigs

Codes:

	A	B	C	D
(a)	5	4	3	1
(b)	3	1	2	4
(c)	5	1	3	4
(d)	3	4	2	1

212. According to the principal of location in jigs and fixtures,how many degress of freedom are to be eliminated to have a body fixed in space
(a) 3 (b) 4
(c) 5 (d) 6

213. Lathe bed is usually made of
(a) structural steel (b) stainless steel
(c) cast iron (d) non-ferrous materials

214. Average cutting speed in machining mild steel by single point tool of H.S.S is
(a) 10 m/min (b) 20 m/min
(c) 30 m/min (d) 40 m/min

215. Tool life is said to be over if
(a) a poor surface finish is obtained
(b) sudden increase in power and cutting force with chattering take place

(c) overheating and fuming due to friction start
(d) all of the above

216. Tool life is most affected by
(a) cutting speed
(b) tool geometry
(c) feed and depth
(d) not using coolant and lubricant

217. The spindle speeds of machine tools are usually designed to follow
(a) arithmetical progression
(b) geometrical progression
(c) harmonical progression
(d) logarithmic progression

218. The spindle speeds in a cutting tool are 160, 229, 328, 496.... The next higher speed will be
(a) 642
(b) 671
(c) 695
(d) 709

219. Flank wear occurs mainly on
(a) nose part, front relief face and side relief face
(b) nose part and top face
(c) cutting edges
(d) all of the above

220. Which of the following properties are essential for a tool material used for high speed machining
(a) red hardness and impact resistance
(b) red hardness and wear resistance
(c) toughness and impact resistance
(d) red hardness, wear resistance and toughness

221. The characteristics that enables one material to cut another is its relative
(a) toughness
(b) hardness
(c) resilience
(d) ductility

222. Most machinable metal is one which
(a) produces discontinuous chips
(b) permits maximum metal removal per tool grind
(c) results in minimum value of shear angle
(d) all of the above

223. The back rake required to machine brass by H.S.S tool is
(a) 10°
(b) 5°
(c) 0°
(d) –5°

224. In electro-discharge machining, the tool is made of
(a) tungsten carbide
(b) properly heat treated alloy steel
(c) diamond
(d) brass or copper

225. Which is false statement about electro-discharge machining
(a) it can machine very hard materials
(b) very good surface finish is obtained
(c) section to be machined should be thick
(d) metal removal rate is very slow

226. In electro-chemical machining operation, the gap between tool and work is kept of the order
(a) no gap, two are in contact with each other
(b) 0.25 mm
(c) 0.75mm
(d) 1.25 mm

227. In spark erosion machining process, the gap between tool and workpiece is filled with
(a) brine solution (b) acid solution
(c) a liquid dielectric (d) an electrolytic solution

228. The machining method which uses abrasive slurry is known as
(a) electro-discharge machining (b) laser machining
(c) plasma arc machining (d) ultrasonic machining

229. A big advantage of electro-chemical machining over electro-discharge machining is that
(a) it is more accurate and precise (b) it consumes less power
(c) its cost is low (d) tool wear is negligible

230. Spark erosion machining method can be used for the machining
(a) conducting materials only (b) non-conducting materials only
(c) semi-conductors only (d) both conducting and non-conducting materials

231. In spark erosion machining process,removal of metal takes places during
(a) charging of the capacitor (b) discharging of the capacitor
(c) all times (d) alternate cycles only

232. Ultrasonic machining method is best suited for
(a) brittle materials (b) stainless steels
(c) plastics (d) non-ferrous alloys

233. With H.S.S tools, highest cutting speed is used while machining
(a) cast iron (b) mild steel
(c) brass (d) aluminium

234. In machining, chips break due to
(a) ductility (b) toughness
(c) work hardening (d) tearing of the work material

235. Best coolant and lubricant for cast iron is
(a) water soluble oils or sulphur based and mineral oils
(b) mineral and fatty oils
(c) soluble oils
(d) compressed air

236. 18-4-1 high speed steel contains following elements in the ratio of 18-4-1
(a) W, Cr, V (b) Cr, V, W (c) W, Mn, Cr (d) W, V, Cr

237. The main function of the cutting fluid is
(a) decrease power consumption (b) cool the tool and workpiece
(c) wash away the chips (d) all of the above

238. The increase in back rake angle would affect the surface finish as follows
(a) improve (b) deteriorate
(c) unchanged (d) improve/deteriorate depending on materials

239. Ceramics tools are made from
(a) tungsten oxide (b) silicon carbide
(c) cobalt (d) aluminium oxide

240. Size of sharper is specified by
(a) length of stroke (b) size of table
(c) maximum size of tool (d) ratio of forward to return stroke

241. Best coolant and lubricant for aluminium is
(a) water soluble oils or sulphur based and mineral oils
(b) mineral and fatty oils or soluble oils
(c) soluble oils
(d) dry

242. Powder metallurgy techniques are used in the production of
(a) high carbon tool steels (b) HSS tools
(c) tungsten carbide tools (d) ceramics

243. High speed steel tool material contains carbon
(a) 0.6-1.0% (b) 2-4%
(c) 4-6% (d) 6-10%

244. The binding materials used in cemented carbide tools is
(a) graphite (b) lead
(c) cobalt (d) carbon

245. Trepanning is an operation of
(a) cutting internal threads
(b) producing a hole by removing metal along the circumference of a hollow cutting tool
(c) making a cone-shaped enlargement of the end of a hole
(d) super finishing

246. The helical grooves which extend to the full length of the drill body are called
(a) lips (b) cutting edges
(c) margins (d) flutes

247. For harder materials, the point angle of drill is
(a) increased
(b) decreased
(c) kept at 118°
(d) point angle has nothing to do with type of material

248. Throw away tips are used because
(a) initial cost is low (b) tool changing is easier
(c) regrinding is not possible (d) all of the above

249. Which is correct order for machinability of metals
(a) magnesium alloys, grey C.I., low carbon steel, monel metal
(b) grey cast iron, low carbon steel, magnesium alloys, monel metal
(c) magnesium alloys, low carbon steel, grey C.I., monel metal
(d) magnesium alloys, grey C.I., monel metal, low carbon steel

250. For particular cutting speed, the tool materials in order of tool life are
(a) H.S.S., cemented carbides, ceramics and oxides
(b) ceramics and oxides, cemented carbides, H.S.S.

(c) H.S.S., ceramics and oxides, cemented carbides
(d) cemented carbides, ceramics and oxides, H.S.S.

251. Tool cutting forces, with increase in cutting speed
(a) more or less remain constant (b) increase linearly
(c) decrease linearly (d) unpredictable

252. End relief angle of a single point tool is the angle
(a) by which the face of the tool is inclined sideways
(b) by which the face of the face is inclined towards back
(c) between the surface of the flank immediately below the point and a line drawn from the point perpendicular to the base
(d) between the surface of the flank immediately below the point and a plane at right angles to the centre line of the point of the tool

253. Cutting force and power involved in a machine tool can be measured by using
(a) pyrometer (b) comparator
(c) transducer (d) dynamometer

254. The operation of milling two sides of a workpiece simultaneously is called
(a) gang milling (b) climb milling
(c) square milling (d) straddle milling

255. The arbor of the milling machine is used to hold
(a) cutting tool (b) spindle
(c) over arm (d) mandrel

256. Feed rate in milling operation is expressed as
(a) mm/tooth (b) mm/r.p.m of the milling cutter
(c) metres/minute (d) revolution per minute

257. Negative rake is usually provided on
(a) H.S.S. tools (b) high carbon steel tools
(c) cemented carbide tools (d) all of the above

258. Side relief angle of a single point tool is the angle
(a) by which the face of the tool is inclined sideways
(b) by which the face of the tool is inclined towards back
(c) between the surfaces of the flank immediately below the point and a line drawn from the point perpendicular to the base
(d) between the surfaces of the flank immediately below the point and a plane at right angles to the centre line of the point of the tool

259. A left hand tool on lathe is used for turning in the direction
(a) from right to left (b) from left to right
(c) across the bed (d) in angular position on compound

260. Counter boring is the operation of
(a) enlarging the end of a hole cylindrically
(b) cone-shaped enlargement of the end of a hole
(c) smoothing and squaring the surface around a hole
(d) sizing and finishing a hole

261. In gang milling
(a) several jobs can be performed in one set up
(b) one job is completed on several milling machines located together
(c) two or more cutters are mounted on the arbor and all of them remove the metal simultaneously
(d) all of the above

262. Spot facing is the operation of
(a) enlarging the end of a hole cylindrically
(b) cone-shaped enlargement of the end of a hole
(c) smoothing and squaring the surface around a hole
(d) sizing and finishing a hole

263. Counter-sinking is the operation
(a) enlarging the end of a hole cylindrically
(b) cone-shaped enlargement of the end of a hole
(c) smoothing and squaring the surface around a hole
(d) sizing and finishing a hole

264. Machinability tends to decrease with
(a) increase in hardness and increase in tensile strength
(b) increase in strain hardening tendencies
(c) increase in carbon content, hard oxide and carbide
(d) decrease in grain-size

265. In the Taylor equation, value of index n is closely related to
(a) workpiece material
(b) cutting tool material
(c) working conditions
(d) temperature at chip tool interface

266. Wear limit on cutting tool is of the order of
(a) 0.3 mm
(b) 0.8 mm
(c) 1.6 mm
(d) 2.5 mm

267. Carbide tools wear out faster at
(a) slow speeds
(b) medium speeds
(c) fast speeds
(d) very fast speeds

268. The broaching operation in which the work moves past the stationary tool is called
(a) pull broaching
(b) push broaching
(c) surface broaching
(d) pot broaching

269. For tool A, Taylor's tool life exponent (n) is 0.45 and constant (K) is 90. Similarly for tool B, n = 0.3 and K = 60. The cutting speed (in m/min) above which tool A will have a higher tool life than tool B is
(a) 26.7
(b) 42.5
(c) 80.7
(d) 142.9

270. A taper hole is inspected using a CMM, with a probe of 2 mm diameter. At a hiehgt, Z = 10 mm from the bottom, 5 points are touched and a diameter of circle (not compensated for probe size) is obtained as 20 mm. Similarly, a 40 mm diameter is obtained at a height Z = 40 mm. The smaller diameter (in mm) of hole at Z = 0 is

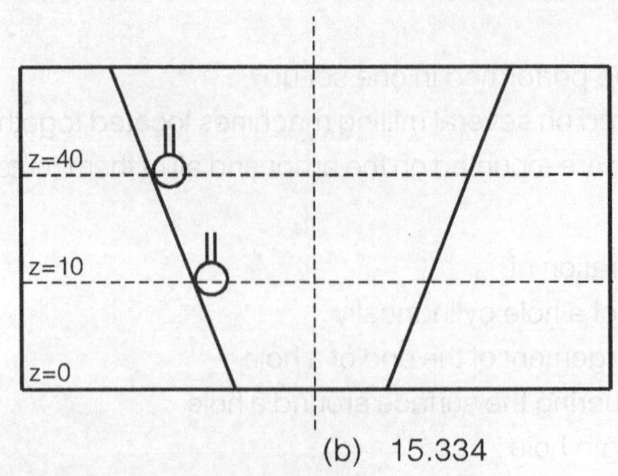

(a) 13.334
(b) 15.334
(c) 15.442
(d) 15.542

Statement for Linked Answer Questions 271 and 272:

In shear cutting operation, a sheet of 5 mm thickness is cut along a length of 200 mm. The cutting blade is 400 mm long (see figure) and zero-shear (S = 0) is provided on the edge. The ultimate shear strength of the sheet is 100 MPa and penetration to thickness ratio is 0.2. Neglect friction.

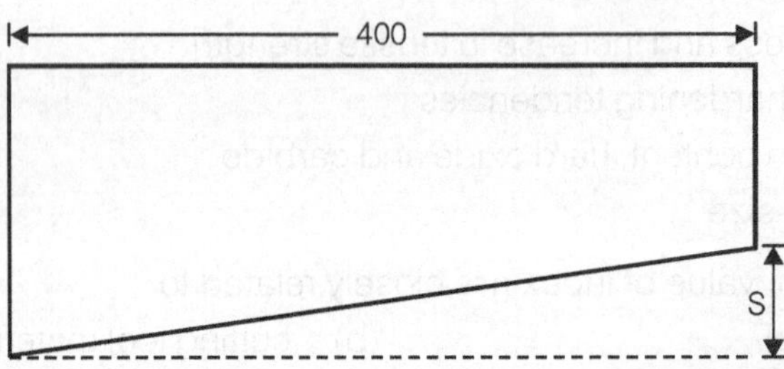

271. Assuming force vs displacement curve to be rectangular, the work done (in J) is
(a) 100
(b) 200
(c) 250
(d) 3300

272. A shear of 20 mm (S = 20 mm) is now provided on the blade. Assuming force vs displacement curve to be trapezoidal, the maximum force (in kN) exerted is
(a) 5
(b) 10
(c) 20
(d) 40

■■■■

ANSWERS

METAL CUTTING

1. (c) **2.** (b) **3.** (c) **4.** (c) **5.** (a) **6.** (b) **7.** (a)
8. (b) **9.** (a) **10.** (a) **11.** (a) **12.** (b) **13.** (a) **14.** (d)
15. (a) **16.** (a) **17.** (d) **18.** (a) **19.** (a) **20.** (b) **21.** (d)
22. (b) **23.** (c) **24.** (b) **25.** (d) **26.** (a) **27.** (b) **28.** (a)
29. (a) **30.** (b) **31.** (b) **32.** (a) **33.** (d) **34.** (a) **35.** (c)
36. (b) **37.** (a) **38.** (a) **39.** (b) **40.** (d) **41.** (a) **42.** (c)
43. (b) **44.** (b) **45.** (c) **46.** (d) **47.** (d) **48.** (a) **49.** (d)
50. (c) **51.** (b) **52.** (c) **53.** (c) **54.** (b) **55.** (a) **56.** (c)
57. (c) **58.** (c) **59.** (b) **60.** (b) **61.** (b) **62.** (d) **63.** (a)
64. (a) **65.** (b) **66.** (c) **67.** (d) **68.** (a) **69.** (a) **70** (b)
71. (d) **72.** (a) **73.** (c) **74.** (a) **75** (b) **76.** (a) **77.** (c)
78. (d) **79.** (a) **80.** (b) **81.** (b) **82.** (c) **83.** (a) **84.** (a)
85. (c) **86.** (a) **87.** (d) **88.** (c) **89.** (d) **90.** (b) **91.** (c)
92. (a) **93.** (a) **94.** (c) **95.** (b) **96.** (a) **97.** (a) **98.** (c)
99. (c) **100.** (b) **101** (b) **102.** (c) **103.** (c) **104.** (a) **105.** (d)
106. (d) **107.** (a) **108.** (b) **109.** (c) **110.** (b) **111.** (d) **112.** (d)
113. (a) **114.** (b) **115.** (d) **116.** (d) **117.** (b) **118.** (b) **119.** (b)
120. (c) **121.** (a) **122.** (c) **123.** (a) **124.** (a) **125.** (c) **126.** (a)
127. (c) **128.** (b) **129.** (c) **130.** (b) **131.** (d) **132.** (d) **133.** (d)
134. (d) **135.** (a) **136.** (c) **137.** (c) **138.** (a) **139.** (d) **140.** (b)
141. (b) **142.** (a) **143.** (d) **144.** (a) **145.** (a) **146.** (b) **147.** (d)
148. (c) **149.** (d) **150.** (d) **151.** (a) **152.** (c) **153.** (a) **154.** (b)
155. (b) **156.** (c) **157.** (b) **158.** (a) **159.** (b) **160.** (d) **161.** (a)
162. (c) **163.** (b) **164.** (d) **165.** (c) **166.** (d) **167.** (d) **168.** (d)
169. (b) **170.** (c) **171.** (d) **172.** (d) **173.** (d) **174.** (a) **175.** (c)
176. (a) **177.** (b) **178.** (b) **179.** (b) **180.** (b) **181.** (b) **182.** (d)

183. (c)	**184.** (d)	**185.** (a)	**186.** (b)	**187.** (c)	**188.** (c)	**189.** (d)
190. (d)	**191.** (d)	**192.** (b)	**193.** (b)	**194.** (c)	**195.** (a)	**196.** (b)
197. (c)	**198.** (c)	**199.** (c)	**200.** (b)	**201.** (d)	**202.** (d)	**203.** (a)
204. (a)	**205.** (c)	**206.** (c)	**207.** (b)	**208.** (d)	**209.** (c)	**210.** (c)
211. (c)	**212.** (c)	**213.** (c)	**214.** (c)	**215.** (d)	**216.** (a)	**217.** (b)
218. (d)	**219.** (a)	**220.** (d)	**221.** (b)	**222.** (b)	**223.** (c)	**224.** (d)
225. (c)	**226.** (b)	**227.** (c)	**228.** (d)	**229.** (d)	**230.** (a)	**231.** (b)
232. (a)	**233.** (d)	**234.** (c)	**235.** (d)	**236.** (a)	**237.** (d)	**238.** (a)
239. (d)	**240.** (a)	**241.** (a)	**242.** (c)	**243.** (a)	**244.** (c)	**245.** (b)
246. (d)	**247.** (a)	**248.** (d)	**249.** (a)	**250.** (b)	**251.** (c)	**252.** (c)
253. (d)	**254.** (d)	**255.** (a)	**256.** (a)	**257.** (c)	**258.** (d)	**259.** (b)
260. (a)	**261.** (c)	**262.** (c)	**263.** (b)	**264.** (d)	**265.** (b)	**266.** (a)
267. (b)	**268.** (d)	**269.** (a)	**270.** (a)	**271.** (a)	**272.** (b)	

CHAPTER 5

Metal Forming

Forming is the process in which the desired size and shape are obtained through the plastic deformation of a material. The stresses induced during the process are greater than the yield strength, but less than the fracture strength, of the material. Forming is developing very fast because metal forming operations such as forging, rolling, drawing, etc., are capable of yielding high productivity compared to other metal working techniques. Metal forming is one of the most important steps in manufacturing of a large variety of products. It involves in obtaining desired shape and size by subjecting the material to large plastic deformation. Equally important is the control of mechanical properties and product quality. Metal forming or metal working processes are divided into two categories i.e. bulk forming and sheet metal forming. Bulk forming refers to processes like forging, rolling, extrusion etc. where there is a controlled plastic flow of material into useful shapes. Forming can be classified in to these three categories:

Cold Working < 0.4 T_m
Warm Working 0.4 – 0.6 T_m
Hot Working > 0.6 T_m
where T_m is melting point of material in Kelvin

Cold working is, when the material is deformed below recrystallization temperature. If the working temperature is higher than the recrystallization is called hot working, if the material is at recrystallization temperature it is called warm working. Following are the characteristics of metal working:

1. Due to work hardening cold working products will be stronger than hot worked products. In hot working since there is scale formation surface finish will be better in cold working.
2. Since in cold working there will not be any shrinkage so better dimensional accuracy will be produced.
3. For hard and brittle materials excessive load will appear in the cold working so the process will be uneconomical and hot working is used for such materials
4. Due to excessive cold working defects like internal cracks may appear in the cold working. Also the process is limited to low strength ductile materials only otherwise the power requirement goes up.

The main objective of metal working analysis is to predict the stresses acting during metal deformation and consequently the forces which must be applied. The working load determines the power requirements and the size of the equipment necessary to perform a particular operation. Measurement of forces also, gives an indication of the efficiency of the process.

In uniaxial tension or pure compression the yield stress can easily be defined and its value read from the stress strain diagram of the material.

1. If the material behaves like that of the elastic-plastic then yield strength can be considered as mean flow stress or nominal stress or the stress at which the material begins to flow and take the shape of

dies. If the material behaves like elastic-work hardening, then average flow stress is taken as nominal stress (σ_0) and it can be calculated as under:

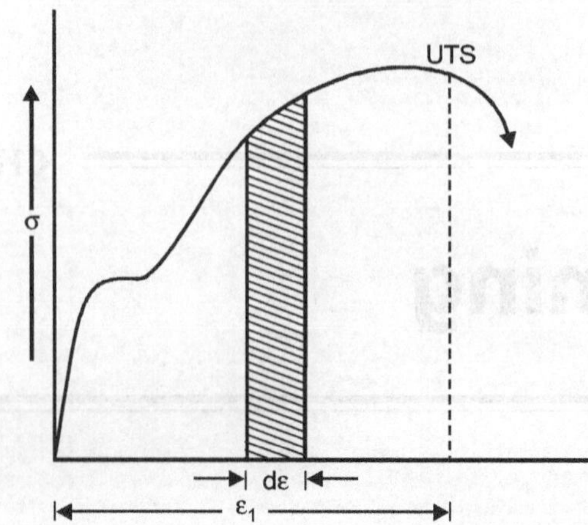

Fig 5. 1 : Wire drawing process

$$\sigma_{fm} = \frac{1}{\varepsilon_1}\int_0^{\varepsilon_1} \sigma_f d\varepsilon = \frac{1}{\varepsilon_1}\int_0^{\varepsilon_1} K\varepsilon^n d\varepsilon \qquad (\text{since } \sigma_f = K\varepsilon^n)$$

$$\sigma_{fm} = \frac{1}{\varepsilon_1}\left[\frac{K\varepsilon^{n+1}}{n+1}\right]_0^{\varepsilon_1} \qquad (\sigma_{fm} \text{ is mean flow stress})$$

$$\sigma_0 = \sigma_{fm} = \frac{K\varepsilon^n}{n+1}$$

In this analysis elastic strains are neglected as compared to plastic strains. At this point it is important to understand the difference between plane strain (von-mises) and the plane stress (tresca hypothesis). If the deformation of material is:

(a) Everywhere parallel to a given plane.

(b) Independent of Z-direction.

Then the deformations are considered as plane strain. If the deformations are not restricted in Z- directions, it is called plane stress case. E.g. wire drawings, extrusion forging in rectangular dies are plane strain case. The stress tensor can be broken into two parts under.

$$\begin{bmatrix} \sigma_{xx} & \tau_{xy} & \tau_{xz} \\ \tau_{yx} & \sigma_{yy} & \tau_{yz} \\ \tau_{zx} & \tau_{zy} & \sigma_{zz} \end{bmatrix} \Rightarrow \begin{bmatrix} \sigma_{xx}-\sigma & \tau_{xy} & \tau_{xz} \\ \tau_{yx} & \sigma_{yy}-\sigma & \tau_{yz} \\ \tau_{zx} & \tau_{zy} & \sigma_{zz}-\sigma \end{bmatrix} + \begin{bmatrix} \sigma & 0 & 0 \\ 0 & \sigma & 0 \\ 0 & 0 & \sigma \end{bmatrix}$$

Where $\begin{bmatrix} \sigma & 0 & 0 \\ 0 & \sigma & 0 \\ 0 & 0 & \sigma \end{bmatrix} \Rightarrow$ hydrostatic component

$\begin{bmatrix} \sigma_{xx}-\sigma & \tau_{xy} & \tau_{xz} \\ \tau_{yx} & \sigma_{yy}-\sigma & \tau_{yz} \\ \tau_{zx} & \tau_{zy} & \sigma_{zz}-\sigma \end{bmatrix} \Rightarrow$ Deviator component

Von-Mises [plane strains] :

In plane strain case it can be represented by a von-mises cylinder. Hydrostatic stress acts along the centre of cylinder and deviator component acts in the radial direction. Hydrostatic component does not cause any plastic deformation. So if a major portion of the stress is converted into hydrostatic component, ductility of the material improves. So hydrostatic pressure is applied on the material in deep drawing operation and this concept is used in hydrostatic extrusion to draw hard and brittle materials. As long as deviator component is within the cylinder, the deformation of the material is within the elastic region. Plastic deformation begins as soon as the resultant stress reaches the surface of the cylinder. Plane stress or tresca hypothesis is represented by hexagon cylinder and hexagon will be inside the circle if yield tensile strength is taken to be same in both the cases. Hexagon will be outside the von-mises circle when yield shear strength is taken to be same in both the cases.

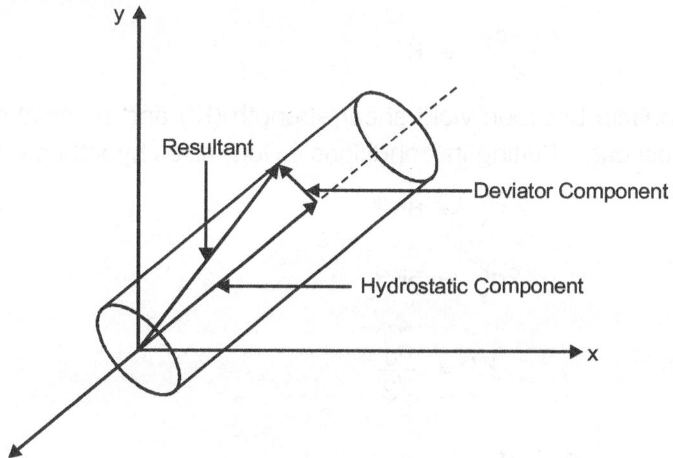

Fig 5.2 : Wire drawing process

For plane strain case

$$\varepsilon_Z = \frac{\partial w_Z}{\partial Z} = 0 \text{ this corresponds to } \sigma_{zz} - \sigma = 0 \text{ where}$$

$$\sigma = \text{Hydrostatic pressure}$$

$$\sigma = \frac{\sigma_{xx} + \sigma_{yy} + \sigma_{zz}}{3}$$

$$\sigma_{zz} - \left(\frac{\sigma_{xx} + \sigma_{yy} + \sigma_{zz}}{3}\right) = 0$$

$$\sigma_{zz} = \frac{\sigma_{xx} + \sigma_{yy}}{2}$$

Putting this value in Von-mises hypothesis we get

$$(\sigma_{xx} - \sigma_{yy})^2 + (\sigma_{yy} - \sigma_{zz})^2 + (\sigma_{zz} - \sigma_{xx})^2 = 6K'^2$$

$$K' = \text{yield shear strength}$$

$$\left(\sigma_{xx}-\sigma_{yy}\right)^2+\left(\sigma_{yy}-\left(\frac{\sigma_{xx}+\sigma_{yy}}{2}\right)\right)^2+\left(\left(\frac{\sigma_{xx}+\sigma_{yy}}{2}\right)-\sigma_{xx}\right)^2=6K'^2$$

$$\left(\sigma_{xx}-\sigma_{yy}\right)^2+\left(\frac{\sigma_{yy}-\sigma_{xx}}{2}\right)^2+\left(\frac{\sigma_{yy}-\sigma_{xx}}{2}\right)^2=6K'^2$$

$$\frac{3}{2}\left(\sigma_{xx}-\sigma_{yy}\right)^2=6K'^2$$

$$\left(\sigma_{xx}-\sigma_{yy}\right)^2=4K'^2$$

or writing in terms of principal stresses σ_1 and σ_2 we get

$$\frac{\sigma_1-\sigma_2}{2}=K'$$

To have a relationship between yield shear strength (K') and nominal stress (σ_0), let us take uniaxial tensile test case. Putting its conditions in von-mises hypothesis we get,

$$2\sigma_{xx}^2=6K'^2$$

$$2\sigma_0^2=6K'^2$$

$$K'=\frac{\sigma_0}{\sqrt{3}}$$

So,
$$\frac{\sigma_x-\sigma_y}{2}=\frac{\sigma_0}{\sqrt{3}}$$

$$\sigma_1-\sigma_2=\frac{2}{\sqrt{3}}\sigma_0$$

$$=1.15\,\sigma_0$$

For plane stress case :

$$\sigma_1-\sigma_2=2K' \text{ (Tresca hypothesis)}$$

From uniaxial tensile test we get

$$\sigma_0=2K'$$

$$K'=\frac{\sigma_0}{2}$$

So
$$\sigma_1-\sigma_2=\sigma_0 \text{ for plane stress}$$

$$K'=\frac{\sigma_0}{2}$$

If flow is restricted in one direction then it is plane strain. If there is no restriction in any direction to flow then it is plane stress.

5.1 WIRE DRAWING

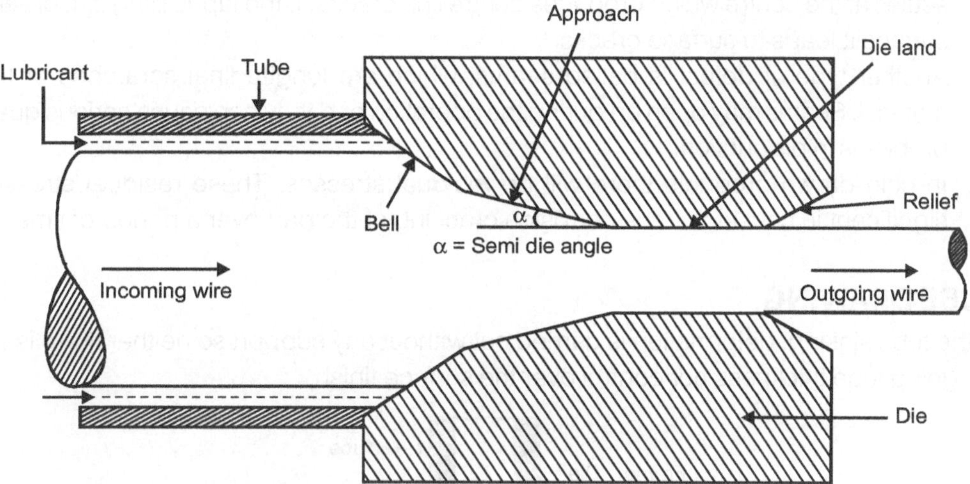

Fig 5. 3 (a) : Wire drawing process

A circular cross section billet is reduced in diameter and increased in length by pulling through a conical die of semi angle. The theory developed has some assumptions:

- A uniform state of stress is assumed to prevail at all points at a plane normal to the axis.
- The redundant work is negligible.
- A cylindrical state of stress is assumed to exist within the contact zone. The axial stress and die pressure can be assumed to be principle stresses.

Before wire enters into the die it should be given a twist or bend to remove oxide layer from it, the wire is then dipped into a lubricant. As the wire is entering into the bell section of die it will be coated with a layer of lubricant. As lubricant hits the die, it looses all its momentum which converts into pressure which enables the lubricant to enter into the deformation zone. For heavy reduction solid lubricants like soap powder, Aluminium, graphite, lime etc. are used. It produces dull surface. Liquid lubricants like soap solution, oil etc. gives bright finish. In the initial stages wire comes in contact with the bell, the shape of the bell causes the hydrostatic pressure to increase and promotes the flow of lubricant into the die. Actual deformation of the material takes place in the approach section of die. By providing die land, although the drawing forces will increase but it is required for die stability.

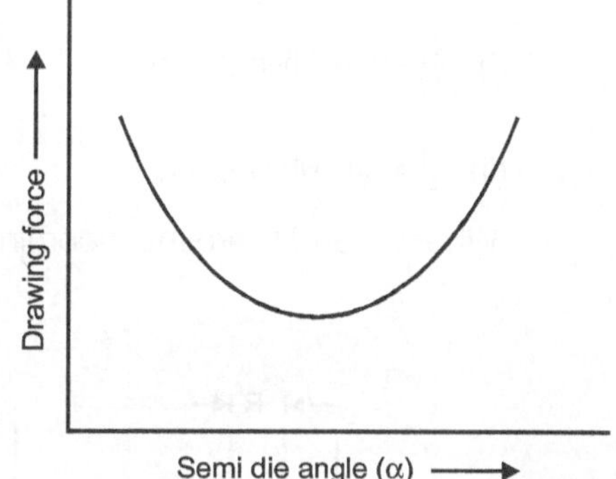

Fig. 5.3 (b) : Wire drawing process

By increasing the semi die angle the contact length between material and die decreases. This decreases the friction and hence drawing forces decreases. When the semi die angle increases beyond a certain value the dead metal zone starts forming within the metal flow region. These regions not only creates obstacle in the metal flow but also starts moving in the backward direction. This increases the drawing force requirement. Dead metal zone produces bright finish to the deformed material. Towards the end, some back relief is provided for the elastic recovery.

5.1.1 Defects of wire drawing

1. If removal of oxide from the wire is not proper, these impurities will percolate down and settles at the centre which produces centre line cracks. If the lubricating film breaks over the surface it leads to surface cracks.
2. Another type of defect is called seams, which are longitudinal scratches or folds in the material. Seams may open up during the operation and they can cause serious quality control problems in production.
3. In cold drawn products there will be residual stresses. These residual stresses can be significant in causing stress corrosion cracking of the part over a period of time.

5.2 TUBE DRAWING

In the tube sinking, deformation is carried out without any support so neither there is a control of out going tube thickness nor control over the surface finish.

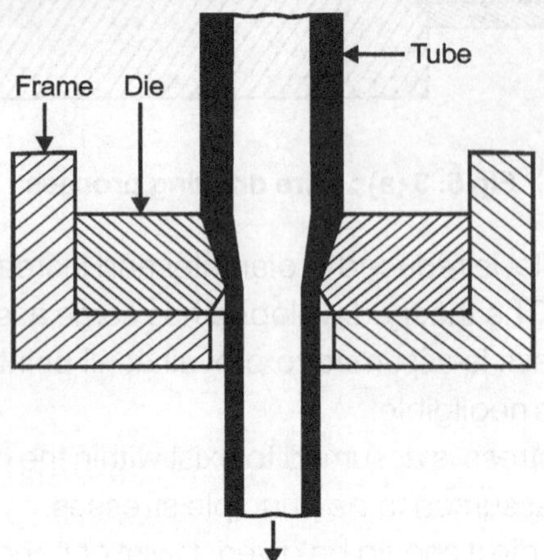

For tube sinking operation

(i) $\frac{t_1}{R} \leq 0.33$ when $t_2 > t_1$

(ii) $\frac{t_1}{R} > 0.33$ when $t_2 < t_1$

Where t_1 and t_2 are the incoming and outgoing tube thickness respectively.

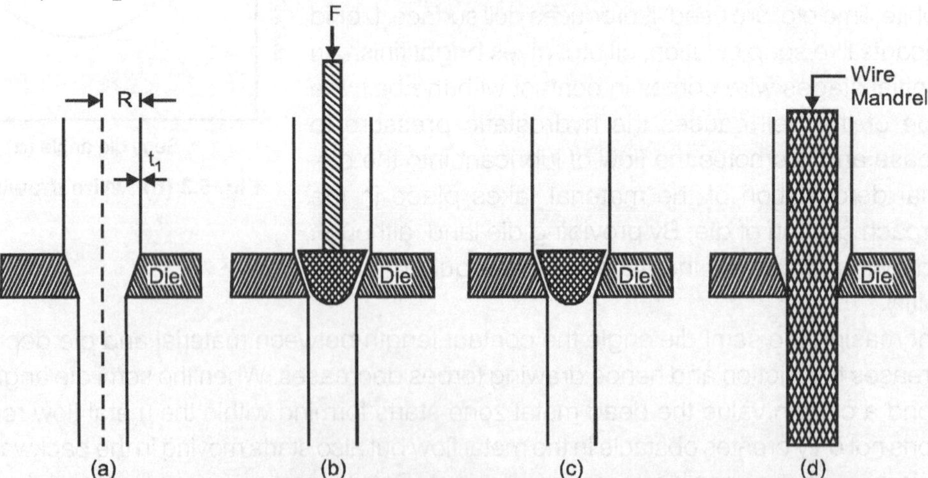

Fig 5.4 (a) : (a) Tube sinking operation. (b) tube drawing with fixed mandrel (c) Tube drawing with floating mandrel (d) tube drawing with moving mandrel

In case of floating mandrel there is a partial control over plastic deformation at the inside so surface will be better but dimensions will not be controlled. In the case of fixed mandrel by carrying the load on the mandrel thickness of tube can be controlled.

In case of tube drawing with moving mandrel the speed of wire is equal to the velocity of out going tube so at the entry to the die, velocity of wire will be higher than the velocity of incoming tube. So relative to wire (in the wire region) tube is going in the backward direction, so friction will appear in the forward direction. On the interface between the tube and die, friction appears in the backward direction. These oppositely acting frictions cancel each other. So for the same amount of reduction drawing pressure will be lowest in case of moving mandrel. So heavy reductions are possible.

5.3 EXTRUSION

Extrusion is defined as the process of pushing the heated billet of metal through an orifice provided into a die, thus forming an elongated part of uniform cross section corresponding to the shape of the die orifice. The pressure is applied hydraulically or mechanically. Intricate shapes in long lengths can be produced by hot extrusion. Depending on the ductility of the material, extrusion may be carried out at room or at elevated temperature. Because a chamber is involved each billet is extruded individually and thus extrusion is a batch or semicontinuous process.

Typical products made from extrusion are railings for sliding doors, tubing for various cross sections, structural and architectural shapes and doors and window frames.

5.3.1 Forward Extrusion or Direct Extrusion

The raw material used is a billet. Billet is heated to its forging temperature and fed into the machine chamber. In this forward extrusion piston and out going material moves in the same direction. Pressure is applied to the billet, forcing the material through the die. In forward direction since billet moves inside the drum so pressure required in overcoming the friction will be more. The length of the extruded part depends on the size of the billet and cross section area of the die.

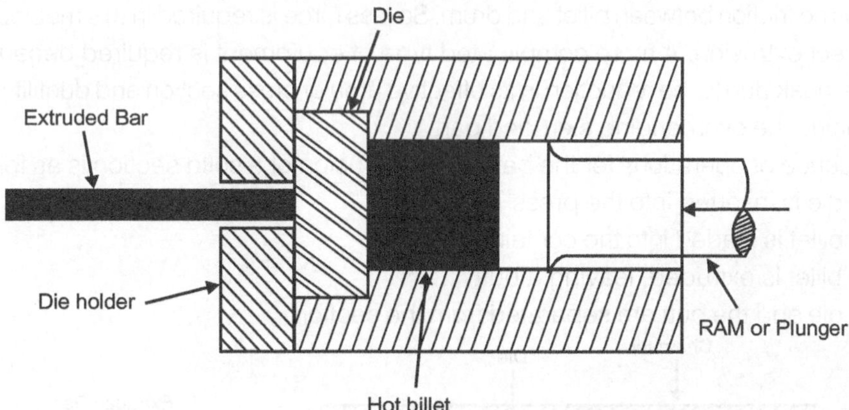

Fig 5.5 : (a) Forward extrusion

A typical sequence of operations for the forward extrusion of a solid section is as follows:

1. The heated billet and the dummy block are loaded into the container
2. The billet is extruded by the force of the ram being pushed against it. This upsets the billet, then forces the metal to flow through the die. During extrusion, a thin shell of material may be left on the container walls. Extrusion is halted in order to leave a thin disk of material (butt) in the container
3. The container is separated from the die, the extruded section with the butt, and the dummy block

4. The discard (butt) is sheared off
5. The shear die, the container, and the ram are returned to their initial (loading) positions

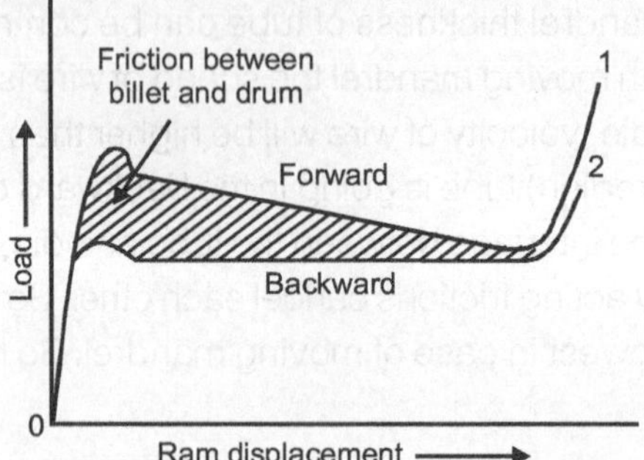

Fig. 5.5(b) Variation of load with displacement in forward and backwarkd extrusion

Typical load versus ram displacement curves shows that the load in forward extrusion initially increases very rapidly as the billet upsets to fill the container. This is followed by a further increase in pressure, and extrusion begins. A somewhat cone-shaped deformation zone then develops in front of the die aperture. After the maximum load has been reached, the extrusion pressure falls as the billet length decreases until a minimum is reached, then rapidly increases again. This last pressure increase occurs because only a disk of the billet remains and the metal must flow radially toward the die aperture. Resistance to deformation increases considerably with decreasing thickness.

5.3.2 Backward Extrusion or Indirect Extrusion

In backward extrusion die is placed in the piston, so out going material will come out in the opposite direction to that of the piston. In this process billet does not move inside the chamber. So there is no friction between billet and drum. So less force is required in this method in comparison to the direct extrusion. A more complicated type of equipment is required because the plunger becomes weak due to the reduction in its effective area of cross section and ductility is experienced in supporting the overhanging extruded part.

The sequence of operations for the backward extrusion of a solid section is as follows:

1. The die is inserted into the press
2. The billet is loaded into the container
3. The billet is extruded, leaving a butt
4. The die and the butt are separated from the section

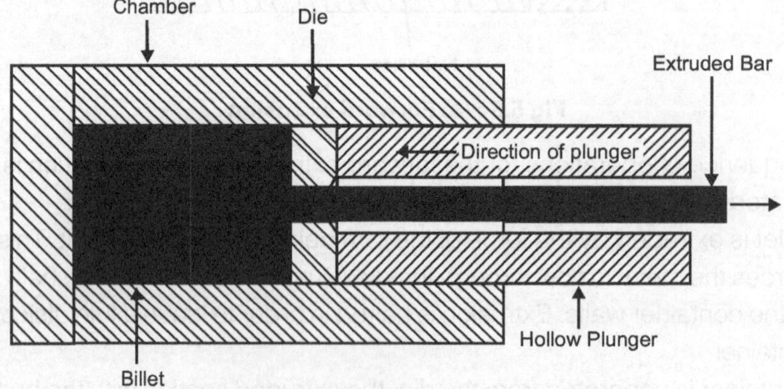

Fig 5.5 : (b) Backward extrusion

Backward extrusion offers a number of advantages, as follows:

1. A 25 to 30% reduction in maximum load relative to direct extrusion
2. Extrusion pressure is not a function of billet length, because there is no relative displacement between the billet and the container. Therefore, billet length is not limited by the load required for this displacement but only by the length and stability of the hollow stem needed for a given container length
3. No heat is produced by friction between the billet and the container; consequently, no temperature increase occurs at the billet surface toward the end of extrusion, as is typical in the direct extrusion of aluminum alloys. Therefore, in backward extrusion, there is a lesser tendency toward cracking of the surfaces and edges, and extrusion speeds can be significantly higher
4. The service life of the tooling is increased, especially that of the inner liner, because of reduced friction and temperatures

The disadvantage of backward extrusion is that impurities or defects on the billet surface affect the surface of the extrusion and are not automatically retained as a shell or discard in the container. As a result, machined billets are used in many cases. In addition, the cross-sectional area of the extrusion is limited by the size of the hollow stem.

5.3.3 Hydrostatic Extrusion

In hydrostatic extrusion the billet is smaller in diameter than the chamber, which is filled with a fluid and the pressure is transmitted to the billet by a ram. Billet does not come in contact with the piston and the material is deformed due to development of hydrostatic pressure within the fluid. Since hydrostatic pressure increases the ductility, very good quality products can be extruded by this process. In this process large hydrostatic pressure applied to the billet, the product in the receiving chamber is maintained under a lower pressure. There is no friction to overcome along the container walls.

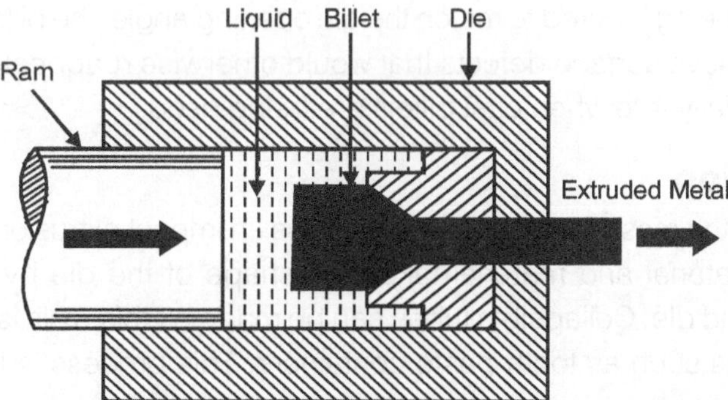

Fig. 5.5(d) Hydrostatic extrusion

In Hydrostatic extrusion, the billet in the container is extruded through the die by the action of a liquid pressure medium rather than by direct application of the load with a ram. The process of pure hydrostatic extrusion differs from conventional extrusion in that the billet is completely surrounded by a fluid, which is sealed off and is pressurized sufficiently to extrude the billet through the die. Hydrostatic extrusion can be done hot, warm, or cold and can be used to extrude brittle materials that cannot be processed by conventional extrusion. Hydrostatic extrusion also allows greater reductions in area (higher extrusion ratios) than either cold or conventional hot extrusion. The primary advantages of simple hydrostatic extrusion over conventional hot or cold extrusion methods are:

1. There is no friction between the billet and the container. Therefore, the pressure at the beginning of extrusion is much lower, and billets of any length can theoretically be extruded
2. Friction at the die can be significantly reduced by a film of pressurized lubricant between the deforming metal and the die surface
3. The lower extrusion pressures and the reduced die friction of hydrostatic extrusion allow the use of either higher extrusion ratios or lower extrusion temperatures
4. The uniform hydrostatic pressure in the container means that billets do not have to be straight; coiled wire can also be extruded

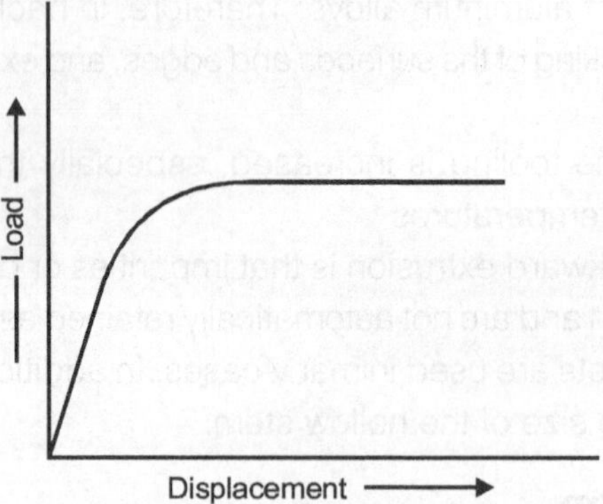

Fig. 5.5(e) Hydrastatic Extrusion

Limitations of the hydrostatic extrusion process include :

1. Containment of the fluid under high pressure (up to 2 GPa, or 290 ksi) requires reliable seals between the container bore surface and both the ram and die. The technology required to achieve dependable seals at these points is widely available. However, also sealing between the billet nose and the die can easily be achieved by chamfering or tapering the billet nose to match the entry angle of the die
2. In addition to being tapered to match the die opening angle, the billet is also usually machined all over to remove surface defects that would otherwise reappear on the extruded product. This is especially true when cast billets are being used.

5.3.4 Impact extrusion

This comes under the mostly cold extrusion process. In impact extrusion a heavy punch is allowed to fall over the material and material takes the shape of the die by flowing in the clearance between punch and die. Collapsible tubes can be made by this. This is used to make collapsible tubes of soft alloys such as tooth paste containers. This process is limited to soft and ductile materials.

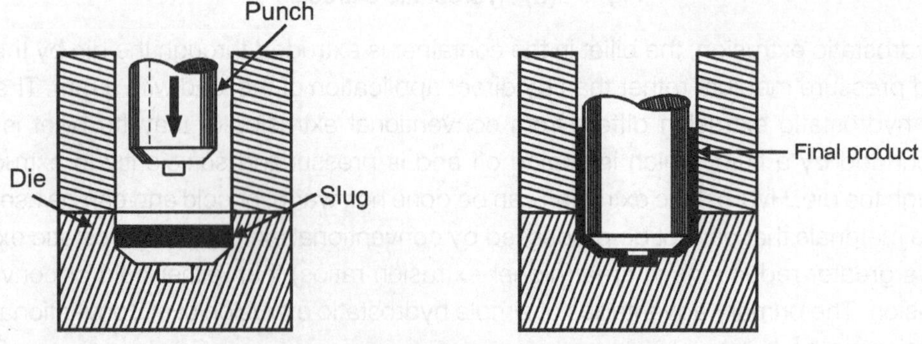

Fig 5.5 : (f) Impact extrusion

5.4 ANALYSIS OF WIRE DRAWING AND EXTRUSION

(A) Johnson's Equation

$$\sigma_d = \sigma_0 [a + b \ln R]$$

where a, b are Johnson's constant, σ_0 is nominal stress and R is extrusion ratio.

R = Extrusion ratio = $\frac{A_i}{A_f}$ (A_i is initial area and A_f is final area)

For hot extrusion

$$\sigma_d = K_1 \ln R$$

K_1 = Extrusion constant which depends upon temperature.

Example : 5.1

A cylindrical billet of 40mm diameter and 100mm length is reduced by backward extrusion to a 15 mm diameter. If the Johnson's equation has a = 0.8 b = 1.5 and strength coefficient of material 750 MPa strain hardening coefficient n = 0.15 Determine

(i) Extrusion ratio (ii) true strain (iii) Extrusion strain (iv) Ram force

Solution :

Given data

$$a = 0.8$$
$$b = 1.5$$
$$K = 750 \text{ MPa}$$
$$n = 0.15$$

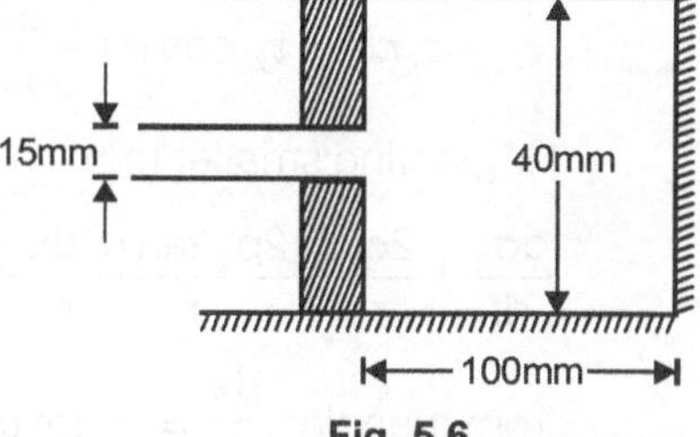

Fig. 5.6

$$\frac{\pi}{4} D_i^2 L_i = \frac{\pi}{4} D_f^2 L_f$$

$$(40)^2(100) = (15)^2 L_f$$

$$\Rightarrow \quad L_f = 711.11 \text{ mm}$$

$$\text{Extrusion strain} = e = \frac{L_f - L_i}{L_i}$$

$$= \frac{711.11 - 100}{100} = 6.11$$

$$\text{True strain} = \varepsilon = \ln[1+e] = \ln[1 + 6.11] = 1.96$$

$$\text{Extrusion ratio} = R = \frac{A_i}{A_f} = 7.11$$

$$\sigma_0 = \frac{K\varepsilon^n}{n+1} = \frac{750 (1.96)^{0.15}}{1.15} = 721 \text{ MPa}$$

So,

$$\sigma_d = \sigma_0 [a + b \ln R] = 721[0.8 + 1.5 \ln 7.11] = 2698.16 \text{ MPa}$$

$$\text{Ram force} = \sigma_d \times \frac{\pi}{4}\left[D_i^2 - D_f^2\right] = 2.9 \text{ MN}$$

(B) Slab Method

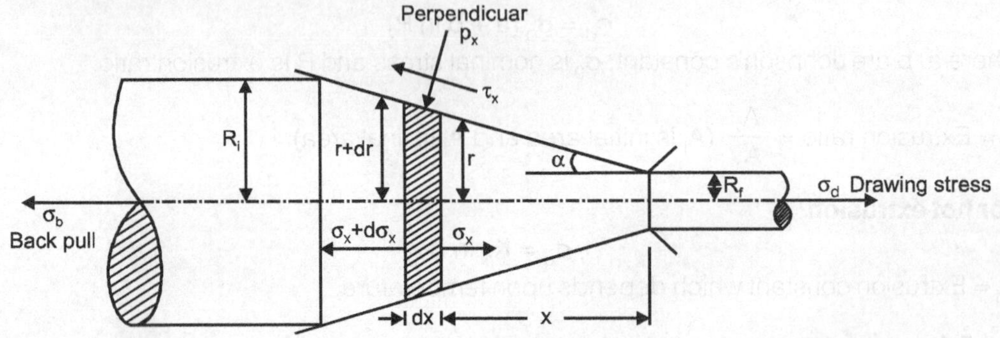

Fig 5.7 : Stresses in drawing

Mass in x direction

$$\Rightarrow (\sigma_x + d\sigma_x)\,\pi\,(r + dr)^2 - \sigma_x \pi r^2 + \tau_x \cos\alpha \left(2\pi r \frac{dx}{\cos\alpha}\right) + p_x \sin\alpha \left(\frac{2\pi r dx}{\cos\alpha}\right) = 0 \qquad ...(1)$$

$$\Rightarrow \pi\left[\sigma_x r^2 + \sigma_x d^2 r + 2rdr\sigma_x + d\sigma_x r^2 + d\sigma_x dr^2 + 2rdrd\sigma_x\right]$$

$$-\sigma_x \pi r^2 + \tau_x \cos\alpha\left(\frac{2\pi r dx}{\cos\alpha}\right) + p_x \sin\alpha\left(\frac{2\pi r dx}{\cos\alpha}\right) = 0 \qquad ...(2)$$

Neglecting smaller terms and divide by r^2dr

$$\frac{d\sigma_x}{dr} + \frac{2\sigma_x}{r} + \frac{2p_x \tan\alpha}{r}\frac{dx}{dr} + \frac{2\tau_x}{r}\frac{dx}{dr} = 0$$

from triangle; $\frac{dx}{dr} = = \cot\alpha$; and $\tau_x = \mu p_x$

$$\text{So } \frac{d\sigma_x}{dr} + \frac{2\sigma_x}{r} + \frac{2p_x}{r} + \frac{2\mu p_x \cot\alpha}{r} = 0$$

Assuming σ_x and p_x are the principle stresses.

$$\sigma_x + p_x = 2K'$$

$$\frac{d\sigma_x}{dr} + \frac{2[2K']}{r} + \frac{2\mu\cot\alpha}{r}(2K' - \sigma_x) = 0$$

$$\frac{d\sigma_x}{dr} = -\frac{2}{r}\left[2K' + \mu\cot\alpha\,[2K' - \sigma_x]\right]$$

let $\mu \cot\alpha = B$

$$\frac{d\sigma_x}{dr} = -\frac{2}{r}\left[2K' + B[2K' - \sigma_x]\right]$$

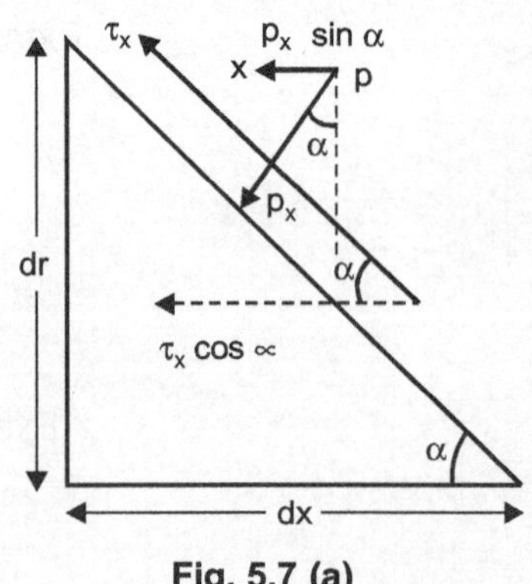

Fig. 5.7 (a)

$$\int \frac{d\sigma_x}{[2K' + B[2K' - \sigma_x]]} = \int -\frac{2}{r}dr$$

$$\frac{-1}{B}\ln\left[2K' + B[2K' - \sigma_x]\right] = -2\ln r + C_1 \qquad ...(3)$$

Boundary conditions

at $r = R_i$; $\sigma_x = \sigma_b$ where σ_b = back pull

Before entering into the die the wire is pulled in the backward direction to keep it straight otherwise it will damage the bell section of die. By increasing the back pull although the drawing stress increases it is required to increase the die life.

$$\frac{-1}{B}\ln\left[2K' + B\left[2K' - \sigma_b\right]\right] = -2\ln R_i + C_1 \quad ...(4)$$

By equating Eq (3) and (4) $\dfrac{[2K' + B(2K' - \sigma_x)]}{[2K' + B(2K' - \sigma_b)]} = \left(\dfrac{r}{R_i}\right)^{2B}$

$$2K' + B(2K' - \sigma_x) = \left(\frac{r}{R_i}\right)^{2B}\left\{2K' + B\left(2K' - \sigma_b\right)\right\}$$

$$\text{So } (2K')\left(\frac{1+B}{B}\right)\left(1-\left(\frac{r}{R_i}\right)^{2B}\right) + \sigma_b\left(\frac{r}{R_i}\right)^{2B} = \sigma_x$$

$$\text{Now} = 2K'\left(\frac{1+B}{B}\right)\left(1-\left(\frac{R_f}{R_i}\right)^{2B}\right) + \sigma_b\left(\frac{R_f}{R_i}\right)^{2B} \quad ...(5)$$

In the wire drawing both von-mises and tresca yield criteria in a spherical coordinate system with spherical symmetry leads to the same result i.e.

$\sigma_1 - \sigma_2 = \sigma_0$

So in the analysis $2K' = \sigma_0$

(Reference: B .Avitzur, "Metal forming: Process and analysis"TMH, 1977)

But it should be noted that in case of tube drawing operation since it is plane strain case

$$K' = \frac{\sigma_0}{\sqrt{3}}$$

In all the tube drawing cases eq. (5) will be used with slight modifications.

Case - I : Tube sinking

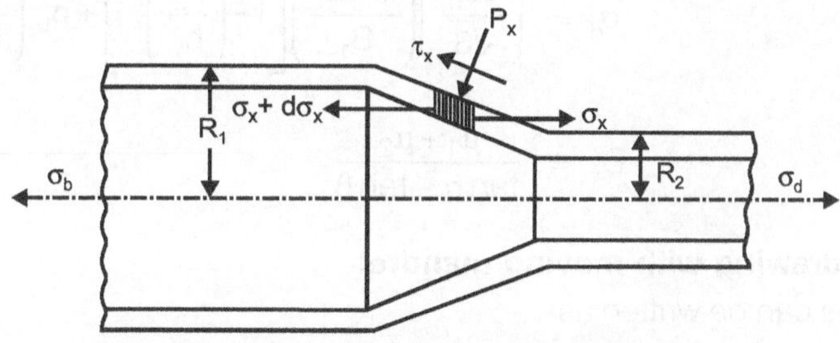

Fig 5.8 : Tube Sinking Analysis

$$\sigma_d = \left(\frac{2\sigma_0}{\sqrt{3}}\right)\left(\frac{1+B}{B}\right)\left[1-\left(\frac{R_2}{R_1}\right)^{2B}\right] + \sigma_b\left(\frac{R_2}{R_1}\right)^{2B} \quad ...(6)$$

$B = \mu \cot \alpha$

Case II : Tube drawing with floating mandrel

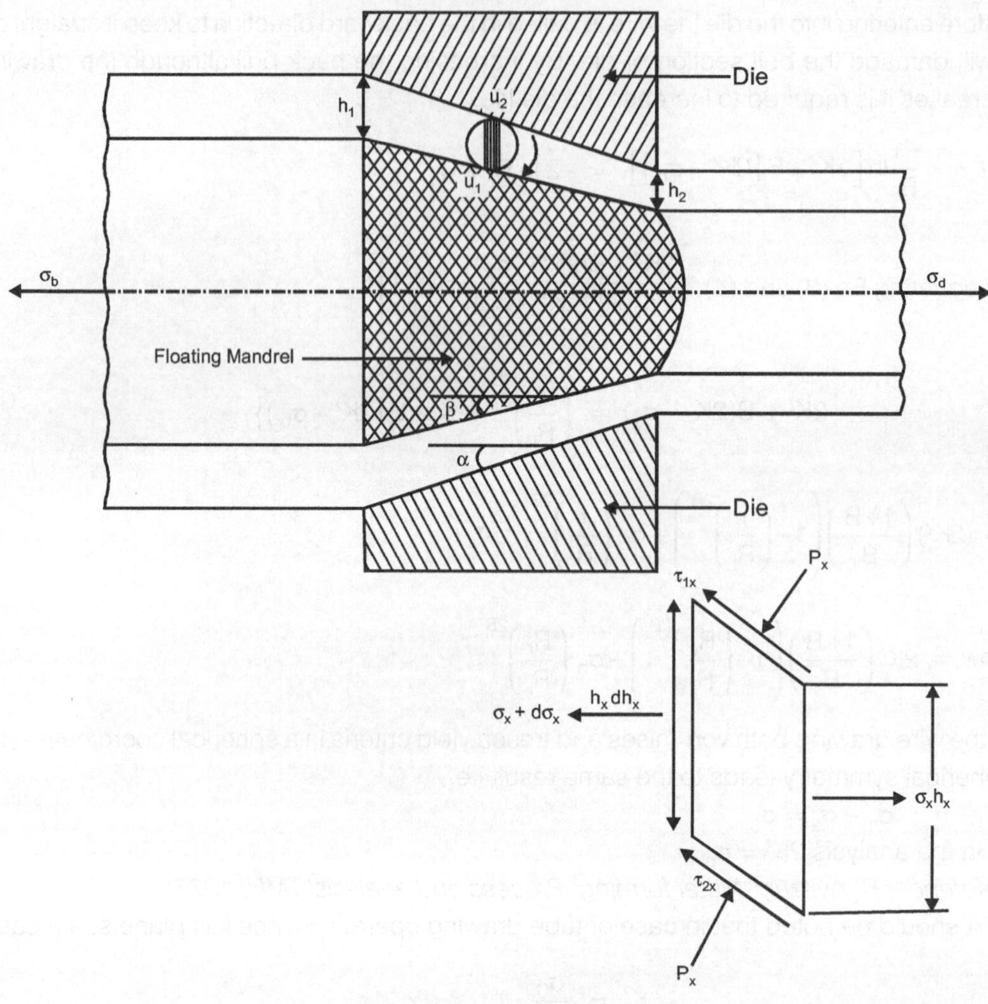

Fig 5.9 : Mandrel Analysis

$$\sigma_d = \left(\frac{2\sigma_0}{\sqrt{3}}\right)\left(\frac{1+B_1}{B_1}\right)\left[1-\left(\frac{h_2}{h_1}\right)^{B_1}\right]+\sigma_b\left(\frac{h_2}{h_1}\right)^{B_1} \quad ...(7)$$

where $$B_1 = \frac{\mu_1+\mu_2}{\tan\alpha-\tan\beta}$$

Case III : Tube drawing with moving mandrel

Similar extrusions can be written as

$$\sigma_d = \left(\frac{2\sigma_0}{\sqrt{3}}\right)\left(\frac{1+B_2}{B_2}\right)\left[1-\left(\frac{A_2}{A_1}\right)^{B_2}\right]+\sigma_b\left(\frac{A_2}{A_1}\right)^{B_2} \quad ...(8)$$

Where $$B_2 = \frac{\mu_1-\mu_2}{\tan\alpha-\tan\beta}$$

A_1 is incoming area at tube and A_2 is exit area at tube

Case IV : Frictionless wire drawing

Case (iii): Now put $\mu = 0$ in initial equation (2)

So, $$\frac{d\sigma_x}{dr} + \frac{2\sigma_x}{r} + \frac{2P_x}{r} = 0$$

$$\sigma_x + P_x = \sigma_0 \qquad \left[\because K' = \frac{\sigma_0}{2}\right]$$

$$\int d\sigma_x = \int -\frac{2\sigma_0}{r} dr$$

$$\sigma_x = -2\sigma_0 \ln(r) + C_1$$

No back pull

Boundary conditions $\quad r = R_i;\ \sigma_x = 0$

$$\sigma_d = \sigma_0 \ln\left(\frac{R_i}{R_f}\right)^2$$

$$\sigma_d = \sigma_0 \ln\left(\frac{\pi R_i}{\pi R_f}\right)^2$$

$$\sigma_d = \sigma_0 \ln\left(\frac{A_i}{A_f}\right) \Rightarrow \sigma_d = \sigma_0 \ln R \qquad ...(9)$$

Drawing force in all the cases $= \sigma_d \times \pi R_0^2$

When $\sigma_d = \sigma_0$ then maximum reduction occurs

So, $$1 = \ln\left(\frac{A_i}{A_f}\right)$$

$$e = \frac{A_i}{A_f}$$

$$\%\text{ reduction} = \frac{A_i - A_f}{A_i} \times 100$$

$$= \left(1 - \frac{A_f}{A_i}\right) \times 100$$

$$= \left(1 - \frac{1}{e}\right) \times 100$$

$$\%\text{ reduction} = \left(\frac{e-1}{e}\right) \times 100 = 63\%$$

For extrusion:

the boundary conditions will be reverse i.e.

$r = R_i;\ \sigma_x = \sigma_b = 0$

$$\sigma_d = 2K'\left[\frac{1+B}{B}\right]\left[1 - \left(\frac{A_f}{A_i}\right)^B\right]$$

When

$$r = R_0;\ \sigma_x = 0$$

$$\sigma_d = 2K'\left[\frac{1+B}{B}\right]\left[\left(\frac{A_i}{A_f}\right)^B - 1\right]$$

$$\sigma_d = 2K'\left[\frac{1+B}{B}\right]\left[R^B - 1\right] = \sigma_0\left(\frac{1+B}{B}\right)(R^B - 1) \qquad \text{...(10)}$$

where R is extrusion ratio.

Example : 5.2

In a drawing operation the initial diameter of the blank is given as 100mm and final diameter of the blank is 50mm. Find the load required for the operation. Given σ_0 = 250Mpa.

Solution :

$$D_i = 100 \text{ mm}$$

$$D_f = 50 \text{ mm}$$

$$\sigma_0 = 250 \text{ MPa}$$

$$\sigma_d = \sigma_0 \ln R = 346.51 \text{ MPa}$$

$$\text{Load} = \frac{\pi}{4} D_f^2 \sigma_d = \frac{\pi}{4}(50^2)346.51 = 680.37 \text{ kN}$$

Example : 5.3

An Aluminium rod of 6.25 mm diameter is drawn into wire of 5.6 mm diameter. Semi die angle is 10.1°. Find the drawing stress, considering the friction to be 0.04 and nominal stress 35 Mpa. Also calculate the maximum reduction that can be given to material.

Solution :

$$d_i = 6.25 \text{ mm}$$

$$d_0 = 5.6 \text{ mm}$$

$$\alpha = 10.1°$$

$$\mu = 0.04$$

$$\sigma_0 = 35 \text{ MPa}$$

$$\sigma_d = ?$$

σ_b is not given so it is zero

$$\sigma_d = \sigma_0\left(\frac{B+1}{B}\right)\left[1 - \left(\frac{r_f}{r_i}\right)^{2B}\right]$$

$$B = \mu \cot \alpha$$

$$B = 0.04 \cot (10.1)$$

$$B = 0.224$$

$$\sigma_d = 35\left(\frac{0.224+1}{0.224}\right)\left[1 - \left(\frac{5.6}{6.25}\right)^{2(0.224)}\right]$$

$$\sigma_d = 10.6 \times \frac{\sqrt{3}}{2}$$

$$\sigma_d = 9.17 \text{ MPa}$$

Maximum reduction

$$\sigma_d = \sigma_0 \text{ (For maximum reduction)}$$

$$1 = \left(\frac{B+1}{B}\right)\left[1-\left(\frac{A_f}{A_i}\right)^B\right]$$

$$1 = \left(\frac{0.224+1}{0.224}\right)\left[1-\left(\frac{A_f}{A_i}\right)^{0.224}\right]$$

$$\left(\frac{A_f}{A_i}\right) = 0.403$$

$$\% \text{ reduction} = \left(\frac{A_i - A_f}{A_i}\right)\times 100 = (1-0.403)\times 100$$

$$\therefore \% \text{ reduction} = 59.6\%$$

Example : 5.4

An alloy is extruded in forward extrusion process from 150 mm diameter billet to 50mm diameter rod, the mass flow stress of material is 250 MPa. The billet is 380 mm long. Semi die angle is 60° and coefficient of friction is 0.1. Determine extrusion load.

Solution :

$$D_i = 150 \text{ mm}; D_f = 50 \text{ mm}$$

$$\sigma_{fm} = \sigma_0 = 250 \text{ MPa}; \alpha = 60°$$

$$L = 380 \text{ mm}; \mu = 0.1$$

$$B = \mu \cot \alpha$$

$$= 0.1 \cot 60° = 0.0577$$

For plane strain

$$K' = \frac{\sigma_0}{\sqrt{3}}$$

$$K' = \frac{250}{\sqrt{3}} = 144.33 \text{ MPa}$$

$$\sigma_d = \sigma_0\left(\frac{1+B}{B}\right)\left(R^B - 1\right)$$

$$\sigma_d = 250\left(\frac{1+0.0577}{0.0577}\right)\left(9^{0.0577} - 1\right)$$

$$\sigma_d = 619.4 \text{ MPa}$$

$$P_f = \frac{4K'L}{D} = \frac{4\,(144.33)\,(380)}{150}$$

$$P_f = 1462 \text{ MPa}$$

$$P = \sigma_d + P_f$$

$$P = 2081.94 \text{ MPa}$$

$$\text{Total load} = 2081.94 \times \frac{\pi}{4} \times 150^2 \times 10^{-6}$$

$$= 36.79 \text{ N}$$

Example : 5.5

Determine the drawing stress to reduce the 10 mm diameter of steel to 8 mm with a semi die angle of 6° and coefficient of friction of 0.09. Consider the flow curve equation of material to be $\sigma_f = 1300\varepsilon^{0.3}$ MPa.

Solution :

Given data

$$\alpha = 6°$$

$$\mu = 0.09$$

$$B = \mu \cot \alpha = 0.09 \cot 6$$

$$= 0.857$$

True strain $$(\varepsilon) = \ln\left(\frac{l_f}{l_i}\right) = \ln\left(\frac{A_i}{A_f}\right)$$

$$\varepsilon = \ln\left(\frac{d_i^2}{d_f^2}\right) = \ln\left(\frac{10}{8}\right)^2 = 0.446$$

Mean flow stress $$\sigma_0 = \sigma_{fm} = \frac{K\varepsilon^n}{n+1}$$

$$= \frac{1300\,(\varepsilon)^{0.3}}{1.3} = 785 \text{ MPa}$$

So the drawing stress $$\sigma_d = \sigma_0\left(\frac{1+B}{B}\right)\left[1-\left(\frac{r_f}{r_i}\right)^{2B}\right]$$

$$\sigma_d = 785\left(\frac{1+0.857}{0.857}\right)\left[1-\left(\frac{4}{5}\right)^{2\times 0.857}\right]$$

$$\sigma_d = 540.5 \text{ MPa}$$

Example : 5.6

After passing through the conical portion, the wire passes through the die land of 2 mm length. The incoming wire diameter is 6 mm and outgoing diameter is 5 mm. The semi die angle is 8° and coefficient of friction is 0.05. If the nominal strength of material is 25 MPa, calculate the pull required to draw the wire.

Solution :

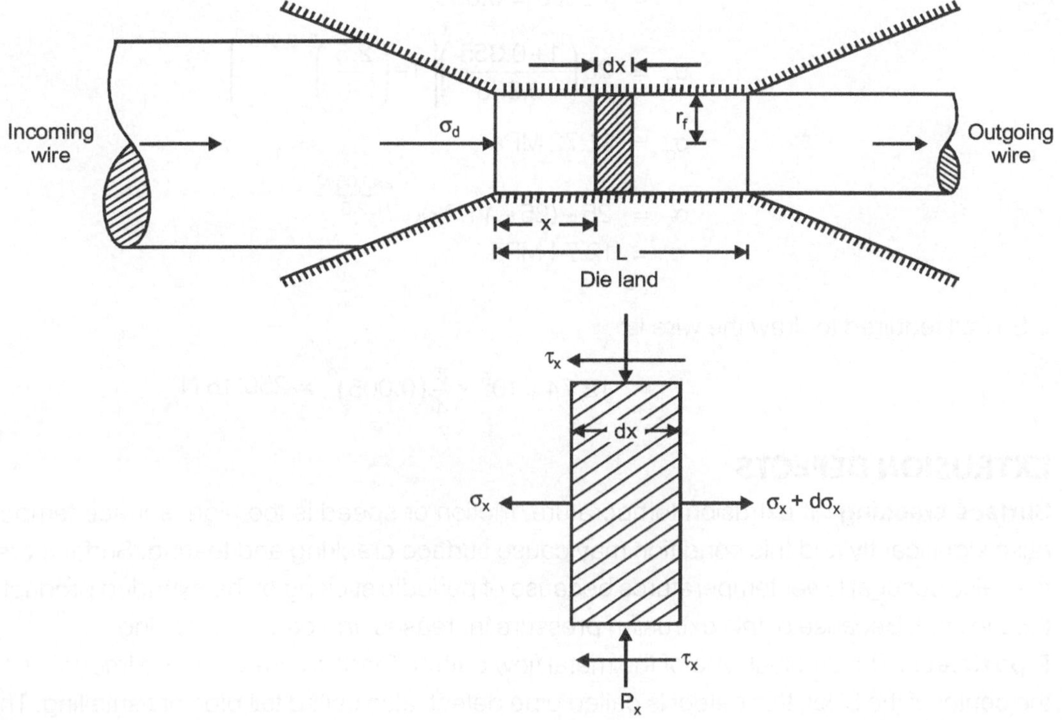

Fig. 5.10

Having the equilibrium of forces in the horizontal direction

$$(\sigma_x + d\sigma_x)\pi r_f^2 - \sigma_x \pi r_f^2 - \tau_x.2\pi r_f dx = 0 \quad ...(11)$$

So $$d\sigma_x = \frac{2\tau_x}{r_f}dx = \frac{2\mu p_x}{r_f}dx$$

Considering σ_x and p_x and also due to spherical coordinate system both von-mises and Tresca will lead to the same result.

$$\sigma_x + P_x = \sigma_0$$

$$P_x = \sigma_0 - \sigma_x$$

$$\int \frac{d\sigma_x}{\sigma_0 - \sigma_x} = \int \frac{2\mu}{r_f}dx$$

$$\ln(\sigma_0 - \sigma_x) = \frac{-2\mu}{r_f}x + c_1 \quad ...(12)$$

Boundary condition : at $x = 0, \sigma_x = \sigma_d$ where

$$\sigma_d = \sigma_0\left(\frac{1+B}{B}\right)\left[1-\left(\frac{r_f}{r_i}\right)^{2B}\right]$$

So $$\sigma_x = \sigma_0 - (\sigma_0 - \sigma_d)e^{-\frac{2\mu x}{r_f}} \quad ...(13)$$

at $x = L, d_x = s_f$

Data given in the problem $\alpha = 8°, \mu = 0.05, \sigma_0 = 25$ MPa

$$r_i = 3 \text{ mm}, r_f = 2.5 \text{ mm}, L = 2 \text{ mm}$$

So,

$$B = \mu \cot \alpha = 0.355$$

$$\sigma_d = 25\left(\frac{1+0.355}{0.355}\right)\left(1-\left(\frac{2.5}{3}\right)^{2\times0.355}\right)$$

$$\sigma_d = 11.72 \text{ MPa}$$

$$\sigma_x = 25-(25-11.72)e^{-\frac{2\times0.05\times2}{2.5}}$$

$$\sigma_t = 12.74 \text{ MPa}$$

So Pull required to draw the wire is:

$$= 12.74\times10^6\times\frac{\pi}{4}(0.005)^2 = 250.16 \text{ N}$$

5.5 EXTRUSION DEFECTS

Surface cracking : If extrusion temperature, friction or speed is too high, surface temperature raise significantly and this condition may cause surface cracking and tearing. Surface cracking may also occur at lower temperatures because of periodic sticking of the extruded product along the die land. Because of this extrusion pressure increases and causes cracking.

Pipe defect : The unusual type of the metal flow pattern tends to draw surface impurities toward the center of the billet, this defect is called pipe defect, also called tail pipe or fishtailing. This can be minimized by modifying the flow pattern to more uniform. These impurities can also be removed by chemical etching of the surface oxides before extrusion.

Internal cracking : Extruded product center can develop various cracks called center cracking, center burst, arrowhead fracture or chevron cracking. These cracks are attributed to a state of hydrostatic tensile stresses at the center line in the deformation zone in the die. Center cracking increases with increase in die angle, increase in amount of impurities, and increase in extrusion ratio and friction.

5.6 FORGING

Forging is shaping of metals by squeezing them between two or more dies to obtain the desired shape. Forging can be done either at room temperature or at elevated temperature. If the forging operation is carried out below the recrystallization temperature, it's known as cold forging and if at temperature above recrystallization is called hot forging. At elevated temperatures there will not be any work hardening.

If forging is done in an open hearth is known as smith forging. In this hot work piece given desired shape by using hand held tools and hammers. Now-a-days power driven hammers are used to impart the repeated blows. The anvil and hammer are mostly flat and the desired shape is obtained by a manipulation of the job between the blows.

Edging : It is performed by a pair of concave dies and the material moves toward the center and it increases the size of stock.

Fullering : It is performed by a pair of convex dies and in this material moves away from the center, this decreases the size of a slab.

There are three types of forging

1. Open die forging

2. Closed die forging
3. Impression die forging.

Open die forging : In open die forging material is kept in the lower die and hammer comes in contact with the material. This results in poor surface finish and also cracks appear at top surface over the forged component. So a layer of material has to be removed from work piece. So in open die forging there is poor material utilization. In this there is less control in determining mechanical properties and difficult to maintain tolerances and surface finish.

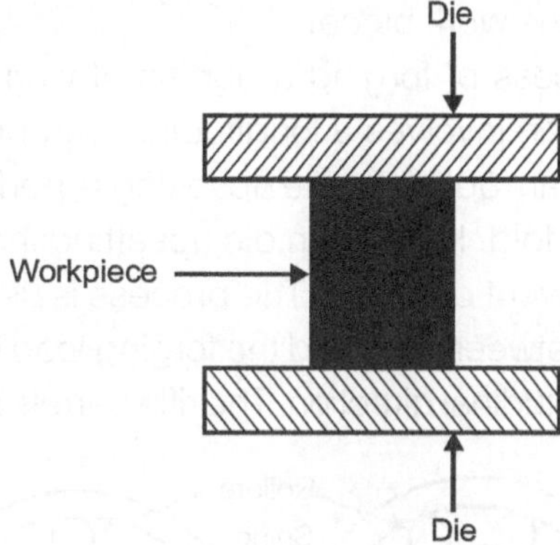

Fig 5.11 : (a) Open die forging

Forgings are made by this process when:

1. The forging is too large to be produced in closed dies
2. The required mechanical properties of the worked metal that can be developed by open-die forging cannot be obtained by other deformation processes
3. The quantity required is too small to justify the cost of closed dies

Closed die forging : In closed die forging operation metal is kept between top and bottom dies. Bottom die is connected with a gutter and the extra material settles down in those gutters called flash. After forging these flash has to be removed. This is the method by which we can make coins. Closed-die forging is the shaping of hot metal completely within the walls or cavities of two dies that come together to enclose the workpiece on all sides. The impression for the forging can be entirely in either die or can be divided between the top and bottom dies. The forging stock, generally round or square bar, is cut to length to provide the volume of metal needed to fill the die cavities, in addition to an allowance for flash and sometimes for a projection for holding the forging. The flash allowance is, in effect, a relief valve for the extreme pressure produced in closed dies. Flash also acts as a brake to slow the outward flow of metal in order to permit complete filling of the desired configuration.

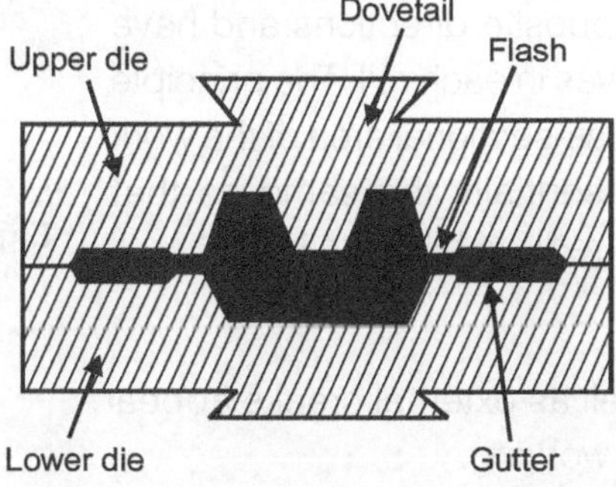

Fig 5.11 : (b) Closed die forging

Impression die forging : In impression die forging the calculated amount of material is kept in the pair of dies and there are no gutters with the lower die and it acquires the shape of the die cavities while being forced between two shaped dies. The thin flash which flow outwards cools rapidly and because of its frictional resistance it subjects the material in the die cavity to high pressures there by encouraging the filling of the die cavity. It provides very good surface finish.

Some other processes are :

Hubbing : Impressions are there in the hammer and by repeated blow to work piece same impressions are develop in the work piece.

Upset forging : It is a process of forging a portion of work. The upsetting operation may be closed or open. The operation involves a longitudinal compression of the bar stock. If the over hanging portion is greater than '3d' close die upsetting is performed otherwise material will bend and it will appear defect like fold. In an open die upsetting the length of the unsupported portion should not exceed 3d to prevent buckling. The process is used to produce bolt head.

Swaging : Material is kept between dies and the forging load is applied to the work through these dies. Metal moves in the longitudinal direction. The rifle barrels are being produced by this process.

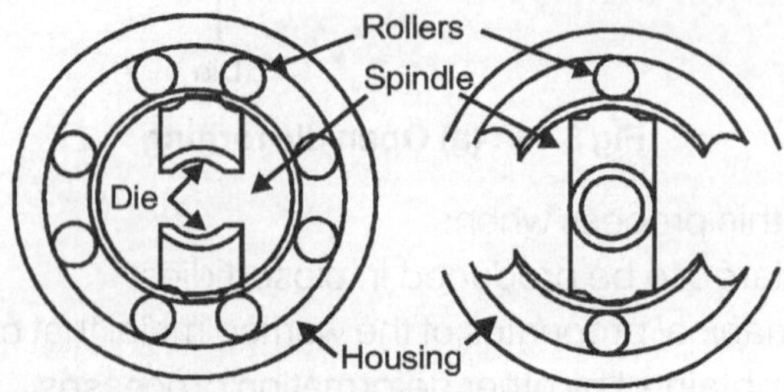

Fig 5.11 : (c) Rotary swaging

Roll forging : Rolls are made up of non uniform radius and when the lump in radius comes in contact with the material, work piece material deforms. Railway wheels are made by this procedure. The process is also used for reducing the diameter of rods. The heated work piece is placed between the dies in an open position and after a half revolution of the rolls the work pieces rolled out it is then put in the smaller group and the operation continued until the desire dimension is achieved. Roll forging (also known as hot forge rolling) is a process for reducing the cross-sectional area of heated bars or billets by passing them between two driven rolls that rotate in opposite directions and have one or more matching grooves in each roll. The principle involved in reducing the cross-sectional area of the work metal in roll forging is essentially the same as that employed in rolling mills to reduce billets to bars.

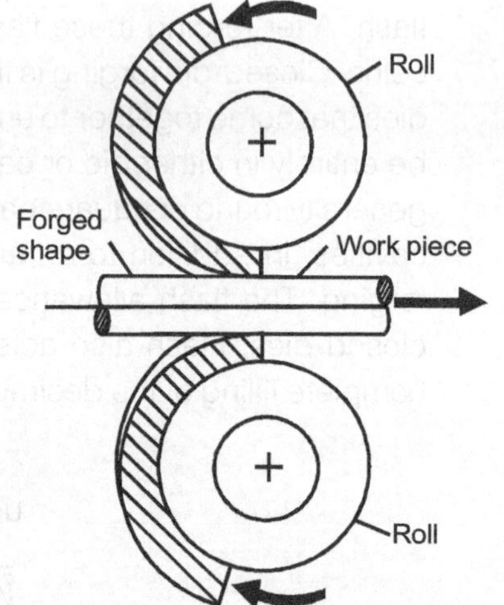

Fig. 5.11 : (d) Principle of roll forgoing

5.6.1 Forging defects

- **Cracks:** Internal as well as external cracks appear due to excessive cold working.
- **Fold:** It is discontinuity produced when two surfaces of metal fold against each other without fusing properly.

- **Barreling :** when hot work comes in contact with the dies there will be a decrease in the temperature of the surface coming in contact with dies. Upon applying the forging load, since central portion will have lower flow stresses, it will move faster than the material near the edges. This result in non cylindrical large components called Barreling.

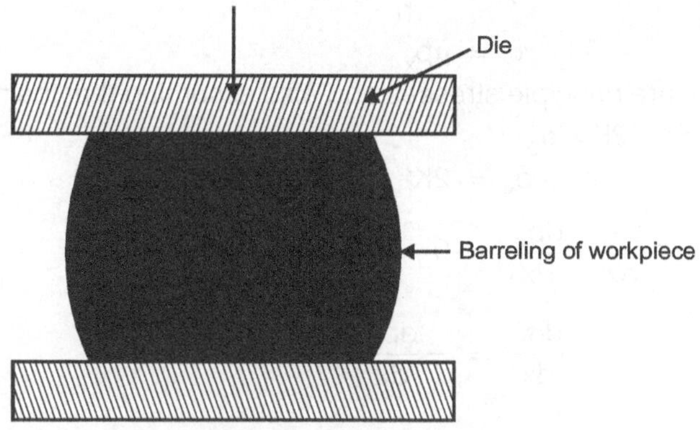

Fig. 5.12

5.6.2 Analysis of forging

(a) Forging between Flat dies : (plane strain compression)

After forging the width of the stock does not change. Height decreases and length increases. Let say h is the final height of billet and w is the width.

Let us have equilibrium of forces in horizontal direction.

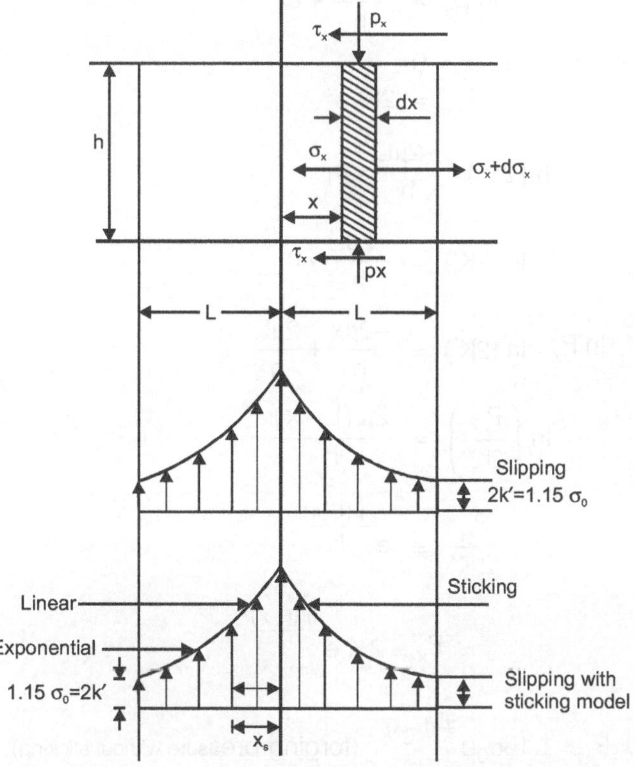

Fig. 5.13 : Friction hill (forging in rectangular dies)

$$(\sigma_x + d\sigma_x)hw - \sigma_x hw - 2w\tau_x dx = 0 \quad ...(14)$$
$$\sigma_x hw + d\sigma_x hw - \sigma_x hw - 2w\tau_x dx = 0$$
$$d\sigma_x hw - 2w\tau_x dx = 0$$
$$d\sigma_x = \frac{2\tau_x dx}{h} \quad ...(15)$$

as $\tau_x = \mu p_x$

Assuming σ_x and p_x are principle stresses and also assuming that the material is behaving as a plane stress case. So $2K = \sigma_0$

$$\sigma_x + p_x = 2K'$$
$$\frac{d\sigma_x}{dx} + \frac{dp_x}{dx} = 0$$
$$\frac{d\sigma_x}{dx} = -\frac{dp_x}{dx}$$

From equation (1)

$$-dp_x = \frac{2\tau_x dx}{h}$$
$$-dp_x = \frac{2\mu p_x}{h} dx$$
$$\int \frac{-dp_x}{p_x} = \int \frac{2\mu}{h} dx$$
$$\ln p_x = \frac{-2\mu x}{h} + c_1$$

Boundary conditions; at $x = L$, $\sigma_x = 0$

So $P_{x_1} = 2K'$

$$\ln(2k') = \frac{-2\mu L}{h} + c_1$$
$$\ln(2K') = \frac{-2\mu L}{h} + c_1$$

So $$\ln P_x - \ln(2K') = \frac{-2\mu x}{h} + \frac{2\mu L}{h}$$
$$\ln\left(\frac{P_{x_1}}{2K'}\right) = \frac{2\mu(L-x)}{h}$$
$$\frac{P_{x_1}}{2K'} = e^{\frac{2\mu(L-x)}{h}}$$
$$P_{x_1} = 2K'e^{\frac{2\mu(L-x)}{h}}$$
$$P_{x1} = 1.15\sigma_0 e^{\frac{2\mu}{h}(L-X)} \quad \text{(forging pressure without sticking)} \quad ...(16)$$

($K' = \sigma'/\sqrt{3}$ since it is plane strain condition)

When the shear stress becomes equal to the yield shear strength, sticking takes place between the die and the work piece material.

Forging load

Because force is symmetric on two sides so

$$F = 2\int_0^L p_x w dx$$

$$\text{Average die pressure} = \frac{F}{2Lw}$$

For sticking

$$\tau_x = K' \text{ [yield strength]}$$

$$\text{at} \quad x = x_s \quad \mu p_x = K' = \tau_x$$

Apply this condition in previous equation

$$\int dp_x = -\int \frac{2K'}{h} dx$$

$$P_{x_2} = -\frac{2K'x}{h} + c_2 \qquad \text{...(17)}$$

Boundary conditions at; $x = x_s$; $p = \frac{K'}{\mu}$

$$\frac{K'}{\mu} = -\frac{2K'x_s}{h} + c_2 \qquad \text{...(18)}$$

From (17) and (18)

$$P_{x_2} = \frac{K'}{\mu} + \frac{2K'}{h}(x_s - x) \quad \text{Forging pressure with sticking} \qquad \text{...(19)}$$

Calculation of sticking length

$$\Rightarrow \quad \frac{K'}{\mu} = 2K'e^{\frac{2\mu}{h}(L-x_s)}$$

$$\Rightarrow \quad \frac{1}{2\mu} = e^{\frac{2\mu}{h}(L-x_s)}$$

$$\Rightarrow \quad \frac{2\mu}{h}(L-x_s) = \ln\left(\frac{1}{2\mu}\right)$$

$$\Rightarrow \quad x_s = L - \frac{h}{2\mu}\ln\left(\frac{1}{2\mu}\right) \qquad \text{...(20)}$$

Forging load with sticking and slipping

$$\text{Load} = 2\left[\int_0^{x_s} P_{x_2} w dx + \int_{x_s}^{L} P_{x_1} w dx\right] \qquad \text{...(21)}$$

This is the load with sticking

(B) Forging between circular dies

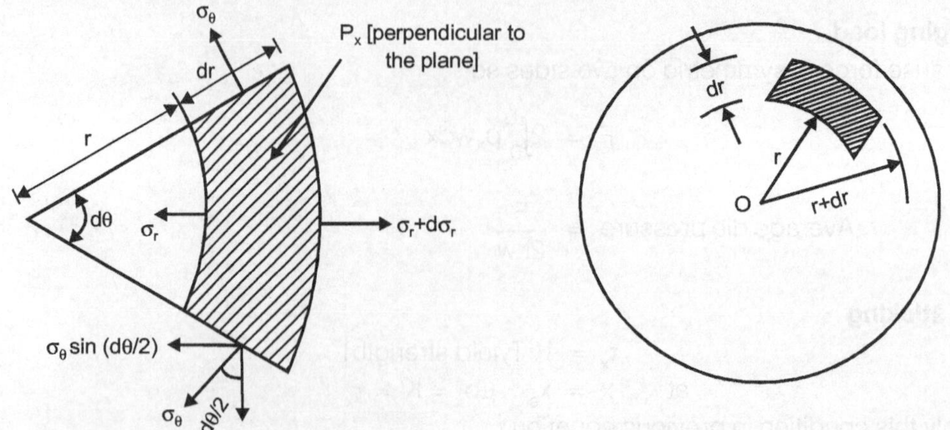

Fig. : 5.14

Considering the equilibrium of forces.

$$(\sigma_r + d\sigma_r)(r + dr)\,d\theta\,h - \sigma_r r d\theta\,h - 2\sigma_\theta \sin\frac{d\theta}{2}\,dr\,h - 2\tau_r\,(rd\theta dr) = 0 \qquad ...(22)$$

Neglect higher order terms and divide by dr hr

$$(\sigma_r r + \sigma_r dr + d\sigma_r r + d\sigma_r dr)\,d\theta\,h - \sigma_r r d\theta h - 2\sigma_0 sn\frac{d\theta}{2}\,drh - 2\tau_r r d\theta dr = 0$$

$$\sin\frac{d\theta}{2} \approx \frac{d\theta}{2}$$

$$\frac{\sigma_r}{r} + \frac{d\sigma_r}{dr} - \frac{\sigma_\theta}{r} - \frac{2\tau_r}{h} = 0$$

$$\frac{d\sigma_r}{dr} + \left[\frac{\sigma_r - \sigma_\theta}{r}\right] - \frac{2\tau_r}{h} = 0 \qquad ...(23)$$

Circumferential strain

$$\sigma_0 = \frac{(r + dr)\,d\theta - rd\theta}{rd\theta} = \frac{drd\theta}{rd\theta}$$

$$\sigma_0 = \frac{dr}{r}$$

This is also known as radial strain

$$\varepsilon_0 = \varepsilon_r$$

If these are equal then $\sigma_\theta = \sigma_r$

(a) Since $\sigma_r = \sigma_\theta = \sigma$, normal pressure (forging) and friction stress in $\tau_r(\tau_{RX})$ and other friction stresses like $\tau_{\theta y}$ and $\tau_{R\theta}$ are zero. So von-mises yield criteria

$$\frac{1}{6}\left\{(\sigma_r - \sigma_\theta)^2 + (\sigma_\theta - \sigma_{yy})^2 + (\sigma_{yy} - \sigma_r)^2\right\} + \tau_{Ry}^2 + \tau_{y\theta}^2 = \frac{\sigma_0^{\ 2}}{3}$$

Putting all the values in the above expression we get

$$(\sigma_r + p_x)^2 + 3\tau_x^2 = \sigma_0^2$$

(b) If σ_r and p_x are the principal stresses then friction can be neglected to zero,

So $\sigma_r + p_x = \sigma_0$

In case of pure torsion if K′ is the yield strength in shear then

$$3K'^2 = \sigma_0^2$$

So
$$K' = \frac{\sigma_0}{\sqrt{3}}$$

So
$$\sigma_r + p_x = \sqrt{3}K'$$

The following analysis is adapted from B. Avitzer, "Metal forming: process and analysis". TMH, 1977.

So the corresponding formulas for circular compression are

$$p_{r_1} = \sigma_0 e^{\frac{2\mu}{h}(R-r)} \quad ...(24)$$

$$p_{r_2} = \frac{K'}{\mu} + \frac{2K'}{h}(R_s - r) \quad ...(25)$$

and sticking radius is

$$R_s = R - \frac{h}{2\mu}\ln\left(\frac{1}{\sqrt{3}\mu}\right) \quad ...(26)$$

Forging load:

$$\text{Without sticking} = \int_0^R p_{r_1} 2\pi r dr \quad ...(27)$$

$$\text{With sticking} = \int_0^{R_s} p_{r_2} 2\pi r dr + \int_{R_s}^{R} p_{r_1} 2\pi r dr \quad ...(28)$$

Example : 5.7

In a forging operation the initial and final diameter of blank is 200 mm and 400 mm. Height is 60 mm. find true strain?

Solution :

Given data is

$$d_1 = 200 \text{ mm}$$
$$d_2 = 400 \text{ mm}$$
$$h_1 = 60 \text{ mm}$$

$$\frac{\pi}{4} d_1^2 h_1 = \frac{\pi}{4} d_2^2 h_2$$

$$d_1^2 h_1 = d_2^2 h_2$$

$$h_2 = 15 \text{ mm}$$

$$\varepsilon = \ln\frac{h_1}{h_2} = \ln\frac{60}{15} = 1.38$$

Example : 5.8

A strip of lead with initial dimensions 24 mm × 24 mm × 150 mm is forged between two flat dies to a final size of 6 mm × 96 mm × 150 mm. If the coefficient of friction is 0.05, determine the maximum forging force. The average yield stress of lead in tension is 7 N/mm².

Solution :

Given,

Initial dimensions of billet 24 × 24 × 150 mm^3

Final dimensions of billet 6 × 96 × 150 mm^3

as width never changes after forging of rectangular billet and all the calculations of forging operation are done on the final dimensions of the billet.

So the dimensions of billet are as

$$\text{thickness of billet } (h) = 6 \text{ mm}$$
$$\text{length of billet } (2L) = 96 \text{ mm}$$
$$\text{width of billet } (w) = 150 \text{ mm}$$
$$\text{coefficient of friction } (\mu) = 0.05$$

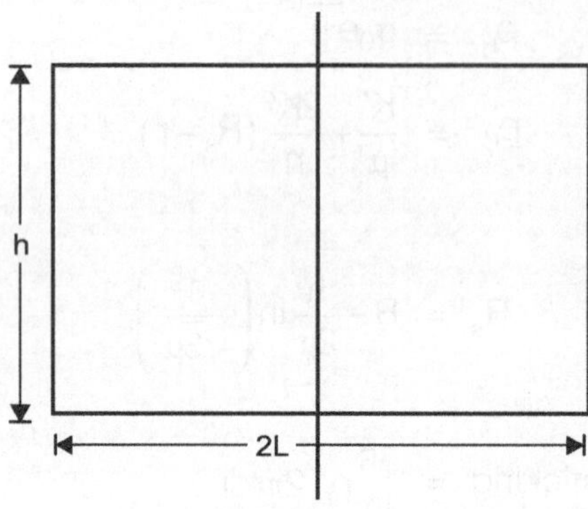

Fig. : 5.15

Step I : Checking for the sticking

As we know sticking length in case of rectangular billet is given by

$$\text{sticking length } (x_s) = L - \frac{h}{2\mu} \ln\left(\frac{1}{2\mu}\right)$$

$$x_s = 48 - \frac{6}{2 \times .05} \ln\left(\frac{1}{2 \times 0.05}\right)$$

$$x_s = -90.155 \text{ mm}$$

negative sign shows that sticking will not occur during the operation.

Step II :

As we know that forging stress when there is no sticking is given by

$$P_{x_1} = 2K' e^{\frac{2\mu}{h}(L-x)}$$

as forging of rectangular billet is a case of plane strain,

So $$2K' = \frac{2\sigma_0}{\sqrt{3}}$$ where σ_0 = average yield stress

$$\sigma_0 = 7 \text{ N/mm}^2 \quad \text{(given)}$$

$$\Rightarrow \qquad P_{x_1} = \frac{2\times 7}{\sqrt{3}} e^{\frac{2\times 0.05}{6}(48-x)}$$

$$P_{x_1} = 8.08e^{\frac{(48-x)}{60}}$$

so,

$$\text{forging load} = 2x\int_0^L P_{x_1}.w.dx$$

$$= 2x\int_0^{48} 8.08\, e^{\left(\frac{48-x}{60}\right)} \times 150\ dx$$

$$= (2\times 8.08\times 150)\int_0^{48} e^{\frac{(48-x)}{60}} dx$$

$$= (2\times 8.08\times 150)\times(-60)\left| e^{\frac{(48-x)}{60}} \right|_0^{48}$$

$$= -(2\times 8.08\times 150\times 60)\left(e^0 - e^{\frac{48}{60}}\right)$$

$$\text{forging load} = 178.24\text{ kN}$$

Example : 5.9

A strip of lead with initial dimensions 24 mm x 24 mm x 150 mm is forged between two flat dies to a final size of 6 mm x 96 mm x 150 mm. If the coefficient of friction is 0.25, determine the maximum forging force. The average yield stress of lead in tension is 7 N/mm^2.

Solution :

Given,

Initial dimensions of billet 24 × 24 × 150 mm^3

Final dimensions of billet 6 × 96 × 150 mm^3

as width never changes after forging of rectangular billet and all the calculations of forging operation are done on the final dimensions of billet

So, the dimensions of billet are

thickness of billet (h) = 6 mm

length of billet (2L) = 96 mm

width of billet (w) = 150 mm

coefficient of friction (μ) = 0.25

Step I : Checking for the sticking

As we know sticking length in case of rectangular billet is given by

$$\text{Sticking length } (x_s) = L - \frac{h}{2\mu}\ln\left(\frac{1}{2\mu}\right)$$

$$x_s = 48 - \frac{6}{2\times 0.25}\ln\left(\frac{1}{2\times 0.25}\right)$$

$$x_s = 39.68\text{ mm}$$

+ve sign shows sticking will occur during the forging operation.

Step II :

As we know that forging stress during sticking zone is given by

$$P_{x_2} = \frac{K'}{\mu} + \frac{2K'}{h}(x_s - x)$$

as forging of rectangular billet is a case of plane strain,

So $$2K' = \frac{2\sigma_0}{\sqrt{3}}$$ where, σ_0 = mean flow stress

$$\sigma_0 = 7\ \text{N/mm}^2\ \text{(given)}$$

$$\Rightarrow \quad P_{x_2} = \frac{\sigma_0}{\sqrt{3}\mu} + \frac{2\sigma_0}{\sqrt{3}.h}(x_s - x)$$

$$P_{x_2} = \left(\frac{7}{\sqrt{3} \times .25}\right) + \left(\frac{2 \times 7}{\sqrt{3} \times 6}\right)(39.68 - x)$$

$$P_{x_2} = 16.16 + 53.45 - 1.347\,x$$

$$P_{x_2} = 69.61 - 1.347\,x$$

So, forging load during sticking $$= 2x\int_0^{x_s} P_{x_2}.w dx$$

$$= (2 \times 150)\,x\int_0^{39.68}(69.61 - 1.347x)\,dx$$

$$= 300x\left|69.61x - \frac{1.347x^2}{2}\right|_0^{39.68}$$

$$= 300x\left|(69.61 \times 39.68 - \frac{1.347}{2}(39.68)^2\right|_0^{39.68}$$

$$= 510.507\ \text{kN}$$

Step : III

Forging stress during Non-sticking zone. As we know forging stress during non-sticking zone is given by

$$P_{x_1} = 2K'e^{\frac{2\mu}{h}(L-x)}$$

$$P_{x_1} = \left(\frac{2 \times \sigma_0}{\sqrt{3}}\right).e^{\frac{2 \times .25}{6}(48-x)}$$

$$P_{x_1} = 8.08\,e^{\frac{(48-x)}{12}}$$

So, Forging load $$= 2x\int_{x_s}^{L} P_{x_1}.w.dx$$

$$= (2 \times w)\int_{39.68}^{48} 8.08\,e^{\left(\frac{48-x}{12}\right)}dx$$

$$= (2 \times 150 \times 8.08) \times (-12) \left| e^{\frac{48-x}{12}} \right|_{39.68}^{48}$$

$$= -(2 \times 150 \times 8.08 \times 12) \left(e^0 - e^{\frac{48-39.68}{12}} \right)$$

$$= 29.098 \text{ kN}$$

So, total forging load = Load (1) + Load (2)

$$= 510.507 + 29.098$$

Total forging load = 539.60 kN

Example : 5.10

A certain disc of lead of radius 150 mm and thickness 50 mm is reduced to a thickness of 25 mm by open die forging. If the coefficient of friction between the job and die is 0.25, determine the maximum forging force. The average shear yield stress of lead can taken as 4 N/mm²

Solution :

Given

Circular billet and open die forging

Initial radius of circular billet (R_i) = 150 mm

Initial thickness of circular billet (h_i) = 50 mm

Final radius of circular billet (R_f) = R_f mm

Final thickness of circular billet (h_f) = 25 mm

Coefficient of friction (μ) = 0.25

average shear yield stress = σ_0 = 4 N/mm²

Calculation of final radius of circular billet after forging operation

As, (volume of billet before forging) = (volume of billet after forging)

$$\pi R_i^2 \times h_i = \pi R_f^2 \times h_f$$

$$\Rightarrow \quad R_f^2 = R_i^2 \times \frac{h_i}{h_f}$$

$$\Rightarrow \quad R_f^2 = 150^2 \times \frac{50}{25}$$

$$\Rightarrow \quad R_f = 212.13 \text{ mm}$$

As we know all the forging calculations are done on the final dimensions of the billet after forging operation.

So, final dimensions of billet are

Radius of billet (R) = R_f = 212.13 mm

thickness of billet (h) = h_f = 25 mm

Step I :

$$\text{Sticking radius } (R_S) = R - \frac{h}{2\mu}\left[\ln\left(\frac{1}{\sqrt{3}\mu}\right)\right]$$

$$= (212.3) - \left(\frac{25}{2\times 0.25}\right)\left[\ln\left(\frac{1}{\sqrt{3}\times 0.25}\right)\right]$$

$$R_S = 170.28 \text{ mm}$$

+ ve sign shows that sticking will occur.

Step II:

Calculation for the stress in the sticking zone; as we know stress in the sticking zone is given by

$$P_{r_2} = \frac{K'}{\mu} + \frac{2K'}{h}(R_S - r)$$

$$K' = \frac{\sigma_0}{2}$$

$$P_{r_2} = \frac{1}{2}\left[\frac{4}{0.25} + \frac{2\times 4}{25}(R_S - r)\right]$$

$$P_{r_2} = 35.245 - 0.16r$$

$$\text{So, forging load during sticking} = \int_0^{R_S} P_{r_2}\, 2\pi r.dr$$

$$= \int_0^{170.28} (35.245 - 0.16r) 2\pi r.dr$$

$$= 2\pi \int_0^{170.28} (35.245r - 0.16r^2)dr$$

$$= 1.556 \text{ MN}$$

Forging load during sticking = 1.804 MN

Step III:

Calculation for the stress in non-sticking zone $(R_S - r)$

$$P_{r_1} = \sigma_0 e^{\frac{2\mu}{h}(R-r)}$$

$$\Rightarrow \quad = 4\times e^{\frac{2\times 0.25}{25}(212.13-r)}$$

$$P_{r_1} = 4\times e^{\frac{(212.13-r)}{50}}$$

So, forging load in $= \int_{R_S}^{R} P_{r_1} 2\pi r.dr$

$$\text{non-sticking region} = \int_{170.28}^{212.13} (4\times 2\pi)\, e^{\frac{(213.13-r)}{50}} r.dr$$

$$= (8\pi)\int_{170.28}^{212.13} r.e^{\left(\frac{213.13-r}{50}\right)} dr$$

$$= (8\pi)\left[-50.r.e^{\left(\frac{212.13-r}{50}\right)} - \int(-50)e^{\left(\frac{212.13-r}{50}\right)} dr\right]_{170.28}^{212.13}$$

$$= (8\pi)\left[-50.r.e^{\left(\frac{212.13-r}{50}\right)} - (-50)^2 e^{\left(\frac{212.13-r}{50}\right)}\right]_{170.28}^{212.13}$$

$$= (8\pi)\left[\left(-50\times 212.13e^0\right)-\left(-50\times 170.28e^{\frac{212.13-170.28}{50}}\right)-50^2\left(e^0-e^{\frac{212.13-170.28}{50}}\right)\right]$$

= 0.309 MN

So, total forging load = Load (1) + Load (2)

= 1.556+0.309=1.865MN

Example : 5.11

A cylinder of height 60 mm and diameter 100 mm is forged at room temperature between two flat dies. Find the die load at the end of compression to a height of 30 mm, using slab method of analysis. The yield strength of the work material is given as 120 N/mm² and the coefficient of friction is 0.05. Assume that volume is constant after deformation. There is no sticking. Also find the mean die pressure.

Solution :

Given

Initial height of cylinder (h_i) = 60 mm

Initial diameter of cylinder (D_i) = 100 mm

Final height of cylinder after forging (h_f) = 30 mm

Final diameter of cylinder after forging (D_f) = D_f mm

Coefficient of friction (μ) = 0.05

Yield strength = 120 N/mm²

Sticking is not present (given)

Step I : Calculation of final diameter of cylinder after forging

as (volume of cylinder before forging) = (volume of cylinder after forging)

$$\frac{\pi}{4}D_i^2 \times h_i = \frac{\pi}{4}\times D_f^2 \times h_f$$

$$\Rightarrow \quad D_f^2 = D_i^2 \times \frac{h_i}{h_f} = 100^2 \times \frac{60}{30}$$

$$D_f = 141.42 \text{ mm}$$

$$\text{Radius } (R_f) = 70.71 \text{ mm}$$

As we know all the forging calculations are done on the final dimensions of the billet.

Step II : Calculation of forging stress

(No sticking) forging stress if sticking is not present is given by

$$P_{r_1} = (2K')e^{\frac{2\mu}{h}(R-r)}$$

As we know forging of circular billet is a plane stress case

So, $2K' = \sigma_0$

where σ_0 = mean flow stress

σ_0 = 120 N/mm² (given)

$$P_{r_1} = (120)\,e^{\left(\frac{70.71-r}{30}\right)}$$

So, forging load is given by $= \int_0^R P_{r_1} \times 2\pi r\,dr$

$$= (2\pi)\times(120)\int_0^{70.71} r.e^{\left(\frac{70.71-r}{300}\right)} dr$$

$$= (240\times\pi)\left[(-300)\,r\,e^{\frac{(70.71-r)}{300}} - \int(-300)e^{\frac{(70.71-r)}{300}}.dr\right]$$

$$= (240\pi)\left[-300r\,e^{\frac{(70.71-r)}{300}} - (300)^2 e^{\frac{70.71-r}{300}}\right]_0^{70.71}$$

$$= (240\pi)\left[(-300\times 70.71 - 300^2) - \left(0 - 300^2\,e^{\frac{70.71}{300}}\right)\right]$$

Forging load = 2.04 MN

$$\text{Mean die pressure} = \frac{\text{Load}}{\text{Total area}}$$

$$= \frac{2.04\times10^6}{\pi R^2} = \frac{2.04\times10^6}{\pi\times(70.71)^2}$$

Mean Die Pressure = 130 MPa

Example : 5.12

A cylindrical specimen made up of steel is 150 mm in diameter and 100 mm high. It is upsetted by open die forging to a height of 50 mm. Assume that coefficient of friction is 0.2. Calculate the force required to forge the material. If flow curve equation is $\sigma_f = 1030\,\varepsilon^{0.17}$

Solution :

Given

Initial diameter of billet (D_i) = 150 mm

Initial height of billet (h_i) = 100 mm

Final height of billet (h_f) = 50 mm

Final diameter of billet (D_f) = D_f mm

Coefficient of friction (μ) = 0.2

Flow curve equation = $\sigma_f = 1030\,\varepsilon^{0.17}$

Step I : Calculation of final diameter of billet

(Volume of billet before forging) = (Volume of billet after forging)

$$\frac{\pi}{4}\times D_i^2\times h_i = \frac{\pi}{4}\times D_f^2\times h_f$$

$$D_f^2 = 150^2\times\frac{100}{50}$$

$$\Rightarrow \quad D_f = 212.13\text{ mm}$$

$$\text{Final radius of billet } (R_f) = \frac{D_f}{2} = 106\text{ mm}$$

As we know all the forging calculations are done on the final dimensions of the billet.
So, final dimensions of billet are

$$\text{Radius of billet (R)} = R_f = 106 \text{ mm}$$
$$\text{Height of billet (h)} = h_f = 50 \text{ mm}$$

Step I : Checking for sticking

as we know sticking radius in case of forging of circular billet is given by

$$\text{Sticking Radius } (R_s) = R - \frac{h}{2\mu}\ln\left(\frac{1}{\sqrt{3}\mu}\right)$$

$$= 106 - \frac{50}{2\times 0.2}\ln\left(\frac{1}{\sqrt{3}\times 0.2}\right) = -26.51 \text{ mm}$$

–ve sign indicates that there will not be any sticking

Step II : Calculation of forging stress (No sticking)

As we know that stress in non-sticking zone is given by

$$P_{r_1} = 2K' e^{\frac{2\mu}{h}(R-r)}$$

as forging of circular billet is plane stress case

So $2K' = \sigma_0$

$$\Rightarrow \quad P_{r_1} = \sigma_0 . e^{\frac{2\mu}{h}(R-r)}$$

as true strain (ε) is given by

$$\varepsilon = \ln\left(\frac{h_i}{h_f}\right) = \ln\left(\frac{100}{50}\right) = 0.693$$

in case of circular billet;

$$\sigma_0 = \sigma_f = 1030\,\varepsilon^{0.17}$$
$$= 1030 \times (0.693)^{0.17}$$
$$\sigma_0 = \sigma_f = 967.74 \text{ MPa (due to compression)}$$

$$\Rightarrow \quad P_{r_1} = 967.74 e^{\frac{2\times .2}{50}(106-r)}$$

$$P_{r_1} = 967.74\, e^{\left(\frac{106-r}{125}\right)}$$

So, forging load $= \int_0^R P_{r_1} \times 2\pi r\, dr$

$$= (967.74 \times 2\pi)\int_0^{106} r.e^{\frac{(106-r)}{125}}\, dr$$

$$= (6080.48)\left[-125r.e^{\frac{(106-r)}{125}} - \int (-125)e^{\frac{106-r}{125}}.dr\right]_0^{106}$$

$$= (6080.48)\left[-125r.e^{\frac{(106-r)}{125}} - (-125)^2.e^{\frac{106-r}{125}}\right]_0^{106}$$

$$= (6080.48)\left[(-125\times106-125^2)-\left(0-125^2e^{\frac{106}{125}}\right)\right] = 46.26\text{ MN}$$

So, Forging load $= 46.26$ MN

Example : 5.13

A circular disc of 200 mm diameter and 70 mm height is forged to 40 mm height. If the flow curve equation of the material is given by $\sigma_f = 200\,(0.01+\varepsilon)^{0.41}$ MPa and coefficient of friction is 0.05. Determine forging load, mean die pressure and maximum pressure.

Solution :

Initial diameter of billet (D_i) = 200 mm

Initial height of billet (h_i) = 70 mm

Final height of billet (h_f) = 40 mm

Final diameter of billet (D_f) = D_f mm

Coefficient of friction (μ) = 0.05

Flow curve equation is

$$\sigma_f = 200\,(0.01+\varepsilon)^{0.41}$$

Step I : Calculation of final diameter of billet

as (volume of billet before forging) = (Volume of billet after forging)

$$\frac{\pi}{4}\times D_i^2\times h_i = \frac{\pi}{4}\times D_f^2\times h_f$$

$$D_f^2 = D_i^2\times\frac{h_i}{h_f} = 200^2\times\frac{70}{40} = 70,000$$

$$D_f = 264.57\text{ mm}$$

$$\text{Radius }(R_f) = \frac{D_f}{2} = 132.28\text{ mm}$$

Step II : Checking for sticking

As we know sticking radius in case of circular billet is given by,

$$\text{Sticking radius }(R_s) = R-\frac{h}{2\mu}\ln\left(\frac{1}{\sqrt{3}\mu}\right)$$

$$= 132.28-\frac{40}{2\times0.05}\ln\left(\frac{1}{\sqrt{3}\times0.05}\right)$$

$$R_s = -846.29\text{ mm}$$

–ve sign indicates that will not be any sticking

Step III :

as true strain (ε) is given by

$$\varepsilon = \ln\left(\frac{h_i}{h_f}\right) = \ln\left(\frac{70}{40}\right) = 0.559$$

So mean flow stress (σ_0) is

$$\sigma_0 = \sigma_f = 200\times(0.01+0.559)^{0.41}$$

$$\sigma_0 = 158.71\text{ MPa}$$

Step IV : Calculation of forging stress in case of no sticking

$$P_{r_1} = 2K'.e^{\frac{2\mu}{h}(R-r)}$$

as forging of circular billet is a case of plane stress.

So, $$2K' = \sigma_0$$

$$P_{r_1} = \sigma_0.e^{\frac{2\mu}{h}(R-r)}$$

$$= 158.71e^{\frac{2\times0.05}{40}(132.28-r)}$$

$$P_{r_1} = 158.71\times e^{\frac{(132.28-r)}{400}}$$

So, $$\text{forging load} = \int_0^R P_{r_1}.2\pi r\,dr$$

$$= (158.71\times 2\pi)\int_0^{132.28} r.e^{\left(\frac{132.28-r}{400}\right)} dr$$

$$= (158.71\times 2\pi)\left[r(-400)e^{\left(\frac{132.28-r}{400}\right)} - \int(-400)\,e^{\frac{132.28-r}{400}}\right]$$

$$= (158.71\times 2\pi)\left[-400r\,e^{\frac{(132.28-r)}{400}} - 400^2 e^{\frac{(132.28-r)}{400}}\right]_0^{132.28}$$

$$= (158.71\times 2\pi)\left[(-400\times 132.28 - 400^2) - (0 - 400^2 e^{\frac{132.28}{400}}\right]$$

$$\text{Forging load} = 9.771 \text{ MN}$$

$$\text{Mean die pressure} = \frac{\text{Load}}{\text{Total Area}}$$

$$= \frac{9.771\times 10^6}{\pi(132.28)^2} = 177.74 \text{ MPa}$$

Maximum die pressure :

As forging pressure is given by

$$P_{r_1} = \sigma_0.e^{\frac{2\mu}{h}(R-r)}$$

For maximum pressure put r = 0

$$\Rightarrow \quad \text{Maximum die pressure} = \sigma_0.e^{\frac{2\mu}{h}.R}$$

$$= 158.71\,e^{\frac{2\times.05}{40}\times 132.28}$$

$$\text{Maximum die pressure} = 220.91 \text{ MPa}$$

Example : 5.14

A circular disc of 200 mm diameter is 100 mm thick is compressed between two dies to a thickness of 50 mm. Determine the maximum die pressure if coefficient of friction is 0.1 and yield strength in compression is 230 MPa.

Solution :

Given

Initial dia of billet (D_i) = 200 mm
Initial thickness of billet (h_i) = 100 mm
Final thickness of billet (h_f) = 50 mm
Coefficient of friction (μ) = 0.1
Yield strength (σ_0) = 230 MPa

Step I : Calculation of final diameter of billet

as (volume of billet before forging) = (volume of billet after forging)

$$\frac{\pi}{4} \times D_i^2 \times h_i = \frac{\pi}{4} D_f^2 \times h_f$$

$$D_f^2 = D_i^2 \times \frac{h_i}{h_f}$$

$$= 200^2 \times \frac{100}{50}$$

$$D_f = 282.84 \text{ mm}$$

$$\text{Radius } (R_f) = 141.42 \text{ mm}$$

As we know all the forging calculations are done on final dimensions of billet

So, Radius of billet = $R = R_f = 141.42$ mm

Height of billet = $h = h_f = 50$ mm

as die pressure is given by $P_r = \sigma_0 . e^{\frac{2\mu}{h}(R-r)}$

For maximum pressure $r = 0$

$$\Rightarrow \quad P_{max} = \sigma_0 . e^{\frac{2\mu}{h} \times R}$$

$$= 230 \times e^{\frac{2 \times 0.1 \times 141.42}{50}}$$

$$P_{max} = 404.95 \text{ MPa}$$

5.7 ROLLING

As shown in the figure, grain structure of the metal changes when it passes through the rolls. As a result of squeezing the grains are elongated in the direction of rolling and velocity of the material at exit is higher than at the entry. After crossing the stress zone the grains start refining. But this is the case only in the hot rolling. In the cold rolling they tend to retain the shape acquired by them during rolling.

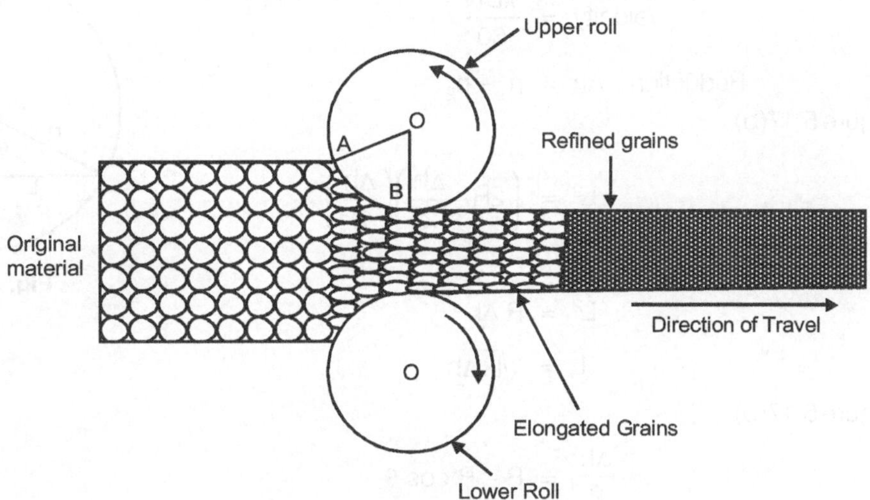

Fig 5.16 : Change of structure during rolling process

The rolls are in contact with the passing metal piece over a sufficient distance represented by the arc AB in the diagram. The angle AOB is called the angle of contact or maximum angle of bite.

5.7.1 Mechanics of Rolling

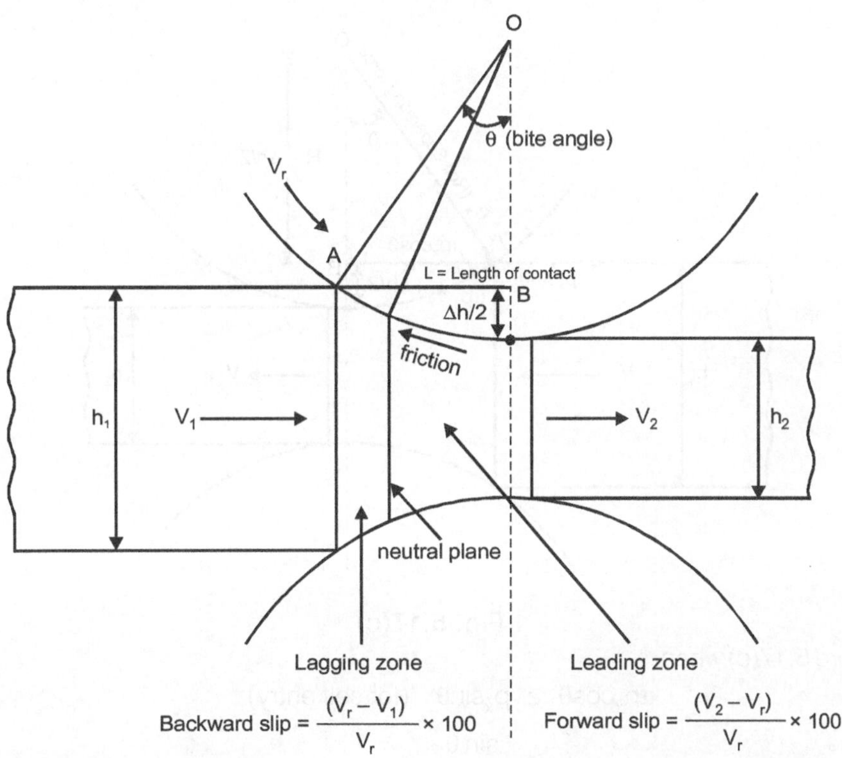

Fig 5.17 : (a) Rolling description

$$\text{Velocity} = \frac{\pi DN}{60}$$

$$\text{Reduction} \quad \Delta h = h_1 - h_2$$

From Figure 5.17(b)

$$L^2 = \left(2R - \frac{\Delta h}{2}\right)\left(\frac{\Delta h}{2}\right)$$

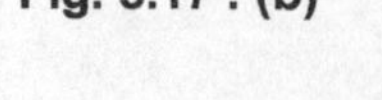

Fig. 5.17 : (b)

Δh^2 is very small

So
$$L^2 = R\,\Delta h$$

$$L = \sqrt{R\,\Delta h} \quad \text{...(29)}$$

From Figure 5.17(b)

$$\frac{\Delta h}{2} = R - R\cos\theta$$

$$\frac{\Delta h}{2} = R(1 - \cos\theta)$$

$$\cos\theta = \frac{R - \frac{\Delta h}{2}}{R}$$

$$\cos\theta = 1 - \frac{\Delta h}{2R}$$

Condition of self entry

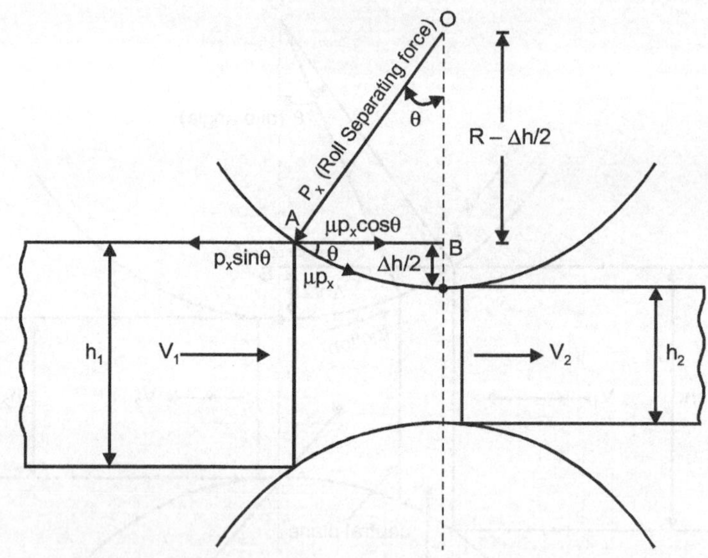

Fig. 5.17(c)

From figure 5.17(c) when

$$\mu p_x \cos\theta \geq p_x \sin\theta \quad \text{(for self entry)}$$

$$\Rightarrow \qquad \mu \geq \frac{\sin\theta}{\cos\theta}$$

$$\Rightarrow \qquad \mu \geq \tan\theta$$

From Fig 5.17(b)
$$\tan\theta = \frac{L}{R-\frac{\Delta h}{2}} = \frac{\sqrt{R\,\Delta h}}{R-\frac{\Delta h}{2}}$$

From the condition of self entry, we get

$$\mu = \frac{\sqrt{R\Delta h}}{R} = \sqrt{\frac{\Delta h}{R}}$$

$$\Delta h = \mu^2 R \quad ...(30)$$

At the entry the velocity of material will be lower than the velocity of rolls, so friction will be in the forward direction. There exists a plane in the deformation zone where the velocity of material is equal to the velocity of roll called neutral plane. Beyond this plane the velocity of rolls and the friction will be in the backward direction. The out going plate thickness will be slightly higher than the minimum gap between the rolls due to elastic recovery.

By increasing the coefficient of friction, the neutral plane shifts towards left. By maintaining the same parameters if the thickness of incoming plate increased, external pressure has to be applied at the entry and this shifts the neutral plane towards right. There will be a situation at which neutral plane coincides with the vertical axis so, there will not be any deformation in the material and the roll will slip over the material. This is called rolling limit and the bite angle at rolling limits is called angle of nip.

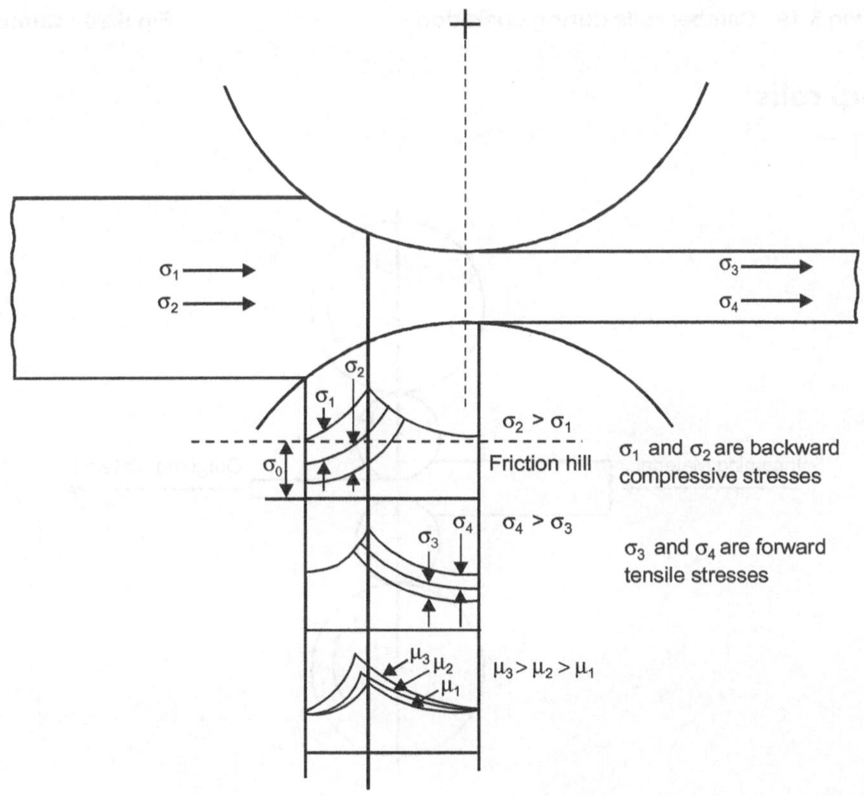

Fig 5.18 : Illustration of friction hill

$$\alpha_n = \text{Angle of nip} = 2\mu \text{ (radians)}$$

$$\alpha_n = 2\mu \times \frac{180}{\pi} \text{ (degrees)}$$

It can be seen from the friction hill that, when the same amount of reduction is given by smaller size rolls lesser amount of rolling load will be required but, there will be a tendency in the roll to deform. This problem can be taken care of by two methods.

5.7.2 Camber rolls

Equal to the deformation of roll, camber is being provided so that when actual deformation is taking place the portion of roll which comes in contact with the material becomes flat. But camber rolls are designed for a particular material and for a particular reduction.

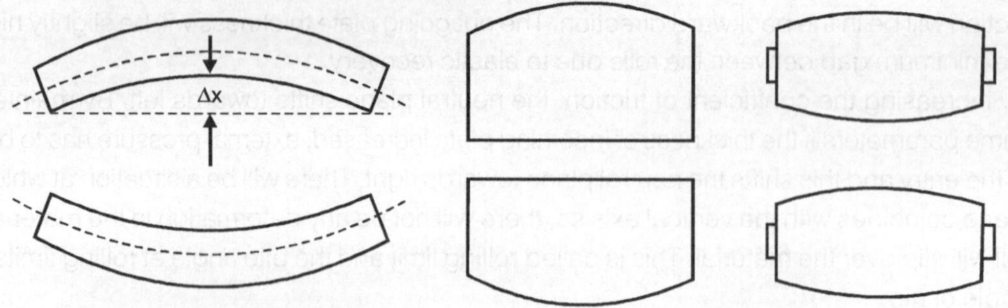

Fig 5.19 : Camber rolls during operation

Fig 5.20 : Camber rolls

5.7.3 Back up rolls

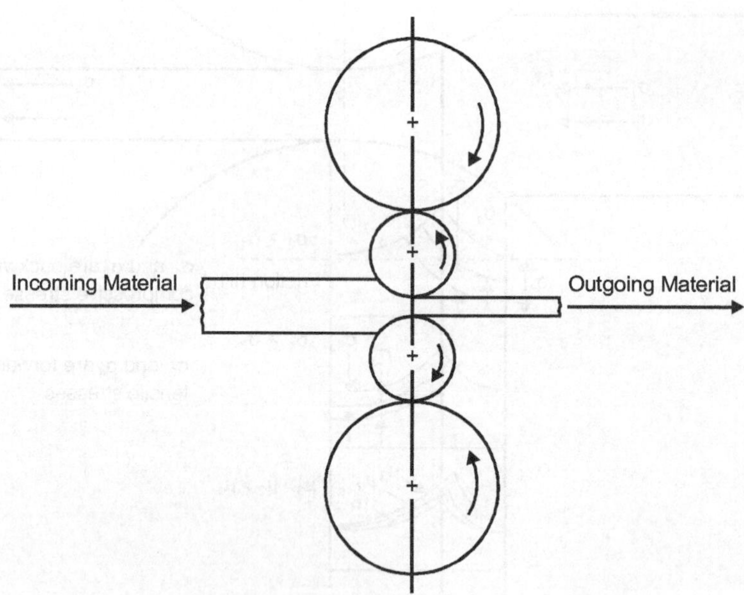

Fig 5.21 : (a) 4 high mill

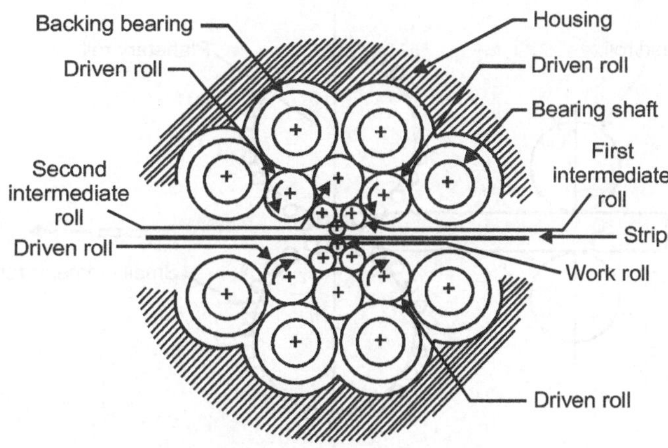

Fig 5.21 : (b) Sandzmier mill

In 4-high mills, bigger rolls are provided for backup and power is given to the smaller rolls. By the variation in % reduction the position of neutral plane will change and hence the direction in which the roll is deforming also changes so in that cluster or sandzmier mills are used.

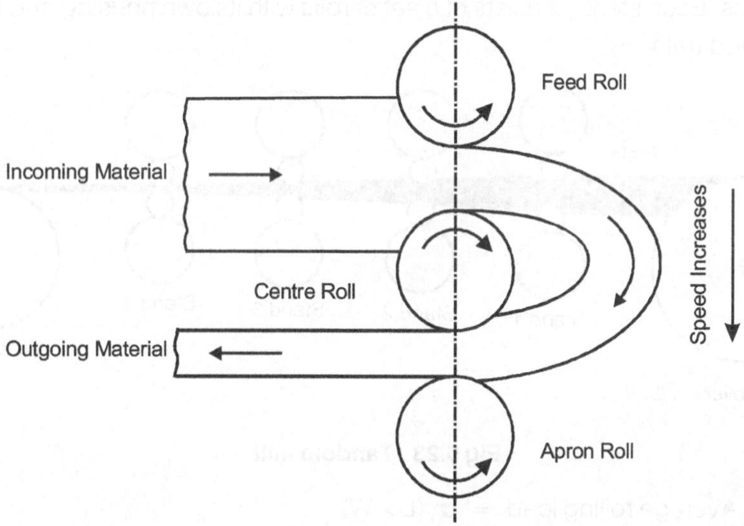

Fig 5.21 : (c) 3 high mills

5.7.4 Three high mill

It is a process of giving 2 stage reduction with the help of 3 rolls. The directions of the rotation of upper and lower rolls are same. But the intermediate roll rotates in a direction opposite to both of these. Although three rolls continuously revolve in the same fixed directions and are never reversed. The work pieces fed in one direction between the upper and middle rolls and reversed direction between the middle and lower rolls.

5.7.5 Planetary Mill

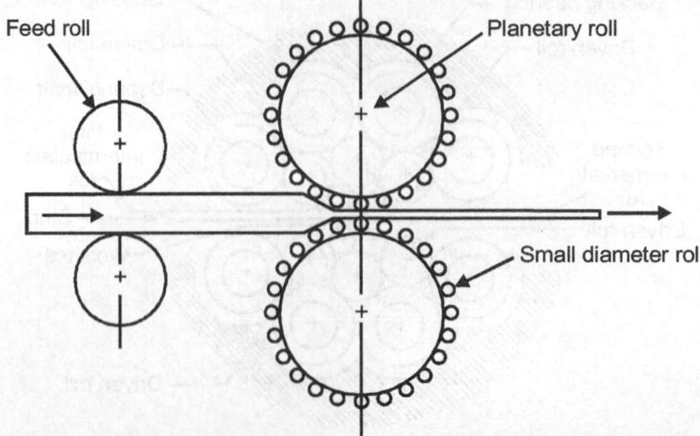

Fig 5.22 : Planetary mill

Over the surface of larger rolls smaller rolls are mounted which rotate in their own axis. Power is given to the larger rolls. Hence there is a rolling limit deformation carried out by a number of pair of rolls in one stage. So a very thick plate can be converted into a thin sheet in one go.

5.7.6 Tandem mill

In tandom rolling the strip is rolled continuously through a number of stands to smaller gauges with each pass. Each stand consists of a set of rolls with its own housing and controls. A group of stands is called train.

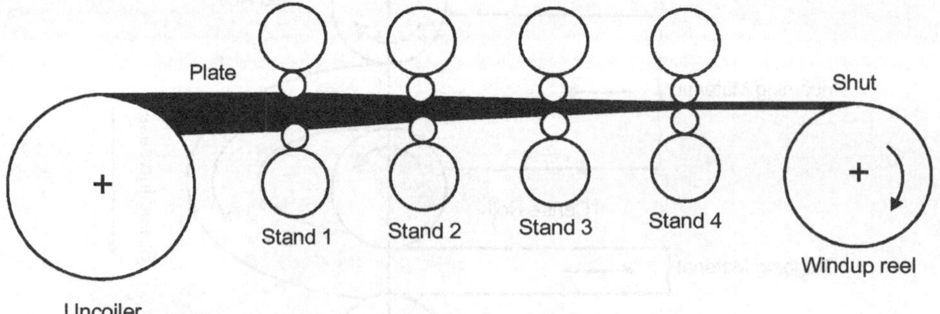

Fig 5.23 : Tandom mill

Average rolling load = σ_0 (L × W)

L = Length of Arc of contact

W = width of sheet

5.7.7 Rolling defects

1. **Wavy edges :** These are the result of roll bending. The strip is thinner along its edges than at its centre. Because the edges elongate more than the centre they buckle because of restraining from expanding freely in the rolling direction.
2. **Spread :** In the rolling of plates and sheet having high width to thickness ratio the width of the material remains constant during rolling. With smaller ratios the width increases considerably in the roll gap. This increase in width is called spreading.

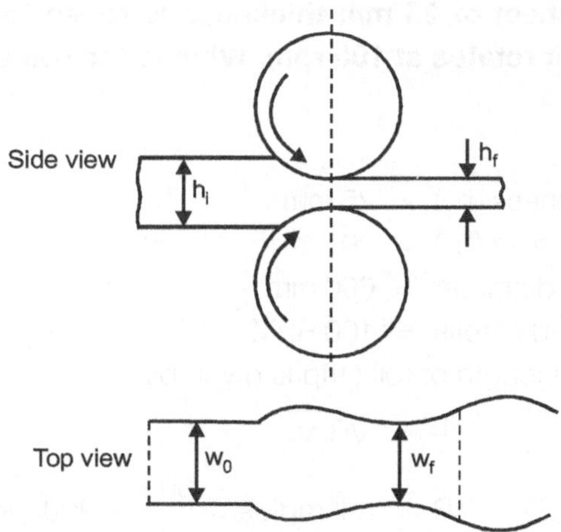

Fig 5.24 : Increase in the width of a strip in flat rolling

3. **Crocodile crack :** Due to weakness at the center sheet bifurcates into two parts. This is called crocodile crack.

5.7.8 Test to calculate the defects

Tensile test specimens are cut 0°, 45° and 90° with the rolling direction. The samples are unloaded before UTS point and the strain ratio is calculated. Higher the value of planer anisotropy more will be the rolling defects like wavy edges. Spread is the increase in width of sheet after rolling.

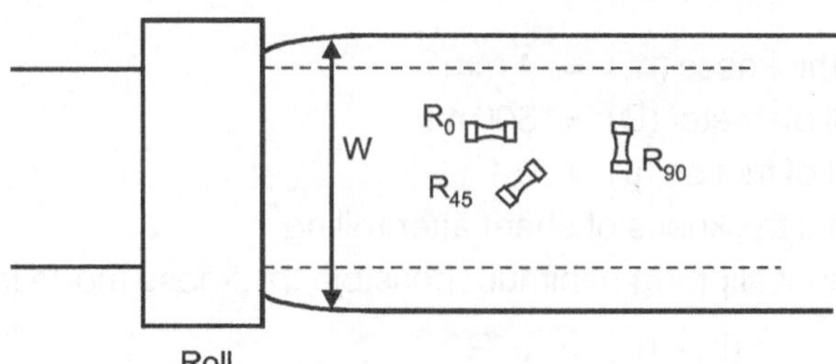

Fig : 5.25

$$\text{Normal anisotropy } (\bar{R}) = \frac{R_0 + 2R_{45} + R_{90}}{4}$$

$$\text{Planar anisotropy} = \Delta R = \frac{R_0 - 2R_{45} + R_{90}}{2}$$

$$\text{where } R = \frac{\text{Width strain}}{\text{Thickness strain}}$$

Formability is the ability of any material to form different shapes. Higher is the value of R (normal anisotropy) more is the formability and more is the value of planar anisotropy larger will be the defects like earing.

When t < 6mm it is called sheet

When t > 6mm it is called plate.

Example : 5.15

In a rolling process, sheet of 25 mm thickness is rolled to 20 mm thickness. Roll is of diameter 600 mm and it rotates at 100 rpm. What is the roll strip contact length?

Solution :

Given

$$\text{Initial thickness } (h_1) = 25 \text{ mm}$$
$$\text{Final thickness } (h_2) = 20 \text{ mm}$$
$$\text{Roll diameter} = 600 \text{ mm}$$
$$\text{Speed of rolls} = 100 \text{ RPM}$$

As we know that contact length of roll strip is given by

$$L = \sqrt{R.\Delta h}$$

where $R = \text{roll radius} = \dfrac{600}{2} = 300 \text{ mm}$

and $\Delta h = h_1 - h_2 = (25 - 20) = 5 \text{ mm}$

Put these values we have

$$L = \sqrt{300 \times 5} = \sqrt{1500} = 38.72 \text{ mm}$$
$$L = 38.72 \text{ mm}$$

Example : 5.16

A 4 mm thick sheet is rolled with 300 mm diameter rolls to reduce thickness without any change in its width. The friction coefficient at the work-roll interface is 0.1. What is the minimum possible thickness of the sheet that can be produced in a single pass?

Solution :

Given

$$\text{initial thickness } (h_1) = 4 \text{ mm}$$
$$\text{roll diameter } (D) = 300 \text{ mm}$$
$$\text{coefficient of friction } (\mu) = 0.1$$

Now let h_2 be the final thickness of sheet after rolling

For single pass without slipping minimum possible thickness mode is

$$(h_1 - h_2) = \mu^2 R$$

where $R = \text{radius of roll} = \dfrac{300}{2} = 150 \text{ mm}$

$$h_2 = h_1 - \mu^2 R \Rightarrow h_2 = 4 - (0.1)^2 \times 150$$
$$h_2 = 4 - 1.5 = 2.5 \text{ mm}$$
$$h_2 = 2.5 \text{ mm}$$

Example : 5.17

The thickness of a metallic sheet is reduced from an initial value of 16 mm to a final value of 10 mm in one single pass rolling with a pair of cylindrical rollers each of diameter of 400 mm. What is the bite angle in degree?

Solution :

Given

$$\text{initial thickness } (h_1) = 16 \text{ mm}$$
$$\text{final thickness } (h_2) = 10 \text{ mm}$$
$$\text{roll diameter} = 400 \text{ mm}$$

we know that

$$\tan\theta = \sqrt{\frac{h_1 - h_2}{R}}$$

where θ = Bite angle

and R = radius of roll = $\frac{400}{2}$ = 200 mm

$$\tan\theta = \sqrt{\frac{16-10}{200}} = 0.1732$$

$$\theta = \tan^{-1}(.1732) = 9.82°$$

$$\text{Bite angle } (\theta) = 9.82°$$

Example : 5.18

In a single pass rolling operation, a 20 mm thick plate with plate width of 100 mm, is reduced to 18 mm. The roller radius is 250 mm and rotational speed is 10 rpm. The average flow stress for the plate material is 300 MPa. What is the power required for the rolling operation in kW is closest to?

Solution :

Given

initial thickness of plate (h_1) = 20 mm

final thickness of plate (h_2) = 18 mm

width of plate (w) = 100 mm

radius of roll (R) = 250 mm

speed of rolling (N) = 10 RPM

Average flow stress for plate material

$$\sigma_0 = 300 \text{ MPa}$$

Now we know that

$$\text{Force} = \text{Stress} \times \text{Area}$$

$$\text{Force} = \sigma_0(L \times W)$$

where $L = \sqrt{R(\Delta h)} = \sqrt{R(h_1 - h_2)}$

Put values of R, h_1 and h_2; we have

$$L = \sqrt{250(20-18)} = \sqrt{250 \times 2} = 22.36 \text{ mm}$$

$$L = 22.36 \text{ mm}$$

Now force = 300 × 22.36 × 100 = 670800 N

Torque per roll, $T = F \times \frac{L}{2}$

Power= $T \times \omega$

$\omega = 2\pi N$

$$\text{Power} = \frac{\pi FLN}{60} = 7.85$$

Power of both rolls= 2×7.85 = 15.7 Kw

5.8 PUNCHING AND BLANKING

Shearing is a metal fabrication process where the flat sheet metal is cut into straight lines. Two blades are forced against each other with a space between them. Generally one of the blades remain stationary. Metal shearing can be performed on sheet, strip, bar, plate, and even angle stock. Bar and angle materials can only be cut to length. However, many shapes can be produced by shearing sheet and plate. Blanking is a metal forming process in which the sheet metal is punched by the die to get the required outer profile of the sheet metal component. During this process the blanking punch penetrates into the sheet metal forcing the profile into the blanking die. The dimensions of the punch is equal to the dimensions of the sheet metal component and the clearance is provided on the punch.

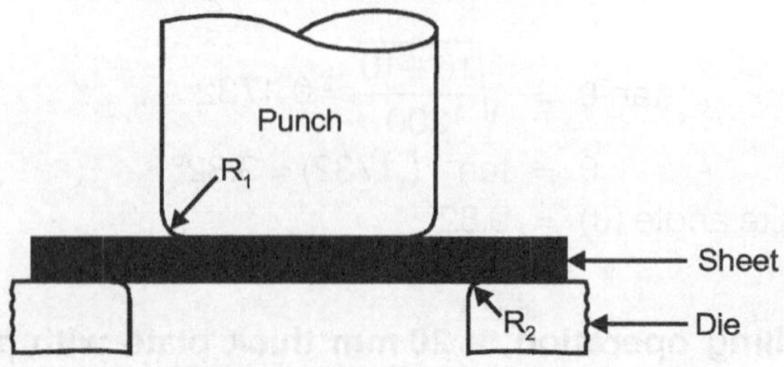

Fig 5.26 : (a) Punching operation

R_1 = punch corner radius

R_2 = Die corner radius

When the pressure on the blank holder is very high, punch and die curvatures are very low and clearance is very small. Sheet metal will shear. When the removed material of the sheet is useful it is called blanking operation and when removed material is a waste it is called punching operation. As Punch coveres roll over depth strains will be primarily elastic. Plastic deformation will begin in the burnish depth. After penetration fracture will start and the material will be sheared. More ductile material, larger will be the fracture angle, which produces large size Burrs. To accommodate these large size burrs clearance has to increase.

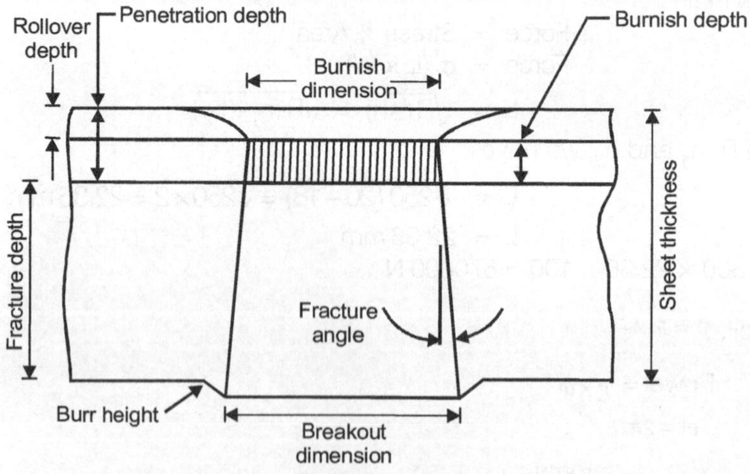

Fig 5.26 : (b) Punched hole

$$\text{Punch load} = \pi D t \times K_1 \left[\frac{\text{Penetration}}{\text{shear on punch}} \right]$$

Where, K_1 is the yield strength in shear, D is diameter of hole/blank and t is sheet thickness.

In punching operation punch is exact so clearance is provided over the die and angle of shear is provided over the punch. In blanking operation die is exact so clearance is provided over the punch and angle of shear is provided over the die.

5.9 DEEP DRAWING

Sheet metal forming (also called press working) involves conversion of flat thin sheet metal blanks into parts of desired shape. Sheet metal forming processes like deep drawing, stretching, bending etc. are widely used to produce a large number of simple to complex components in automotive and aircraft industries, household appliances etc. Most of these parts are manufactured using one or more of the sheet metal forming processes. These processes are also called sheet metal stamping operations. In deep drawing, which is also called cup drawing or radial drawing, flat thin sheets (blanks) are formed into cup shaped components by pressing the central portion of the sheet into die opening using a punch to draw the metal into the desired shape.

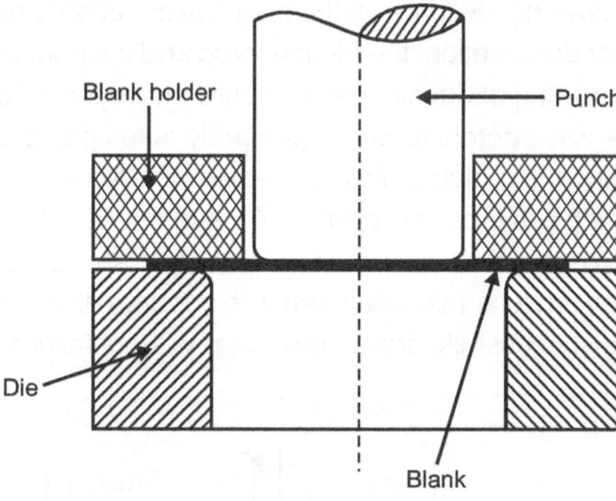

Fig. 5.27(a)

The blank may be circular or rectangular, or of a more complex outline. Blank holder is loaded by a blank holder force, which is necessary to prevent wrinkling and to control the material flow into the die cavity. The punch is pushed into the die cavity, simultaneously transferring the specific shape of the punch and the die to the blank. During the forming stage the material is drawn out of the blank holder-die region and the material is subjected to compressive and tensile stresses in this portion. When a very high blank holding force is applied, the deep drawing process becomes a stretching process.

In stretch forming or stretching, the flange of the flat blank is securely clamped. In some cases, the material is fixed under the blank holder and the remaining part of the blank is formed over a tool. Normally a rigid punch is used to form the part. Stretching is characterized by bi-axial tensile stresses leading to significant thickness reduction in the deformed portion.

Bending is the plastic deformation of metals about a linear axis (called the bending axis) with little or no change in the surface area. The bending process is used not only to form parts such as angle sections, flanges, seams and corrugations, but also to impart stiffness to the part by increasing its moment of inertia. It is the most common type of deformation that occurs in almost all sheet forming operations.

In deep drawing operation, bending of sheet takes place over the die curvature. The bend zone experiences localized strains which are tensile on the outside of the neutral axis and compressive on the inside.

In deep drawing, the majority of the deformation occurs in the flange of the cup. The metal is subjected to three different types of stress systems. These stresses have influence over thickness variation in drawn cup. The primary deformation occurs in the flange of the deforming cup which undergoes radial tension and circumferential compression. Because of this, thickness increases in the flange portion. The second deformation zone is the bending around the die radius while the third deformation zone is the uni-axial stretching (plane strain) in the cup wall, which causes thinning of the metal. In the cup bottom, which is subjected to biaxial tension, the thickness is more or less equal to the initial sheet thickness.

The draw ratio (DR) of a deep drawing process is calculated as the ratio of blank diameter to the cup diameter. For achieving very high draw ratio, redrawing and ironing or annealing between draws is performed. The maximum draw ratio that can be obtained under perfect deep drawing conditions is called limiting draw ratio (LDR). The LDR is considered a good measure of drawability of a material.

In conventional deep drawing the force on the punch required to produce a cup is the summation of the force required for deformation, the friction force and the ironing force (if present). When the punch force in deep drawing of ductile metals increases beyond a certain limit, failure occurs in the cup wall due to tensile stretching and it generally takes place at the transition between the punch profile region and the punch cylindrical region. The force needed to cause deformation is applied through a punch at the central portion of the blank. This force is transmitted through the cup wall to the flange. As the blank size increases, there is an increase in the total force and if the total force exceeds beyond a certain value, fracture occurs in the cup wall. This puts a limit on the blank size that can be successfully drawn into a cup without failure.

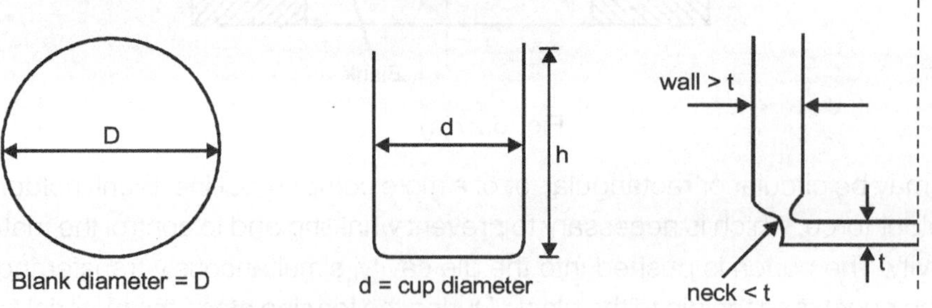

Fig 5.27 : Deep drawing operation

$$\frac{D}{d} = \text{Draw ratio}$$

$$\frac{\pi}{4}D^2 = \frac{\pi}{4}d^2 + \pi dh$$

$$\text{Limiting draw ratio} = \frac{D_{max}}{d}$$

Die-corner radius = 10 t

Punch corner radius = 6 t

Clearance = (1.2 to 1.4) t

Blank Holding Pressure = 2% of yield strength

Thickness in the drawn cup at the punch corner will be lower because maximum load is appearing on the material in this region. Since compressive hoop stress is appeares in the flange region so thickness of the drawn cup on wall region increases. As the diameter of blank increases thickness of the drawn cup in the punch corner region will decrease. Once this thickness reaches a particular value fracture will appear in the drawn cup. The maximum diameter of the blank that can be drawn safely without any fracture is called limiting draw ratio (LDR) which is an indicator of formability of material. In any material larger the value of, work hardening exponent 'n', higher will be the formability of material. Larger thickness sheets will exhibit larger value of LDR.

5.9.1 Redrawing

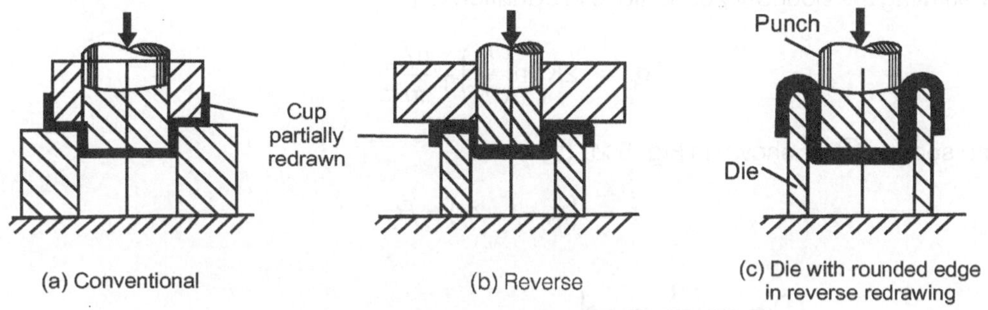

(a) Conventional (b) Reverse (c) Die with rounded edge in reverse redrawing

Fig 5.28 : Redrawing operation

Redrawing is a process of drawing already drawn component to increase LDR. Beverage cans are produced in three redrawing stages and two ironing stages. When the redrawing is performed near LDR, Step rings will appear in the drawn component. In conventional redrawing operation the material bends in the opposite direction around the blank holder and the die corners. In the reverse drawing operation the material bends in only one direction namely along the outer and the inner die corners. Since some amount of strain hardening takes place during the initial operation, annealing is normally advised before commencing the redrawing operation.

5.9.2 Analysis of deep drawing

There will be three loads acting during deep drawing operation i.e. in the Flange, bending and pulling over die curvature. Total load will be addition of these three loads.

Assuming that σ_r and σ_θ are the principle stresses and since σ_θ is compressive so

$$\Rightarrow \quad \frac{d\sigma_r}{dr} + \left(\frac{\sigma_r + \sigma_\theta}{r}\right) = 0$$

$$\Rightarrow \quad \frac{d\sigma_r}{dr} = \frac{-2K'}{r}$$

Flange

Cup wall

Bottom

p_H Blank holder

μp_H

μp_H

Die

r_d

R = Radius of blank

Fig : 5.29 (a)

by integrating

$$\Rightarrow \qquad \int d\sigma_r = \int \frac{-2K'}{r} dr$$

$$\Rightarrow \qquad \sigma_r = -2K' \ln r + c_1 \qquad ...(31)$$

Boundary conditions at r = R, $\sigma_r = \dfrac{\mu p_H}{\pi R t}$

Also $\sigma_r + \sigma_\theta = 2K'$

Substituting the boundary conditions in equation (31)

$$\sigma_r = 2K' \ln\left(\frac{R}{r}\right) + \frac{\mu p_H}{\pi R t} \qquad ...(32)$$

Upto section ① as shown in Fig. 5.29 (b)

$$r = (r_p + t + r_d) = R_d$$

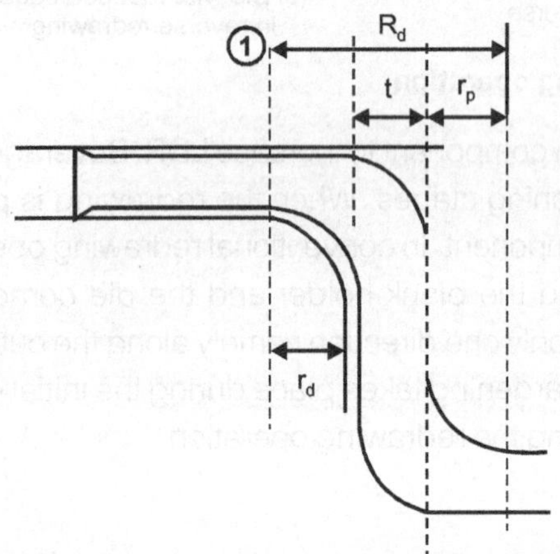

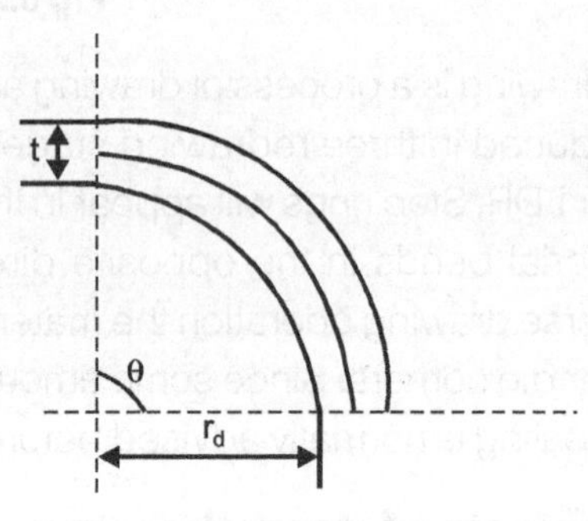

Fig : 5.29 (b)

Drawing pressure up to ① (flange)

$$\sigma_r = \frac{\mu p_H}{\pi R t} + 2K' \ln\left(\frac{R}{r}\right)$$

Punch pressure required to bend the sheet

$$\text{Maximum strain} = \frac{(r_d + t)\theta - \left(r_d + \frac{t}{2}\right)\theta}{\left(r_d + \frac{t}{2}\right)\theta}$$

$$\left(r_d + \frac{t}{2}\right) = r_n = \text{radius of neutral plane.}$$

$$\text{Maximum strain} = \frac{t}{2r_n}$$

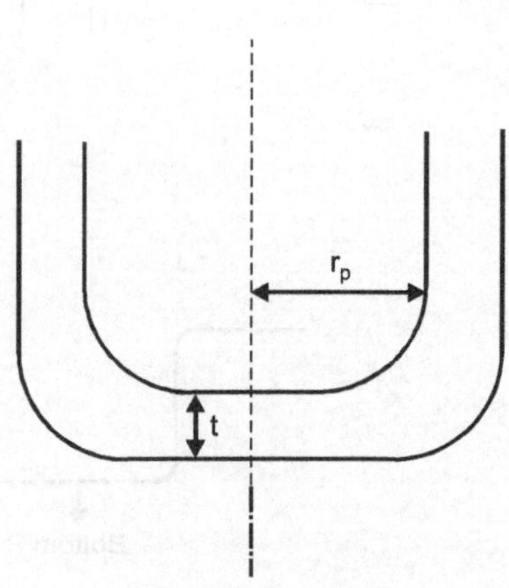

Fig. 5.29 : (c)

$$\text{Average strain} = \frac{0+\frac{t}{2r_n}}{2} = \frac{t}{4r_n}$$

$$\text{Pressure required for bending} = \frac{t}{4r_n}\times\sigma_0$$

Total pressure required by the punch up to bending of sheet

$$\sigma_w = \frac{\mu p_H}{\pi R t} + 2K\ln\left(\frac{R}{R_d}\right) + \frac{\sigma_0 t}{4r_n} \quad \text{...(33)}$$

Since sheet metal is pulled over the die curvature and if the angle of contact between sheet and die is θ, total pressure required by the punch to draw the cup is

$$\sigma_w = \left[\frac{\mu p_H}{\pi R t} + 2K\ln\left(\frac{R}{R_d}\right) + \frac{\sigma_0 t}{4r_n}\right]e^{\mu\theta} \quad \text{...(34)}$$

$$\text{Punch load} = \sigma_w \times 2\pi r_p t \quad \text{...(35)}$$

Example : 5.19

If a fully work hardened Aluminium blank of 100 mm diameter and 1.2 mm. Sheet has a mean flow stress of 350 MPa. It is deep drawn to a cylindrical cup of 50 mm mid wall diameter. The blank holding force is 30 kN, coefficient of friction is 0.1. Take the die curvature radius as 8 times the sheet thickness. Determine the final cup height and appropriate load.

Solution :

Given data

$$R = 50 \text{ mm}$$
$$t = 1.2 \text{ mm}$$
$$\sigma_{fm} = \sigma_0 = 350 \text{ MPa}$$
$$d_p = 50 \text{ mm}$$
$$p_H = 30 \text{ kN}, \mu = 0.1$$
$$r_d = 8t$$
$$R_d = 25 + \frac{1.2}{2} + (8\times 1.2)$$
$$R_d = 35.2 \text{ mm}$$
$$r_n = r_d + \frac{t}{2}$$
$$r_n = 9.6 + 0.6$$
$$r_n = 10.2 \text{ mm}$$

Height can be found by

$$\pi R^2 = \pi r_d^2 + 2\pi r_d h$$
$$(50^2) = (25)^2 + 2(25)h$$
$$h = 37.5 \text{ mm}$$

$$\sigma_w = \left[\frac{\mu p_H}{\pi R t} + 2K\ln\left(\frac{R}{R_d}\right) + \frac{\sigma_0 t}{4r_n}\right]e^{\mu\theta}$$

$$\Rightarrow \quad \sigma_w = \left[\frac{0.1\times30\times10^3}{\pi\times50\times1.2}+350\ln\frac{50}{35.2}+\frac{1.2}{4\times10.2}\times350\right]e^{\frac{0.1\times90\times\pi}{180}}$$

$$\Rightarrow \quad \sigma_w = 174.38$$

$$\text{punch load} = \sigma_w\times2\pi r_p t$$

$$= 174.38\times2\pi\times25\times1.2$$

$$= 32.85\text{ KN}$$

Example : 5.20

A Cylindrical cup without flange is to be drawn from a 2 mm thick sheet. The cup shall have 15 mm diameter and 40 mm height. Reduction ratio is the first and subsequent draws may not exceed 40% and 15% respectively. Determine the blank size and the number of draws necessary.

Solution :

Approximate blank diameter can be calculated by equating the surface area (assuming the variation in thickness to be insignificant)

$$\frac{\pi}{4}D^2 = \frac{\pi}{4}d^2+\pi dh$$

where D is diameter of blank, d is diameter of cup and h is cup height

$$\frac{\pi}{4}(D^2) = \frac{\pi}{4}(15)^2+\pi(15)(40)$$

$$D = 51.23\text{ mm}$$

$$\text{So the overall ratio} = \frac{D}{d} = 3.41$$

Let us say d_1 is the diameter of cup after first draw and it will be the blank diameter for 2nd draw and so on. So overall draw ratio can be written as

$$\underbrace{\frac{D}{d_1}}_{1\text{stage}}\times\underbrace{\frac{d_1}{d_2}\times\ldots}_{n-1\text{stages}} = 3.41$$

Since 40% reduction can be given in the first draw, so

$$d_1 = 0.6D$$

and 15% reduction can be given in the subsequent draws. So

$$d_2 = 0.85\,d_1 \text{ and so on.}$$

So

$$\frac{1}{0.6}\times\left(\frac{1}{0.85}\right)^{n-1} = 3.41$$

$$(n-1)\ln\left(\frac{1}{0.85}\right) = \ln(2.049)$$

$$= 0.717$$

$$n-1 = 4.41$$

$$n = 5.41$$

So total 6 draws are necessary.

5.9.3 Measuring of formability

There are some tests by which we can measure the formability.

1. **Cupping test**

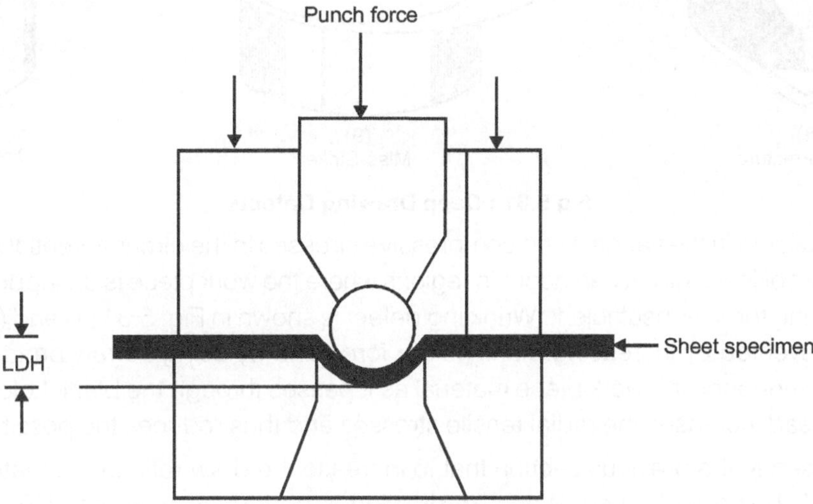

Fig 5.30 : Cupping test

When punching, after some time necking will appear so before the neck has appeared how much the amount of height it is moved is known as Limiting Dome Height (LDH), a measure of formability. Sheet will be restricted to move by draw bead. Higher is LDH more will be formability of shut material.

2. **Formability limit diagrams**

In deep drawing process when we take circles on the space of upper portion and deform the material, there it will form an ellipse and the major and minor axis will form. By using that a graph is formed between major and minor strains of deformed circle, that is called Formability limit diagrams.

5.9.4 Common defects in deep drawing

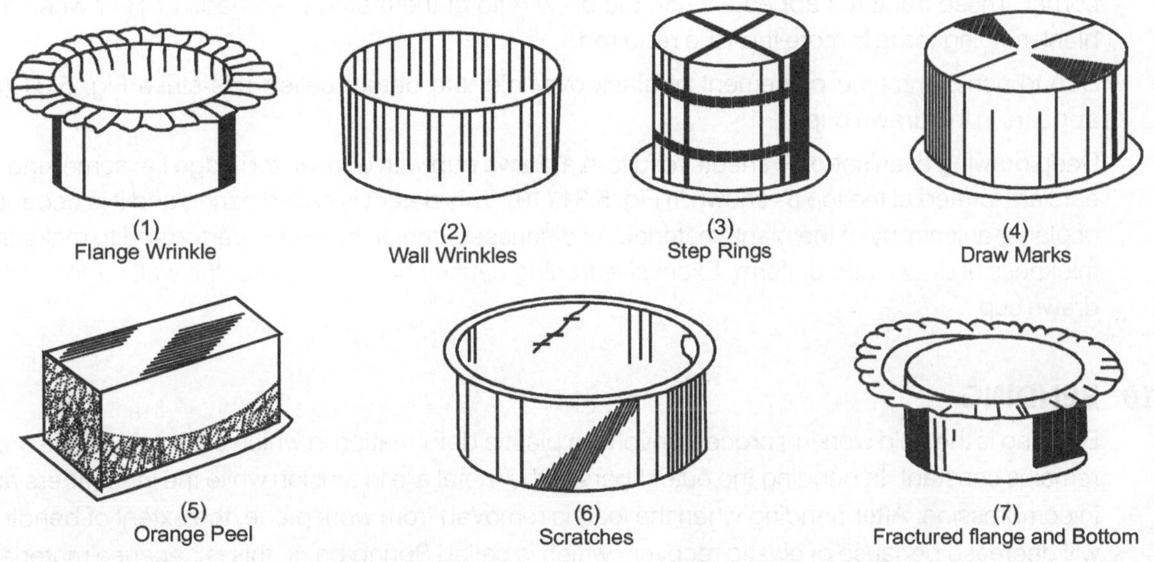

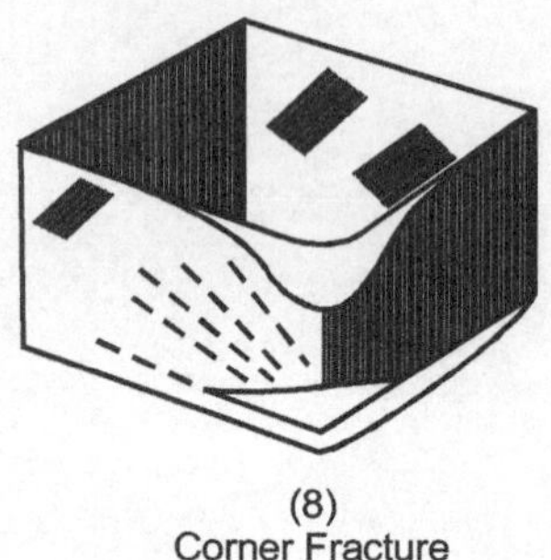

(8)
Corner Fracture

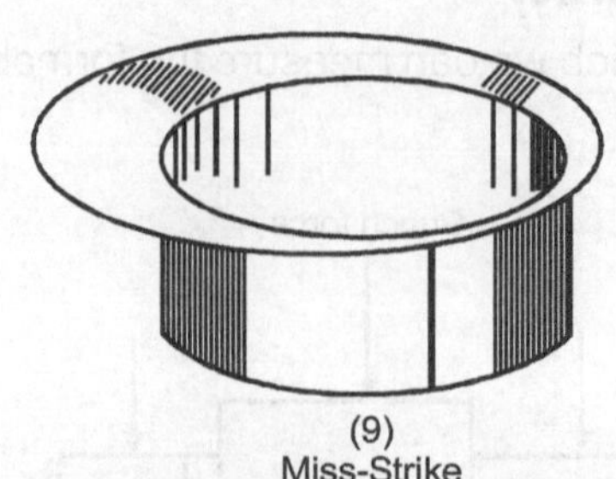

(9)
Miss-Strike

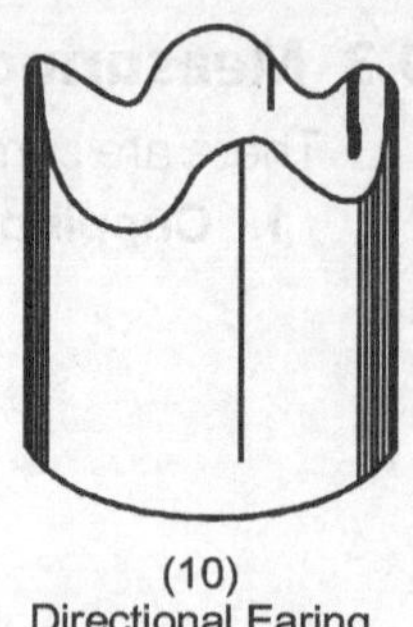

(10)
Directional Earing

Fig 5.31 : Deep Drawing Detects

Wrinkling occurs in the flange when compressive stresses in the circumferential direction reach a critical point of instability. It can occur in regions where the work piece is unsupported or when the blank holding force is insufficient. Wrinkling defect is shown in Fig. 5.31 (1) and (2). The wrinkling can be prevented by increasing blank holder force and by using a draw bead. The draw bead bends and unbends the work piece material as it passes through the blank holder. This bending over the bead increases the radial tensile stresses and thus reduces the possibility of wrinkling.

As discussed in the previous section that to increase the draw ratio of the material Re-drawing operation is being performed. When Re-drawing is performed near the LDR value, step rings Fig. 5.31 (3) appears in the drawn cup. These step rings appears at the position of neck. To avoid this defect, redrawing operation is performed at a lower value of draw ratio.

Due to the improper finish of the die surface, draw marks Fig. 5.31 (4) appears on the drawn cup. Surface of the die generally deteriorates generally due to improper lubrication between the blank and the die. It should be noted that higher will be the draw ratio when the punch is made rough and die is made smooth.

Orange peel Fig. 5.31 (5) appears in the drawn cup due to coarse grain structure of the drawn material and it can be avoided by refining the grain structure of material. Heavy scratches over the drawn cup Fig. 5.31 (6) may appear due to appearance of yield point phenomenon in the material. So the sheet material must be consumed within the strain aging time of the material.

Fracture [Fig. 5.31 (7) and (8)] occurs when the sheet metal is subjected to strains exceeding the safe strain limits of the material. For ductile sheets this fracture usually occurs near the punch corner. These fractures appears when the draw ratio of the material exceeds LDR or when the blank holding force is more then the required.

Due to non-symmetric placement of blank over die, the defect called mis-strike Fig. 5.31 (9) appears in the drawn cup.

Deep drawing of anisotropic sheets results in a drawn cup with uneven top edge i.e. some kind of ears are formed at the top as shown in Fig. 5.31 (10). This defect is called earing and it is because of planar anisotropy of the blank material. As discussed earlier, ironing is performed to make the thickness of drawn cup uniform. Excessive ironing causes burnishing over the wall region of the drawn cup.

5.10 BENDING

Bending is the cold working process involving plastic deformation in which the total surface area remains constant. In bending the outer fibers of the metal are in tension while the inner fibers are in compression. After bending when the load is removed from work piece, the extent of bending will decrease because of elastic recovery which is called Spring back, this is because materials have finite modulus of elasticity. Bending can be defined as a process of fabrication of a sheet metal by uniform straining of material around a straight axis which lies in the neutral plane and normal to the lengthwise direction of the sheet or strip. These processes involve a punch and die arrangement. The punch hits the blank and pushes it inside the die cavity which gives the blank as shape of the die. The shape of punch and die play an important role in giving shape to the blank. Sheet metal bending process takes place with the help of a press. Apart from using punch and dies we also use grippers that hold the part during and in between the process. There are various types of process that are used in bending:

(a) **Air bending :** Air bending is one of the most common methods that are used for sheet metal bending process. In this process the punch only touches the blank that is sufficient to give it the required shape and the angle. In other words the punch doesn't traverse till the end of the die cavity. After the punch travels back, the sheet metal springs back and is of lesser bend than the punch. In air bending process there is no need to change the dies and other necessary equipments to achieve different bend angles as the bend angles depend upon the punch stroke. Also smaller forces are required to bend the sheets but it does require accurate control of punch stroke to achieve required bend angles.

(b) **Bottoming :** In bottom bending process, the punch bottoms the blank into the die cavity. The amount of spring back is very less in this method. The forces involved are higher (approximately 50-60% more) than air bending.

(c) **Coining :** Coining is bending process where we bottom the blank and apply compressive stress to increase the amount of plastic deformation of the bent region. This helps in reducing the springback. The inner radius of the workpiece should be 0.75 times the material thickness. Very high forces are required and hence large capacity machines are used. 5 to 8 times the tonnage of the air bending are required. This increases the machine cost rapidly but extremely high precision bend is achieved due to elimination of springback.

(d) **Other bending processes:** In V-Bending process the clearance between the punch and die is constant. It is taken as the sheet thickness itself. U-bending is a process where we get two parallel bending axis in the same operation. A backing pad is used to force the sheet contacting with the sheet bottom. Around 30% of force is requires at the pad where the sheet bottom comes in contact with the pad. The type of bending process which includes bending of only some portion of the sheet is called flanging. The length of flange and the angle of bend can be adjusted accordingly.

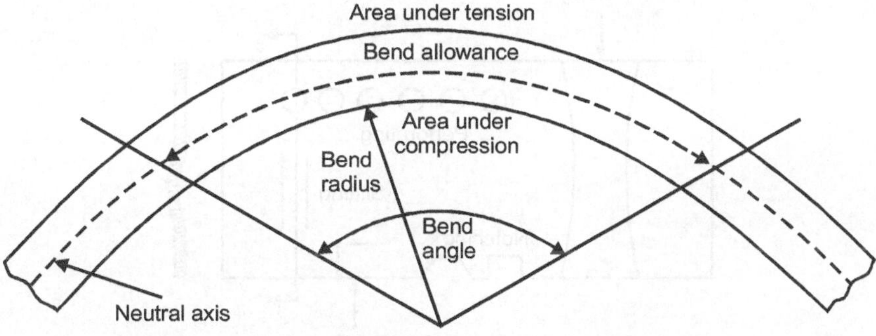

Fig 5.32 : Bending operation

Bend allowance : The length of neutral axis in the curvature region of bend is called bend allowance. In order to estimate the required flat work piece length to make a bend, calculation of bend allowance is essential.

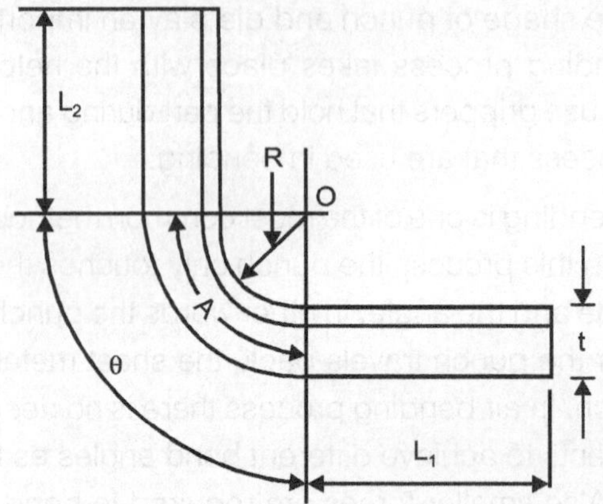

Fig 5.33 : Schematic of bend allowance.

Bend allowance = θ (R + Kt)
t = sheet thickness
θ = angle in radians
R = bend radius
K = stretch factor
= 0.33 when R < 2t
= 0.5 when R ≥ 2t

It is the length of neutral axis in the bend region.

Notching : It is complete removal of small portion of material on the edge of sheet metal.
Lancing : It is creating a tab on the edge without removal of material.
Slitting : It is creating a tab in the middle of sheet.
Nibbling : By using a small punch and die assembly the process of creating a profile is called nibbling.
Perforating : It is a process of producing large number of holes in the sheet metal.
Parting : shearing the sheet into two or more pieces.

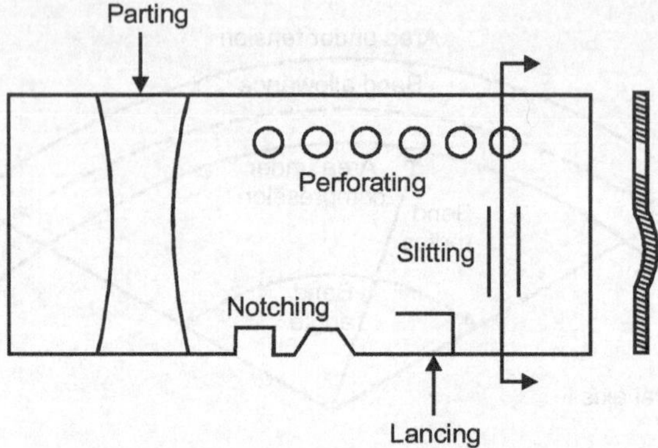

Fig 5.34 : (a) Various operations in forming

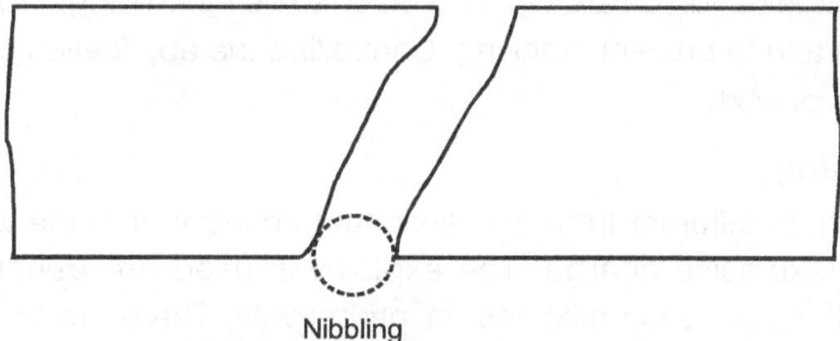

Fig 5.34 : (b)

Types of dies:

1. **Simple die :** On these dies one operation is being performed on one station in a single stroke.
2. **Compound die :** When multiple operations can be performed on a single station in single stroke is called compound die.
3. **Progressive die :** When multiple operations can be performed on multiple stations in a single stroke such dies are called progressive dies.

POINTS TO PONDER

High Energy Rate of Forming

High energy rate forming, is a metal forming process in which the actual forming of the work occurs in a fraction of a second. Explosive forming, Electromagnetic forming, Electrohydraulic forming are the major high rate forming processes. They are called High Energy Rate Forming (HERF) processes because in these forming processes large amount of energy is applied for a very short interval of time. All the high rate forming processes have the common feature of rapid release of energy to the work piece over a few microseconds, thus accelerating the workpiece away from the energy source to velocities considerably greater than that attained in conventional forming processes (typically 20-300 m/s). Since the forming of the metal occurs so quickly, desirable materials for (HERF) will be ductile at high deformation speeds. There are several advantages of using these forming processes, like die costs are low, easy maintenance of tolerances, possibility of forming most metals, and material does not show spring-back effect. The limitations of these processes is the need for skilled personnel.

Advantages:

The energy of deformation is delivered at a much higher rate than in conventional practice. Larger energy is applied for a very short interval of time. High particle velocities are produced in contrast with conventional forming process. The velocity of deformation is also very large and hence these are also called High Velocity Forming (HVF) processes. Many metals tend to deform more readily under extra fast application of force. Large parts can be easily formed by this technique.. The strain rate dependence of strength increases with increasing temperature. The yield stress and flow stress at lower plastic strains are more dependent on strain rate than the tensile strength. High rates of strain cause the yield point to appear in tests on low carbon steel that do not show a yield point under ordinary rates of strain.

Limitations:

Highly skilled personnel are required from design to execution. Transient stresses of high magnitude are applied on the work. Not suitable to highly brittle materials. Source of energy (chemical explosive or electrical) must be handled carefully. Dies need to be much bigger to withstand high energy rates and shocks and to prevent cracking. Controlling the application of energy is critical as it may crack the die or work.

(a) Explosive Forming

Explosive forming, is different from conventional forming in that the punch or diaphragm is replaced by an explosive charge. The explosives used are generally high – explosive chemicals(TNT,RDX), gaseous mixtures, or propellants. There are two techniques of high – explosive forming: stand – off technique and the contact technique.

Standoff Technique: The sheet metal work piece blank is clamped over a die and the assembly is lowered into a tank filled with water. The air in the die is pumped out. The explosive charge is placed at some predetermined distance from the work piece. On detonation of the explosive, a pressure pulse of very high intensity is produced. A gas bubble is also produced which expands spherically and then collapses. When the pressure pulse impinges against the work piece, the metal is deformed into the die with as high velocity as 120 m/s.

Contact Technique: **The explosive charge in the form of cartridge is held in direct contact with the work piece while the detonation is initiated. The detonation builds up extremely high pressures (up to 30,000MPa) on the surface of the work piece resulting in metal deformation, and possible fracture. The process is used often for bulging tubes**

Applications. Explosive forming is mainly used in the aerospace industries but has also found successful applications in the production of automotive related components. The process has the greatest potential in limited – production prototype forming and for forming large size components for which conventional tooling costs are prohibitively high.

(b) Electromagnetic forming

The process is also called magnetic pulse forming, and is mainly used for swaging type operations, such as fastening fittings on the ends of tubes and crimping the terminal ends of cables. The electrical energy stored in a capacitor bank.The tubular work piece is mounted on a mandrel having the die cavity to produce shape on the tube. A primary coil is placed around the tube and mandrel assembly. When the switch is closed, the energy is discharged through the coilproducing a varying magnetic field around it. In the tube a secondary current is induced, which creates its own magnetic field in the opposite direction.The coil is firmly held and hence the work piece collapses into the diecavity due to magnetic repelling force, thus assuming die shape.

Process parameters:

Work piece size, Electrical conductivity of the work material, Size of the capacitor bank.The strength of the current, which decides the strength of the magnetic field and theforce applied.Rigidityand Insulationon the coil.

Advantages:

Suitable for small tubes. Operations like collapsing, bending and crimping can be easily done.Electrical energy applied can be precisely controlled and hence the process is accuratelycontrolled.The process is safer compared to explosive forming.Wide range of applications. Since there is no physical contact between the work piece anddie as compared to the use of a punch in conventional formingprocess, the surface finish can be improved.High production rates are possible. The attribute thatessentially control the production rate would be the time takenfor the capacitor bank to get charged.It is an environmentally clean process as no lubricants arenecessary.

Limitations:

Applicable only for electrically conducting materials.Not suitable for large work pieces. Rigid clamping of primary coil is critical.Shorter life of the coil due to large forces acting on it.

Applications:

Crimping of coils, tubes, wires. Bending of tubes into complex shapes. Bulging of thin tubes.The electromagnetic forming technology has unique advantages inthe forming, joining and assembly of light weight metals such as aluminium because of the improved formability and mechanical properties, straindistribution, reduction in wrinkling, active control of spring back, minimizationof distortions at local features, local coining and simple die.

(c) Electrohydraulic forming

The characteristics of this process are similar to those of explosive forming. The major difference, however, is that a chemical explosive is replaced by a capacitor bank, which stores the electrical energy. The capacitor is charged through a charging circuit. When the switch is closed, a sudden electrical discharge in the form of sparks is produced between electrodes. Water breaks it down into oxygen and hydrogen which then explodes producing a shock wave that pushes the material onto the die surface at a high rate s and this discharge produces a shock wave in the water medium. This shock wave deforms the work plate and collapses it into the die. The energy released is much lesser than that released in explosive forming.

Process Characteristics:

Standoff distance must be optimum. The energy of the pressure pulse depends on the size of capacitor. Water is used as Transfer medium. The die cavity must be evacuated to prevent adiabatic heating of the work due to a sudden compression of air.

Advantages:

Better control of the pressure pulse as source of energy is electrical- which can be easilycontrolled. Safer in handling as compared to the explosive materials. Small to medium size work pieces can be formed. The process does not depend on the electrical properties of the work material.

Limitations:

Suitable only for smaller works. Need for vacuum makes the equipment more complicated. Proper SOD is necessary for effective process.

Aging: A change in material property or properties with time.

Anisotropy: Variations in one or more physical or mechanical properties with direction.

Bendability: The ability of a material to be bent around a specified radius without fracture.

Bend Radius: The radius measured on the inside of a bend which corresponds to the curvature of a bent specimen or the bent area in a formed part.

Bend Test: Evaluation of a sheet metal's response to a bending operation, such as around a fixed radius tool.

Biaxial Stretchability: The ability of sheet material to undergo deformation by loading in tension in two directions in the plane of the sheet.

Blank: The piece of sheet metal, produced in cutting dies, that is to be subjected to further press operations. A blank may have a specific shape developed to facilitate forming or to eliminate a trimming operation subsequent to forming.

Blank Holder: The part of a forming die which holds the blank by pressure against a mating surface of the die to control metal flow and prevent wrinkling. The blank holder is sometimes referred to as "Holddown" or binder area. Pressure is applied by mechanical means, springs, air, or fluid cushions. The pressure exerted by the blank holder against the blank. This is normally adjustable to control metal flow during the drawing.

Blanking: The act of cutting a blank.

Bulge Test: A test wherein the blank is clamped securely around the periphery and by means of hydrostatic pressure the blank is expanded. The blank is usually grinded so that the resulting strains can be measured. This test is usually performed on large blanks of 8-12 inches diameter.

Coining: A process of cold forming metals, in which the metal is shaped between two dies in such a manner as to fill the depression of both sides in relief by displacement of the material.

Cupping: A press forming operation in which a cup-shaped part is produced from sheet metal. Cupping is often the first operation in the production of a complex deep-drawn part.

Cup Test: A formability test in which a sheet blank is drawn or stretched into a hemispherical or cylindrical shape. Variations in blank diameter, cup height or forming load are utilized to evaluate sheet formability.

Deep Drawing: In drawing operation, where a part is produced from a blank by the action of a punch in which the sheet is pulled into a die cavity and the flange of the blank is compressed in the circumferential direction. The area directly under the punch remains undeformed.

Drawability: The ability of a sheet to form cup-shaped parts by the action of a punch under conditions that cause the flange or a portion of the flange to be drawn into the die cavity.

Draw Bead: A ridge constructed around a portion of a die cavity to partially restrain metal flow. A groove in the mating blankholder allows die closing. Sometimes called Die Bead.

Erichsen Test: A test in which a piece of sheet metal, restrained except at the center, is deformed by a spherical ended plunger until fracture occurs. The height of the cup in millimeters at fracture is a measure of the ductility.

Forming Limit Diagram: An empirical curve showing the biaxial strain levels beyond which failure may occur in sheet metal forming. The strains are given in terms of major and minor strains measured from deformed circles, previously imprinted into the undeformed sheet.

Fukui Cup Test: A cupping test combining stretchability and drawability in which a roundnosed punch draws a circular blank into a conical shaped die until fracture occurs at the nose. Various parameters from the test are used as the criterion of formability.

Ironing: Forming between dies whose side-wall clearance is less than the sheet metal thickness such that a thinning action occurs. This usually takes place in the side walls of deep-drawn parts where the inflow of metal from the flange tends to thicken or wrinkle; subsequent ironing produces a more uniform wall thickness.

■■■■

PRACTICE QUESTIONS

1. **Assertion(A) :** In a two high rolling mill there is a limit to the possible reduction in one pass
   Reason(R) : the reduction possible in the second pass is less than that of first pass
   (a) both A and R are individually true and R is correct explanation of A
   (b) both A and R are individually true and R is not correct explanation of A
   (c) A is true but R is false
   (d) A is false R is true

2. Rolling very thin strips of mild steel requires
   (a) large diameter rolls (b) small diameter rolls
   (c) high speed rolling (d) rolling without a lubricant

3. In rolling process roll separating force can be decreased by
   (a) reducing roll diameter
   (b) increasing roll diameter
   (c) providing back up rolls
   (d) increasing the friction between rolls and metal

4. Consider the following charteristic of a rolling process
   1. Shows work hardening effect
   2. Surface finish is not good
   3. Heavy reduction in areas can obtained
   Which of these characteristics associated with hot rolling?
   (a) 1 and 2 (b) 1 and 3
   (c) 2 and 3 (d) 1,2 and 3

5. **Assertion (A) :** Rolling require high friction which increases force and time consumption
   Reason (R) : To prevent damage to the surface of rolled products lubricant should be used
   (a) both A and R are individually true and R is correct explanation of A
   (b) Both A and R are individually true and R is not correct Explanation of A
   (c) A is true but R is false
   (d) A is false R is true

6. Which of the following assumptions are correct for rolling
   1. the material is plastic
   2. the arc of contact is circular with radius greater than radius of roll
   3. coefficient of friction is constant over the arc of contact and acts in one direction throughout the arc of contact
   Select the correct answer using the codes given below
   (a) 1 and 2 (b) 1 and 3
   (c) 2 and 3 (d) 1,2 and 3

7. A strip is to be rolled from thickness of 30mm to 15 mm using a two high mill having rods of millimeter 300 mm the coefficient of friction for unaided bite should nearly be
   (a) 0.35 (b) 0.5
   (c) 0.25 (d) 0.07

8. In one setting of rolls in a 3-high rolling mill one gets
   (a) one reduction in thickness
   (b) two reductions in thickness

(c) three reductions in thickness
(d) two or three reductions in thickness depending upon setting

9. In a rolling operation the maximum draft is given by (r = radius of coil, μ is coefficient of friction)
(a) μR
(b) $\mu\sqrt{R}$
(c) μ^2R^2
(d) μ^2R

10. Flash and Gutter are provided in drop forging dies at the following stage
(a) blocking
(b) preforming (Edging)
(c) finishing
(d) fullering

11. In rolling a strip between two rolls, the position of the neutral point in arc of contact does not depends on
(a) amount of reduction
(b) diameter of rolls
(c) coefficient of friction
(d) material of rolls

12. A wide strip is rolled to a final thickness of 6.35 mm with a reduction of 30% the roll radius is 50cm and coefficient of friction is 0.2 the point of contact is
(a) 0.5 radians
(b) 0.0738 radians
(c) 0.0123 radians
(d) 0.0235 radians

13. The thickness of a metallic sheet is reduced from initial value of 16mm to a final value of 10 mm in one single pass rolling with a pair of cylindrical rollers each of diameter 400 mm the bite angle in degrees will be
(a) 5.936
(b) 7.936
(c) 8.936
(d) 9.936

14. Consider the following statements
Assertions (A) : While rolling metal sheet in rolling mill the edges are sometimes not straight and flat but are wavy.
Reason(R) : Non uniform mechanical properties of flat material rolled out result in waviness of edges
(a) both A and R are individually true and R is correct explanation of A
(b) both A and R are individually true and R is not correct explanation of A
(c) A is true but R is false
(d) A is false R is true

15. A 4mm thick sheet is rolled with 300 mm diameter rolls to reduce thickness without any change in width the friction coefficient at the work roll interface is 0.1 the minimum possible thickness of the sheet that can be produced in single pass is
(a) 1.0 mm
(b) 1.5 mm
(c) 2.5mm
(d) 3.7mm

16. Which one of the following is advantage of forging
(a) good surface finish
(b) low tooling cost
(c) close tolerance
(d) improves physical property

17. Drop forging is used to produce
(a) small components
(b) large components
(c) identical components in large number
(d) minimum size components

18. Consider the following statements

Assertions (A) : In drop forging besides the provision for flash provision is also made in forging die for additional space called gutter

Reason(R) : The gutter helps to restrict the flow of metal thereby helping to fill thin ribs and bases in the upper die

(a) both A and R are individually true and R is correct explanation of A
(b) both A and R are individually true and R is not correct explanation of A
(c) A is true but R is false
(d) A is false R is true

19. Match **List-I** (Products) with **List-II** (Processes) and select the correct answers with the codes given below

	List-I		**List-II**
A.	MS angles and channels	**1.**	Welding angles and channels
B.	Carburettors	**2.**	Forging
C.	Roof trusses	**3.**	Casting
D.	Gear wheels	**4.**	Rolling

Codes:

	A	B	C	D
(a)	1	2	3	4
(b)	4	3	2	1
(c)	1	2	4	3
(d)	4	3	1	2

20. Which of the following process is most commonly used for forging of bolt heads of hexagonal shape

(a) closed die drop forging
(b) open die upset forging
(c) close die press forging
(d) open die progressive forging

21. Match **List-I** with **List-II** and select the correct answers with the codes given below

	List-I		**List-II**
P.	Yield point elongation	**1.**	Distortion of dimension
Q.	Spring back	**2.**	Leuder's bands
R.	Wrinkling	**3.**	Plastic deformation
S.	Residual stress	**4.**	Compressive stress

Codes:

	P	Q	R	S
(a)	3	4	2	1
(b)	4	3	1	2
(c)	2	3	4	1
(d)	2	1	4	3

22. Consider the following operations in forging a hexagonal bolt from a round bar stock whose diameter is equal to bolt diameter

1. flattering
2. upsetting
3. swaging
4. cambering

The correct sequence of these operations is

(a) 1, 2, 3, 4
(b) 2, 3, 4, 1
(c) 2, 1, 3, 4
(d) 3, 2, 1, 4

23. In open die forging a disc of diameter 200 mm and height 60mm is compressed without any barreling effect the final diameter of disc is 400 mm the true strain is

(a) 1.986 (b) 1.686

(c) 1.386 (d) 0.602

24. In forging operation fullering is done to

(a) draw out the material (b) bend the material

(c) upset the material (d) extrude the material

25. Consider the following steps involved in hammer forging a connecting rod from bar stock

1. blocking 2. triming

3. finishing 4. fullering

5. edging

The correct sequence of these operations is

(a) 1, 4, 3, 2 and 5 (b) 4, 5, 1, 3 and 2

(c) 5, 4, 3, 2 and 1 (d) 5, 1, 4, 2 and 3

26. A forging method for reducing the diameter of the bar and in the process making it long is termed as

(a) fullering (b) punching

(c) upsetting (d) extruding

27. The process of removing the burs or flash from a forged component in drop forging is called

(a) swaging (b) perforating

(c) trimming (d) fetting

28. Match **List-I** (Types of forging) with **List-II** (operations) and select the correct answers with the codes given below

	List-I		List-II
A.	A drop forging	**1.**	Metal is gripped in dies & Pressure is applied on heated end
B.	Press forging	**2.**	Squeezing action
C.	Upset forging	**3.**	Metal is placed Between rollers & pushed
D.	Roll forging	**4.**	Repeated hammer blows

	A	B	C	D
(a)	4	1	2	3
(b)	3	2	1	4
(c)	4	2	1	3
(d)	3	1	2	4

29. The forging defect due to hindrance to smooth flow of metal in component called lap occurred because

(a) the corner radius provided is too large

(b) the corner radius provided is too small

(c) draft is not provided

(d) the shrinkage allowance is inadequate

30. Consider the following statements

Assertions (A) : Forging dies are provided with a taper or draft angles on vertical surfaces

Reason (R) : It facilities complete filling of die cavity and favorable grain glow

(a) both A and R are individually true and R is correct explanation of A

(b) both A and R are individually true and R is not correct explanation of A

(c) A is true but R is false

(d) A is false R is true

31. Match **List-I** (Forging/technique) with **List-II** (Processes) and select the correct answers using the codes given below

List-I	**List-II**
A. Smith Forging	**1.** Materials only upset to get the desired Shape
B. Drop Forging	**2.** Carried out manually the blank in open dies
C. Press Forging	**3.** done in close impression dies by Hammers in the blows
D. Machine Forging	**4.** done in close impression dies by continous squeezing process

Codes:

	A	B	C	D
(a)	2	3	4	1
(b)	4	3	2	1
(c)	2	1	4	3
(d)	4	1	2	3

32. Cold forging results in improved quality due to which of the following
1. better mechanical properties of the process
2. unbroken grain flow
3. smoother finishes
4. high pressure

select the correct answers using the codes given below
(a) 1, 2 and 3 (b) 1, 2 and 4
(c) 2, 3 and 4 (d) 1, 3 and 4

33. In wire drawing process, the bright shining surface on the wire is obtained if one:
(a) ones not use a lubricant (b) low tooling cost
(c) uses thick paste lubricant (d) use thin fluid lubricant

34. Which one of the following factors promotes the tendency for wrinkling in the process of drawing?
(a) increase in the ratio of thickness to blank diameter of work material
(b) decrease in the ratio of thick ness to blank diameter of work material
(c) decrease in the holding force on the blank
(d) use of solid lubricants

35. Which one of the following lubricants is most suitable for drawing mild steel wires?
(a) sodium stearate (b) water
(c) lime-water (d) kerosene

36. In drawing operation if D_i = initial diameter and D_0 = outgoing diameter, then what is the degree of drawing equal to?

(a) $\frac{D_i - D_o}{D_i}$ (b) $\frac{D_0 - D_i}{D_0}$

(c) $\frac{D_i^2 - D_0^2}{D_i^2}$ (d) $\sqrt{\frac{D_i^2 - D_0^2}{D_i^2}}$

37. In a wire drawing operation, diameter of a steel wire is reduced from 10 mm to 8 mm. The mean flow of the material is 400 MPa .The ideal force required for drawing (ignoring friction and redundant work) is

(a) 4.48 kN (b) 8.97 kN
(c) 20.11 kN (d) 31.41 kN

38. A cup of 10 cm height and 5cm diameter is to be made from a sheet metal of 2 mm thickness. The number of deductions necessary will be (if the LDR of the material is 1.9):

(a) one (b) two
(c) three (d) four

39. Match **List-I** with **List-II** and select the correct answer using the codes given below the lists:

List-I	**List-II**
A. Drawing	**1.** Soap solution
B. Rolling	**2.** Camber
C. Wire drawing	**3.** Pilots
D. Steel metal	**4.** Crater operations
	5. Ironing progressive dies

Codes:

	A	**B**	**C**	**D**
(a)	2	5	1	4
(b)	4	1	5	3
(c)	5	2	3	4
(d)	5	2	1	3

40. Consider the following statements: earing in a drawn cup can be due to non –uniform

1. speed of the press
2. clearence between tools
3. material properties
4. blank holding

which of these statements are correct?

(a) 1, 2 and 3 (b) 2, 3 and 4
(c) 1, 3 and 4 (d) 1, 2 and 4

41. A shell of 100 mm diameter and 100 mm height with the corner radius of 0.4 mm is to be produced by cup drawing. The required blank diameter is:

(a) 118 mm (b) 161 mm
(c) 224 mm (d) 312 mm

42. Which of the following types of stresses is /are involved in the wire drawing operation?

(a) Tensive only
(b) compressive only
(c) a combination of tensile and compressive stresses
(d) a combination of tensile, compressive and shear stresses

43. In the deep drawing of cups, blanks show a tendency to wrinkle up around the periphery(flange). The most likely cause and remedy of the phenomenon are, respectively

(a) buckling due to circumferential compression; increase blank holder pressure
(b) high blank holder pressure and high friction; reduce blank holder pressure and apply lubricant
(c) high temperature causing increase in circumferential length; apply coolant to blank
(d) buckilng due to circumferential compression; decrease blank holder pressure

44. A cylindrical vessel with flat bottom can be deep drawn by

(a) shallow drawing (b) single action deep drawing
(c) double action deep drawing (d) triple action deep drawing

45. Which of the following methods can be used for manufacturing 2 meter long seamless metallic tubes?

1. drawing
2. extrusion
3. rolling
4. spinning

Which of the following statements are correct?

(a) 1 and 2
(b) 2 and 3
(c) 1, 3 and 4
(d) 2, 3 and 4

46. Consider the following statements if forward extrusion process

1. The ram and the extruded product travel in the same direction
2. The ram and the extruded product travel in the opposite direction
3. The speed of travel of the extruded product is the same as that of the ram
4. The speed of the travel of the extruded product is greater than that of the ram

Which of the following statements are correct

(a) 1 and 3
(b) 2 and 3
(c) 1 and 4
(d) 2 and 4

47. The following operations are performed while preparing the billets for extrusion process

1. Alkaline cleaning
2. Phosphate coating
3. Picking
4. Lubricating with reactive soap

the correct sequence of these operations is

(a) 3, 1, 4, 2
(b) 1, 3, 2, 4
(c) 1, 3, 4, 2
(d) 3, 1, 2, 4

48. Which of the following statements are the salient features of hydrostatic extrusion?

1. It is suitable for soft and ductile material
2. It is suitable for high-strength super-alloys
3. The billet is inserted in to the extrusion chamber and pressure is applied by ram to extrude the billet through the die
4. The billet is inserted in to the extrusion chamber where it is surrounded by suitable liquid. The billet is extruded through the die by applying pressure to the liquid

Select the correct answer using the codes given below:

(a) 1 and 3
(b) 1 and 4
(c) 2 and 3
(d) 2 and 4

49. A brass billet is to be extruded from its intial diameter of 100 mm to a final diameter of 50 mm. The working temperature of 700°C and the extrusion constant is 250 MPa .The force required for extrusion is

(a) 544 MN
(b) 2.72 MN
(c) 1.36 MN
(d) 0.36 MN

50. Magnesium is extuded and not rolled because

(a) It has a low melting point
(b) It has low density
(c) Its reactivity with roll material is high
(d) It has a close-packed hexagonal structure

51. Which of the following materials is used in the manufacture of extrusion nozzles?
(a) Grey cast iron (b) Malleable cast iron
(c) White cast iron (d) Nodular cast iron

52. Extrusion force does not depend upon the
(a) extrusion ratio (b) type of extrusion process
(c) material of the die (d) working temparature

53. What does hydrostatic pressure in extrusion process improve
(a) ductility (b) type of extrusion process
(c) brittleness (d) tensile strength

54. Which of the following are the characteristics of extrusion process?
1. faster speed of production
2. good surface finish
3. dimensions can be maintained with close tolerances
4. mechanical properties of the extruded products are inferior to the that of the rolled products

Select the correct answer using the following codes given below:
(a) 1 and 2 (b) 1, 2 and 3
(c) 3 and 4 (d) 1 and 4

55. Assertion(A) : Indirect extrusion operation can be performed either by moving ram or by moving the container
Reason(R) : Advantage in indirect extrusion is less quantity of scarp compared to direct extrusion
(a) Both A and R are individually true and R is the correct explanation of the A
(b) both A and R are individually true but R is not the correct explanation of A
(c) A is true but R is false
(d) A is false but R is true

56. Which of the following is the correct statements?
(a) Extrusion is used for the manufacture of seamless tubes
(b) Extrusion is used for reducing the diameter of round bars and tubes by rotating dies which open and close rapidly on the work
(c) Extrusion is used to improve fatigue resistance of the metal by setting up compressive stresses on its surface
(d) Extrusion comprises pressing the metal inside a chamber to force it out by high pressure through an office which is shaped to provide the desired form of the finished part

57. Which one the following processes necessarily mandrel of requisite diameter to form the internal hole?
(a) Hydrostatic extrusion (b) Tube drawing
(c) Swaging (d) Wire drawing

58. Assertion (A): Greater force on the plunger is required in the case of the direct extrusion than indirect one
Reason (R): In case of direct extrusion,the direction of the force applied on the plunger and the direction of the movement of the extruded metal are the same
(a) both A and R are individually true and R is correct explanation of A
(b) both A and R are individually true but R is not the correct explanation of A
(c) A is true but R is false
(d) A is false but R is true

59. The bending force required for V-bending, U-bending and edge bending will be in the ratio of:

(a) 1 : 2 : 0.5 (b) 2 : 1 : 0.5
(c) 1 : 2 : 1 (d) 1 : 1 : 1

60. What is the bend allowance in the following Fig. if the bend radius is 100 mm.

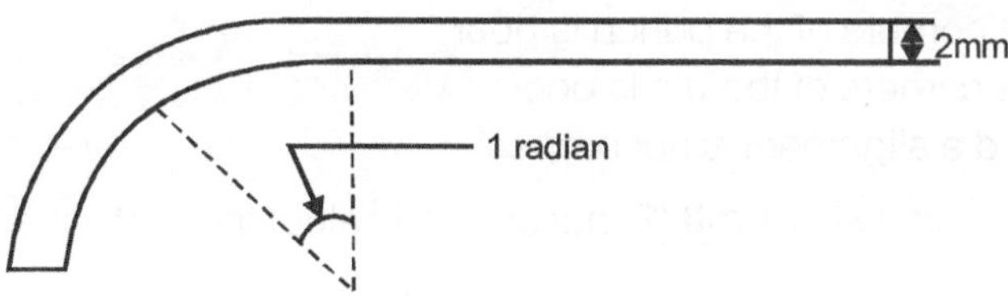

(a) 99 mm (b) 100 mm
(c) 101 mm (d) 102 mm

61. Which of the following is a continuous bending operation oils are used to produce long sections of formed shapes from coil or strip stock?

(a) Strength forming (b) Roll forming
(c) Roll bending (d) Spinning

62. Match the correct combination for the following metal working processes

	Processes		**Associated state of stressTension**
P.	Blanking	1.	Tension
Q.	Stretch forming	2.	Compression
R.	Coining	3.	Shear
S.	Deep drawing	4.	Tension and compression
		5.	Tension and shear

(a) P-2,Q-1,R-3,S-4 (b) P-3,Q-4,R-1,S-5
(c) P-5,Q-3,R-3,S-1 (d) P-3,Q-1,R-2,S-4

63. **Assertion (A):** In sheet metal blanking operation, clearance must be given to die:
Reason (R): The blank should be of required dimensions

(a) both A and R are individually true and R is correct explanation of A
(b) both A and R are individually true but R is not the correct explanation of A
(c) A is true but R is false
(d) A is false but R is true

64. In the blanking operation to produce steel washer,the maximum punch load used is 2×10^5N. The plate thickness is 4 mm and percentage penetration is 25%. The work done during this shearing operation is

(a) 200 J (b) 400 J
(c) 600 J (d) 800 J

65. The force requirement in a blanking operation of low carbon steel sheet is 5.0 kN. The thickness of the sheet is 't' and diameter of the blanked part is 'd'. For the same work material, if the diameter of the blanked part is increases 1.5 d and thick ness reduced to 0.4 t , the new blanking force in kN is

(a) 3.0 (b) 4.5
(c) 290.7 (d) 8.0

66. Best position of crank for blanking operation in a mechanical press is:
(a) top dead center (b) 20 degrees below top dead center
(c) 20 degrees before bottom dead center (d) bottom dead center

67. Which of the following statements are correct for deep drawing.
(a) Clearance at the corners of the punch is poor
(b) The finish at the corners of the punch is poor
(c) The finish at the corners of the die is poor
(d) The punch and die alignment is not proper

68. Match **List-I** (Press-part) with **List-II** (Function)and select the correct answer using the following codes given below

	List-I		**List-II**
A.	Punch plate	1.	Assisting withdrawl of punch
B.	Stripper	2.	Advancing the work-piece through correct distance
C.	Stopper	3.	Ejection of the work piece from die cavity
D.	Knock out	4.	holding the small punch in the proper position

Codes:

	A	B	C	D
(a)	4	3	2	1
(b)	2	1	4	3
(c)	4	1	2	3
(d)	2	3	4	1

69. Consider the following statements related to piercing and blanking
1. Shear on the punch reduces the maximum cutting force
2. Shear increases the capacity of the press needed
3. Shear increases the life of punch
4. The total energy needed to make the cut remains unaltered due to provision of shear

which of the following above statements are correct ?
(a) 1 and 3 (b) 1 and 4
(c) 2 and 3 (d) 3 and 4

70. A metal disc of 20mm diameter is to be punched from a sheet of 2 mm thickness. The punch and the dia clearance is 3%. The required punch dia meter is
(a) 19.88 mm (b) 19.84 mm
(c) 20.06 mm (d) 20.12 mm

71. What is the process of removing metal by a milling cutter which is rotated against the direction of travel of the work piece, called?
(a) down milling (b) up milling
(c) end milling (d) face milling

72. In deciding the clearance between the punch and die in press work in shearing, the following rule is helpful
(a) punch size controls the hole size die size controls blank size
(b) punch size controls both hole size and blank size
(c) die size controls both hole size and blank size
(d) die size controls hole size and punch size controls blank size

73. Which is the force required to be exerted by the punch in order to shear out the blank (from a circular disk) of 20 mm diameter and 1.5 mm thickness and having a shear strength equal to 300 MPa?

(a) 9.0 kN (b) 18.0 kN
(c) 28.3 kN (d) 45.0 kN

74. Steel balls are manufactured by
(a) casting (b) machining
(c) cold heading (d) sintering

75. Match **List-I** (Metal forming process) with **List-II** (Associated feature) and select the correct answer using the following codes given below the lists:

	List-I		List-II
A.	Blanking	**1.**	Shear angle
B.	Flow forming	**2.**	Coiled stock
C.	Roll forming	**3.**	Manderel
D.	Embossing	**4.**	Closed matching dies

Codes:

	A	B	C	D
(a)	1	3	4	2
(b)	3	1	4	2
(c)	1	3	2	4
(d)	3	1	2	2

76. Match **List-I** (Parts) with **List-II** (Manufacturing processes) and select the correct answer using the codes given below the lists:

	List-I		List-II
A.	Seamless tubes	**1.**	Roll
B.	Accurate and forming and smooth tubes	**2.**	Shot
C.	Surfaces having peenin higher hardness	**3.**	Forging & fatigue strength
D.	Embossing	**4.**	Cold forming

Codes:

	A	B	C	D
(a)	1	4	2	3
(b)	2	3	1	4
(c)	1	3	2	4
(d)	2	4	1	3

77. The collapsible tooth paste tubes are manufactured
(a) direct extrusion (b) piercing
(c) impact extrusion (d) indirect extrusion

78. Match **List-I** (Type of modeling) with **List-II** (Mechanism involve and select the correct answer using the codes given below the lists:

	List-I		List-II
A.	Compression moulding	**1.**	Mould cavity must be heated to curve the plastic forced in to
B.	Injected moulding	**2.**	Similar to hydraulic extrusion
C.	Jet moulding	**3.**	Analogous to hot-pressing of powdered metals
D.	Extrusion moulding	**4.**	Analougus to die casting of metals

Codes:

	A	B	C	D
(a)	2	4	1	3
(b)	3	1	4	2
(c)	2	1	4	3
(d)	3	4	1	2

79. Magnetic forming is an example of:
(a) cold forming (b) hot forming
(c) high energy rate forming (d) roll forming forming

80. Consider the following statements
1. Metal forming decreases harmful effects of impurities and improves mechanical strength
2. Metal working process is a plastic de formation process
3. Very intricate shapes can be produced by forging process as compared to casting process
Which f the statements given below are correct?
(a) 1, 2 and 3 (b) 1 and 2 only
(c) 2 and 3 only (d) 1 and 3 only

81. Which is incorrect statement about results of cold working
(a) it increases corrosion resistance
(b) it increases strength, elasticity and hardness with a corresponding decrease in ductility
(c) handling of parts is easier
(d) dimensions tolerances and finish are high

82. In the metal forming processes, the stresses encountered are
(a) less than the yield strength of the material
(b) less than the fractures strength of the material and greater than yield strength
(c) less than the limit of proportionality
(d) less than the elastic limit

83. Rotary swaging is used for
(a) manufacturing bolts and rivets
(b) manufacturing seamless tubes
(c) improving fatigue resistance
(d) reducing diameter of round bars and tubes by rotating die which open and close rapidly on the work

84. Which is incorrect statement about results of hot working
(a) power requirements are low (b) surface finish is good
(c) grain refinement is possible (d) porosity in the metal is largely eliminated

85. Which of the following materials can't be forged
(a) wrought iron (b) cast iron
(c) mild steel (d) high carbon steel

86. Swaging
(a) causes a steadily applied pressure instead of impact force
(b) is used to force the end of a heated bar into a desired shape
(c) is a forging operation in which two halves of a rotating die open and close rapidly while impacting the end of the heated tube or shell
(d) is forging method for reducing the diameter of a bar and in the process making it longer

87. The increase in hardness due to cold working is called
(a) cold hardening (b) hot hardening
(c) work hardening (d) age-hardening

88. The operation of removing the burr or flash from the forged parts in drop forging is known as
(a) lancing (b) trimming
(c) coining (d) burring

89. Metallic cans are usually mass produced by the following process
(a) embossing (b) coining
(c) spinning (d) deep drawing

90. Spinning operation is carried out on
(a) hydraulic press (b) mechanical press
(c) lathe (d) milling machine

91. Production of contours in flat blanks is termed as
(a) blanking (b) rubbing
(c) perforating (d) punching

92. Thread rolling is some what like
(a) cold extrusion (b) cold rolling
(c) cold forging (d) plug rolling

93. Upset forging
(a) causes a steadily applied pressure instead of impact force
(b) is used to force the end of a heated bar into a desired shape
(c) is a forging operation in which two halves of rotating die open and close rapidly while impacting the end of the heated tube or shell
(d) is a forging method for reducing the diameter of a bar and in the process making it longer

94. Which of the following process is different from the rest of the processes
(a) shot peening (b) cold extrusion
(c) sand blasting (d) drop forging

95. It is required to reduce a slab directly to strip in one pase. Which of the following rolling mills can do this function?
(a) three high mill (b) four high mill
(c) planetary mill (d) continuous rolling mill

96. In four high rolling mill the bigger rollers are called
(a) guide rolls (b) back up rolls
(c) mail rolls (d) support rolls

97. Notching is the operation of
(a) removal of excess metal from the edge of a strip to make it suitable for drawing without wrinkling
(b) cutting of the excess metal at edge which was required for gripping purpose during press working operation.
(c) cutting in a single line across a part of the metal strip to allow bending or forming in progressive die operation while the part remains attached to the strip.
(d) punching in which punch is topped as soon as the metal fracture is complete and metal is not removed but held in hole

98. Lancing is the operation of
(a) removal of excess metal from the edge of a strip to make it suitable for drawing without wrinkling
(b) cutting of the excess metal at edge which was required for gripping purpose during press working operation

(c) cutting in a single line across a part of the metal strip to allow bending or forming in progressive die operation while the part remains attached to the strip
(d) punching in which punch is stopped as soon as the metal fracture is complete and metal is not removed but held in hold.

99. Flange wrinkling is the defect found in
(a) blanking (b) spinning
(c) bending (d) deep drawing

100. Injection moulding is the ideal method of processing
(a) plastics (b) thermo-setting plastics
(c) thermoplastics (d) non-ferrous materials

101. The ductility of a material with work hardening
(a) increases
(b) decreases
(c) remains unaffected
(d) may increase/decrease depending on its tensile strength

102. Blanking and piercing operation can be performed simultaneously in
(a) simple die (b) progressive die
(c) compound die (d) combination die

103. In progressive dies
(a) two or more cutting operations can be performed simultaneously
(b) cutting and formation operations are combined and carried out in single operations
(c) workpiece moves from one station to other with separate operation performed at each station
(d) all of the above

■■■■

ANSWERS

METAL FORMING

1. (b)	**2.** (b)	**3.** (a)	**4.** (c)	**5.** (b)	**6.** (a)	**7.** (a)
8. (b)	**9.** (d)	**10.** (b)	**11.** (d)	**12.** (b)	**13.** (d)	**14.** (c)
15. (b)	**16.** (d)	**17.** (c)	**18.** (a)	**19.** (d)	**20.** (a)	**21.** (a)
22. (a)	**23.** (c)	**24.** (a)	**25.** (b)	**26.** (a)	**27.** (c)	**28.** (c)
29. (b)	**30.** (c)	**31.** (a)	**32.** (a)	**33.** (d)	**34.** (c)	**35.** (c)
36. (c)	**37.** (b)	**38.** (b)	**39.** (d)	**40.** (b)	**41.** (c)	**42.** (a)
43. (a)	**44.** (d)	**45.** (a)	**46.** (c)	**47.** (a)	**48.** (d)	**49.** (b)
50. (c)	**51.** (c)	**52.** (c)	**53.** (a)	**54.** (d)	**55.** (c)	**56.** (d)
57. (b)	**58.** (b)	**59.** (a)	**60.** (c)	**61.** (b)	**62.** (d)	**63.** (d)
64. (a)	**65.** (a)	**66.** (b)	**67.** (b)	**68.** (a)	**69.** (a)	**70** (a)
71. (b)	**72.** (a)	**73.** (c)	**74.** (c)	**75** (c)	**76.** (a)	**77.** (c)
78. (b)	**79.** (c)	**80.** (a)	**81.** (a)	**82.** (b)	**83.** (d)	**84.** (b)
85. (b)	**86.** (c)	**87.** (c)	**88.** (b)	**89.** (d)	**90.** (c)	**91.** (a)
92. (b)	**93.** (b)	**94.** (c)	**95.** (c)	**96.** (b)	**97.** (a)	**98.** (c)
99. (d)	**100** (b)	**101.** (b)	**102.** (c)	**103.** (c)		

CHAPTER 6

Welding

Welding is the process of joining similar or dissimilar metals with or without the application of heat and pressure. Weldability of a material depends upon various factors like the metallurgical changes that occur due to welding, changes in hardness in and around the weld, extent of oxidation and the extent of cracking tendency of the joint.

6.1 METALLURGY IN WELDING

Metal is heated over a range of temperature up to fusion and followed by cooling to ambient temperature. Because of differential heating, the material far away from the weld bead will be simply warmed out, but as the weld area is approached, progressively higher temperatures are obtained, resulting in a corresponding complex microstructure. The heating and cooling also results in setting up internal stresses and plastic strain in the weld. So at high temperatures physical, chemical and metallurgical properties are liable to change.

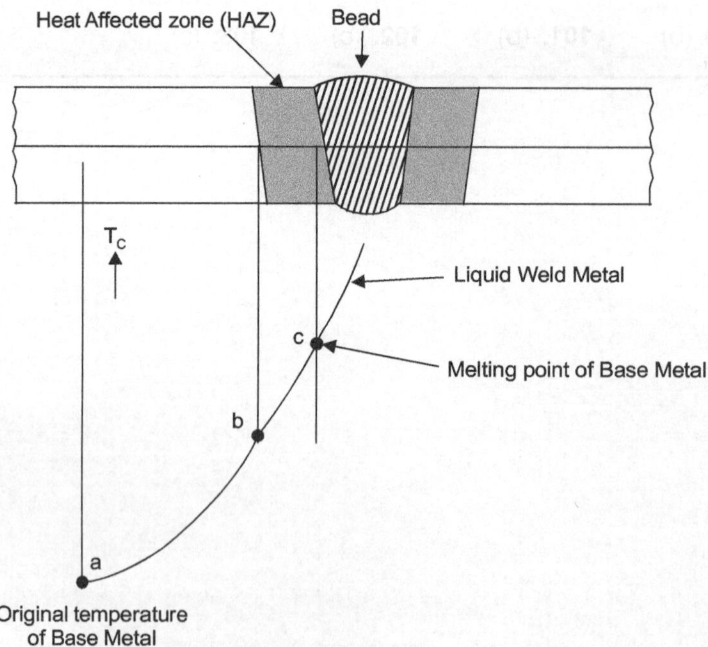

Fig. 6.1 : Graph showing the change of thermal conditions

Depending on the welding process used and the metal being welded, there exists wide variation in the maximum liquid temperature and in the slope of temperature gradient curve for the unfused metal. Three distinct zones can be identified in a weld joint which are :

1. Base metal
2. Heat effected zone.
3. Weld metal.

A joint produced without the filler metal is called autogenous and its weld zone is composed of the resolidified base metal. A joint made with a filler metal called weld metal. After the application of heat and the introduction of the filler metal into the weld area, the molten weld joint is allowed to cool to ambient temperatures. Since central portion of the bead will be cooled slowly, long columnar grains will develop (see TTT diagram discussion in chapter 2) and in the outward direction grains will become finer and finer (Fig. 6.2). So the ductility and toughness of bead metal at the centre will be higher. Thus the original structure consisting of ferrite and pearlite in steels is changed to another micro structure. The composition of the first crystals which form in a molten alloy may be quite different from the composition of liquid but as the freezing proceeds, the crystals readjust their composition to that of the initial liquid alloy in order to satisfy the condition of equilibrium depending upon the type of phase diagram exhibit by alloy.

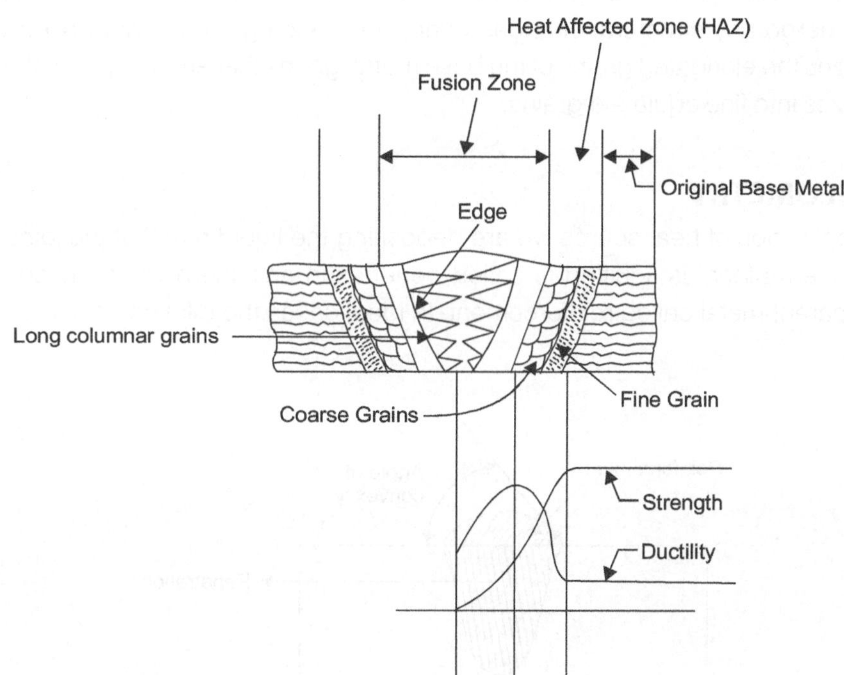

Fig. 6.2 : Regions in weld bead

The weld metal when it is in the molten state has a good capacity to dissolve gases which come into contact with it, like oxygen, nitrogen and hydrogen. As the metal starts cooling, capacity of dissolving gases goes on decreasing, thus causing additional volume of gases to be evolved. Due to the appearance of mushy zone during solidification, a portion of these gases may be trapped into the bead called porosity. This decreases the strength of weldment.

Grain growth in heat affected zone

Large regular shaped crystals

Long elongated (columnar) crystals

Fig. 6.3 : Structure of the weld metal

As shown in Fig. 6.3 in the structure of weld section the columnar crystals are formed near fusion faces due to directional cooling of weld towards the centre. Since the inner part of the weld cools more uniformly it results an enlarged but regular crystal structure. Surface of weld beyond the slag coating is being cooled very fast, so small and slightly chilled crystal structure will be formed. Cooling rates can be controlled by preheating of the weld before welding.

The heat affected zone is within the base metal itself. It has a microstructure different from that of the base metal prior to welding, because it is subjected to elevated temperature for a substantial period of time during the welding. The portions of the base metal that are far from the heat source do not undergo any structural changes during the welding. The heat applied during welding recrystalizes the elongated grains of the base metal, grains that are away from the weld metal will recrystalizes into fine equiaxed grains.

6.2 BEAD GEOMETRY

By the application of heat source we are depositing the liquid metal at the joint. The depth, up to which the material is liquified is called penetration and there will be raised portion of the plane of parent metal called reinforcement. In ideal cases the dilution should be as close to be as 100%.

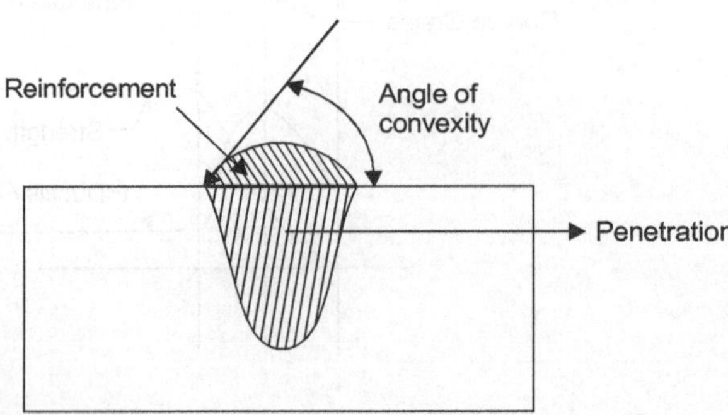

Fig. 6.4 : Bead geometry

A_p = Area of penetration

A_R = Area of Reinforcement

$$\text{Dilution} = \frac{A_P}{A_P + A_R}$$

6.3 CLASSIFICATION OF WELDING PROCESS

1. Arc Welding
2. Resistance Welding
3. Solid State Welding
4. Radiant Energy Welding
5. Thermit Welding
6. Allied process
7. Gas Welding

6.4 ARC WELDING

6.4.1 Shielded Metal Arc Welding (SMAW)

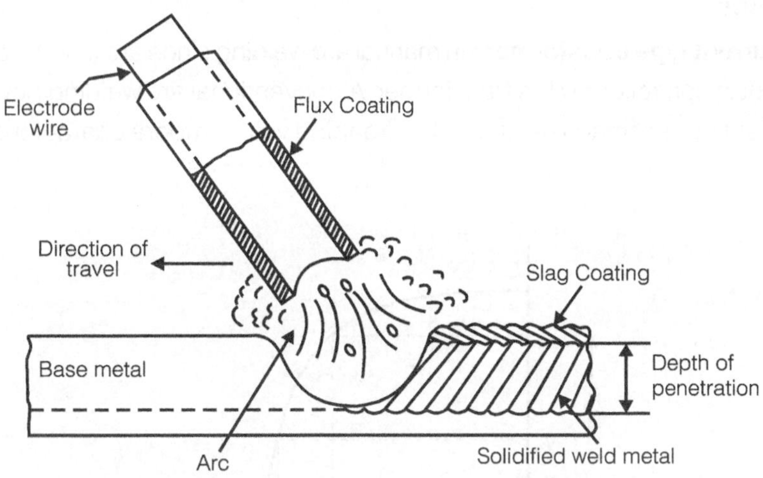

Fig. 6.5 : Welding using SMAW process

This is also called as stick electrode welding or coated electrode welding or manual arc welding. It uses coated electrodes of 2.5 to 6.35 mm diameter and 300 to 450 mm length held in an electrode holder. The power source used is of constant current type. Initially the arc is being created between wire electrode and work piece, flux by taking energy from the arc gets melted and a mixture of liquid metal and molten flux gets simultaneously settled in the groove. Since the density of flux is lower, it comes over the surface of bead and provides slag coating. It has following functions.

(a) Since flux gives alloys to the bead material, bead becomes stronger than even the parent metal. The region in the parent metal near the bead where the temperature has gone beyond the recrystalization temperature is called heat affected zone and it is the weakest portion in the weld metal.

(b) It provides the heat treatment to the bead and as a result of that at the center of the bead there will be coarse grains and in the outward direction grains become finer and finer.

(c) It protects the bead from the attack of atmosphere gases.

There are two movements of electrode holder.

(a) The rate at which electrode wire is getting consumed with the same rate the electrode has to be fed towards the molten pool. It is because arc length decides the arc voltage and arc current. If lower is the current settings, the temperature of the molten pool will be lower, as a result of that the liquid flux (slag) trying to come out, material will be solidified and slag will be trapped with in the bead. This defect is called slag inclusion. Also variable current leads to improper bead geometry.

(b) Movement of electrode along the longitudinal direction of bead.

If both these movements are controlled manually it is called manual arc welding. If first movement is controlled by machine it is called **semi-automatic arc welding** and if both movements are controlled by machine it is called **automatic arc welding.** The temperature in the core of the arc ranges between 6000 – 7000°C. SMAW can be performed both on AC and DC source with drooping characteristics.

6.4.2 Transformer

Constant current type transformer : In manual arc welding since arc length can not be controlled, so arc current is controlled by the transformer. A conventional arc welding power source is known as the constant current machine. It has the drooping volt – ampere characteristic curve as shown in Fig. 6.6

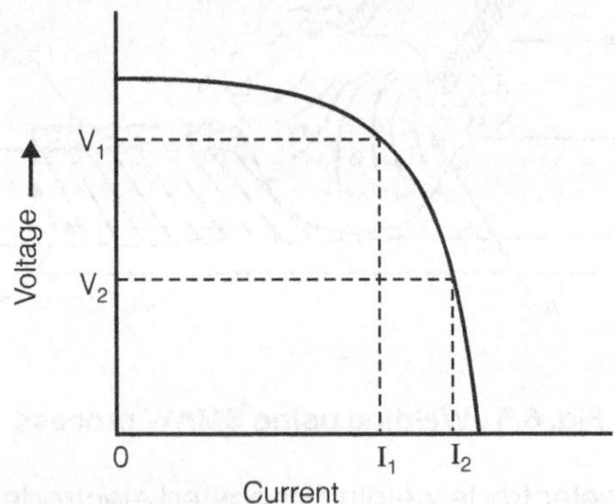

Fig. 6.6 : Characteristics of constant current type transformer

The constant current curve shows that the welding power source produces maximum output voltage with no load and as load increases, the output voltage decreases. As it can be observed in Fig 6.6 that for a major change in arc voltage, there is insignificant change in arc current.

Constant voltage transformer : A constant voltage welding power source has essentially a flat volt ampere characteristics curve though usually with a slight droop. The curve may be shifted up or down to change the voltage. These characteristics are used only for continuous electrode wire welding such as gas welding semiautomatic and automatic welding processes.

If the load in the circuit changes, the power source automatically adjusts its current output to satisfy the requirement and maintains essentially the same voltage across the output terminals. The constant voltage welding power system has its greatest advantage when the current density of the electrode wire is high.

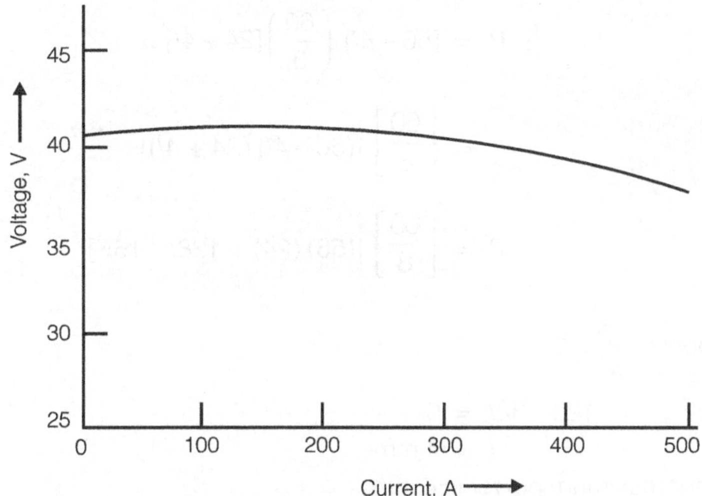

Fig. 6.7 : Characteristics of constant voltage transformer

$$V = (OCV) - \left[\frac{OCV}{SCC}\right] I$$

This is for semi automatic arc welding.

$$OCV = \text{Open circuit voltage.}$$
$$SCC = \text{Short circuit current.}$$

For stable arc in constant voltage transformer $V_{arc} = V_{transformer}$

For stable arc in constant current transformer $I_{arc} = I_{transformer}$

Example : 6.1

The arc length voltage characteristic is given by expression V = 24 + 4*l* (*l* = arc length in mm). The volt ampere characteristic of a power source can be approximated by a straight line with open circuit voltage 80V and short circuit current is 600A. Determine optimum arc length for maximum power.

Solution :

$$OCV = 80V$$
$$SCC = 600A$$
$$V = 80 - \left[\frac{80}{600}\right] I$$

Stable

$$V_t = V_a$$
$$I = F(l)$$
$$24 + 4l = 80 - \left[\frac{80}{600}\right] I$$
$$\left[\frac{80}{600}\right] I = 80 - 24 - 4l$$
$$I = [56 - 4l] \left[\frac{60}{8}\right]$$
$$P = VI$$

$$P = [56 - 4l]\left(\frac{60}{8}\right)[24 + 4l]$$

$$= \left[\frac{60}{8}\right][(56 - 4l)(24 + 4l)]$$

$$P = \left[\frac{60}{8}\right][(56)(24) + 128l - 16l^2]$$

For maximum power $\frac{dp}{dl} = 0$

$$128 - 32l = 0$$
$$l = 4 \text{ mm}$$

The arc length for maximum power is 4 mm.

Example : 6.2

Volt-amp characteristics of a power source is given by equation $I_t^2 = -600(V - 60)$

1. **If the arc characteristic is $I = 20(V - 16)$, determine the power of stable arc.**
2. **If arc length voltage are represented by $V_a = 30 + 5.5l$, determine optimum arc length for maximum power.**

Solution :

1. Arc characteristics is given by

$$I_a = 20(V - 16)$$

Volt-amp characteristics of power source is given by

$$I_t^2 = -600(V - 60),$$

as current-voltage variation is parabolic, so the power source is dropper or constant current source.

Hence, for the stability of arc;

$$I_t = I_a;$$

(i.e. current across transformer is equal to current across arc)

or $$I_t^2 = I_a^2$$

$$-600(V - 60) = \{20(V - 16)\}^2$$
$$-600V + 36{,}000 = 400V^2 + 102{,}400 - 12{,}800V$$
$$400V^2 - 12200V + 66400 = 0$$
$$4V^2 - 122V + 664 = 0$$

Solve this equation, we get roots of this equation as;

$$V = \frac{122 \pm \sqrt{122^2 - 4 \times 4 \times 664}}{2 \times 4} = \frac{122 \pm 65.26}{8}$$

$\Rightarrow$ $V = 23.40$ volts, 7.09 volts

But at $V = 7.09$ volts; $I_a = 20(V - 16)$

$= 20(7.09 - 16) = -ve$

So neglect $V = 7.09$ volts

Hence voltage (v) = 23.40 volts

Current of arc at this voltage = $I = 20\,(v - 16)$

$\Rightarrow \quad I = 20\,(23.40 - 16) = 148$ Amps.

So, power of arc = $V \times I$

$= 23.40 \times 148 = 3463.2$ W

or Power = 3.463 kW Ans.

2. Arc length voltage is given by

$$V = 30 + 5.5l$$

Volt-amp characteristic of power source is given by

$$I_t^2 = -600\,(V - 60);$$

On putting value of V in the above equation, we get;

$$I_t^2 = -600\,(30 + 5.5l - 60)$$

$$I_t^2 = -600\,(5.5l - 30)$$

$$I_t = \{-600\,(5.5l - 30)\}^{1/2}$$

As, power (P) is given by

$$P = V \times I$$

$$P = (30 + 5.5l) \times \{-600\,(5.5l - 30)\}^{1/2}$$

For maximum power, $\dfrac{dp}{dl} = 0$

$$\frac{dp}{dl} = \left\{(30+5.5l)\times\frac{1}{2}\{-600\,(5.5l-30)\}^{-\frac{1}{2}}\times(-600\times5.5)\right\}+\left\{-600(5.5l-30)^{\frac{1}{2}}\right\}\times5.5$$

equate $\dfrac{dp}{dl} = 0$

$$\Rightarrow \frac{(30+5.5l)\,\{-600\,(5.5l-30)\}^{-\frac{1}{2}}(-600\times5.5)}{2} = \{-600\,(5.5l-30)\}^{\frac{1}{2}}\times5.5$$

$$\Rightarrow \frac{(30+5.5l)\times(-600\times5.5)}{2} = -600\,(5.5l-30)\times5.5$$

$$\Rightarrow \frac{(30+5.5l)}{2} = (5.5l-30)$$

$\Rightarrow 30 + 5.5l = 11l - 60$

$\Rightarrow 90 = 5.5l$

$\Rightarrow l = 16.36$ mm

Hence, optimum arc length for maximum power = I_{opt} = 16.36 mm

Example : 6.3

Arc length voltage characteristics can be represented by

$$V = 20 + 4l$$

If the arc length in the welding operation varies between 4mm to 6mm and the welding current between 450 Amp. to 550 Amp, assuming a linear power source characteristic, calculate.

1. **Open circuit voltage (OCV)**
2. **Short circuit current (SCC)**

Solution :

Power source characteristic is linear. (given)

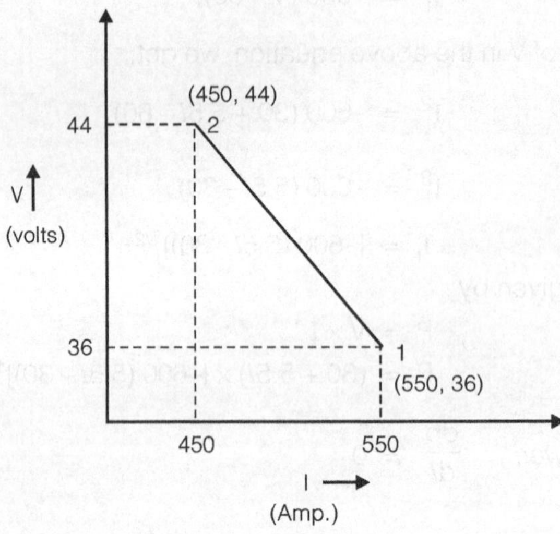

Fig. 6.8

as arc length voltage characteristic is;

$$V = 20 + 4l$$

at arc length (l_1) = 4 mm

$$V_1 = 20 + (4 \times 4) = 36 \text{ volts}$$

at arc length (l_2) = 6 mm

$$V_2 = 20 + (4 \times 6) = 44 \text{ volts}$$

equation of line can be written as;

$$(V - V_1) = \left(\frac{V_2 - V_1}{I_2 - I_1}\right)(I - I_1)$$

$$(V - 36) = \left(\frac{44-36}{450-550}\right)(I - 550)$$

$$V - 36 = \frac{-8}{100}(I - 550)$$

$$\Rightarrow \quad V - 36 = \frac{-8}{100}I + 44$$

$$\Rightarrow \quad V = 80 - \left(\frac{8}{100}\right)I \quad ...(1)$$

as linear power source characteristic is given by

$$V = OCV - \left(\frac{OCV}{SCC}\right)I \quad ...(2)$$

Compare (1) and (2); we get

$$OCV = 80V$$

$$\frac{OCV}{SCC} = \frac{8}{100} \Rightarrow SCC = 1000 \text{ Amp.}$$

$\Rightarrow$

$$OCV = 80 \text{ volts}$$

$$SCC = 1000 \text{ Amp.}$$

Example : 6.4

A direct current welding machine with a linear power source characteristic provides open circuit voltage of 80 V and short circuit current of 800 A. During welding with the machine, the measured arc current is 500 A corresponding to an arc length of 5.0 mm and the measured arc current is 460 A corresponding to an arc length of 7.0 mm. What is the linear voltage (E) – arc length (L) characteristic of the welding arc.

Solution :

Power source characteristic is linear.

Open circuit voltage (OCV) = 80V

Short circuit current (SCC) = 800A

For the linear power source characteristics, voltage (E) is given by:

$$E = OCV - \left(\frac{OCV}{SCC}\right)I$$

$\Rightarrow$

$$E = 80 - \frac{80}{800} \times I$$

$$E = 80 - \left(\frac{1}{10}\right)I \quad ...(1)$$

Let linear voltage (E) - arc length (l) characteristic of welding arc be

$$E = a + bl \quad ...(2)$$

a, b are constants

For arc current (I) = 500 A, arc length (l) = 5 mm;

From (1)

$$E = 80 - \frac{1}{10} \times 500$$

From (2)

$$E = 80 - 50 = 30V$$

$$E = a + 5b$$

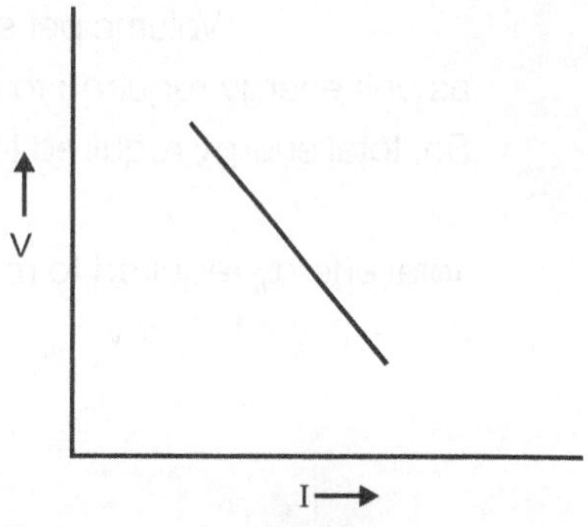

Fig. 6.9

Equate these values of E, we get

$$a + 5b = 30 \quad ...(3)$$

Similarly

$$\text{For arc current (I)} = 460 \text{ A},$$

$$\text{arc length } (l) = 7 \text{ mm}$$

From (1) $$E = 80 - \frac{1}{10} \times 460$$

$$E = 34V$$

From (2) $$E = a + 7b$$

Equate these values of E, we get

$$a + 7b = 34 \quad \text{...(4)}$$

On solving equation (3) and (4)

$$a + 5b = 30 \quad \text{...(3)}$$

$$a + 7b = 34 \quad \text{...(4)}$$

Subtract equation (3) from (4); we get

$$2b = 4 \Rightarrow b = 2$$

Put this value of b = 2 in equation (3)

$$a + (5 \times 2) = 30$$

$$\Rightarrow \quad a = 20$$

So, equation (2) becomes

$$E = 20 + 2l$$

Example : 6.5

In arc welding of a butt joint, the welding speed is to be selected such that highest cooling rate is achieved. Melting efficiency and heat transfer efficiency are 0.5 and 0.7, respectively. The area of the weld cross section is 5 mm² and the unit energy required to melt the metal is 10 J/mm³. If the welding power is 2kW. What is the welding speed in mm/s.

Solution :

Given,

Area of weld cross-section (A) = 5 mm²

Unit energy required to melt the metal = 10 J/mm³

Welding power = 2 kW = 2000 W

Melting efficiency (η_m) = 0.5

Heat transfer efficiency (η_H) = 0.7

Let welding speed = V mm/s

So, Volume welded per second = (Area of cross-section of weld) × (Welding speed)

Volume per second = 5 × V = 5 V mm³/s

as unit energy required to melt the metal = 10 J/mm³

So, total energy required to melt the metal = 10 × volume

= 10 × 5 VJ/s

Total energy required to melt the metal = 50V J/s

as melting efficiency (η_m) = 0.5, so the total energy supplied to melt the metal

$$= H_m = \left(\frac{50\,V}{0.5}\right)\frac{J}{s} \quad \text{...(1)}$$

as welding power = 2000 W

and heat transfer efficiency (η_H) = 0.7

So, total energy available $H_T = \eta_H \times 2000 = 2000 \times 0.7$ W

$$H_T = 1400\text{ W} \quad \text{...(2)}$$

as (1) and (2) are same, so equate these,

$$\frac{50V}{0.5} = 1400$$

$\Rightarrow$ $V = 14$ mm/s i.e. welding speed = 14 mm/s

Example : 6.6

Calculate the melting efficiency in case of arc welding of steel with a potential of 20 V and current of 200 A. The travel speed is 5 mm/s and the cross-sectional area of the joint is 20 mm^2. Heat required to melt steel may be taken as 10 J/mm^3 and the heat transfer efficiency as 0.85.

Solution :

Given,

Potential (V) = 20 volts

Current (I) = 200 Amp.

Weld speed (v) = 5 mm/s

Cross-section area of joint (A) = 20 mm^2

Heat required to melt steel = 10 J/mm^3

Heat transfer efficiency (η_T) = 0.85

Let Melting efficiency = η_m

as power is given by

$$P = V \times I$$

$$\Rightarrow \quad P = 20 \times 200 = 4000 \text{ W}$$

as heat transfer efficiency (η_T) = 0.85

So, total energy available at bead = $P \times \eta_T$

$= 4000 \times 0.85$

$= 3400$ W ...(1)

Volume welded per second is given as,

Volume welded per second = (cross-section area of weld) × (weld speed)

$= 20 \times 5 = 100$ mm^3/s

as, heat required to melt steel = 10 J/mm^3

So, total heat required to melt steel = 10 × 100 = 1000 J/s

as melting efficiency = η_m

So total heat to be supplied to melt steel = $\dfrac{1000}{\eta_m}$ ...(2)

as (1) and (2) are same,

so $$3400 = \frac{1000}{\eta_m}$$

$\Rightarrow$ $\eta_m = 0.2941$

i.e. Melting efficiency = 29.41%

Example : 6.7

The cross-section of weld bead is shown in Fig.6.10 The profile of the bead and the fusion zone are taken circular for convenience. Bead width and radii of curvature of circular profiles are given in the figure. Calculate

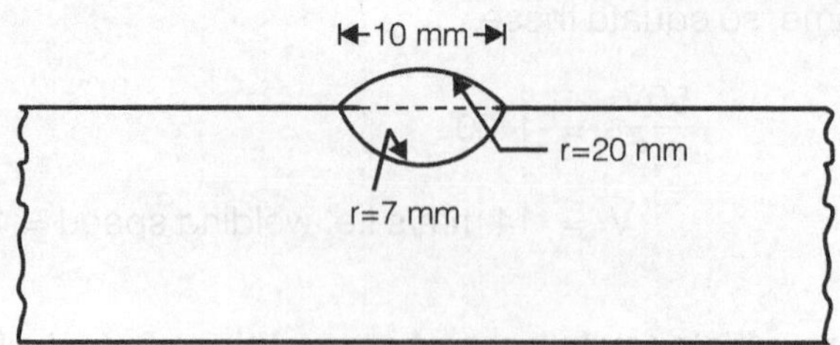

Fig. 6.10

(a) the bead height,

(b) penetration, and

(c) dilution (%)

All dimensions in Fig. are in mm.

Solution :

As reinforcement is a part of circle of radius 20 mm.

As we know, perpendicular drawn from the centre on the chord bisect the chord as from the properties of circle we know

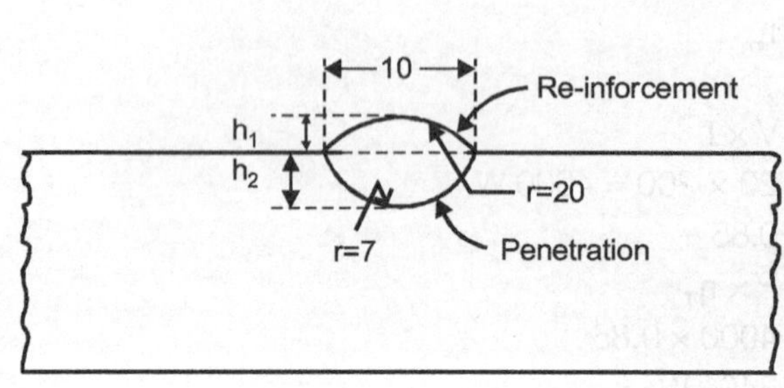

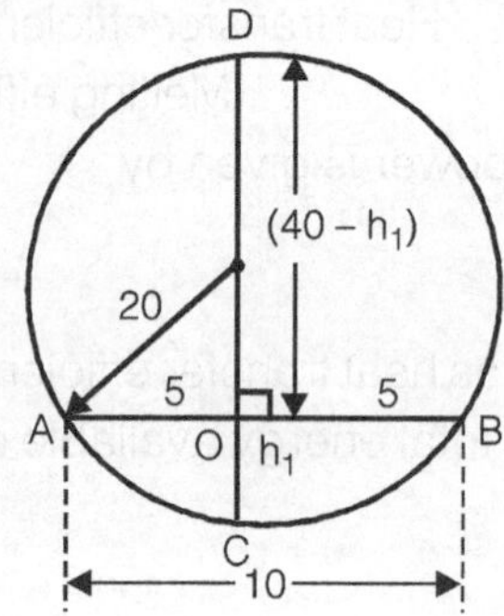

Fig. 6.11

$$OA \times OB = OC \times OD$$

$$5 \times 5 = h_1 \times (40 - h_1)$$

$$\Rightarrow \quad h_1^2 - 40h_1 + 25 = 0$$

Roots of this equation are

$$\Rightarrow \quad h_1 = \frac{40 \pm \sqrt{40^2 - 4 \times 25}}{2}$$

$$h_1 = \frac{40 \pm 38.72}{2}$$

$$h_1 = 39.36 \text{ mm}, 0.64 \text{ mm}$$

$$\Rightarrow \quad h_1 = 0.64 \text{ mm}$$

Similarly, penetration is also a part of the circle of radius 7 mm.

From the properties of circle,

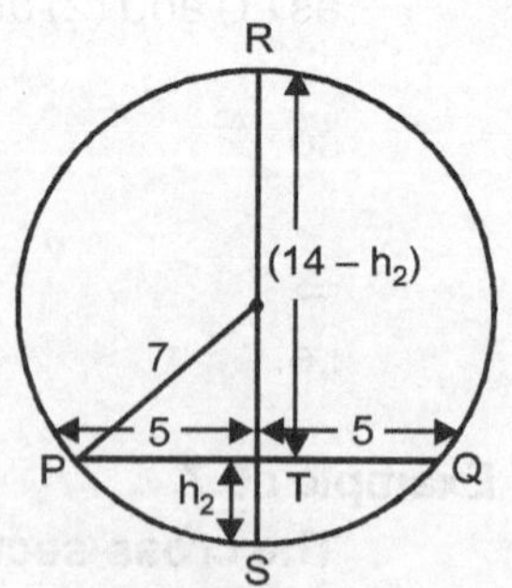

Fig. 6.12

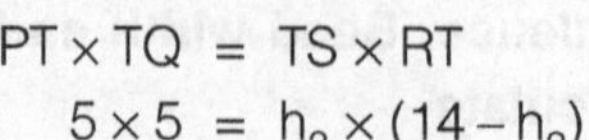

$$PT \times TQ = TS \times RT$$

$$5 \times 5 = h_2 \times (14 - h_2)$$

$$\Rightarrow \qquad h_2^2 - 14h_2 + 25 = 0$$

So, roots of this equation are

$$h_2 = \frac{14 \pm \sqrt{14^2 - 100}}{2}$$

$$= \frac{14 \pm 9.79}{2}$$

$$h_2 = 11.89 \text{ mm}, 2.10 \text{ mm}$$

$$\Rightarrow \qquad h_2 = 2.10 \text{ mm}$$

(a) Bead height = $h_1 + h_2 = 0.64 + 2.10 = 2.74$mm

(b) Penetration = $h_2 = 2.10$ mm

(c) % dilution = $\left(\frac{A_P}{A_P + A_R}\right) \times 100$

$$A_P = \text{Penetration area}$$

$$A_R = \text{Reinforcement area}$$

As penetration is a part of circle of radius 7 mm.

From ΔOBC;

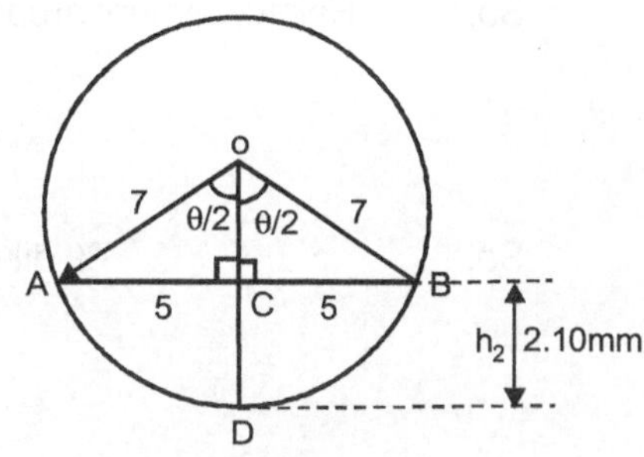

Fig. 6.13

$$\sin\frac{\theta}{2} = \frac{BC}{OB} = \frac{5}{7}$$

$$\Rightarrow \qquad \frac{\theta}{2} = \sin^{-1}\left(\frac{5}{7}\right)$$

$$\Rightarrow \qquad \theta = 91.16°$$

So, area of sector $= \frac{\theta}{360°} \times \pi r^2$

$$= \frac{91.16}{360} \times \pi \times 7^2 = 38.98 \text{ mm}^2$$

area of ΔOAB;

$$= \frac{1}{2} \times AB \times OC$$

$$= \frac{1}{2} \times 10 \times 4.9 \qquad \{OC = OD - DC = 7 - 2.1 = 4.9 \text{ mm}\}$$

$$= 24.5 \text{ mm}^2$$

So, penetration area = Area of sector (OADB) – Area of ΔOAB

$$A_P = 38.98 - 24.5 = 14.48 \text{ mm}^2$$

$$A_P = 14.48 \text{ mm}^2$$

Similarly

Reinforcement is a part of circle of radius 20 mm.

From ΔPST;

$$\sin\frac{\theta}{2} = \frac{ST}{PS} = \frac{5}{20}$$

$$\Rightarrow \quad \frac{\theta}{2} = \sin^{-1}\left(\frac{5}{20}\right)$$

$$\theta = 28.95°$$

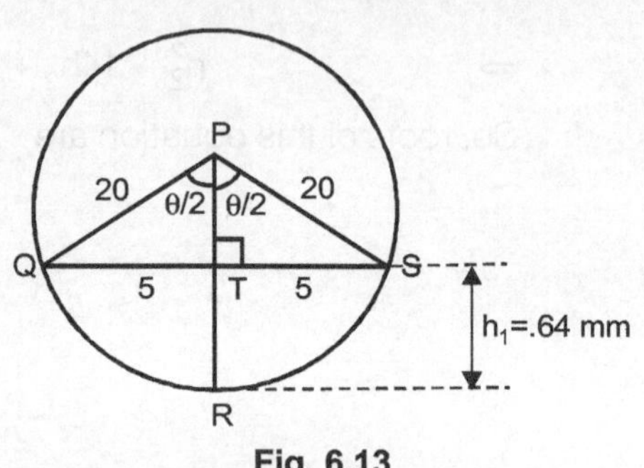

Fig. 6.13

So, area of sector (PQRS) $= \frac{\theta}{360°} \times \pi r^2$

$$= \frac{28.95}{360°} \times \pi \times 20^2$$

$$= 101.05 \text{ mm}^2$$

$$\text{area of } \Delta PQS = \frac{1}{2} \times QS \times PT$$

$$= \frac{1}{2} \times 10 \times 19.36 \qquad \{PT = PR - RT = 20 - 0.64 = 19.36 \text{ mm}\}$$

$$= 96.8 \text{ mm}^2$$

So; Reinforcement area (A_R) = (Area of sector PQRS) – (Area of ΔPQS)

$$= 101.05 - 96.8$$

$$A_R = 4.25 \text{ mm}^2$$

So, $$\% \text{ dilution} = \left(\frac{A_P}{A_P + A_R}\right) \times 100$$

$$= \left(\frac{14.48}{14.48 + 4.25}\right) \times 100$$

$$\% \text{ dilution} = 77.30\%$$

Example : 6.8

Welding of C40 steel plate of 10 mm thickness required a current of 160 Amps, while it was 360 amps when the plate thickness was increased by 50%. Estimate the welding current for 8 mm thickness of same material.

Solution :

Given,

Current requirement for 10mm thick plate = 160 Amps.

Current requirement for 15mm thick plate = 360 Amps.

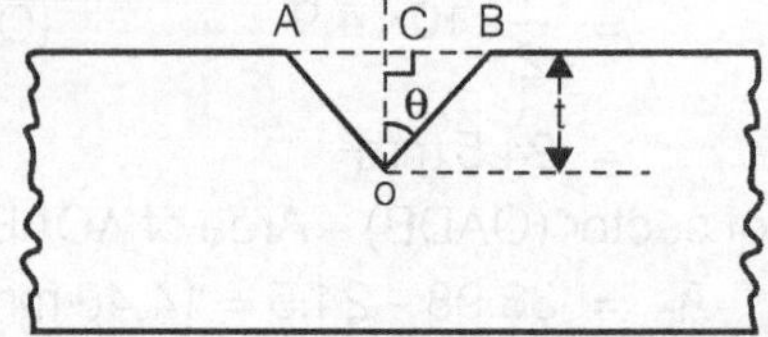

Fig. 6.15

$$\text{area of weld} = \frac{1}{2} \times AB \times OC$$

$$= \frac{1}{2} \times (t \times \tan\theta) \times 2 \times t$$

$$= t^2 \tan\theta$$

Let welding speed $= v$

So, Volume of weld $= (t^2 \tan\theta)\, v$

as volume of weld $\propto$ (current)^2

$$v\,(t^2 \tan\theta) \propto I^2$$

$$\Rightarrow \quad I \propto t$$

$$\Rightarrow \quad I = a + bt$$

at t = 10 mm, I = 160A

$$\Rightarrow \quad 160 = a + 10b \qquad ...(1)$$

at t = 15 mm, I = 360 A

$$\Rightarrow \quad 360 = a + 15b \qquad ...(2)$$

Subtract (1) from (2),

$$200 = 5b$$

$$\Rightarrow \quad b = 40$$

From equation (1)

$$160 = a + (10 \times 40)$$

$$\Rightarrow \quad a = 160 - 400 = -240$$

$$\Rightarrow \quad I = -240 + 40t$$

So; at thickness (t) = 8mm

$$I = -240 + (40 \times 8)$$

$$I = 80 \text{ Amps.}$$

$\Rightarrow$ Welding current = 80 Amps.

6.4.3 Submerged Arc Welding (SAW)

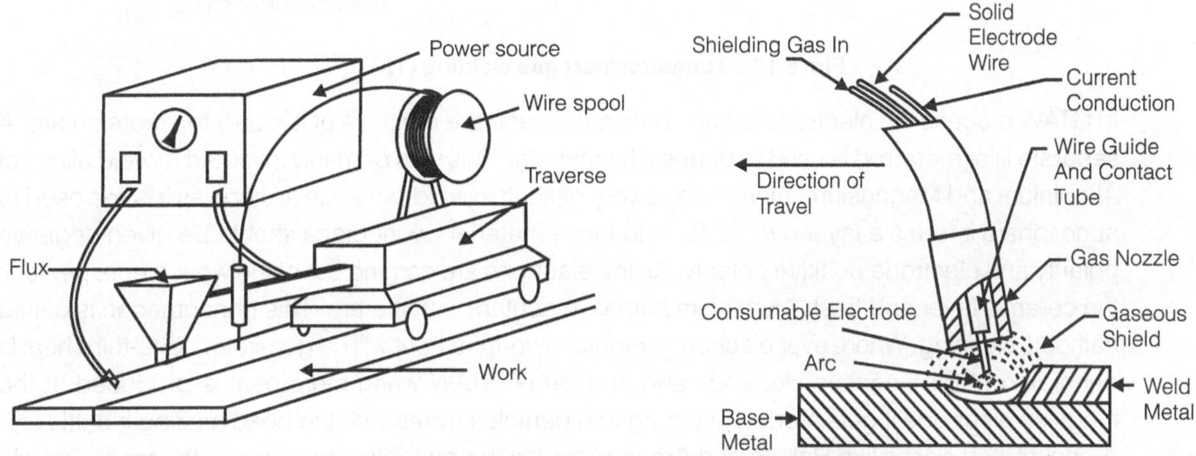

Fig. 6.16 : (a) and (b) Welding using SAW process

This is semi-automatic version of SMAW process which can produce long weld runs. In this type of welding electrode is in the form of spool of copper coated wire and granular flux is used. Copper is coated to increase the conductivity of wire. Powdered flux is initially poured into the welding area and arc is maintained within the pool of flux. This minimizes the Spatter of liquid metal and suppresses the intense ultra violet radiation. Flux also acts as a thermal insulator by deep penetration of heat into the work piece. Flux is fed on weld zone by gravity through flow nozzle, a long continuous weld can be performed. It is mainly used in the down hand welding position in a semi – automatic welding process and the feeding of electrode towards the weld pool is controlled by machine. The welds obtained by this process are stronger than the base plate, if done carefully.

The metals most widely welded through this process are low carbon steels, low alloy steels, stainless steels and high alloy steels. It is extensively used in joints in thick plates in pressure vessels, ships, bridges, structural work, welded pipes and nuclear reactors. It can not be used for plates less than 5mm thick. The process is also used to make LPG cylinders.

6.4.4 Gas Tungsten Arc Welding (GTAW) or Tungsten Inert Gas Welding (TIG)

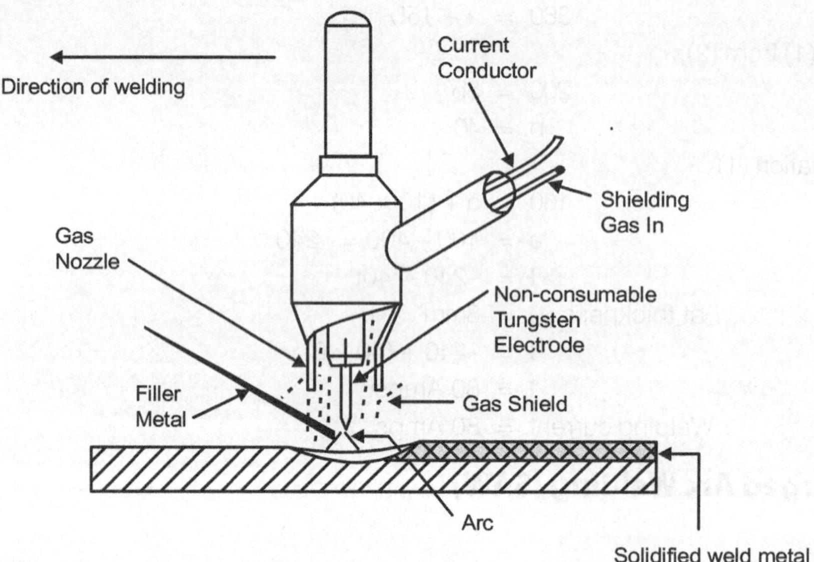

Fig. 6.17 : Tungsten inert gas welding (TIG)

In GTAW process the electrode is non consumable and the purpose of it is only to create an arc. A separate filler metal rod is used to deposit the material. This was primarily invented to weld alloys of Aluminium and Magnesium. Aluminium is very difficult to weld because as soon as it is exposed to atmosphere it forms a layer over it. To weld these materials work piece should be given negative polarity and electrode positive polarity. As the electrons are coming out of the work piece, peels of the ceramic layer and fresh Aluminium comes in contact with the arc. This phenomenon is called cathodic cleaning. Where ever electron bombards two-third heat will be generated, one-third heat is generated at the negative electrode and in case of GTAW when more heat is produced at the tungsten electrode it gets eroded and tungsten particles mixes with the bead making it brittle.

A pool of inert gases like Helium and Argon provides the protective covering to the molten metal. This reverse polarity is also used to weld very thin sheets because when the development of heat over work piece material is more, holes will be produced in the material and there will not be any welding. The torches for carrying current beyond 100 amperes are usually water cooled. Arc initiation is normally done by touching the electrode on a graphite block. The gas nozzle of a GTAW torch is its weakest part and is fastened to the torch body by threaded connection. These nozzles are generally made of ceramic material.

This process is used to weld in all positions and is extensively used for welding aluminium, magnesium, stainless steels, copper, Nimonic alloys, Monel, inconel etc. it is especially used in aircraft industry, chemical plant and nuclear plant.

Comparison of Different Electrode polarities:

	DC, electrode positive	DC, electrode negative	AC
1. Penetration	shallow	Deep	Intermediate
2. Heat generation	2/3rd at electrode, 1/3rd at workpiece	1/3rd at electrode, 2/3rd at workpiece	50% on both
3. Metal deposition rate	High	Low	Intermediate
4. Thickness of work to be welded	Thin sheets	Thick sheets	Intermediate
5. Stable smaller arc	Easier	Easier	Difficult
6. Arc blow	Severe	Severe	Insignificant

As the thickness of work decreases negative polarities should be given for work otherwise due to development of excessive heat holes will develop in the sheet. To control the metal deposition rate GTAW process is used. If the thickness further decreases, the operating current should decrease.

Electrode Penetration in GTAW : If negative polarity is given to electrode it should be grinded to a point to increase the density of current which leads to deeper penetration. If the electrode is given positive polarity, diameter of tip is increased to minimize the electrode erosion.

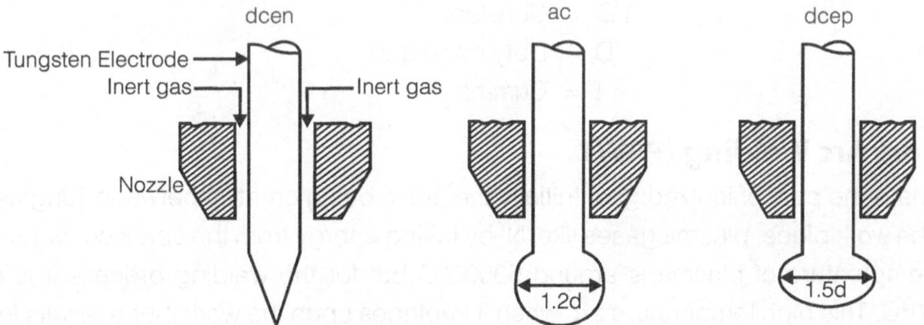

Fig. 6.18 : Shapes of tungsten electrodes

The selection of the electrode should be good to get the best results. The extension of the cup should be approximately equal to the inside diameter of the cup or slightly more in case of fillet welds. Otherwise the shielding effect will be reduced. Instead of GTAW in Gas metal Arc Welding (GMAW) tungsten electrode is replaced by a consumable electrode which is in the form of spool.

6.4.5 Arc Blow

Due to fixed polarity magnetic lines forms in the work piece material. Upon welding at the center of the work magnetic lines will be equally distributed on both sides. So arc will be straight. But while welding near the edges, since most of the magnetic lines will be concentrated in the material, arc will be deflected towards the workpiece. This phenomenon is called Arc Blow which results in severe spatter and improper bead geometry.

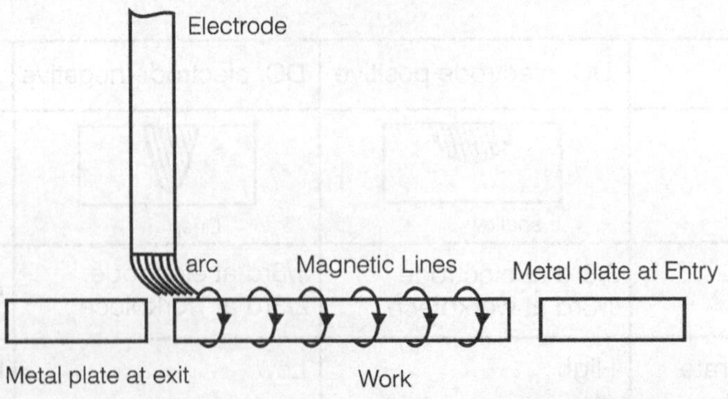

Fig. 6.19 : Arc blow due to magnetic field

Arc Blow phenomenon will be more intensified while welding with DC with bare electrodes. To avoid this phenomenon metal plates are kept at the entry and at the exit called Tab in and Tab out. These plates also minimize the end defects. Arc Blow phenomenon may also appear when large piece of iron is present at the welding site.

6.4.6 Duty Cycle

It is the time span in successive 10 min interval during which current can be drawn from the transformer, after that there should be a compulsory break otherwise the temperature of its essential components will increase. Usually 60% duty cycle is the standard industrial rating. A power source with such a rating can supply its rated output for 6 minutes in every ten minute interval of its operation.

$$I^2D = \text{Constant}$$

where D = Duty cycle and

I = Current

6.4.7 Plasma Arc Welding (PAW)

Plasma is the pool of ionized gas. Initially the arc is being created between Tungsten electrode and the work piece, plasma gases like 'N' by taking energy from the spark converts into plasma. The temperature of plasma is around 50000°C but for the welding process it is restricted to 20000°C. This high temperature arc when it impinges upon the work piece results in reuniting of electrons and ions to form atomic and then molecular gas, releasing heat in the process which is thus utilized for welding. Due to increase source of heat penetration will through and without spreading the heat affected zone it is called keyhole phenomenon, this phenomenon will ensure 100% penetration and gives wine glass weld bead. More intense the source of heat, deeper will be the penetration and maximum can be the welding speed and hence lesser will be the heat affected zone. A separate filler rod is used to deposit the material. When plasma torch is used for cutting it is called metal melting process and not oxidation process. Plasma arc is concentrated because it is fed through a small nozzle. The energy is obtained from a dc power source of the constant current type having an open circuit voltage of 70 – 80 volts and duty cycle of 60%. There are two types plasma arc welding. In the transferred – arc method the work piece is welded as part of the electrical circuit. The arc transfers from the electrode to the work piece. In the non transferred type the arc occurs between the electrode and the nozzle and the heat is carried to the work piece by plasma gas. The thermal transfer mechanism is similar to that for Oxyfuel flame. The major application of the process is in welding stainless steel, titanium, having very high melting points and super alloys. Commercially it is used in aeronautical industry, precision instrument industry and jet engine manufacturing. This plasma arc welding has serious draw backs. For example the intense arc results in extensive ultra violet and infra – red radiation which can harm the skin.Tungsten electrode

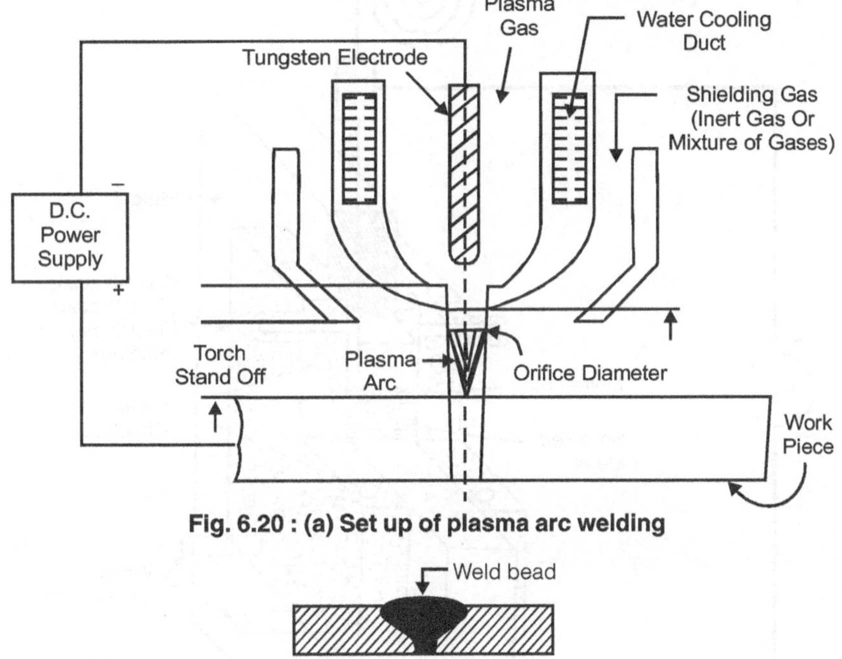

Fig. 6.20 : (a) Set up of plasma arc welding

Weld bead

Fig. 6.20 : (b) Wine glass bead structure in PAW

6.4.8 Electro slag/Electro gas Welding

Initially the arc is being created between starting block and the Copper coated electrode wire followed by addition of flux. Welding progresses in the vertical direction and due to that temperature of the molten pool keeps on increasing. A pair of water cooled cu shoes is being provided on the sides to avoid any spillage of liquid metal to the sides. Once the temperature of the molten pool reaches a particular value arc is switched off and the molten pool is so hot that it starts consuming the wire without any arc. Temperature is controlled in this fashion to minimize the heat affected zone. If the heat effected zone increases it may cause distortion of the plate.

The point when molten metal is not able to consume any more wire, arc is restarted again. Restarting the arc is easy on electro gas welding. In electro gas welding an inert gas is used in place of slag or flux. The duty cycle of the transformer is 100% and electrode should be given positive polarity. Both AC and DC power sources are used with a rating of 1000 amperes at an open circuit voltage of 55 volts and duty cycle of 100 percent.

The process is used to join two thick plates along the edges. The major application is in **ship building**. Welding is done with joint in vertical position. It can be used for welding plates of thickness 20mm to 40mm. It is also used extensively in the heavy ferrous industry. This finds extensive use in the construction of pressure vessels, press frames, water turbine and heavy plate fabrication industries.

Welding current depends on welding voltage and electrode feed rate, it increases with it. Increase in welding current increases the welding speed, but higher welding current may cause cracking. Welding voltage affects the depth of penetration and stable operation of the process. Excessive voltage causes overheating of the metal, gassing of the slag pool, and even sparking.

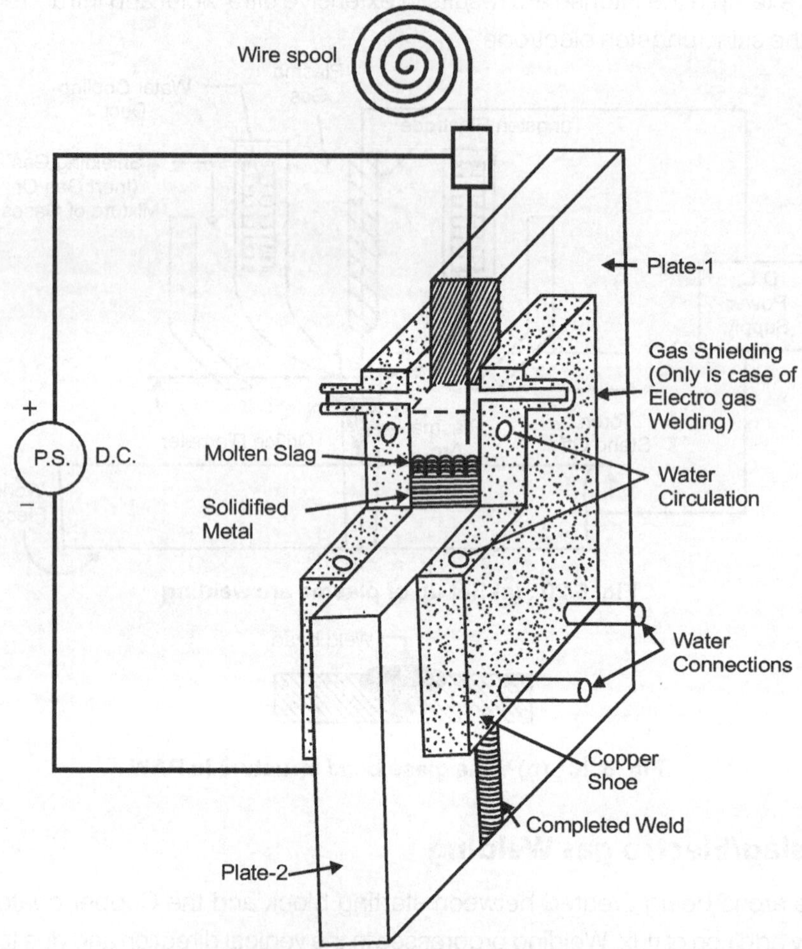

Fig. 6.21 : Electro gas welding

6.5 RESISTANCE WELDING

Some of the important resistance welding processes are :

A. Spot
B. Projection
C. Seam
D. HFRW/HFIW
E. Flash butt welding

The resistance welding is produced by means of electrical resistance across the two components to be joined. The heat generated in this process is given by

$$H = I^2RT$$

H = heat generated in joules

I = current in amperes.
R = resistance in ohms.
t = time of current flow in seconds

The welds produced by resistance welding are normally without the addition of any filler material, electrodes and shielding gases.

6.5.1 Spot welding

A pair of sheet is held between two copper electrodes. Indentation is produced over the sheet by applying the pressure on the electrodes. Due to indentation there will not be any air gap between electrode and the sheet. Step down transformers are used in resistance welding, in the secondary winding of which there is single turn. When a high current is passed through conductors, maximum resistance will be there at the interface between the two sheets due to presence of air gap. As a result of that material in both the sheets will be liquified and a spot (nugget) will be formed. Kickless cables are used to supply the current for the safety of operator. Rocker arm type machines are used for smaller parts and press type are used for larger work pieces.

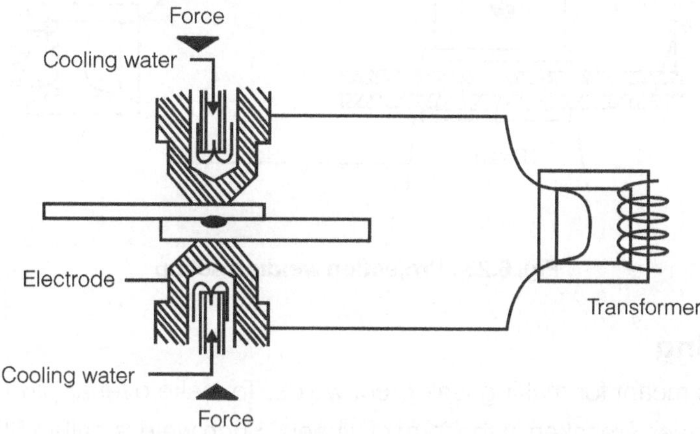

Fig. 6.22 : (a) Resistance welding set up

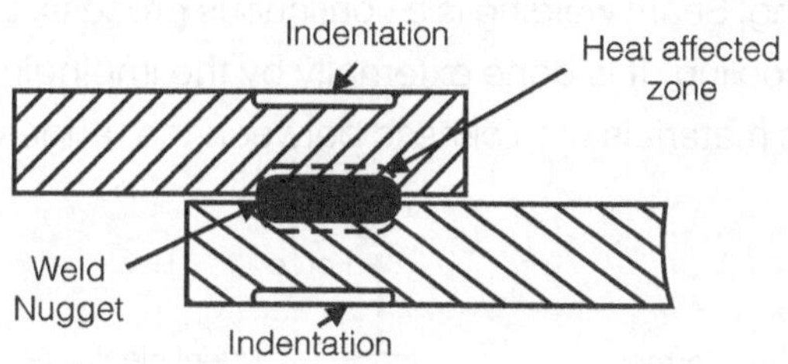

Fig. 6.22 : (b) Spot welding nugget

The diameter of the nugget is $d_n = 6\sqrt{t}$

Height of the nugget is $h_n = 2t -$ indentation

Spot welding is mainly used for lap welding of thin sheets particularly in the welding of automobile and refrigerator bodies and high quality work in automobile engines.

6.5.2 Projection welding

To make number of spot welds in one set up projection welding is used. These projections are made by EMBOSSING process by the flow of current. The projection materials become soft and the pressure on the electrode is suddenly increased and projections are allowed to collapse on the other sheet. During this stage the sheet which has projections tends to be hotter. So projections are provided in the thicker sheet or the sheet which has higher thermal conductivity. There are so many projections: round button or dome type, ring type, elongated projection, shoulder projection, radius projection and cross wires. Equipment used for projection welding is similar to that for spot welding except that the rod electrodes are replaced by flat copper plates. Projection welding reduces the amount of current and pressure needed to achieve a good bond between the two pieces, with consequential less chance of shrinkage and distortion in the areas around the weld zone.

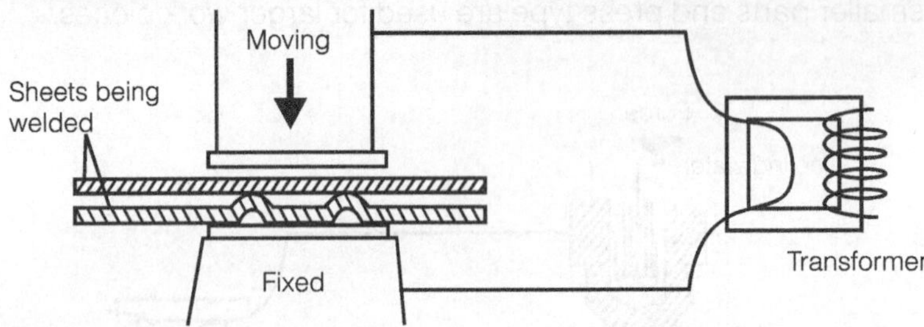

Fig. 6.23 : Projection welding set up

6.5.3 Seam Welding

Spot welding is meant for making leak proof welds. To make overlapping spot weld (leak proof welding) electrodes are taken in the form of wheels. Such weld is called Stitch Weld. To produce stitch welds at higher speeds seam welding is used. In seam welding the electrodes used in the form of rollers or wheels, due to shunting of current through the already made welds, the current required is higher than in normal spot welding. Pressure is applied to fuse metal properly into the nugget as in spot welding. Seam welding is a continuous process. Electrodes get overheated and hence need constant cooling. It is done externally by the impinging water jets. Seam welding is capable of welding thin materials of thickness between the ranges of 2.5 to 4.7 mm.

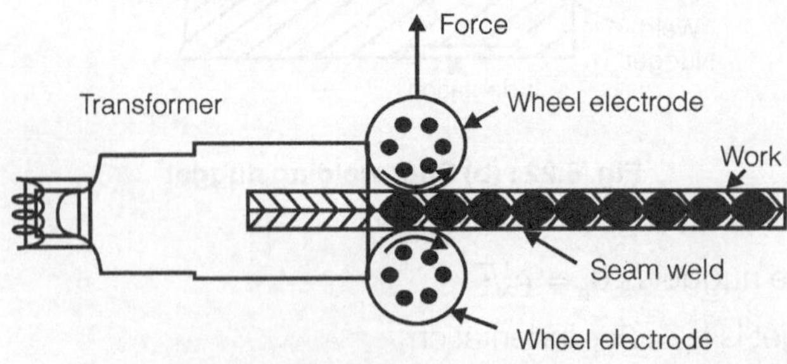

Fig. 6.24 : Seam welding set up

Example : 6.9

Two sheets of 1.5 mm thickness are to be spot welded by passing a current of 10,000 A for 0.1 sec. The maximum indentation allowed is 10% of sheet thickness. The density of spot weld nugget is 8 gm/cm³. If 1380 Joules are required to melt 1 gm of steel. Calculate the efficiency if effective resistance to be 200 micro ohm.

Solution :

Given,

$$\text{Current (I)} = 10{,}000\text{ A}$$
$$\text{Resistance (R)} = 200\ \mu\Omega = 200 \times 10^{-6}\ \Omega$$
$$\text{time } (t_1) = 0.1\text{ s}$$

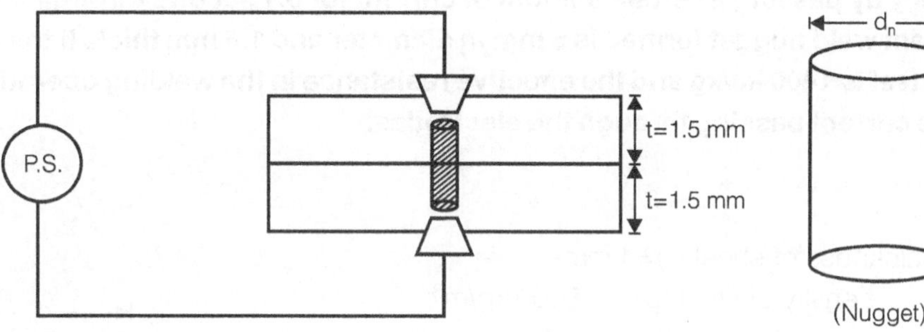

Fig. 6.25 : Seam welding set up

So, total heat supplied $(H_s) = I^2Rt_1$

$$= (10{,}000)^2 \times (200 \times 10^{-6}) \times 0.1$$
$$H_s = 2000\text{ J}$$

as, we know, diameter of nugget $(d_n) = 6 \times \sqrt{t} = 6 \times \sqrt{1.5}$

$$d_n = 7.34\text{ mm}$$

Height of nugget $(h_n) = 2 \times (\text{thickness} - \text{indentation})$

$$= 2 \times \left(1.5 - \frac{10 \times 1.5}{100}\right)$$
$$= 2 \times 1.5 \times 0.9$$
$$h_n = 2.7\text{ mm}$$

So, volume of nugget $(V_n) = \frac{\pi}{4} \times d_n^2 \times h_n$

$$V_n = \frac{\pi}{4} \times 7.34^2 \times 2.7$$

$\Rightarrow$ $V_n = 114.24\text{ mm}^3$

or $V_n = 114.24 \times 10^{-3}\text{ cm}^3$

as density of nugget = 8 gm/cm³

So, mass of nugget = density × volume

$$= 8 \times 114.24 \times 10^{-3}$$

mass of nugget = 0.913 gm

as 1380 joules are required to melt 1 gm of steel

So total heat required to melt = 1380 × 0.913

= 1261.28 Joules

So, $$\text{efficiency } (\eta) = \frac{\text{Heat Required to melt}}{\text{Heat Supplied to melt}} \times 100$$

$$= \frac{1261.28}{2000} \times 100$$

$$\eta = 63.06\%$$

So, efficiency = 63.06%

Example : 6.10

Spot welding of two 1 mm thick sheets of steel (density = 8000 kg/m³) is carried out successfully by passing a certain amount of current for 0.1 second through the electrodes. The resultant weld nugget formed is 5 mm in diameter and 1.5 mm thick. If the latent heat of fusion of steel is 1400 kJ/kg and the effective resistance in the welding operation is 200 μΩ, what is the current passing through the electrodes.

Solution :

Given,

thickness of sheet (t) = 1 mm

density of steel (ρ) = 8000 kg/m³

time (t_1) = 0.1 s

diameter of nugget (d_n) = 5 mm

thickness of nugget (h_n) = 1.5 mm

Latent heat of fusion of steel = 1400 kJ/kg

Resistance (R) = 200 μΩ

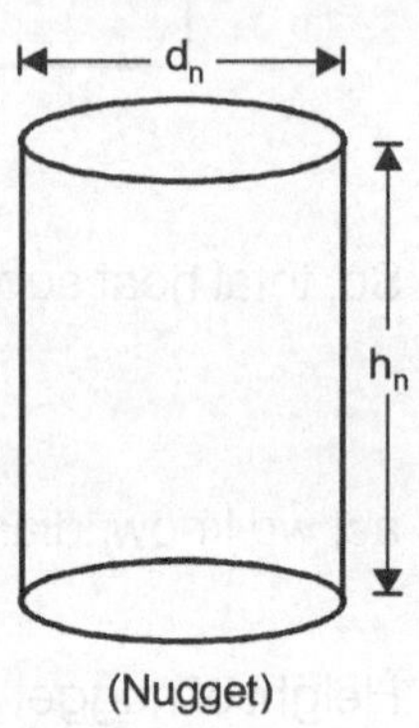

Fig. 6.26 : Seam welding set up

Volume of Nugget is given by,

$$\text{Volume} = \frac{\pi}{4} \times d_n^2 \times h_n = \frac{\pi}{4} \times 5^2 \times 1.5$$

Volume of nugget = 29.45 mm³

Volume of nugget = 29.45×10^{-9} m³

Mass of Nugget = density × volume of nugget

$= 8000 \times 29.45 \times 10^{-9}$

Mass of Nugget = 2.356×10^{-4} kg

as Latent heat of fusion = 1400 kJ/kg

So, total heat of fusion = 1400 × mass

$= 1400 \times 2.356 \times 10^{-4}$ kJ = 0.329867 kJ

Total heat = 329.867 J ...(1)

Total heat is also given as $H = I^2Rt_1$

$H = I^2 \times 200 \times 10^{-6} \times 0.1$ ...(2)

Equation (1) and (2);

$$I^2 \times 200 \times 10^{-6} \times 0.1 = 329.867$$

$$I^2 = 16.493 \times 10^6$$

$\Rightarrow$ $I = 4061.2$ Amp.

$\Rightarrow$ Current passing through the electrodes = 4061.2 Amp.

Example : 6.11

In an arc welding process, the voltage and current are 25 V and 300 A respectively. The arc heat transfer efficiency is 0.85 and welding speed is 8 mm/sec. What is the net heat input (in J/mm).

Solution :

Given,

Voltage (V) = 25 volts

Current (I) = 300 Amp.

Arc heat transfer efficiency (η_T) = 0.85

Welding speed (v) = 8 mm/s

Let net heat input = H_1 J/mm

Total heat supplied i.e. Power = $V \times I$

$= 25 \times 300 = 7500$ W

as heat transfer efficiency is 0.85 so net

heat input at the bead is $= \eta_T \times 7500$

$= 0.85 \times 7500$

Heat input = 6375 Watts ...(1)

Heat input is also given as

Heat input = H_1 × Welding speed = $H_1 \times 8$ ...(2)

From (1) and (2)

$H_1 \times 8 = 6375$

$\Rightarrow$ $H_1 = 796.875$ J/mm

So Net heat input = 796.875 J/mm

Example : 6.12

Two metallic sheets, each of 2.0 mm thickness, are welded in a lap joint configuration by resistance spot welding at a welding current of 10 kA and welding time of 10 millisecond. A spherical fusion zone extending up to the full thickness of each sheet is formed a properties of the metallic sheets are given as :

ambient temperature = 293 K

melting temperature = 1793 Kdensity = 7000 kg/m^3

latent heat of fusion = 300 kJ/kgspecific heat = 800 J/kgK

Assume :

1. **contact resistance along sheet-interface is 500 micro-ohm and along electrode-sheet interface is zero;**
2. **no conductive heat loss through the bulk sheet materials; and**
3. **the complete weld fusion zone is at the melting temperature.**
   What is the melting efficiency (in %) of the process?

Solution :

Given,

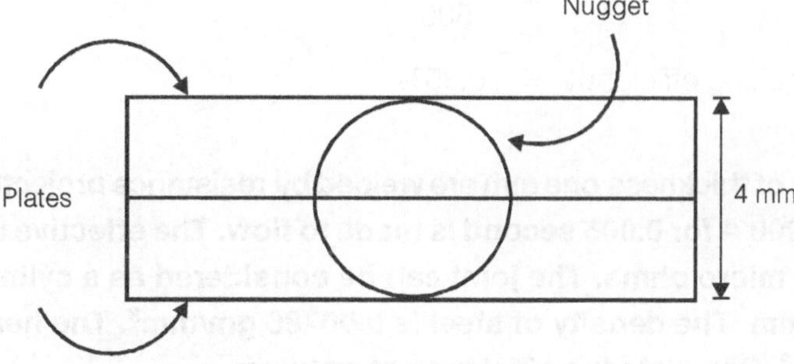

Fig. 6.27 : Seam welding set up

Thickness of each plate (t) = 2 mm

Welding current (I) = 10 kA = 10,000 A

Welding time (t_1) = 10 millisecond = 10×10^{-3} s

Spherical fusion zone is formed upto full thickness of each plate

ambient temperature (T_{amb}) = 293 k

Melting temperature (T_{MP}) = 1793 k

density (ρ) = 7000 kg/m³

Latent heat of fusion (L) = 300 kJ/kg = 300,000 J/kg

Specific heat (c) = 800 J/kg k

Contact Resistance (R) = 500 micro ohm = $500 \times 10^{-6}\ \Omega$

Total heat supplied (H) = I^2Rt_1

$= (10000)^2 \times 500 \times 10^{-6} \times 10 \times 10^{-3}$

$= 500$ J

As nugget is of spherical shape with radius of 2mm.

$$\text{Volume of nugget} = \frac{4}{3}\pi r^3 = \frac{4}{3}\times\pi\times 2^3$$

$$= 33.51 \text{ mm}^3$$

$\Rightarrow$ Volume of nugget = 33.51×10^{-9} m³

Mass of nugget (m) = density × volume

$= 7000 \times 33.51 \times 10^{-9}$ kg

$m = 2.345 \times 10^{-4}$ kg

Total heat required to form nugget is sum of heat required to raise the temperature from ambient to melting point (sensible heat) and heat required to change phase (Latent heat)

$\Rightarrow$ total heat required = sensible heat + latent heat

$= mC\,\Delta T + mL$

$= m[C(T_{MP} - T_{amb}) + L]$

$= 2.345 \times 10^{-4}\,[800\,(1793 - 293) + 300{,}000]$

Total heat required to melt = 351.75 J

$$\Rightarrow \text{melting efficiency} = \frac{\text{total heat required to melt}}{\text{total heat supplied}}\times 100$$

$$= \frac{351.75}{500}\times 100$$

melting efficiency = 70.35%

Example : 6.13

Two steel sheets of thickness one mm are welded by resistance projection welding technique. A current of 30,000 A for 0.005 second is made to flow. The effective resistance of joint can be taken as 100 micro ohms. The joint can be considered as a cylinder of diameter 5 mm and height 1.5 mm. The density of steel is 0.00786 gm/mm³. The heat needed for welding steel is 10 J/mm³. Calculate the efficiency of welding.

Solution :

Given,

thickness of each sheet (t) = 1 mm

current (I) = 30,000 A

time (t_1) = 0.005 s

Resistance (R) = 100 micro ohms

= $100 \times 10^{-6}\Omega$

Welding is done by resistance projection welding process.

Nugget or spot generated is of cylindrical shape.

diameter of nugget (d_n) = 5 mm

Height of nugget (h_n) = 1.5 mm

density of steel (ρ) = 0.00786 gm/mm³

Heat needed for welding steel = 10 J/mm³

Total heat supplied for weld = I^2Rt_1

= $(30{,}000)^2 \times 100 \times 10^{-6} \times 0.005$

Total heat supplied for weld = 450 J

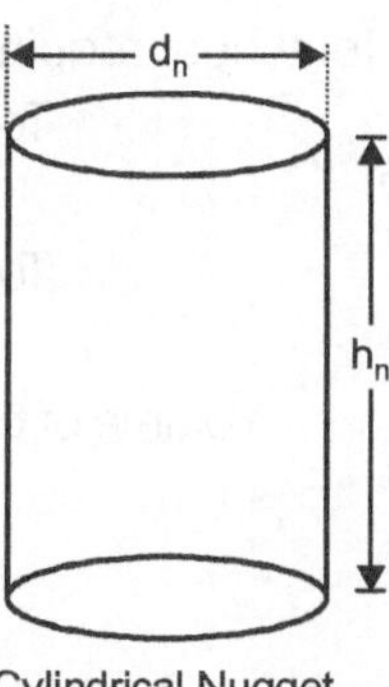

Fig. 6.28 : Seam welding set up

Volume of cylindrical nugget = $\frac{\pi}{4} \times d_n^2 \times h_n$

$$= \frac{\pi}{4} \times 5^2 \times 1.5 = 29.45 \text{ mm}^3$$

So, heat needed for welding steel = 10 × volume

= 10 × 29.45 J

Heat needed = 294.5 J

So, efficiency of welding = $\frac{\text{Heat needed}}{\text{Heat supplied}} \times 100$

$$= \frac{294.5}{450} \times 100$$

Efficiency = 65.44%

Example : 6.14

What are the common defects found in welding? Explain their causes and remedies.

A spot welding obtained with a current of 35,000 A for a period of 0.01 sec. If the effective resistance of the joint be taken as 120 micro ohms, and if the joint can be considered as a cylinder of 6 mm diameter and 2.5 mm in height, calculate the heat distributed to the surroundings.

Take heat required for melting steel as 10 J/mm³.

Solution :

Given,

Current (I) = 35,000 A

time (t_1) = 0.01 s

Resistance (R) = 120 μΩ= $120 \times 10^{-6}\ \Omega$

Nugget formed is cylindrical with

diameter (d_n) = 6 mm

height (h_n) = 2.5 mm

Total heat required for melting steel = 10 J/mm^3

Total heat supplied = I^2Rt_1

$= (35000)^2 \times 120 \times 10^{-6} \times 0.01$

Total heat supplied = 1470 J

Volume of cylindrical nugget $= \frac{\pi}{4} \times d_n^2 \times h_n$

$= \frac{\pi}{4} \times 6^2 \times 2.5 = 70.68 \text{ mm}^3$

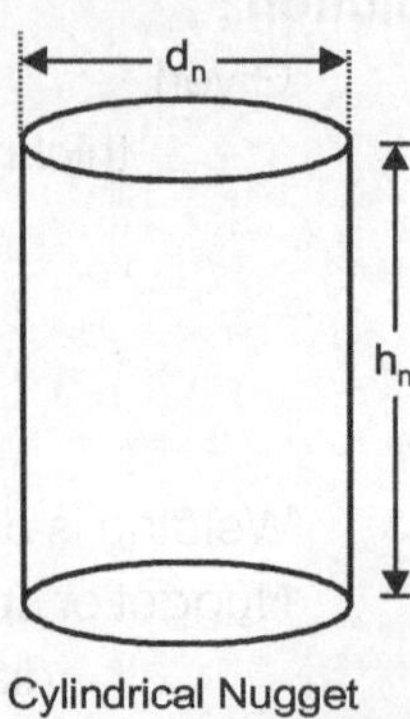

Fig. 6.29 : Seam welding set up

So heat required for melting steel = 10 × 70.68 = 706.8 J

So, heat lost surrounding per nugget = 1470 – 706.8 = 763.2 J

Example : 6.15

For resistance spot welding of two aluminium sheets, each 2 mm thick, a current of 5000 A was passed for 0.15 sec. The total resistance was estimated to be 75 μΩ and the nugget diameter and thickness were measured to be 5 mm and 2.5 mm respectively. What would be the proportion of heat energy utilized for welding if the melting energy per unit volume for aluminium is taken as 2.9 J/mm^3?

Solution :

Given,

thickness of each plate (t) = 2 mm

current (I) = 5000 A

time (t_1) = 0.15 s

Resistance (R) = 75 μΩ = $75 \times 10^{-6}\ \Omega$

Nugget diameter (d_n) = 5 mm

Nugget thickness (h_n) = 2.5 mm

Melting energy required per unit volume = 2.9 J/mm^3

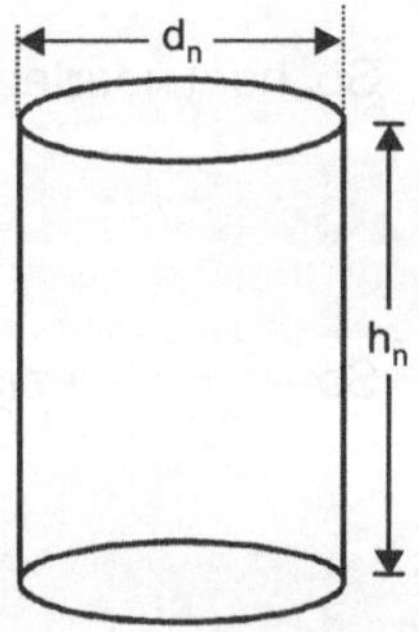

Fig. 6.30 : Seam welding set up

Volume of nugget $= \frac{\pi}{4} \times d_n^2 \times h_n$

$= \frac{\pi}{4} \times 5^2 \times 2.5 = 49.087 \text{ mm}^3$

So, melting energy required = 2.9 × volume

$= 2.9 \times 49.087$ J

Total energy required to melt = 142.353 J

Total energy supplied = I^2Rt_1

$= (5000)^2 \times 75 \times 10^{-6} \times 0.15$

Total energy supplied = 281.25 J

So, proportion of heat utilized $= \frac{\text{Heat Required}}{\text{Heat Supplied}} \times 100$

$= \frac{142.353}{281.25} \times 100 = 50.61$

⇒ Proportion of Heat utilized = 50.61%

Example : 6.16

Two 8mm thick steel plates are placed 5 mm apart and welded by a butt joint. Welding is carried out at 20 volts and at speed of 5 mm/s. Heat transfer efficiency is 0.80. If the heat required to melt steel is 10 J/mm^3 and melting efficiency is 0.625, then the weld current (in Amp) will be

Solution :

Given, thickness of each plate (t) = 8 mm

distance between plates (d) = 5 mm

voltage (V) = 20 volts

Speed of weld (v) = 5 mm/s

Heat transfer efficiency (η_T) = 0.80

Heat required to melt metal = 10 J/mm^3

Melting efficiency (η_m) = 0.625

Let weld current = I Amp.

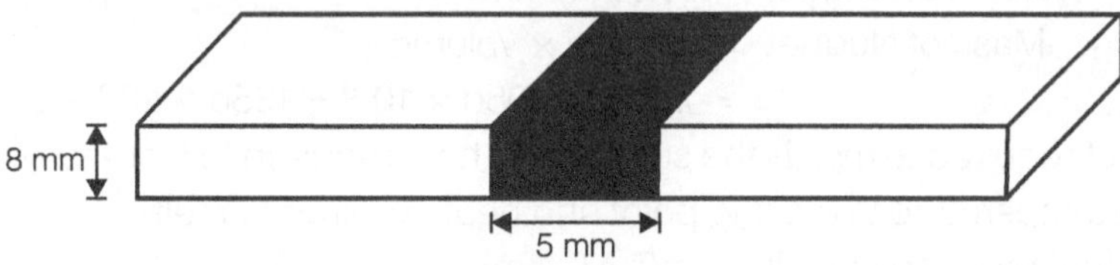

Fig. 6.31

$$\text{Total heat energy supplied} = V \times I$$

$$= 20 \times I \ \frac{\text{Joules}}{\text{S}}$$

as heat transfer efficiency is 0.80

So, total heat energy supplied to bead = 0.80 × 20 × I

Heat supplied = 16I Watt

Volume of metal melted per second = (Area) × (Velocity of weld)

$$= (8 \times 5) \times (5) = 200 \text{ mm/s}$$

As heat required to melt = 10 J/mm^3

So, total heat required to melt = 10 × 200

= 2000 J/s

as melting efficiency = 0.625

$$\eta_{\text{melting}} = \frac{\text{Heat required}}{\text{Heat supplied}}$$

$$0.625 = \frac{2000}{16I}$$

$\Rightarrow$ I = 200 Amp.

$\Rightarrow$ Weld Current = 200 Amp.

Example : 6.17

In spot welding (pulsed laser) of aluminium plates (density = 2700 kg/m^3, specific heat (c) = 896 J/kg k, melting temperature = 933 k, latent heat of melting = 398 kJ/kg) at a temperature

of 30°C, a pulse with energy of 0.5 J is focussed onto an area of 0.05 mm². If the entire energy is coupled into the material. What will be the depth of weld? Assume the cross-section area of the weld circular and is uniform throughout its depth and only heat conduction in the direction of penetration.

Solution :

Given,

density of plate = 2700 kg/m³

specific heat (c) = 896 J/kg K

Melting temperature (T_{MP}) = 933 K

Ambient temperature (t_{amb}) = 273 + 30 = 303 K

$$\text{Latent heat of melting (L)} = 398\frac{kJ}{kg} = 398 \times 1000\ \frac{J}{kg}$$

Let depth of spot = dmm

as nugget formed has circular cross-section throughout its depth i.e. it is cylindrical

Volume of Nugget = (cross-section area) × depth

Volume of Nugget = (0.05 d) mm³

Mass of Nugget = density × volume

$= 2700 \times 0.05d \times 10^{-9} = 135d \times 10^{-9}$ kg

Amount of heat required to melt is the sum of total heat required to increase temperature of metal from ambient temperature to melting point and heat required to melt.

⇒ Heat required to melt = mC ΔT + mL

$= m\{C(T_{MP} - T_{amb}) + L\}$

$= 135d \times 10^{-9}\{896(933 - 303) + 398 \times 1000\}$

Heat required to melt = 0.1299d J

Now Heat required to melt = Heat energy supplied

0.1299d = 0.5

d = 3.849mm

⇒ depth of weld = 3.849mm Ans.

6.5.4 High Frequency Resistance Welding (HFRW)

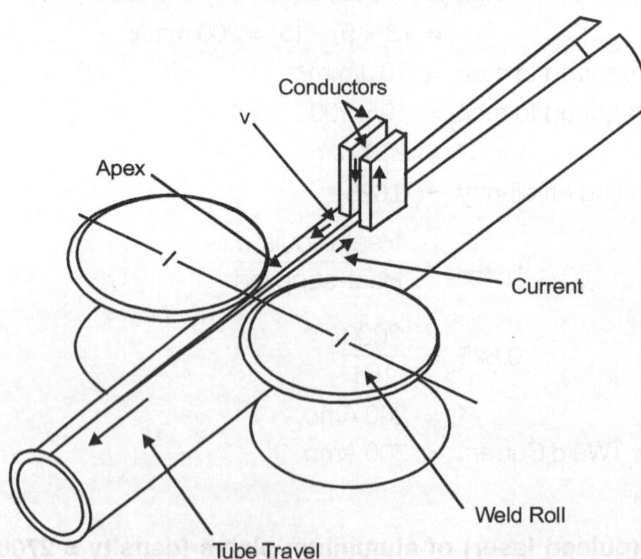

Fig. 6.32 : High frequency resistance welding process of tube making

Initially the steel is cut into the strips and as it approaches towards the concave rolls the sheet will bend in the form of a pipe. Mandrel is placed at the center so that sheet bends exactly in a cylindrical fashion. A pair of conductors is kept on both sides of edges and a high frequency current is passed. Due to skin effect current flows through low inductance rather than low resistance i.e. through air gap. This result in the heating of edges and when the pressure is applied by the rolls there will be a continuous welding. Since rubbing takes place between the blocks and work piece material, after welding some kilometers of pipe the blocks should be replaced. This process used to make iron pipes.

6.5.5 Flash Butt Welding

Two workpieces are held together with the help of electrodes as show in figure 6.33 and a very high current around 100000 A is passed through these conductors. By the production of flash, oxide layer from the ends will be removed and material will also become soft. Current is stopped and axial pressure is increased to make a joint. Weld is formed through plastic deformation. In this two clamps will be present to hold the work pieces, one is fixed and the other is movable so that they can adjust. After applying high force weld is achieved and metal is expelled forming a rough fin or flash around the joint. This can be removed by machining.

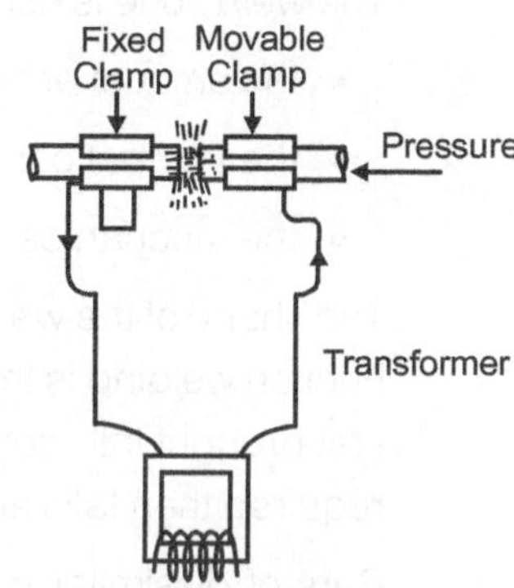

Fig. 6.33 : Principle of flash butt welding

The process is extensively used for welding mild steels, medium carbon steels and alloy steels as well as non ferrous metals like Aluminium alloys and titanium. This process is suitable for end to end or edge to edge joining of sheets of similar or dissimilar metals 0.2mm to 25mm thick.

6.6 SOLID STATE WELDING PROCESSES

6.6.1 Friction Welding

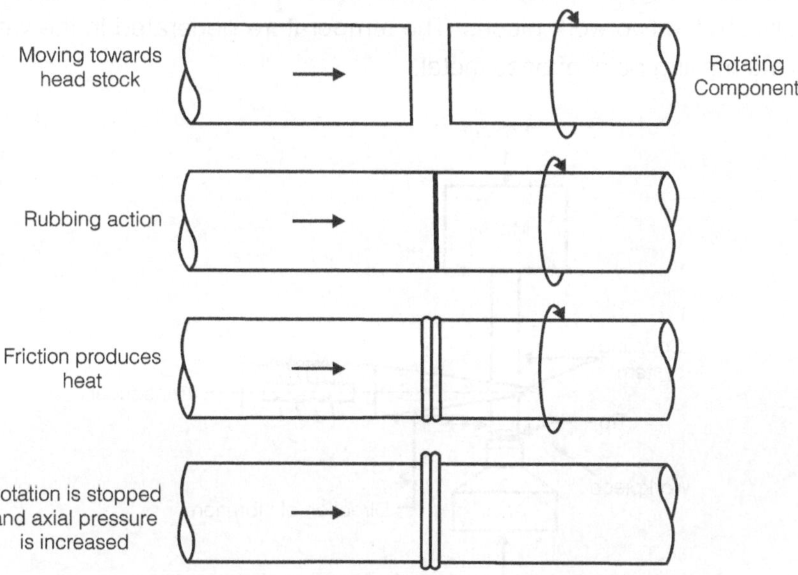

Fig. 6.34 : Steps in friction welding

The principle of this operation is changing mechanical energy into heat energy. Two samples are joined together along their axis by this process. One part is held in the chuck and the other is in the tail stock and the spindle is rotated at very high speed (around 5900 r.p.m). Both pieces are allowed to rub against each other and as a result of that heat will be produced. The material near edges will become softer and once it reaches particular temperature, breaks are applied and axial pressure is increased to cause the forging action and hence welding. No joint preparation is required for most of the non ferrous metals, since the rotating action is self cleansing. For some non ferrous metals, preparatory cleansing of the joint faces is important.

The weld zone is usually confined to a narrow region whose size depends upon

- the amount of heat generated
- the thermal conductivity of the materials and
- the mechanical properties of the materials at elevated temperatures.

The shape of the weld joint depends upon the rotational speed and on the axial pressure applied. Friction welding is the high speed process suited to production welding. When the interfaces are first brought into contact, maximum power is required, breaking up the surface film. The power required then falls and remains nearly constant while the joint is raised to welding temperature.

Bars of dissimilar metals and varying cross section can be joined by this process. The major application of the process is in welding drill to the shank, welding engine valves to stem, axle to its hub etc.

6.6.2 Ultrasonic Welding

Heat affected zone is minimum in ultrasonic welding. High frequency vibrations are being produced by a transducer in the range of 20 to 60 KHz and are transported to the work with the help of sonotrode. Load is applied on the sonotrode tip with the help of pneumatic, hydraulic or a spring actuated device. Higher the frequency of vibration higher is the rate at which energy is transmitted. Due to ultrasonic energy oxide layer at the interface will be broken and there will be a fresh metal contact between the two work pieces. The temperature generated in the weld zone is usually 30-50% of the melting point of base metal.

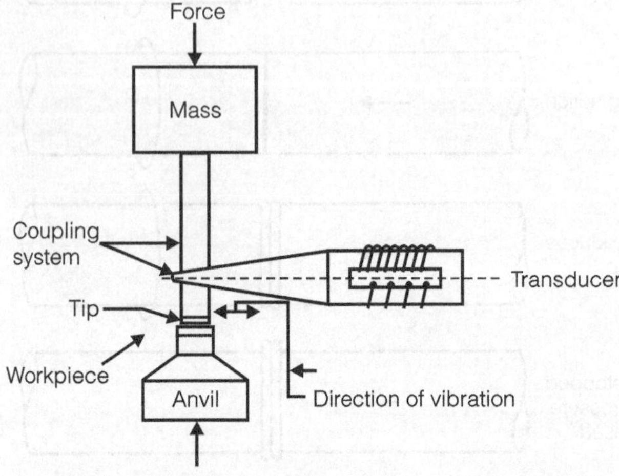

Fig. 6.35 : Components of Ultrasonic welding machine.

Ultrasonic welding equipment consists of two main parts, power source and a transducer. The power source converts the 50 Hz mains supply to a high frequency electric power and that is converted by the transducer to magnetic flux and then the kinetic motion which is amplified through a velocity transformer. This set up can be used for spot and seam welds. The major application of process is in armature winding, aluminum wire connectors in transducers and diodes.

6.6.3 Explosive Welding

Two plates can be joined together along the surface by this process. Two plates to be welded are placed at an inclination to each other, the angle vary from 1°-10°. The thicker plate is called the target plate and the thinner plate is called flyer plate has a buffer plate of PVC or rubber, between it and the explosive. The charge is exploded by a detonator placed at one end of flyer plate. When the charge explodes, the flyer plate moves towards the target plate and Stand off distance is being provided for the initial momentum to develop in the flyer plate. Because of the movement with higher velocity pressure is produced at the interface. Under such high velocity and pressure the metal flows ahead of the joining front acting like a fluid jet resulting in a bond of the interlocking type. This interlocking is an essential aspect of an explosion welding and cause of strength. Flyer plate is kept at an angle to give direction to the weld and objective of buffer plate is to protect the flyer plate from explosive.

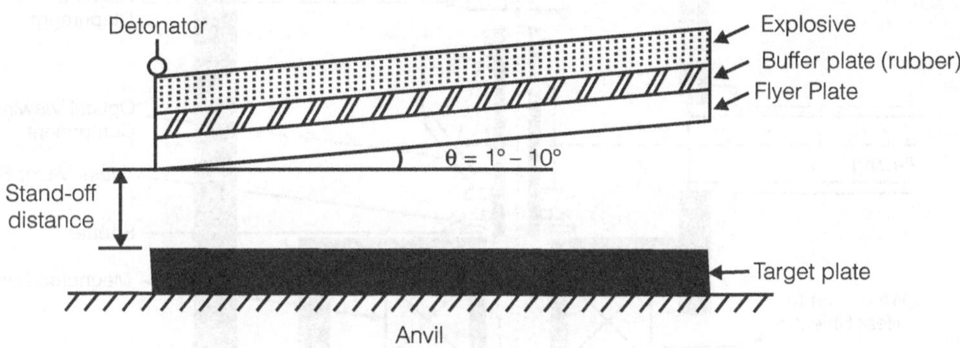

Fig. 6.36 : Schematic of Explosive welding

The impact velocity depends on the ratio of the explosive to that of the weight of the flyer plate and also on the contact angle. The maximum velocity can be decided by the sound in the target plate material because at supersonic velocities the wave in the target cannot propagate ahead of the bonding front. Minimum velocity is decided by magnitude at which the projected material becomes sufficiently plastic on impact to form a divided jet. The application of this process is in tube to tube sheet joints in heat exchangers, weld to pipe joints, blocking of leaking tubes in boilers, also when two plates are to be joined along the surface etc. This is also used for cladding a plate or slab with dissimilar metal.

6.7 RADIANT ENERGY WELDING

6.7.1 Electron Beam Welding

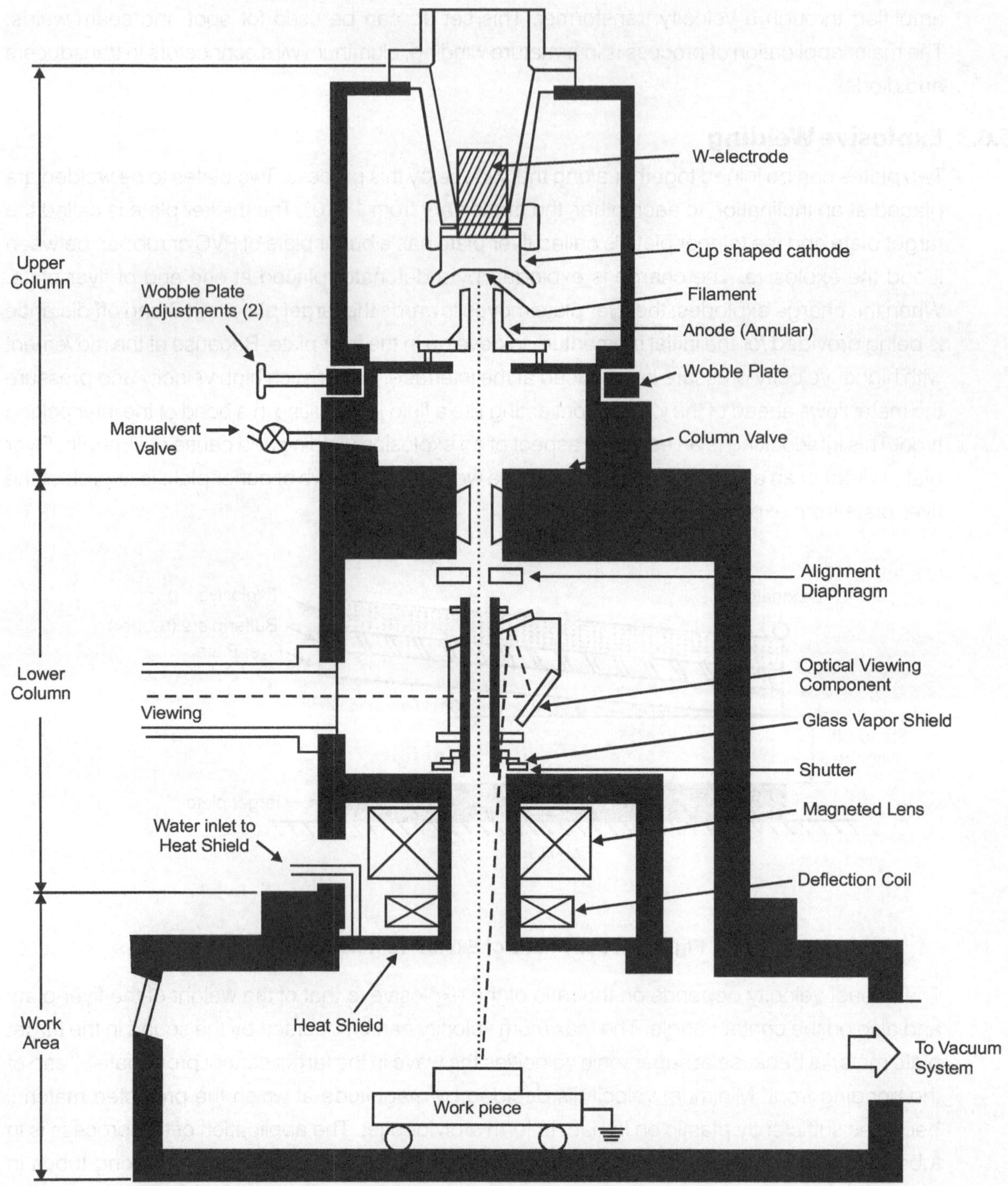

Fig. 6.37 : Electron Beam Welding

In electron beam welding a beam of electrons is used to melt the metal where it has to be welded. Electrons are generated with the help of tungsten electrode and cup shaped cathode directs the electron towards annular anode. Since there are holes in anode, major portion of electrons will comes on the other side. These electrons are converged to beam by creating magnetic field with the help of field electrodes. This electron beam has highly concentric energy and can melt the metal within no time. Work piece experiences deep penetration and very insignificant heat affected zone. This phenomenon is called Key Hole phenomena.

Entire assembly is kept inside a chamber and vacuum is maintained in it. It is because high energy electrons will collide with air particles, particles will ionize and electrons will loose all their energy. As high energy electrons hit the work piece material, X- Rays will be produced, so inside surface of the container is lead lined. To make the electrons move uninterrupted it is essential to use the vacuum throughout the EBW gun column and the chamber. The electron beam is required to reach the work piece placed in the welding chamber to affect the necessary welding. If the electrons are projected into the normal atmosphere they lose all their kinetic energy in collision with atoms and molecules of oxygen and nitrogen.

Depending on the accelerating voltage the EBW guns are rated as low voltage and high voltage types with the range of voltages between 15-30 kv and 150kv respectively. This process is widely used in space industries, nuclear missile, air craft industries, cluster gears welding, intricate valve arrangements made of corrosion resistant alloys for automotive industries as well as in pressure capsules. The process produces very deep penetration and almost ideal weld is achieved. Other major applications are in welding Ti, Be, stainless steels, super alloys etc.

6.7.2 Laser Beam Welding

In laser welding a concentrated coherent light beam impinges at the desired spot to melt and weld the metal. Atoms from the ground level are pumped to high energy level with the help of primarily Xenon lamp. From the high energy level atom comes automatically to metastable state and once there is sufficient population inversion, atoms come to the ground level by emitting a bundle of energy called photon. These photons will be directed back and forth in the lasing tube. 98% reflecting lens means photons will not be allowed to come out, unless they are horizontal. Laser beam can be converged to a point. Laser emission is obtained when the upper level is sufficiently populated at the expense of lower level. Such a state is known as population inversion and the method obtained is called pumping. Laser beam welding is more versatile than the electron beam welding because welding is possible in air, through a transparent medium and along a profile.

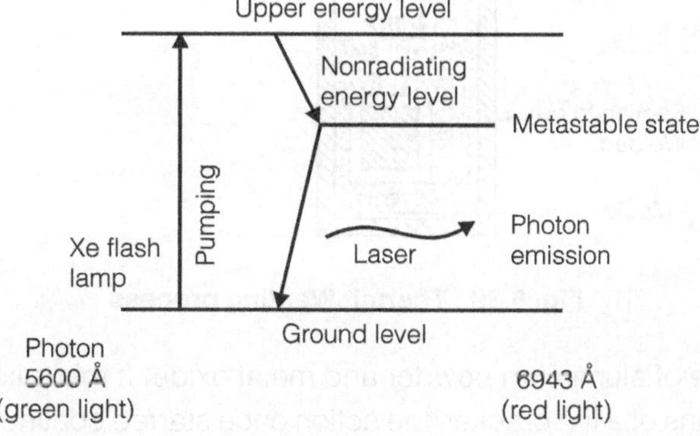

Fig. 6.38 : (a) Principle of laser emission

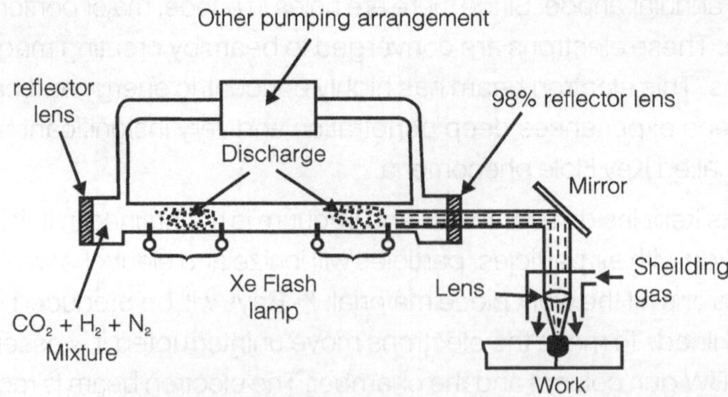

Fig. 6.38 : (b) Set up of CO_2 Laser beam welding

The most useful laser for welding is the laser in which the lasing medium is mixture of CO_2, nitrogen, and helium in the ratio of 1: 1: 10 at a pressure of 20 to 50mm of mercury. A laser consists of a glass tube in which the lasing gas mixture flows. There is one electrode at each of the two ends between which a high voltage discharge is set up. There is one reflector at each of the ends. The space between the two reflectors is called laser cavity. The laser beam emitted through the semi reflecting surface is focused to the desired spot. For industrial purposes the lasing material often used is ruby. Ruby is aluminium oxide with chromium atoms to the extent of 0.05%. The application of this is in radio engineering, electronics, welding copper and Aluminium alloys etc.

6.8 THERMIT WELDING PROCESS

$8Al + 3Fe_3O_4 \rightarrow 4Al_2O_3 + 9Fe + \text{Heat}$

(Ratio of Al and Fe according to atomic weight is 1:3)

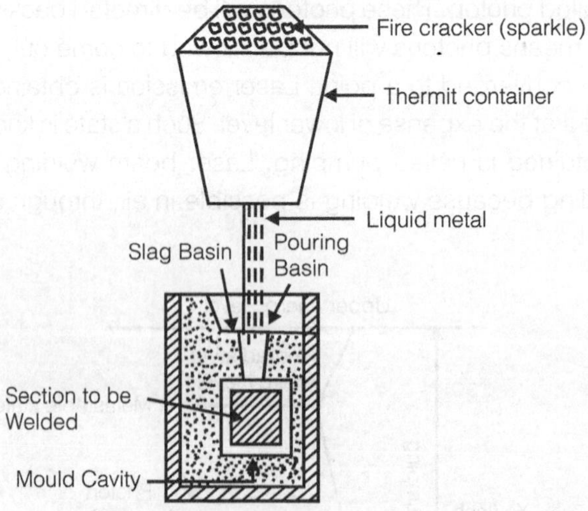

Fig. 6.39 : Thermit Welding process

Thermit is a mixture of aluminium powder and metal oxide. If this mixture is placed in a crucible and ignited by means of a fire cracker, the action once started continuous throughout the mass of the mixture, giving out great heat because reaction is exothermic. The aluminium is a strong reducing agent, and combines with the oxygen from the iron oxide, the iron oxide being reduced to iron and intense heat will be released. The aluminium oxide floats to the top of the molten metal as a slag which are contained in the mould. The high temperature of the iron results in excellent fusion taking place with the parts to be welded. A small percentage of additions like manganese or other alloying elements is done to increase the strength of joint.

Thermit welding involves aligning the parts to be joined, but with a gap between them is usually filled with wax, around which a sand or ceramic mold is built. If the parts are thick the mold cavity may be preheated to improve welding and dry the mold. Drying the mould is very important; otherwise superheated steam trapped in the mold can cause explosions. This process is used for joining tracks on site. This is also used in welding cable conductors, reinforcing bars and for heavy repairs such as those of broken necks and ship sterns.

6.9 ALLIED PROCESSES

If the parent material cannot withstand high temperature, or the parts to be joined are delicate or intricate or metals with different properties, thickness etc. then to weld with those processes is difficult, so to weld those, allied processes are introduced. In this additional metal is deposited on a substrate either by spraying or by some welding process are used :

There are three types of allied process

1. Soldering (flux is $ZnCl_2$)
2. Brazing (flux is Borax)
3. Adhesive bonding (Thermosetting plastic)

Soldering : The two surfaces are cleaned and a flux is applied over them. This flux dissolves whatever amount of oxide is remaining on the surface and also protects the surface from further oxidation. After the application of the flux the pieces are heated and the solder is then applied on the gap. The material flows into the interface of mating surface by capillary action. On cooling, it solidifies and provides a joint of adequate strength. Soldering is done by thoroughly cleaning the pieces with the help of wire brush, emery cloth, file or even steel wool. On cooling, the joint is cleaned by hot water to avoid corrosive action of the flux residue. Soldering of lead pipe is known as **wiping.**

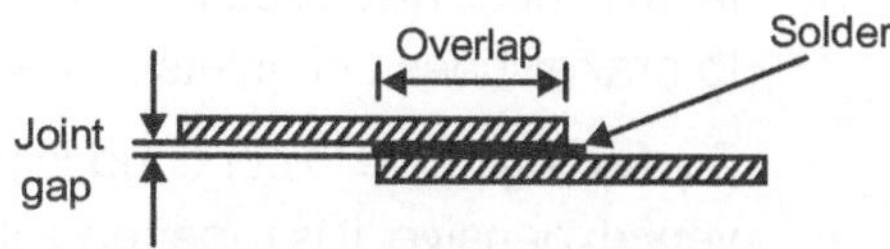

Fig. 6.40 : Soldering process

Soldering is used extensively in the electronics industry. Solders are not used for load bearing members because of the low strength than brazing. This process is used to join Copper with such precious metals such as silver and gold.

Brazing : The process is performed above 450° C but below solidus temperature of the base metal. The surfaces to be joined are grinded and then cleaned by applying some chemicals like Carbon Tetrachloride. A brazing flux (borex) is then applied over the surface to dissolve any remaining metal oxide. A brazing material like brass is poured at the joint. The molten metal fills by capillary action. Upon cooling and solidification of the filler metal, a strong joint is obtained. Joints to be brazed are made with small clearance, so that brazing flux is applied, to dissolve solid metal oxide which is still present and to prevent further oxidation. Brazing fluxes are generally borax, boric acid, borates, fluorides and chlorides. Surfaces to be brazed must be clean in order to obtain proper wetting and spreading characteristics of the molten filler metal in the joint and to develop maximum bond strength. Residual flux left on the brazed joint can be removed by washing with hot water followed by air drying. It is used to braze dissimilar metals with good joint strength. Intricate, light weight shapes can be joined rapidly and with little distortion. Brazing can be automated and used for mass production.

Torch
Base metal
Flux
Filler metal

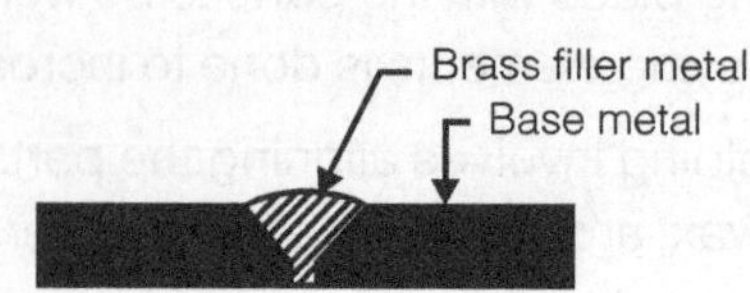

Fig. 6.41 : Brazing process

Adhesive bonding : In adhesive bonding a metal is joined to another metal or non metal by the use of an adhesive which consists of the thermosetting type. The pieces to be joined are thoroughly cleaned by chemicals or by mechanical means. Adhesives are applied to the cleaned surfaces by brushing, spraying, roller coating or dipping. The layer of adhesive applied depends upon the metal being bonded, type of adhesive, solvent used and the strength aimed. There is no metallic bond, but the bond is created by chemical action between the adhesive and metal being joined. The surface, before applying adhesive is degreased and suitably roughened. Usually the adhesive is applied in a liquid solvent and light pressure applied and joint heated to about for an half an hour for curing. Adhesive joints can be made to resist water or gas pressure and adhesive bonds have high damping capacity and thus high fatigue resistance.

This is used in manufacturing of railway cars, microwave reflectors, refrigerators, and storage tanks. This is also used in processes which include fastening of stiffeners to aircraft skin, attaching to brake shoes and joints in the aircraft wing and tail assemblies.

Surfacing : Due to continuous use of rolls in the rolling mill, rolls get eroded and since these are very expensive, it is repaired either by surfacing or cladding. In surfacing the material is liquified and with the help of compressed neutral gases, liquified metal is sprayed over the rolls so that desired properties can be achieved, which include corrosion resistance, wear resistance, dimension control and metallurgical needs.

Different types of surfacing are : Cladding, Hardfacing, Buildup, and Buttering. In Cladding operation, electrode is in the form of thin sheet and with the help of submerged arc welding and gas arc welding it is coated over the rolls. In Hardfacing a metal is deposited over another surface to increase the hardness of another surface and to make it resistant to abrasion, impact, erosion, galling and cavitation. Abrasion resistance is the most important application in hardfacing. Builtup overlay is the rebuilding of worn out parts to restore them to original shape and dimensions. Buttering is the process of depositing one or more layers of a material between those metallurgically non-compatible materials which individually have compatibility with the material forming the buttering layer.

6.10 GAS WELDING

Oxygen cylinder valves are made of Brass and Acetylene cylinder valves are made by Steel. Brass valve does not corrode so easily and that is why brass valves are used in oxygen cylinder. Acetylene is a very dangerous gas because it can explode under its own weight. So calcium silicate is filled in the cylinder and then acetone is poured. Acetylene is absorbed in acetone. The absorbing capacity of acetone increases by increasing pressure. So the pressure of acetylene cylinder is higher. Only stainless steel valves will be used in acetylene cylinders because acetylene by reacting with copper produces copper acetylene which is an explosive.

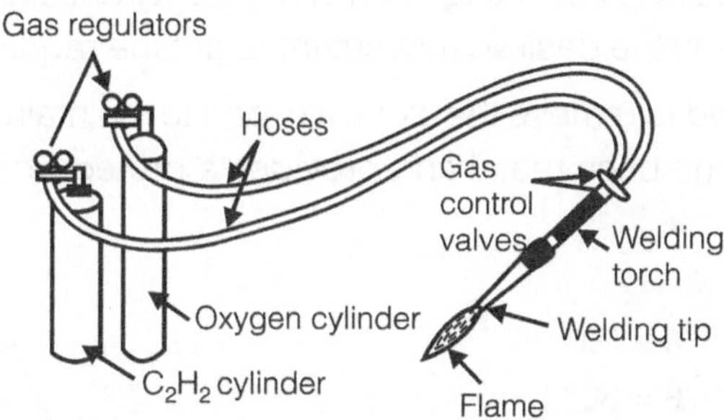

Fig. 6.42 : (a) Basic equipment used in oxyfuel gas welding.

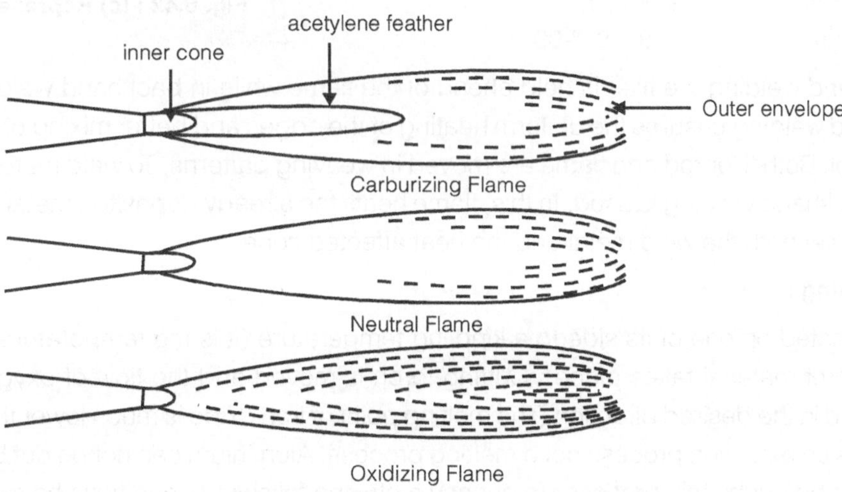

Fig. 6.42 : (b) Different types of flames

carburizing flames contains unburnt Carbon and after solidification of weld the bead will have pitted surface over it and the weld bead appear to be very hard and brittle, because lot of free Carbon will be present in the flame and it is used to weld high Carbon steels and cast Irons. It will not produce any sound. Because it contains unburnt carbon its temperature is lower than a neutral or an oxidizing flame. If the excess unburnt Carbon finds its way to the molten metal the weld puddle appears to be boiling.

Neutral flame produces hissing sound and the flame is used to weld low Carbon steels and Aluminium. This consists of nearly one to one ratio of acetylene and oxygen by volume. It consists of two parts namely the inner cone and the outer envelope. It has the clear luminous hissing cone indicating that the combustion is complete. Such a flame makes the hissing sound and it is useful to weld metals. The temperature of neutral flame is 3150°C.

Oxidized flames are the hottest flames and it produces roaring sound. This flame has an excess of oxygen over the acetylene. It consists of a very short pointed white inner cone and a shorter outer envelope. The reduction of length of the inner cone is a measure of excess oxygen. This is the hottest flame produced. These flames are used to weld alloys of Copper and Zinc. In welding these metals, the oxidizing flame produces a base metal oxide layer to protect the evaporation of low point alloying elements. The temperature of these flames is around 3480°C.

The welding procedures requires the ignition of oxy-acetylene flame and the manipulation of the torch to move the flame in the desired movement to get the required weld.

A welding flux is required to remove the oxide film and to maintain a clean surface. The flux melts at the melting point of the base metal and provides a protective layer against reaction with the atmospheric gases.

Thickness of plate

t ≤ 1 mm	θ = 20°
= 1-3 mm	θ = 30°
= 3-5 mm	θ = 40°
= 5-7 mm	θ = 50°
= 7-10 mm	θ = 60°
= 10-15 mm	θ = 70°
> 15 mm	θ = 80°-90

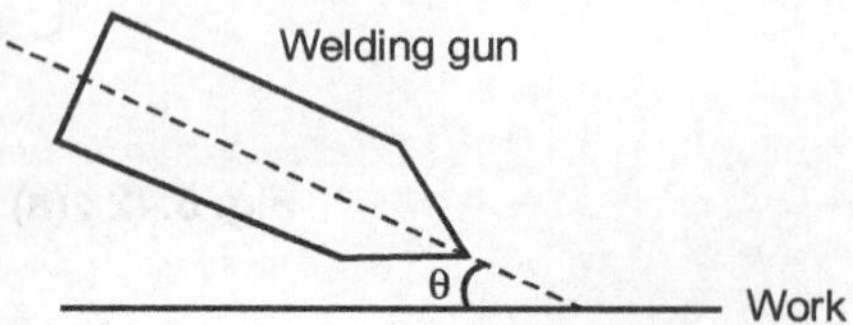

Fig. 6.42 : (c) Representing tip angle

In forehand welding the filler is held ahead of the flame while in backhand welding it follows it. Forehand welding ensures the uniform heating of the edges and better mixing of the metal in the weld pool. Both filler rod and flame are moved in weaving patterns. To weld material thicker than 5mm backhand welding is used. In this, flame heats the already deposited metal and that serves to heat treat both the weld metal and the heat affected zone.

Gas cutting :

Iron is heated on one of its side to a kindling temperature (it is the temperature at which rapid oxidation of material takes place). Cutting torch is moved and the flow of oxygen is suddenly increased in the desired direction. The melting point of iron oxide is much lower than iron. So gas cutting is an oxidation process not a melting process. Aluminium can not be cut by this process. Steel can be cut by this process. To cut any metal the following conditions have to be full filled

1. The kindling temperature should be lower than the melting temperature of the metal
2. The melting point of the parent metal should have higher temperature than the metal oxide which is formed by reaction with oxygen and it should be fluid in molten state so as to blow out easily.
3. It should have low thermal conductivity so that the material can be rapidly raised to its kindling temperature.

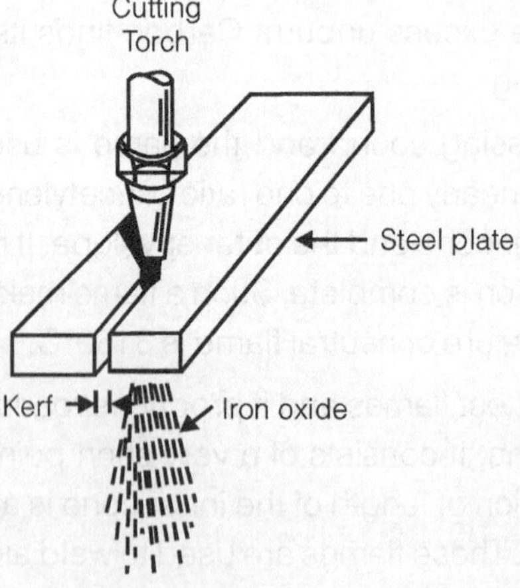

Fig. 6.43 : Flame cutting of steel plate with oxyacetylene torch

6.11 FORCES IN WELDING

The modes of the metal transfer which is operative at an intense during welding is dependent upon the cumulative effect of the different forces acting on the molten droplet growing at the tip of the electrode. The individual role of each force may vary. Some forces help in detaching the molten droplet while others help in retaining it at the electrode tip.

Gravity : This force will try to detach the molten droplet from the tip of the electrode and attach to the work piece. But in case of over head welding it is an attaching force. Numerically it is equal to the weight of the detaching droplet.

$$F_g = mg$$

Surface Tension : What ever is the welding position surface tension is an attaching force.

Drag : In gas welding drag force will appear due to flow of inert gases and it is always detaching force.

Electromagnetic pinch force or Lorentz force : Electrode can be considered as a number of parallel conductors and when the current flows in same direction an attractive force will develop. This will try to detach the molten droplets from the tip of electrode. In over head welding current settings has to be higher. Larger the current settings finer will be droplet size of liquid metal and higher will be the pinch force. This is required to move the liquid metal against gravity. Also in overhead welding work should be given +ve polarity to minimize dilution.

$$\text{Electromagnetic pinch force} \propto \frac{I^2}{200}$$

6.12 TYPES OF METAL TRANSFER

Short circuit transfer : In short circuit there is a periodic bridging of electrode and molten pool resulting in arc getting extinguished. Consequently heavy flow of current occurs which results in increased heating of the bridge. Decreased viscosity and surface tension, increased electromotive and hydrodynamic forces result in the transfer of molten metal from the electrode to the weld pool. With this the bridge breaks and the voltage jumps to open circuit voltage and the arc is reignited.

Dip transfer : This is also a short circuit mode of metal transfer but in this the electrode is fed into the weld pool at a fast pace so that the electrode is dipped into the molten metal before droplet is detached. Dip transfer also appears in electroslag welding. Due to lower temperature of molten pool there are high chances of slag inclusion.

Globular Metal Transfer : When the current settings are slightly higher molten droplet size increases and the detachment primarily takes place by gravity and groove will be filled drop by drop as a result of that bead density will be low. The detached globule travels under the action of gravity and hydrodynamic forces directly towards the weld pool and is referred to as drop transfer. This type of transfer encountered where the arc length is medium to long. The droplet temperature is also higher than in the case of short – circuit transfer.

Spray Transfer : This type of transfer takes place at higher current settings which results in bead structure. As the current density is raised the droplet growth rate increases proportionately to the increase in temperature and electromagnetic forces in the form of pinch effect become significant. Droplets are pinched off before they attain the size permitted by the surface tension and that results in spray mode transfer. Drop spray is found to give least spatter and smoke with higher deposition efficiency than other variants of spray modes. Spray mode transfer is associated with GMAW with medium to long lengths.

Repelled Transfer : In D.C arc welding because of fixed polarity magnetic lines setup in the work piece and the tendency of arc is to be deflected towards the magnetic lines, the welding process will have lot of spatter. In case of arc blow metal deviate from its intended path and produces severe and pattern called repelled transfer. To avoid this problem of arc blow metal pieces are kept at the entry and exit called tab inns and tab outs. So while performing welding at the ends magnetic lines will continue in this material and there will not be any arc blow. This mode of transfer is usually encountered in welding of steel with medium arc length and low medium welding current.

Explosive Transfer : Sometimes the droplet is shattered either while it is still at the electrode tip or soon after detachment. This type of metal transfer is known as explosive transfer and is attributed to the formation of gas bubbles within the liquid drop at the electrode tip. The bubbles are due to the formation of CO in the case of steels and some absorbed gases in the case of non – ferrous metals.

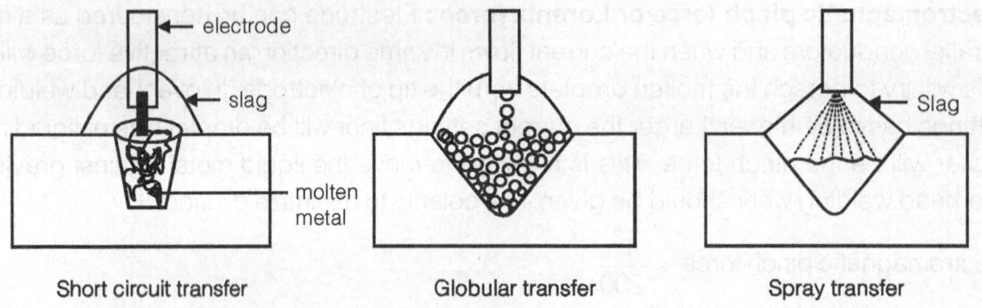

Fig. 6.44 : Different types of metal transfer

6.13 TYPES OF FLUXES

1. **Cellulose :** The mixture contains

   20 - 60% TiO_2
   10 - 20% cellulose
   15 - 30% quartz
   3 - 100 ppm is H_2
   0 - 15% carbonates
   5 - 10% ferromanganese

   These fluxes are having high hydrogen content and by reaction it produces H_2 gas and it is exothermic reaction. Thus deeper penetration will be produced but as the gas is coming out it leaves either large number of fine holes or some time cracks called hydrogen embrittlement. The metal deposited by electrodes with these coatings is fine grained and contains hardly any oxygen. These coatings are used for positional welding, particularly where a good degree of penetration is required.

2. **Rutile :**

   40 - 60% Natural TiO_2
   15 - 30 ppm hydrogen
   10 - 14% Thermomanganese
   0 - 5% Organic cellulose
   15 - 20% Quartz
   0 - 15% Carbonates

Rutiles are characterized by heavy alloy content due to which it produces thick coating of slag over it. These are also called general purpose electrodes. These electrodes give medium penetration welds of good appearance with very high mechanical properties. Advantage is for high arc stability and positional welding properties. Most of the general purpose coated electrodes used for welding low carbon structural steels are of this type.

3. **Acidic coating :** The mixture contains Fe_3O_4, MnO_2, Quartz, Silicates, Carbonates, Ferromanganese and very high hydrogen content.
   The slag produced have acid reaction and thus, dissolve basic oxide like MnO, consequently a large amount of manganese is transferred to the also. The manganese present in the slag lowers the viscosity, which tends to improve the appearance of the weld bead and makes it possible to deposit weld metal in all positions.

4. **Basic coatings :** The mixture contains
   40 - 50% $CaCO_3$
   20 - 40% Fluorspar and feld spar
   0 - 5% quartz
   0 - 10% Rutile
   5 - 10% Ferroalloys
   Although it is easy to weld by basic coated electrodes but due to the presence of $CaCO_3$, flux is having a tendency to capture moisture from the surrounding. So the electrode has to be baked in an oven before use and it has to be consumed as fast as possible. In these coatings there is no use of minerals which combine with water. This ensures lowest possible content of hydrogen in the weld metal. That is why they are also known as low – hydrogen electrodes. The hydrogen electrodes have superior weld properties with highest ductility of any of the weld deposits. The welds are therefore resistant to cracking.

6.14 ISO DESIGNATION

Basis for coding :

1. A prefix letter.
2. Letters to denote the type of covering.
3. A code number of four digits.
4. In some case a suffix, consisting of one or more letters, to indicate special characteristics.
   Example representation can be shown as,
   E R 4 3 1 2 H J X

1. Here E = extrusion is used to make the electrode
2. R = Type of flux used
   A – Acidic
   B – Basic.
   R – Rutile.
   C – Cellulose.
3. In code number the first two digits give the mechanical properties of the weld deposit and last two digits give the performance characteristics of the coverings.
   4 and 3 are mechanical properties of the joint
   4-Yield strength and ultimate tensile strength
   3-% of elongation.

1-Welding position.
2-Welding current and voltage settings.
4. H - Hydrogen Controlled electrode.
J-Iron powdered electrode.
X-Radio graphic quality of weld.

6.15 WELDING POSITIONS

The angle which the weld root makes with the horizontal is called weld slope and by drawing normal over the bead the angle this line makes with the vertical is called weld rotation.

Weld position	Symbol	Slope	Rotation
Flat	F	0 - 5°	0 - 10°
Horizontal	H	0 - 5°	30 - 90°
Vertical	V	80 - 90°	0 - 180°
Down head	D	80 - 90°	0 - 180°
Over head	O	0 - 15°	115 - 180°

6.16 TYPES OF JOINTS

1. **Butt joint :** Butt weld can be formed by welding plates end to end. The angle will be 135°-180°
2. **Lap joint :** A lap joint is the one in which plates overlap each other. The angle will be 0°-5°
3. **T-joint :** The angle of T-joint is 5°-90°.
4. **Edge joint :** The angle of edge joint is 30°-135°
5. **Corner joint :** The angle is 5°-30°
6. **Cruciform joint :**

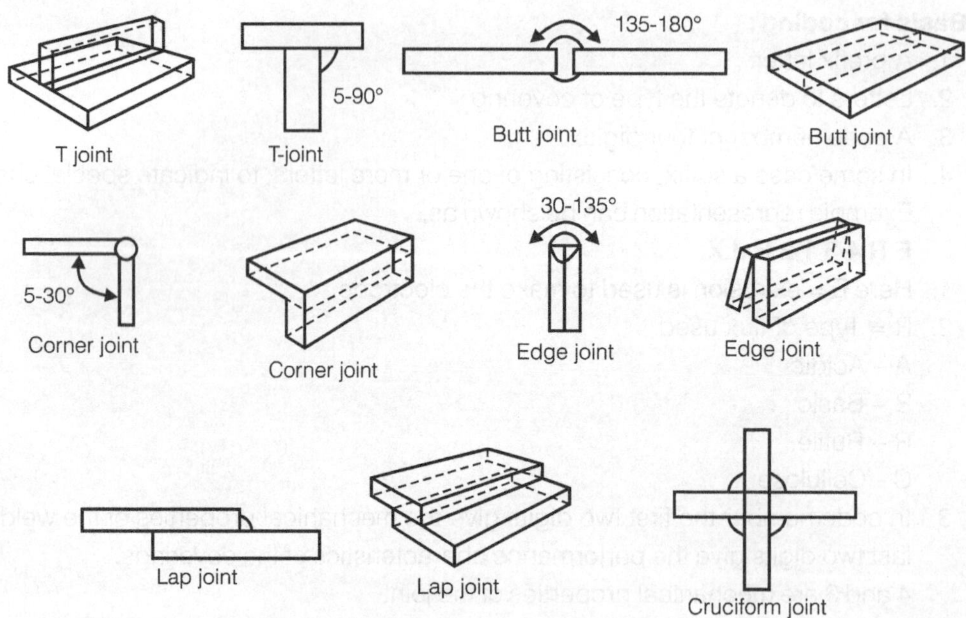

Fig. 6.45 : Different types of welds

These are type of joints which can be prepared through different welding processes. Joints are prepared with an incident angle with root face and root gap as required for the plate thickness. The plates can be tack welded in position to prevent movement due to expansion and contraction during welding and tack welds must be strong enough to prevent movement of the plates.

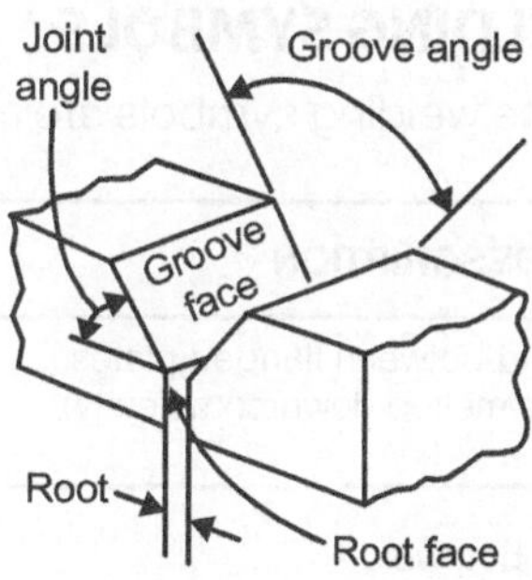

Fig. 6.46 : Nomenclature for edge preparation

6.17 TYPES OF WELD

1. Fillet weld
2. Groove weld
3. Spot
4. Seam
5. Surfacing
6. Plug.

The root gap is provided to achieve full penetration while root face avoids burn – through due to excessive heat concentration at sharp edges. The groove angle is dependent upon proper fusion of the sides with the weld metal from the electrode.

Fillet weld :

$$A = l \cos 45° \times W$$

It always fails in the minimum cross section, due to shear.

So $\quad$ force $= K' \times A$

K' = Yield strength in shear

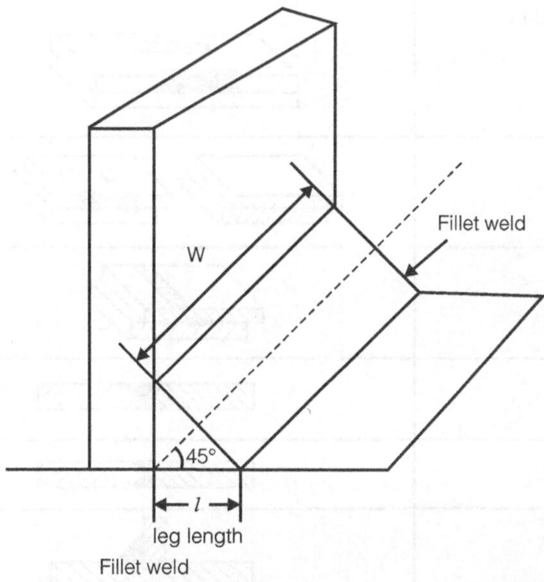

Fig. 6.47

Using intense source of heat energy or by giving work piece positive polarity of penetration extends in another work material, after solidification there will be a joint between the two materials. This is called plug weld.

6.18 WELDING SYMBOLS

Some welding symbols are represented as below.

DESCRIPTION	SECTIONAL REPRESENTATION	SYMBOL
1. Butt weld between flanged plates (flanges melted down completely)		
2. Square butt weld		
3. Single-V butt weld		
4. Single-bevel butt weld		
5. Single-V butt weld with broad Root face		
6. Single-bevel butt weld with broad Root face		
7. Single butt weld		
8. Single-J butt weld		
9. Backing or sealing run		
10. Fillet weld		
11. Plug weld (circular or Elongated hole, Completely Filled)		
12. Spot weld (resistance or Arc welding or projection weld)		
13. Seam weld		
14. Flat (flush) single-V butt weld		
15. Convex double-V butt weld		
16. Concave fillet weld		
17. Flat (flush) single-V butt weld with flat (flush) backing run		

Fig. 6.48 : Representation of welding symbols and supplementary welding symbols.

Representation of the weld :

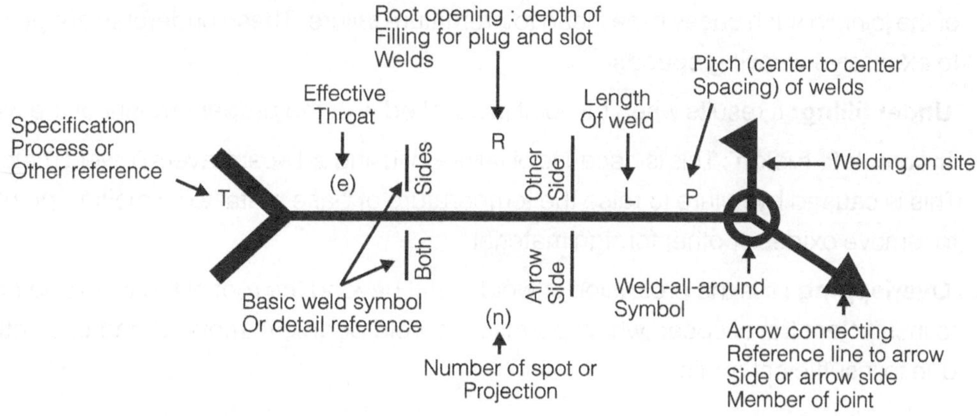

Fig. 6.49 : (a) Various representations of symbols.

Information below the line refers to the same side of arrow and information above the line refers to the opposite side. Flag indicates that welding is performed on sight. Initially there was a V-groove, but there was some defect in the bead. So the deposited material was removed (Back Gouging) and V was prepared on the other side. To minimize the distortion sealing run was performed and its reinforcement was grinded and the process was performed using GTAW process on sight.

55°
GTAW
BG
65°

Fig. 6.49 : (b)

6.19 DEFECTS IN WELDING

The ideal weld should be such that adequate fusion exists between the filler metal and edge preparation together with good penetration. The major discontinuities are

1. **Porosity :** Porosity is caused due by the entrapment of gas bubbles by the freezing dendrites during the cooling of the wed pool and occur essentially only in the weld metal, or by chemical reactions during welding, or by contaminants. These are generally in the shape of spheres or of elongated pockets. The porosity may be distributed through out weld zone, or may be concentrated in a certain region. This can be reduced by proper selection of filler metals, by preheating the weld area, by proper cleaning the weld zone, and by reducing the welding speeds.
2. **Slag inclusions:** Inclusions may be caused by compounds such as oxides, fluxes, and electrode coating materials, which are trapped in the weld zone. If shielding is not provided properly the contaminants from atmosphere will contribute to such inclusions. This can be reduced by cleaning the weld bead, by providing enough shielding gas.
3. **Incomplete fusion and penetration :** It is usually caused by insufficient heat and too fast travel of torch or electrode. This can be avoided by raising the temperature of the base metal, cleaning the weld area, providing enough shielding gas. Incomplete penetration occurs when the depth of the welded joint is insufficient. This can be reduced by increasing the heat input, reducing the speed, changing joint design and ensuring proper fit.

4. **Under cutting :** It is the melting or burning away the base metal at the toe of the weld as sharp recess or notch. Undercut can be as stress raiser and can reduce the fatigue strength of the joint. In such cases it may lead to premature failure. These undercuts are generally due to excessive weaving speeds.
5. **Under filling:** It results when the joint is not filled with the proper amount of the weld metal.
6. **Incomplete fusion :** This is because of failure in fusing adjacent layers of weld metal together. This is caused by failure to raise the temperature of base metal to its melting point or failure to remove oxides or other foreign material.
7. **Overlapping :** it is the protrusion of weld metal beyond the root of the weld and occurs due to insufficient travel speed which permits the weld puddle to move ahead of electrode, also due to insufficient angle.

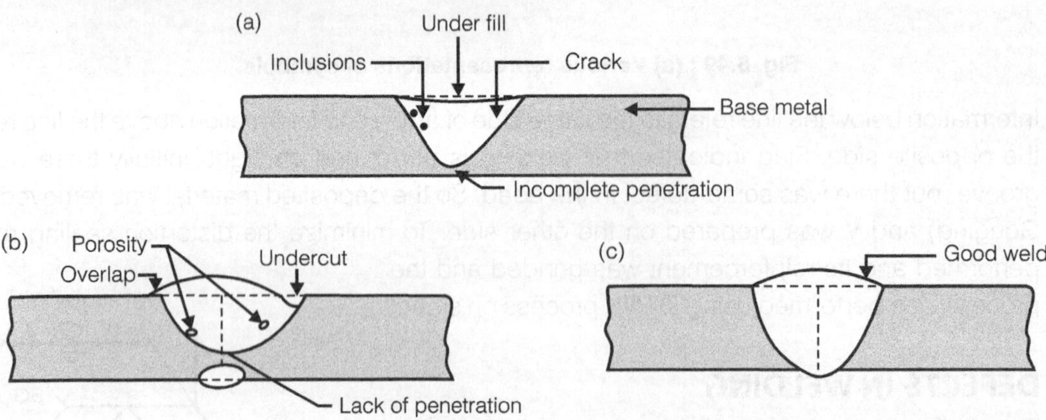

Fig. 6.50 : (a) Different welding defects

8. **Cracks :** Cracking is the most serious defect in a weld since it can lead to premature failure particularly in a dynamically loaded component. The typical types of cracks are longitudinal, transverse, crater, underbead and toe cracks. Cracks occurs due to; temperature gradients that causes thermal stresses in weld zone, variations in the compositions of the weld zone, hydrogen embrittlement, inability of the weld metal to contract during cooling and embrittlement of grain boundaries.

   There are two types of cracks, hot crack and cold crack. Hot cracks occur while the joint is still at elevated temperatures. Cold cracks develop after the weld metal has solidified.

   This can be prevented by changing the joint design, by changing the parameters, procedures and the sequence of the welding process, preheating the component, and by avoiding rapid cooling of the weld components.
9. **Lamellar tears :** The work piece when tested at its thickness direction it is weaker alignment of nonmetallic impurities and inclusions. This is mainly in rolled plates. In welding such components these lamellar tears will appear, because of shrinkage of the restrained components during cooling. This can be avoided by providing for shrinkage of the members.

10. **Surface damage :** This occurs due to spatter during the welding. Because of that small droplet of metal deposited on adjacent surfaces. Such surface discontinuities are objectionable. By using proper welding techniques and procedures surface damage can be avoided.

11. **Stresses and Distortion in welding :** In the welding process we have a molten pool of metal which consists partly of the parent metal melted or fused from the side of the joint, and partly of the electrode or filler rod. As welding proceeds this pool travels along and heat is lost by conduction and radiation resulting in cooling of the joint. The cooling takes place with varying rapidity depending on many factors such as size of work, quality of weld metal being deposited, thermal conductivity of the parent metal and the melting point and specific heat of the weld metal.

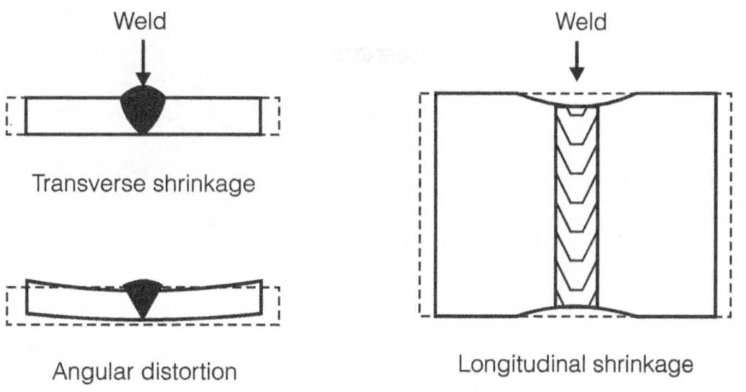

Fig. 6.50 : (b) Distortion of parts after welding

As the weld proceeds there are areas surrounding the weld in varying conditions of expansion and contraction, and thus a varying set of forces will be set up in the welded parent metal. When the weld has cooled, these forces still remain, they are called residual stresses and they are not due to any external load but due to internal forces. These stresses cause elastic and plastic deformation. The process of removal of these residual stresses is called stress relieving.

Stresses are set up during manufacturing due to rolling and forging. Some of these stresses will be relieved during welding because of heating and because of that distortion reduces. If the welded plates are free to move distortion occurs and if the plates are rigid the stresses remain as residual stresses.

Depositing the weld metal with slightly higher current, gives wider and deeper layer. On cooling the plate distorts and bends in upward direction with the weld on the inside of the bend. If the quantity of heat supplied is small, the contraction forces will be small and no distortion occurs. But because of the higher heat higher temperature will be set up on the upper side of the plate, therefore the upper side contracts more than the lower side and due to the pull of the contracting line of the weld metal distortion occurred. An increase in the weld speed, distortion also increases and greater the number of layers of weld metal deposited, the greater will be the distortion.

Distortion is reduced by step welding. In this the line of weld is broken up into short lengths, each length is ending where the other began. By this the expansion and contraction area next to each other neutralizes.

Peening is the processes of lightly hammering the weld and the surrounding parent metal in order to relieve stresses present and to consolidate the structure of the weld metal. It may be carried out while the weld is still hot or immediately the weld has cooled.

Preheating and post heating is done to reduce the internal stresses and also to harden the areas in the HAZ. To minimize angular distortion very small amount of material is deposited on the opposite face called sealing run. It also closes the root gap. The reinforcement of the sealing run is undesirable. So it has to be grinded. This is called dressed flush.

In electroslag or electrogas welding when temperature of molten pool is not controlled, the thick plates will bend, which called rotational distortion. In welding cast iron, preheating must is necessary, otherwise because of lack of ductility it will easily fracture before it distort.

■■■■

PRACTICE QUESTIONS

1. The correct sequence of the given materials in ascending order of their weldability is:

(a) Ms, copper, cast iron, aluminium
(b) Cast iron, Ms, aluminium
(c) Copper, cast iron, Ms, aluminium
(d) Aluminium, copper, cast iron, Ms

2. Match the following:

	Work material		**Type of joining**
P.	Aluminium	**1.**	Submerged arc welding
Q.	Die steel	**2.**	Soldering
R.	Copper wire	**3.**	Thermit welding
S.	Titanium sheet	**4.**	Atomic hydrogen welding
		5.	Gas tungsten arc welding
		6.	Laser beam welding

(a) P-2, Q-5, R-1, S-3
(b) P-6, Q-3, R-4, S-4
(c) P-4, Q-1, R-6, S-2
(d) P-5, Q-4, R-2, S-6

3. Consider the following statements:

The size of the heat affected zone will increase with:

1. increase starting temperature
2. increase welding speed
3. increase thermal conductivity of base metal
4. increase in base metal thickness

Which of the following given above are correct?

(a) 1, 2 and 4
(b) 1, 2 and 3
(c) 1 and 3
(d) 2 and 3

4. Two plain carbon steel specimens having 0.8% carbon contents are welded. If we observe the weldment under Metallurgical Microscope from centre towards either side, the following structures are observed at different zones:

1. Fine pearlite
2. Coarse pearlite
3. Martensite

Select the correct answer using the codes given below:

(a) 1, 2, 3
(b) 1, 3, 2
(c) 2, 1, 3
(d) 3, 1, 2

5. Fabrications weldability test is used to determine

(a) Mechanical properties required for satisfactory performance of welded joint.
(b) Susceptibility of welded joint for cracking
(c) Suitability for joint design
(d) Appropriate machining process

6. Which one of the following is not a fusion welding process.

(a) Gas welding
(b) Arc welding
(c) Brazing
(d) Resistance welding

7. Consider the following statements:
Assertions(A): Aluminium has poor weldability
Reason(R): Aluminum has high thermal conductivity and high affinity to oxygen.
(a) Both A and R are individually true and R is the correct explanation of A.
(b) Both A and R are individually true but R is not the correct explanation of A.
(c) A is true but R is false.
(d) A is false but R is true.

8. Consider the following statements: MIG welding process uses
1. Consumable electrode
2. Non-consumable electrode
3. D.C power supply
4. A.C power supply
Of these statements:
(a) 2 and 4 are correct
(b) 2 and 3 are correct
(c) 1 and 4 are correct
(d) 1 and 3 are correct

9. Which of the following gases can be used for GMAW as a shielding gas?
1. Argon
2. Oxygen
3. Carbon dioxide
Select the correct answer using the code given below:
(a) 1, 2 and 3
(b) 1 and 2
(c) 2 and 3
(d) 1 and 3

10. Consider the following statements:
Assertions (A): The electrodes of ac arc welding are coated with sodium silicate, where as electrodes used for dc arc welding are coated with potassium silicate binders.
Reason (R): Potassium has a lower ionisation potential than sodium.
(a) Both A and R are individually true and R is the correct explanation of A
(b) Both A and R are individually true but R is not the correct explanation of A.
(c) A is true but R is false
(d) A is false but R is true.

11. An arc welded joint is shown in the figure. The part labeled 'B' in the figure is known as:
(a) Weld preparation
(b) Penetration
(c) Reinforcement
(d) Slag

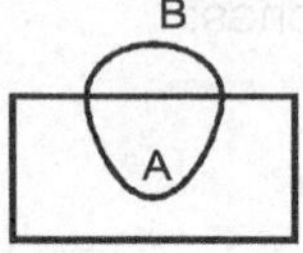

12. In an arc welding process, the voltage and current are 25 V and 300 A respectively. The arc heat transfer efficiency is 0.85 and welding speed is 8 mm/sec. The net heat input (in J/mm) is.
(a) 64
(b) 797
(c) 1103
(d) 79700

13. The voltage-current characteristics of a DC generator for arc welding is a straight line between an open-circuit voltage of 80 V and short-circuit current of 300A. The generator settings for maximum arc power will be:
(a) 80V and 150A
(b) 40V and 300A
(c) 40V and 150A
(d) 80V and 300A

14. Which of the following statements is correct?
(a) No flux is used in gas welding of mild steel
(b) Borax is the commonly used flux coating on welding electrodes.

(c) Laser beam welding employs a vaccum chamber and thus avoids use of a shielding method.
(d) AC can be used for GTAW process

15. Arc blow is more common in
(a) AC welding
(b) DC welding with straight polarity
(c) DC welding with bare electrodes
(d) AC welding with bare electrodes

16. Pinch effect in welding is the result of
(a) Expansion of gases in the arc
(b) Electromagnetic forces
(c) Electric force
(d) Surface tension of the molten metal

17. In manual arc welding, the equipment should have drooping characteristics in order to maintain the
(a) Voltage constant when arc length changes
(b) Current constant when arc length changes
(c) Temperature in the arc constant
(d) Weld pool red-hot

18. In arc welding DC reverse polarity is used to bear greater advantage in
(a) overhead welding
(b) flat welding of lap joints
(c) edge welding
(d) flat welding of butt Joints

19. Match **List-I** (Ingredients) with **List-II** (Welding functions) and select the correct answer using the codes given below the lists:

List-I	**List-II**
A. Silica	**1.** Arc stabilizer
B. Potassium silicate	**2.** Deoxidizer
C. Ferro silicon	**3.** Fluxing agent
D. Cellulose	**4.** Gas forming material

Codes:

	A	B	C	D
(a)	3	4	2	1
(b)	2	1	3	4
(c)	3	1	2	4
(d)	2	4	3	1

20. Match **List-I** (Welding defects) with **List-II** (Causes) and select the correct answer using the Codes given below the lists:

List-I	**List-II**
A. Spatter	**1.** Damp electrodes
B. Distortion	**2.** Arc blow
C. Slag inclusion	**3.** Improper cleaning in multipass welding
D. Porosity	**4.** Poor joint selection

Codes:

	A	B	C	D
(a)	4	2	3	1
(b)	4	2	1	3
(c)	2	4	1	3
(d)	2	4	3	1

21. In arc welding of a butt joint, the welding speed is to be selected such that highest cooling rate is achieved. Melting efficiency and heat transfer efficiency are 0.5 & 0.7, respectively. The area of the weld cross section is 5mm^2 and the unit energy required to melt the metal the is 10 J/mm^3. If the welding power is 2kW, the welding speed in mm/s closest to

(a) 4 (b) 14
(c) 24 (d) 34

22. Which of the following are the major characteristics of submerged are welding?

1. High welding speeds
2. High deposition rates
3. Low penetration
4. Low cleanliness

Select the correct answer using the code given below:

(a) 2 and 3 (b) 1, 2 and 3
(c) 3 and 4 (d) 1 and 2

23. Consider the following statements:

1. In arc welding, 65% to 75% heat is generated at the node.
2. Duty cycle in case of arc welding is the cycle of complete welding of work piece from the beginning
3. Arc blow is more common with DC welding

Which of the statements given above are correct?

(a) 1, 2 and 3 (b) 1 and 2
(c) 2 and 3 (d) 1 and 3

24. In MIG welding at high current, the metal is transferred into the form of which of the following?

(a) A fine spray of metal (b) Molten drops
(c) Weld pool (d) Molecules

25. The coating material of an arc welding electrode contains which of the following

1. Deoxidizing agent
2. Arc stabilizing agent
3. Slag forming agent

Select the correct answer using the code given below:

(a) 1, 2 and 3 (b) 1 and 2 only
(c) 2 and 3 only (d) 1 and 3 only

26. **Assertion(A):** Submerged arc welding is not recommended for high carbon steels, tool steels, aluminium, magnesium etc...

Reason(R): This is because of unavailability of suitable fluxes, reactivity at high temperatures and low sublimation temperatures.

(a) Both A and R are individually true and R is the correct explanation of A
(b) Both A and R are individually true and R is not the correct explanation of A
(c) A is true but R is false
(d) A is false but R is true

27. The maximum heat in resistance welding is at the :

(a) tip of the positive electrode
(b) tip of the negative electrode
(c) top surface of the plate at the time of electric contact with the electrode
(d) interface between the two plates being Joined

28. Two 1 mm thick steel sheets are to be spot welded at a current of 5000 A. Assuming effective resistance to be 200 micro-ohms and current flow time of 0.2 seconds, heat generated during the process will be
(a) 0.2 Joules
(b) 1 Joules
(c) 5 Joules
(d) 1000 Joules

29. In resistance welding, heat is generated due to the resistance between
(a) Electrode and work piece
(b) Asperities between touching plates
(c) Two dissimilar metals being in contact
(d) Inter atomic forces

30. Two metallic sheets, each of 2.0 mm thickness, are welded in a lap joint configuration by resistance spot welding at a welding current of 10 KA and welding time of 10 milliseconds, A spherical fusion zone extending up to the full thickness of each sheet is formed a properties of the metallic sheets are given as:
ambient temperature = 233 K
melting temperature = 1793 K
density = 7000 kg/m^3
specific heat = 800 J/kgK
latent heat of fusion = 300 kJ/kg
Assume:
1. contact resistance along sheet-interface is 500 micro-ohm and along electrode-sheet Interface is zero.
2. no conductive heat loss through the bulk sheet materials; and
3. the complete weld fusion zone is at the melting temperature. The melting efficiency(in %) of the process is

(a) 50.37
(b) 60.37
(c) 70.37
(d) 80.37

31. Spot welding of two 1mm thick sheets of steel (density = 8000 kg/m^3) is carried out successfully by passing a certain amount of current for 0.1 second through the electrodes. The resultant weld nugget formed is 5mm in diameter and 1.5 mm thick. If the latent heat of fusion of steel is 1400 kJ/kg and the effective resistance in the welding operation is 200 ohms, the current passing through the electrodes is approximately
(a) 1480 A
(b) 3300 A
(c) 4060 A
(d) 9400 A

32. Tool material not suited to resistance welding is
(a) aluminium oxide
(b) stellite
(c) high speed steel
(d) masonite

33. In oxy-acetylene gas welding, for complete combustion, the volume of oxygen required per unit of acetylene is
(a) 1
(b) 1.5
(c) 2
(d) 2.5

34. In gas welding of mild steel using an oxy-acetylene flame, the total amount of acetylene consumed was 10 liters. The oxygen consumption from the cylinder is
(a) 5 liters
(b) 10 liters
(c) 15 liters
(d) 25 liters

35. In oxyacetylene gas welding, temperature at the inner cone of the neutral flame is around
(a) 3500°C
(b) 3200°C
(c) 2900°C
(d) 2550°C

36. Oxyacetylene reducing flame is used while carrying out the welding on
(a) mild steel (b) high carbon steel
(c) grey cast iron (d) alloy steels

37. Assertion(A): If neutral flame is used in oxy-acetylene welding, both oxygen and acetylene cylinders of same capacity will be emptied at the same time.
Reason(R): Neutral flame uses equal amounts of oxygen and acetylene.
(a) Both A and R are individually true and R is the correct explanation of A
(b) Both A and R are individually true and R is not the correct explanation of A
(c) A is true but R is false
(d) A is false but R is true

38. Consider the following statements:
1. In gas welding, the torch should be held at an angle of 30° to 45° from the horizontal plane
2. In gas welding, the angle of the torch depends upon the thickness of metal to be formed.
3. Drag in gas cutting is the time difference between heating of the plate and starting the oxygen gas for cutting.

Which of the statements given above are correct
(a) 1, 2 and 3 (b) 1 and 2
(c) 2 and 3 (d) 1 and 3

39. Consider the following statements in respect of oxyacetylene welding
1. The joint is not heated to a state of fusion
2. No pressure is used
3. Oxygen is stored in steel cylinder at a pressure of 14 MPa
4. When there is an excess of acetylene used, there is a decided change in appearance of flame.

Which of the statements given above are correct
(a) 1, 2 and 3 (b) 2, 3 and 4
(c) 1, 3 and 4 (d) 1, 2 and 4

40. The metal powder used in thermit welding of steel is
(a) Al (b) Cu
(c) Pb (d) W

41. In atomic hydrogen welding, hydrogen acts as
(a) A heating agent
(b) One of the gases to generate the flame
(c) An effective shielding gas protecting the weld
(d) A lubricant to increase the flow characteristics of weld metal

42. In which one of the following welding techniques is vaccum environments required?
(a) Ultrasonic welding (b) Laser beam welding
(c) Plasma arc welding (d) Electron beam welding

43. High speed electron beam welding is focused on the weld spot using:
(a) Vaccum lens (b) Insert gas lens
(c) Optical lens (d) Magnetic lens

44. Which one of the following welding processes consists of minimum heat affected zone(HAZ)?
(a) Shielded Metal arc welding(SMAW) (b) Laser Beam Welding(LBW)
(c) Ultrasonic Welding(USW) (d) Metal Inert Gas Welding(MIG)

45. Consider the following statements in respect of the laser beam welding:

1. It can be used for welding any metal or their combinations because of very high temperature of the focal points.
2. Heat affected zone is very large because of quick heating
3. High vaccum is required to carry the process

Which of the statements given below is/are correct?

(a) 1 and 2 only (b) 2 and 3 only
(c) 1 only (d) 1, 2 and 3

46. In a fillet welded joint, the weakest area of the weld is:

(a) toe (b) root
(c) throat (d) face

47. Match **List-I** (Filler rod materials) with **List-II** (Joining process) and select the correct answer using the codes given below the lists:

List-I	**List-II**
A. Mild steel	**1.** MIG welding
B. Bronze	**2.** Soldering
C. Brass	**3.** Brazing
D. Lead and tin alloy	**4.** Thermit welding
	5. Braze welding

Codes:

	A	B	C	D
(a)	1	5	3	2
(b)	4	3	2	5
(c)	4	3	5	2
(d)	1	3	5	4

48. The strength of a brazed joint

(a) decreases with increase in gap between the joining surfaces
(b) increase with increase in gap between the two joining surfaces
(c) decreases up to certain gap between the two joining surfaces beyond which it increases
(d) increases up to certain gap between the two joining surfaces beyond which it decreases

49. Match the **List-I** (Welding problems) with **List-II** (Causes) and select the correct answers using the codes given below the lists:

List-I	**List-II**
A. Cracking of base metal	**1.** Excessive stress weld metal
B. Cracking of bead	**2.** High joint base metal rigidity
C. Porosity	**3.** Failure to remove slag from previous deposit
D. Inclusion	**4.** Oxidation
	5. Excessive hydrogen, oxygen, nitrogen in the welding atmosphere

Codes:

	A	B	C	D
(a)	2	1	5	3
(b)	3	4	2	1
(c)	2	4	5	3
(d)	3	1	4	2

50. Hot cracks occur in the weld and fusion zone as the metal solidifies. Which of the following are the cases for hot cracks?

1. Presence of sulphur and phosphorous in the base metal.
2. High carbon or alloy content of the base metal
3. Moisture in the joint or electrode
4. Joint restraint

Select the correct answer using the code given below:

(a) 1, 2 and 4
(b) 1, 2 and 3
(c) 3 and 4
(d) 1, 2, 3 and 4

51. Which of the following effects are possible due to residual stresses in welding?

1. Reduce dimensional stability
2. Welding cracking
3. Effect on fatigue strength
4. Reduction in the creep strength

Select the correct answer:

(a) 1 and 2 only
(b) 2 and 3 only
(c) 1 and 3 only
(d) 1, 2, 3 and 4

52. The main criterion for selection of electrode diameter in arc welding is

(a) material to be welded
(b) thickness of material
(c) voltage used
(d) current used

53. Which of the following is preferred for welding of non-ferrous metals by arc welding?

(a) A.C low frequency
(b) A.C high frequency
(c) D.C
(d) all the above

54. The material used for coating the electrode is called:

(a) binder
(b) slag
(c) deoxidizer
(d) flux

55. Following gases are used in tungsten inert gas welding

(a) CO_2 and H_2
(b) argon and neon
(c) helium and neon
(d) argon and helium

56. T-joint weld is used

(a) where longitudinal shear is present
(b) where severe loading is encountered and the upper surface of both pieces must be in the same plane.
(c) to join two pieces of metal in the same manner as revet joint metals
(d) to join two pieces perpendicularly

57. Copper is

(a) easily spot welded
(b) very difficult to be spot welded
(c) as good for spot welding as any other material.
(d) preferred to be welded by spot welding.

58. Two sheets of same material but different thickness can be butt welded by

(a) adjustment of the current
(b) time duration of current
(c) pressure applied
(d) changing the size of one electrode

59. Projection welding is
(a) multi-spot welding process (b) continuous spot welding process
(c) used to form mesh (d) used to make cantilevers

60. Grey cast iron is best welded applied
(a) TIG (b) MIG
(c) submerged work (d) oxy-acetylene welding

61. Seam welding is
(a) multi-spot welding process (b) continuous spot welding process
(c) used to form mesh (d) used for welding cylindrical objects

62. Laser welding finds widest applications in
(a) heavy industry (b) structural work
(c) process industry (d) electronic industry

63. In arc welding, temperature of the following order may be generated
(a) 1000°C (b) 1500°C
(c) 5500°C (d) 8000°C

64. Long arc lengths in welding would require the voltages and currents respectively as follows:
(a) high, high (b) low, low
(c) high, low (d) low, high

65. Fluxes are used in welding in order to protect the molten metal and the surfaces to be joined from
(a) oxidation (b) carburizing
(c) distortion and warping (d) unequal temperature districution

66. Metal deposited on to the workpiece from the electrode
(a) falls because of gravity
(b) is attracted towards the workpiece due to the positive polarity of the workpiece
(c) is attracted towards the workpiece due to the negative polarity of the workpiece
(d) is attracted due to electromagnetic action.

67. The phenomenon of weld decay occurs in
(a) cast iron (b) brass
(c) bronze (d) stainless steel

68. Weld spatter is
(a) flux (b) electrode coating
(c) welding defect (d) welding technique

69. Arc blow occurs in
(a) gas welding (b) gas cutting
(c) DC arc welding (d) AC arc welding

70. Low hydrogen electrodes are baked prior to use in order that
(a) proper strength is obtained (b) welding is free from blow
(c) welding is free from moisture pick up (d) current required is minimum

71. In straight polarity welding
(a) electrode holder is connected to the negative and work to positive
(b) electrode holder is connected to the positive and work to negative
(c) work is positive and holder is earthed
(d) holder is positive and work is earthed.

72. In thermit welding, the iron oxide and aluminium are mixed in the proportion of
(a) 1 : 1 (b) 3 : 1
(c) 1 : 3 (d) mixture is of differet oxides
(e) none of the above

73. TIG welding is best suited for welding
(a) mild steel (b) stainless steel
(c) carbon steel (d) aluminium

74. In the manual TIG welding, the angle of the electrode holder with the direction of welding normally is
(a) 30° (b) 45°
(c) 60° (d) 70°

75. Arc stability is better with
(a) AC welding
(b) DC welding
(c) both AC and DC welding
(d) specially designed wave forms

76. Two stainless steels foils of 0.1 mm thickness are to be joined. Which of the following process would be best suited
(a) gas welding (b) electro slag welding
(c) TIG welding (d) MIG welding

77. Arc length in arc welding should be equal to
(a) half the diameter of electrode rod (b) rod diameter
(c) twice the rod diameter (d) 2.5 times the rod diameter

78. In coated electrode in arc welding
(a) both rod and coating melt simultaneously
(b) coating melts first and then rod melts
(c) which melts first depends on polarity
(d) rod melts first and then coating melts

79. The advantages of electroslag welding are
(a) ability to weld metals of great thickness in a single pass without calling for joint preparation
(b) high welding speed
(c) little distortion and good stress distribution across the weld
(d) protection from contamination

80. In arc welding, if arc is too short, it will result in
(a) formation of large globules in an irregular pattern because of wandering of arc, leading to poor fusion with base metal
(b) electrode sticking to the base metal and base metal not melting and bead resting on top of the work, leading to poor fusion and gas and slag holes
(c) arc extinction
(d) operator hazard

81. In braze welding the filter metal is (an other type of fusion welding)
(a) Distributed by capillary action
(b) Melted and deposited at the point where the weld is to be made. The process is performed at the melting point of filler material

(c) Both of the above
(d) Not required

82. Too high welding current in arc welding would result in
(a) excessive piling up of weld metal, poor penetration, wasted electrodes
(b) excessive spatter, under cutting along edges, irregular deposits, wasted electrodes.
(c) too small bead, weak weld and wasted electrodes
(d) excessive piling up of weld metal, overlapping without penetration of edges, wasted electrodes

83. The temperature of the inner luminous cone of neutral flame is the order of
(a) 1000°C (b) 2000°C
(c) 2500°C (d) 3500°C

84. The maximum flame temperature occurs at
(a) the tip of flame (b) the inner cone
(c) next to the inner cone (d) at the outer cone

85. Distortion in welding occurs due to
(a) use of excessive current (b) improper clamping methods
(c) use of wrong electrodes (d) oxidation of weld pool

86. Oxygen to acetylene ratio in case of neutral flame by weight is
(a) 0.8 : 1.0 (b) 1 : 1
(b) 1.2 : 1 (d) 2 : 1

87. The main advantage of MIG welding over TIG welding is that
(a) former can be used to weld hard to weld metal
(b) former permits use of large currents thereby allowing higher deposition
(c) welding rate is very fast
(d) welding is completely automatic

88. Which of the following materials is best cut by oxy-cutting process
(a) brass (b) bronze
(c) stainless steel (d) mild steel

89. The acetylene cylinder is filled with a material saturated with
(a) calcium carbide (b) calcium oxide
(c) acetone (d) acetylene

90. The commonly used flux for brazing is
(a) resin (b) NH_4CL
(c) borax (d) soft iron

91. Two pipes of inner diameter 100 mm and outer diameter 110 mm each are joined by flash-butt welding using 30 V power supply. At the interface, 1 mm of material melts from each pipe which has a resistance of 42.4 Ω. If the unit melt energy of 64.4 MJm^{-3}, then time required for welding (in s) is
(a) 1 (b) 5
(c) 10 (d) 20

■■■■

ANSWERS

WELDING

1. (d)	**2.** (d)	**3.** (c)	**4.** (c)	**5.** (b)	**6.** (c)	**7.** (a)
8. (d)	**9.** (d)	**10.** (a)	**11.** (c)	**12.** (b)	**13.** (c)	**14.** (d)
15. (c)	**16.** (b)	**17.** (b)	**18.** (c)	**19.** (c)	**20.** (d)	**21.** (b)
22. (d)	**23.** (d)	**24.** (a)	**25.** (a)	**26.** (c)	**27.** (d)	**28.** (d)
29. (b)	**30.** (c)	**31.** (c)	**32.** (a)	**33.** (d)	**34.** (d)	**35.** (b)
36. (c)	**37.** (c)	**38.** (b)	**39.** (b)	**40.** (a)	**41.** (a)	**42.** (d)
43. (d)	**44.** (b)	**45.** (c)	**46.** (c)	**47.** (a)	**48.** (d)	**49.** (a)
50. (d)	**51.** (d)	**52.** (b)	**53.** (c)	**54.** (d)	**55.** (d)	**56.** (a)
57. (b)	**58.** (d)	**59.** (a)	**60.** (d)	**61.** (b)	**62.** (d)	**63.** (c)
64. (c)	**65.** (a)	**66.** (d)	**67.** (d)	**68.** (c)	**69.** (c)	**70** (c)
71. (a)	**72.** (b)	**73.** (d)	**74.** (a)	**75** (b)	**76.** (c)	**77.** (b)
78. (d)	**79.** (a)	**80.** (b)	**81.** (b)	**82.** (b)	**83.** (d)	**84.** (b)
85. (a)	**86.** (b)	**87.** (b)	**88.** (d)	**89.** (c)	**90.** (b)	**91.** (c)